Ecological Studies, Vol. 187

Analysis and Synthesis

Edited by

M.M. Caldwell, Logan, USA
G. Heldmaier, Marburg, Germany
R.B. Jackson, Durham, USA
O.L. Lange, Würzburg, Germany
H.A. Mooney, Stanford, USA
E.-D. Schulze, Jena, Germany
U. Sommer, Kiel, Germany

Ecological Studies

Volumes published since 2001 are listed at the end of this book.

J. Nösberger S.P. Long R.J. Norby M. Stitt G.R. Hendrey H. Blum (Eds.)

Managed Ecosystems and CO$_2$

Case Studies, Processes, and Perspectives

With 76 Figures, 5 in Color, and 37 Tables

 Springer

Prof. Dr. Josef Nösberger
Institute of Plant Sciences
ETH
Universitätstrasse 2
8092 Zurich, Switzerland

Prof. Dr. Stephen P. Long
Department of Plant Biology
University of Illinois
379 Edward R. Madigan Laboratory
1201 W. Gregory
Urbana, IL 61801, USA

Prof. Dr. Richard J. Norby
Oak Ridge National Laboratory
Bldg. 1062, Bethel Valley Road
Oak Ridge, TN 37831-6422
USA

Prof. Dr. Mark Stitt
Max Planck Institute
of Molecular Plant Physiology
Am Mühlenberg
14476 Golm
Germany

Prof. Dr. George R. Hendrey
Queens College
City University of New York
65-30 Kissena Blvd
Flushing NY 11367, USA

Dr. Herbert Blum
Institute of Plant Sciences
Swiss Federal Institute
of Technology (ETH)
8092 Zurich, Switzerland

Cover illustration. Background: One ring of the Swiss FACE with a grass and clover sward (photo Manuel Schneider). Upper left corner: Arbuscular mycorrhizal fungus showing spores of *Glomus mosseae* and its hyphal network (photo Adrian Leuchtmann). Petri dish: Colony-forming soil bacteria grown on a nutrient agar (photo Michel Aragno). Lower right corner: Net ecosystem CO_2 exchange measurement in the Swiss FACE experiment (photo Manuel Schneider).

ISSN 0070-8356
ISBN-10   3-540-31236-6 Springer Berlin Heidelberg New York
ISBN-13   978-3-540-31236-9 Springer Berlin Heidelberg New York

Springer is a part of Springer Science+Business Media
springer.com

© Springer-Verlag Berlin Heidelberg 2006
Printed in Germany

Editor: Dr. Dieter Czeschlik, Heidelberg, Germany
Desk editor: Dr. Andrea Schlitzberger, Heidelberg, Germany
Cover design: *design & production* GmbH, Heidelberg, Germany
Typesetting and production: Friedmut Kröner, Heidelberg, Germany

31/3152 YK – 5 4 3 2 1 0 – Printed on acid free paper

Preface

There are times when it is difficult to acknowledge the obvious. It is obvious that human life itself depends on the continuing capacity of ecosystems to provide their broad spectrum of benefits. The conditions of the major ecosystems and their continuing capacity to support us have been assessed by the United Nations Development Programme, the United Nations Environment Programme, the World Bank and the World Resource Institute (2002) and their findings were published in *World resources 2000–2001: people and ecosystems: the fraying web of life*. The report takes stock of the global condition of various classes of ecosystems and their capacities to continue to provide what we need. These assessments considered environmental impacts not only the quantity and quality of outputs of market products, such as food, fiber and timber, but also on the biological bases for plant and ecosystem production, including soil and water quality, biodiversity and changes in land use over time. Humanity has focused for too long on how much man can take from our ecosystems, with little attention to the impacts of large-scale environmental assaults such as burning fossil fuels, changes in land-use and intensification of agriculture. These assaults have led to a higher concentration of atmospheric carbon dioxide ($[CO_2]$), ozone and other greenhouse gases. Through such assessments, we have come to realize that solid, scientific knowledge is a prerequisite to evaluating the capacity of the biosphere to adapt to the global-scale environmental changes that society is creating, exploring the possibilities to mitigate potential negative consequences and making sound management decisions. The examples presented in this volume demonstrate that our knowledge about the effects of increasing atmospheric $[CO_2]$ on agroecosystems, forests and grassland is rapidly improving; our ability to influence these changes has simply not grown at the same pace.

The present analyses show that experimental results obtained at the level of selected tissues and individual plants sometimes can be misleading when scaled-up to the ecosystem level. Attempts to understand the larger, ecological effects of increasing [CO$_2$] involve exposing today's ecosystems to expected future levels of [CO$_2$]. Prior to the work described in this volume, most information about plant responses to elevated [CO$_2$] was derived from experimental studies in greenhouses, controlled environmental chambers, transparent field enclosures or open-top chambers. While all these methods provide an atmosphere with enriched [CO$_2$], they also significantly alter other aspects of the environment surrounding the plant and, consequently, the responses of test plants to the applied treatment. Thus, the development of the *free air carbon dioxide enrichment* (FACE) technologies presented a unique possibility to measure the integrated response of an intact ecosystem with a focus on stand-level mechanisms over decadal time-scales. Moreover, FACE experiments offer a distinctive platform that facilitates multidisciplinary approaches to address the essential features of a plant stand and how they might influence the response to increasing [CO$_2$]. These features include: the closed canopy which constrains growth and developmental responses, varying source-sink balances during the growth cycle of plants that influence the rate of photosynthesis, and full occupancy of the soil by the root system, affecting symbionts and other soil biota and constraining nutrient cycles. The longer time-scale that can be addressed with FACE enables studies of changes in soil food-webs and soil carbon content. A modern and integrated treatment of these aspects is the motivation for the production of this volume.

Our discussions of FACE start with a description of the origin and continuing development of FACE technologies and give an analysis of their strengths and weaknesses. The second section presents case studies from several ecosystems, including crops, grassland and plantation forests. The focus of these case studies is largely ecological. However, they also illustrate the consequences of their findings for plant breeding and management of these ecosystems in the future. The third section deals with underlying mechanistic processes at different levels of organisations – from genomics to the ecosystem. They include recent cutting-edge work, statistical meta-analysis and conceptual models that make substantial headway in synthesizing broad questions. The closing chapter reflects upon the perspectives on the future of FACE studies and identifies significant gaps in the current data that need to be filled to ensure a full understanding of the response of ecosystems to increasing CO$_2$ concentrations.

The need to understand responses of ecosystems to a CO$_2$-enriched world, combined with the role of the biosphere in mediating the rate of change in [CO$_2$], requires a coordinated and comprehensive investigation of the impacts of atmospheric changes on those ecosystems that cover the greatest part of the global land area or have the greatest potential to alter [CO$_2$]. The questions that need to be addressed range from soil chemistry, soil food-webs and rhizosphere biology, through to net primary production, allocation of photosynthate to different plant parts and the responses at the community and ecosystem levels in a wide range of plant stands. This work has utilized a rich arsenal of scientific techniques from genomics and micrometeorology to ecosystem modeling.

The contributors to this volume hope the book will stimulate further innovation both in experimental techniques and in the application of field research to plants and ecosystems and, thus, enhance our knowledge of the processes and functions of managed ecosystems. This is no small task.

We thank ETH Zurich for supporting a workshop at which the preparation of this volume was initiated.

April 2006

Josef Nösberger
Stephen P. Long
Richard J. Norby
Mark Stitt
George R. Hendrey
Herbert Blum

Reference

United Nations Development Programme, United Nations Environment Programme, World Bank, World Resources Institute (2002) World resources 2000–2001: people and ecosystems: the fraying web of life. World Resources Institute, Washington, D.C.

Contents

Part B   Case Studies

3   The Effects of Free-Air [CO₂] Enrichment
**   of Cotton, Wheat, and Sorghum**   47
            B.A. KIMBALL

4   **SoyFACE: the Effects and Interactions**
          of Elevated [CO_2] and [O_3] on Soybean   71
          D.R. Ort, E.A. Ainsworth, M. Aldea, D.J. Allen,
          C.J. Bernacchi, M.R. Berenbaum, G.A. Bollero, G. Cornic,
          P.A. Davey, O. Dermody, F.G. Dohleman, J.G. Hamilton,
          E.A. Heaton, A.D.B. Leakey, J. Mahoney, T.A. Mies,
          P.B. Morgan, R.L. Nelson, B. O'Neil, A. Rogers,
          A.R. Zangerl, X.-G. Zhu, E.H. DeLucia, and S.P. Long

7 **Responses of an Arable Crop Rotation System
 to Elevated [CO$_2$]** 121
H.J. Weigel, R. Manderscheid, S. Burkart, A. Pacholski,
K. Waloszczyk, C. Frühauf, and O. Heinemeyer

8   **Short- and Long-Term Responses of Fertile Grassland
         to Elevated [CO$_2$]** 139
         A. Lüscher, U. Aeschlimann, M.K. Schneider,
         and H. Blum

9   **Impacts of Elevated CO$_2$ on a Grassland Grazed by Sheep:
         the New Zealand FACE Experiment** 157
         P.C.D. Newton, V. Allard, R.A. Carran, and M. Lieffering

**10   Responses to Elevated [CO₂] of a Short Rotation, Multispecies
       Poplar Plantation: the POPFACE/EUROFACE Experiment . 173**
G. Scarascia-Mugnozza, C. Calfapietra, R. Ceulemans,
B. Gielen, M.F. Cotrufo, P. DeAngelis, D. Godbold,
M.R. Hoosbeek, O. Kull, M. Lukac, M. Marek, F. Miglietta,
A. Polle, C. Raines, M. Sabatti, N. Anselmi, and G. Taylor

Part C Processes

**20   The Potential of Genomics and Genetics to Understand
           Plant Response to Elevated Atmospheric [CO$_2$]**   351
           G. TAYLOR, P.J. TRICKER, L.E. GRAHAM, M.J. TALLIS,
           A.M. RAE, H.TREWIN, and N.R. STREET

**21   The Impact of Elevated Atmospheric [CO$_2$] on Soil C
           and N Dynamics: A Meta-Analysis**   373
           K.-J. VAN GROENIGEN, M.-A. DE GRAAFF, J. SIX, D. HARRIS,
           P. KUIKMAN, and C. VAN KESSEL

Part D   Perspectives

Contributors

AESCHLIMANN, U.
Institute of Plant Sciences, Swiss Federal Institute of Technology (ETH),
8092 Zürich, Switzerland

AINSWORTH, E.A.
USDA/ARS Department of Plant Biology, University of Illinois, 190 ERML,
1201 W. Gregory Drive, Urbana, Illinois 61801, USA

ALDEA, M.
Department of Plant Biology, University of Illinois, 190 ERML265 Morrill
Hall, 505S. Goodwin Ave., Urbana, Illinois 61801, USA

ALLARD, V.
INRA-Agronomie, 234 avenue de Brézet, 6300 Clermont-Ferrand, France

ALLEN, D.J.
USDA/ARS Department of Plant Biology, University of Illinois, 190 ERML,
1201 W. Gregory Drive, Urbana, Illinois 61801, USA

ANSELMI, N.
Department of Plant Protection, University of Tuscia,
Via S. Camillo De Lellis, 01100 Viterbo, Italy

ARAGNO, M.
Laboratoire de Microbiologie, Institut de Botanique, Case postale 2,
2007 Neuchâtel, Switzerland, e-mail: michel.aragno@unine.ch

BERENBAUM, M.R.
Department of Entomology and Plant Biology, University of Illionois,
265 Morrill Hall, 505 S. Goodwin Ave., Urbana, Illinois 61801, USA

BERNACCHI, C.J.

Illinois State Water Survey, Center for Atmospheric Sciences,
2204 Griffith Drive, Champaign, Illinois 61820, USA

BERNHARDT, E.S.

Nicholas School of the Environment and Earth Sciences, Duke University,
Durham, North Carolina 27708, USA

BINDI, M.

Department of Agronomy and Land Management (DISAT),
University of Florence, P. le delle Cascine 18, 50144 Firenze, Italy,
e-mail: marco.bindi@unifi.it

BLUM, H.

Institute of Plant Sciences, Swiss Federal Institute of Technology (ETH),
8092 Zürich, Switzerland

BODDY, E.L.

Environmental Science, University of Wales at Bangor, Bangor,
Gwynedd LL57 2UW, UK

BOLLERO, G.A.

Department of Crop Science, University of Illinois, AW101 Turner Hall,
1102 S. Goodwin Ave., Urbana, Illinois 61801, USA

BURKART, S.

Institute of Agroecology, Federal Agricultural Research Centre (FAL),
Bundesallee 50, 38116 Braunschweig, Germany

CALFAPIETRA, C.

Department of Forest Environment and Resources, University of Tuscia,
Via S.Camillo De Lellis, 01100 Viterbo, Italy

CARRAN, R.A.

AgResearch Grasslands, Private Bag 11008, Palmerston North, New Zealand

CEULEMANS, R.

Department of Biology, University of Antwerpen, Universiteitsplein 1,
2610 Wilrijk, Belgium

CORNIC, G.
Laboratory D'Écophysiologie Végétale, Université Paris XI Orsay,
91405 Paris, France

COTRUFO, M.F.
Department of Environmental Sciences, II University of Naples,
via Vivaldi 43, 81100 Caserta, Italy

DAVEY, P.A.
Department of Plant Biology, University of Illinois, 190 ERML, 1201 W. Gregory Dr., Urbana, Illinois 61801, USA

DEANGELIS, P.
Department of Forest Environment and Resources, University of Tuscia,
Via S.Camillo De Lellis, 01100 Viterbo, Italy

DE GRAAFF, M.-A.
Department of Plant Sciences, One Shields Avenue, University of California–Davis, Davis, California 95616, USA

DELUCIA, E.H.
Department of Plant Biology, University of Illinois, 265 Morrill Hall,
505 S. Goodwin Ave., Urbana, IL 61801, USA

DERMODY, O.
Department of Plant Biology, University of Illinois, 265 Morrill Hall,
505 S. Goodwin Ave., Urbana, Illinois 61801, USA

DOHLEMAN, F.G.
Department of Crop Science, University of Illinois, AW101 Turner Hall,
1102 S. Goodwin Ave., Urbana, Illinois 61801, USA

ELLSWORTH, D.S.
Nicholas School of the Environment and Earth Sciences, Duke University,
Durham, North Carolina 27708, USA

FARRAR, J.F.
Environmental Science, University of Wales at Bangor, Bangor,
Gwynedd LL57 2UW, UK

FERRIS, H.
Department of Nematology, University of California, Davis, California 95616,
USA

FINZI, A.C.
Nicholas School of the Environment and Earth Sciences, Duke University,
Durham, North Carolina 27708, USA

FOX, T.C.
Department of Plant Sciences, University of California, Davis,
California 95616, USA

FRÜHAUF, C.
Institute of Agroecology, Federal Agricultural Research Centre (FAL),
Bundesallee 50, 38116 Braunschweig, Germany

GIELEN, B.
Department of Biology, University of Antwerpen, Universiteitsplein 1,
2610 Wilrijk, Belgium

GIUNTOLI, A.
IBiMET, National Research Council, Florence, Italy

GODBOLD, D.
School of Agricultural and Forest Science, University of Wales Deiniol Road,
Bangor LL57 2UW, UK

GRAHAM, L.E.
School of Biological Sciences, Bassett Crescent East, University of Southampton, Southampton SO16 7PX, UK

GRAY, A.M.
Environmental Science, University of Wales at Bangor, Bangor,
Gwynedd LL57 2UW, UK

GUNDERSON, C.A.
Environmental Sciences Division, Oak Ridge National Laboratory,
Bldg. 1062, Bethel Valley Road, Oak Ridge, Tennessee 37831-6422, USA

HAMILTON, J.G.

Department of Plant Biology, University of Illinois, 265 Morrill Hall,
505 S. Goodwin Ave., Urbana, Illinois 61801, USA

HANSON, P.J.

Environmental Sciences Division, Oak Ridge National Laboratory,
Bldg. 1062, Bethel Valley Road, Oak Ridge, Tennessee 37831-6422, USA

HARRIS, D.

Stable Isotope Facility, University of California–Davis, Davis,
California 95616, USA

HARTWIG, U.A.

Academia Engiadina, 7503 Samedan, Switzerland,
e-mail: ueli.hartwig@academia-engiadina.ch

HASEGAWA, T.

National Institute for Agro-Environmental Sciences, 3-1-3 Kannondai,
Tsukuba, Ibaraki 305-8604, Japan

HEATON, E.A.

Department of Crop Science, University of Illinois, AW101 Turner Hall,
1102 S. Goodwin Ave., Urbana, Illinois 61801, USA

HEINEMEYER, O.

Institute of Agroecology, Federal Agricultural Research Centre (FAL),
Bundesallee 50, 38116 Braunschweig, Germany

HENDREY, G.R.

Queens College, City University of New York, 65–30 Kissena Blvd, Flushing,
New York 11367, USA, e-mail: ghendrey@qc.edu

HILL, P.W.

Environmental Science, University of Wales at Bangor, Bangor,
Gwynedd LL57 2UW, UK, e-mail: p.w.hill@bangor.ac.uk

HOFMOCKEL, K.S.

Nicholas School of the Environment and Earth Sciences, Duke University,
Durham, North Carolina 27708, USA

HOOSBEEK, M.R.
Department of Environmental Sciences, Wageningen University,
Duivendaal 10, 6700 AA Wageningen, Netherlands

HYMUS, G.J.
Department of Biological Sciences, University of Essex, Colchester CO4 3SQ,
UK

JASTROW, J.D.
Environmental Research Division, Argonne National Laboratory, Argonne,
Illinois, USA

JONES, D.L.
Environmental Science, University of Wales at Bangor, Bangor,
Gwynedd LL57 2UW, UK

KAMMANN, C.
Institute for Plant Ecology, University of Giessen, 35392 Giessen, Germany

KARNOSKY, D.F.
Michigan Technological University, 1400 Townsend Drive, Houghton,
Michigan 49931-1295, USA, e-mail: karnosky@mtu.edu

KIM, H.Y.
Chonnam National University, 300 Yongbong-dong, Buk-gu,
Gwangju 500-757, Korea

KIMBALL, B.A.
United States Arid-Land Agricultural Research Center, Agricultural Research
Service, United States Department of Agriculture, 21881 North Carbon Lane,
Maricopa, Arizona 85239, USA, e-mail: bkimball@uswcl.ars.ag.gov

KOBAYASHI, K.
The University of Tokyo, Graduate School of Agricultural and Life Sciences,
1-1-1 Yayoi, Bunkyo-ku, Tokyo 113-8657, Japan,
e-mail: aclasman@mail.ecc.u-tokyo.ac.jp

KUIKMAN, P.
Alterra, Soil Sciences Center, P.O. Box 47, 6700 AA Wageningen,
The Netherlands

Kull, O.

Institute of Botany and Ecology, University of Tartu, Riia 181, 51014 Tartu, Estonia

Leakey, A.B.D.

Institute of Genooomic Biology, University of Illionois, 04 ASL, 1207 W. Gregory Dr., Urbana, Illinois 61801, USA

Lichter, J.

Nicholas School of the Environment and Earth Sciences, Duke University, Durham, North Carolina 27708, USA

Lieffering, M.

AgResearch Grasslands, Private Bag 11008, Palmerston North, New Zealand

Long, S.P.

Department of Plant Biology, University of Illinois, 379 Edward R. Madigan Laboratory, 1201 W. Gregory Drive, Urbana, Illinois 61801, USA, e-mail: slong@uiuc.edu

Lukac, M.

School of Agricultural and Forest Science, University of Wales Deiniol Road, Bangor LL57 2UW, UK

Lüscher, A.

Agroscope FAL Reckenholz, Reckenholzstrasse 191, 8046 Zurich, Switzerland, e-mail: andreas.luescher@fal.admin.ch

Magliulo, E.

ISAFOM, National Research Council, Naples, Italy

Mahoney, J.

Environmental Sciences Department, Brookhaven National Laboratory, Upton, New York 11973, USA

Manderscheid, R.

Institute of Agroecology, Federal Agricultural Research Centre (FAL), Bundesallee 50, 38116 Braunschweig, Germany

MAREK, M.

Institute of Landscape Ecology, Academy of Sciences, Pooiei 3b, 60300 Brno, Czech Republic

MATAMALA, R.

Nicholas School of the Environment and Earth Sciences, Duke University, Durham, North Carolina 27708, USA

MIES, T.A.

Department of Crop Sciences, University of Illinois, 190 ERML, 1201 W. Gregory Drive, Urbana, Illinois 61801, USA

MIGLIETTA, F.

IBIMET-CNR, Institute of Biometeorology, National Research Council, Via Caproni 8, 50145 Florence, Italy

MIURA, S.

Hokkaido Tenpoku Agricultural Experiment Station, Midorigaoka, Hama-Tonbetsu, Hokkaido 098-5738, Japan

MOORE, D.

Nicholas School of the Environment and Earth Sciences, Duke University, Durham, North Carolina 27708, USA

MOORE, J.C.

Department of Biological Sciences, University of Northern Colorado, Greeley, Colorado 80639, USA

MORGAN, P.B.

Department of Plant Biology, University of Illinois, 190 ERML, 1201 W. Gregory Dr., Urbana, Illinois 61801, USA

NELSON, R.L.

USDA/ARS, Department of Crop Science, University of Illinois, AW101 Turner Hall, 1102 S. Goodwin Ave., Urbana, Illinois 61801, USA

NEWTON, P.C.D.

AgResearch Grasslands, Private Bag 11008, Palmerston North, New Zealand, e-mail: paul.newton@agresearch.co.nz

NÖSBERGER, J.
Institute of Plant Sciences, ETH, Universitätstrasse 2, 8092 Zurich,
Switzerland, e-mail: josef.noesberger@ipw.agrl.ethz.ch

NORBY, R.J.
Oak Ridge National Laboratory, Bldg. 1062, Bethel Valley Road, Oak Ridge,
Tennessee 37831-6422, USA, e-mail: rjn@ornl.gov

OKADA, M.
National Agricultural Research Center for Tohoku Region, 4 Akahira,
Shimokuriyagawa, Morioka, Iwate 020-0198, Japan

O'NEIL, B.
Department of Entomology, University of Illinois, 320 Morrill Hall,
505 S. Goodwin Ave., Urbana, Illinois 61801, USA

OREN, R.
Nicholas School of the Environment and Earth Sciences, Duke University,
Durham, North Carolina 27708, USA

ORT, D.R.
USDA/ARS, Department of Plant Biology, University of Illinois, 190 ERML,
1201 W. Gregory Drive, Urbana, Illinois 61801, USA, e-mail: d-ort@uiuc.edu

OSBORNE, C.P.
Department of Biological Sciences, University of Essex, Colchester CO4 3SQ,
UK

PACHOLSKI, A.
Institute of Agroecology, Federal Agricultural Research Centre (FAL),
Bundesallee 50, 38116 Braunschweig, Germany

PHILLIPS, D.A.
Department of Plant Sciences, University of California, Davis, California
95616 USA, e-mail: daphillips@ucdavis.edu

PIPPEN, J.S.
Nicholas School of the Environment and Earth Sciences, Duke University,
Durham, North Carolina 27708, USA

POLLE, A.

Forstbotanisches Institut, Georg August Universität Göttingen, Büsgenweg 2, 37077 Göttingen, Germany

PREGITZER, K.S.

Ecosystem Science Center, School of Forestry and Environmental Science, Michigan Technological University, Houghton, Michigan 49931, USA

RAE, A.M.

School of Biological Sciences, Bassett Crescent East, University of Southampton, Southampton SO16 7PX, UK

RAINES, C.

Department of Biological Sciences, University of Essex, Colchester CO4 3SQ, UK

ROGERS, A.

Environmental Sciences Department, Brookhaven National Laboratory, Upton, New York, 11973-5000, USA, e-mail: arogers@bnl.gov

SABATTI, M.

Department of Forest Environment and Resources, University of Tuscia, Via S.Camillo De Lellis, 01100 Viterbo, Italy

SADOWSKY, M.J.

Department of Soil, Water and Climate; and BioTechnology Institute, University of Minnesota, St. Paul, Minnesota 55108, USA

SCARASCIA-MUGNOZZA, G.

Institute of Agroenvironmental & Forest Biology, National Research Council, Villa Paolina, 05010 Porano, Italy, e-mail: gscaras@unitus.it

SCHLESINGER, W.H.

Nicholas School of the Environment and Earth Sciences, Duke University, Durham, North Carolina 27708, USA, e-mail: schlesin@duke.edu

SCHNEIDER, M.K.

Swiss Federal Institute of Aquatic Science and Technology Eawag, 8600 Dübendorf, Switzerland

SIX, J.
Department of Plant Sciences, One Shields Avenue,
University of California–Davis, Davis, California 95616, USA

STREET, N.R.
School of Biological Sciences, Bassett Crescent East, University of Southampton, Southampton SO16 7PX, UK

TALLIS, M.J.
School of Biological Sciences, Bassett Crescent East, University of Southampton, Southampton SO16 7PX, UK

TARNAWSKI, S.
Laboratoire de Microbiologie, Institut de Botanique, Case postale 2,
2007 Neuchâtel, Switzerland

TAYLOR, G.
School of Biological Sciences, Bassett Crescent East, University of Southampton, Southampton SO16 7PX, UK, e-mail: G.Taylor@soton.ac.uk

THOMAS, R.B.
Nicholas School of the Environment and Earth Sciences, Duke University, Durham, North Carolina 27708, USA

TREWIN, H.
School of Biological Sciences, Bassett Crescent East, University of Southampton, Southampton SO16 7PX, UK

TRICKER, P.J.
School of Biological Sciences, Bassett Crescent East, University of Southampton, Southampton SO16 7PX, UK

TSCHAPLINSKI, T.J.
Environmental Sciences Division, Oak Ridge National Laboratory,
Bldg. 1062, Bethel Valley Road, Oak Ridge, Tennessee 37831-6422, USA

VACCARI, F.
IBiMET, National Research Council, Florence, Italy

VAN GROENIGEN, K.-J.

Laboratory for Soil Science and Geology, Wageningen University, PO Box 37, 6700 AA Wageningen, the Netherlands, e-mail: kees.janvangroenigen@wur.nl

VAN KESSEL, C.

Department of Plant Sciences, One Shields Avenue, University of California–Davis, Davis, California 95616, USA

WALOSZCZYK, K.

Institute of Agroecology, Federal Agricultural Research Centre (FAL), Bundesallee 50, 38116 Braunschweig, Germany

WEIGEL, H.J.

Institute of Agroecology, Federal Agricultural Research Centre (FAL), Bundesallee 50, 38116 Braunschweig, Germany, e-mail: hans.weigel@fal.de

WULLSCHLEGER, S.D.

Environmental Sciences Division, Oak Ridge National Laboratory, Bldg. 1062, Bethel Valley Road, Oak Ridge, Tennessee 37831-6422, USA

ZANGERL, A.R.

Department of Entomology, University of Illinois, 320 Morrill Hall, 505 S. Goodwin Ave., Urbana, Illinois 61801, USA

ZHU, X.-G.

Department of Plant Biology, University of Illinois, 190 ERML, 1201 W. Gregory Drive, Urbana, Illinois 61801, USA

Abbreviations

$\phi_{CO_2,max}$	maximum apparent quantum efficiency of CO_2 uptake
A	net leaf CO_2 uptake
A'	daily integral of leaf photosynthetic CO_2 uptake
AA	amino acids
AHL	*N*-acyl homoserine lactone
AMF	arbuscular mycorrhizal fungi
ANPP	aboveground primary productivity
APAR	absorbed photosynthetically active radiation
APP	aboveground primary production
AQUA	absolute quantification of proteins
A_{sat}	light-saturated CO_2 uptake
BAI	basal area increment
BNL	Brookhaven National Laboratory
BPP	belowground primary production
$c[CO_2]$	current concentration of atmospheric carbon dioxide
$c[O_3]$	current tropospheric ozone
C_a	external CO_2 concentration
CCER	canopy CO_2 exchange rates
CFD	computational fluid dynamic
C_i	intercellular CO_2 concentration
C_{mic}	soil microbial biomass
C_{new}	amount of C taken up by the soil during the experiment
C_{old}	old soil C pool
C_{org}	organic matter
C_{total}	total soil C pool
CWSI	crop water stress index
D^2H	stem volume index
DAPG	2,4-diacetylphloroglucinol
DAT	days after transplanting
DMP	dry matter production
DOC	dissolved organic C

DOY	day of year
$e[CO_2]$	elevated concentration of atmospheric carbon dioxide
$e[O_3]$	elevated concentration of tropospheric ozone
EST	expressed sequence tag
ET	evapotranspiration
FACE	free air carbon dioxide enrichment
FACTS	forest-atmosphere carbon transfer and storage
FAO	Food and Agriculture Organization of the UN
fAPAR	fractions of absorbed photosynthetically active radiation
FATE	free air temperature enhancement
G	soil heat flux
GCM	general circulation climate model
g_m	leaf mesophyll conductance
GPP	gross primary production
g_s	stomatal conductance
H	sensible heat flux
HGI	height growth in the year
ICAT	isotope-coded affinity tags
IR	infrared
IRGA	infrared gas analyzer
J_{max}	maximum rate of electron transport
J_{PSII}	daily integral of photosystem II electron transport
l	stomatal limitation to photosynthesis
LAI	leaf area index
LAR	leaf area ratio
LD	linkage disequilibrium
L_e	latent heat flux
LM_A	leaf mass per unit area
LPI	leaf plastochron index
LUE	light use efficiency
MRT	mean residence time
MS	mass spectrometry
N_A	N expressed on a leaf area basis
N_{area}	N content per unit leaf area
NDVI	normalized difference vegetation index
NEP	net ecosystem production
N_M	whole-canopy N concentration
NMR	nuclear magnetic resonance
NNI	N nutrition index
NO_x	oxidized nitrogen
NPP	net primary production
N_{sym}	N assimilated from symbiotic N_2 fixation
OTC	open-top chambers

PAR	photosynthetically active radiation
PAWC	plant available water content
PBL	planetary boundary layer
PCA	principal components analysis
PGPR	plant growth-promoting rhizobacteria
PI	panicle initiation
PID	proportional-integral-differential
PLFA	phospholipid fatty acids
PNL	progressive nitrogen depletion
PNL	progressive nitrogen limitation
PPI	photosynthetic photon irradiance
ppm	for CO_2: parts per million by volume in the atmosphere
Q_b	between class heterogeneity
Q_m	heterogeneity, explained by the regression model
Q_p	photosynthetic photon irradiance
Q_t	total heterogeneity
QTL	quantitative trait loci
QTN	quantitative trait nucleotides
Q_w	within-class heterogeneity
QY	maximum apparent quantum yield of CO_2 uptake
RAS	superfamily of GTPases
RGR	relative growth rate
RH	relative humidity
R_L	respiration in the light
R_M	maintenance respiration
R_N	nighttime respiration / nitrogen uptake respiration
R_n	net radiation
RNY	relative N yield
RNYT	relative N yield total
R_S	in situ soil respiration
r_s	stomatal resistance
R_{SA}	autotrophic soil respiration
R_{SH}	heterotrophic soil respiration
R_T	fine-root respiration
RT-PCR	reverse transcriptase polymerase chain reaction
Rubisco	ribulose-1,5-bisphosphate carboxylase/oxygenase
RuBP	ribulose-1,5-bisphosphate
RY	relative yield
RYT	relative yield total
S:R	shoot-to-root ratio
SIR	substrate induced respiration
SLA	specific leaf area
SNP	single nucleotide polymorphisms

SOM	soil organic matter
SRF	short rotation forestry
SSR	simple sequence repeat
T_a	air temperature
T_{air}	air temperature
T_L	leaf temperature
TNC	total non-structural carbohydrate
T-RFLP	terminal, restriction fragment length polymorphism
UWW	under water weight
$V_{c,max}$	maximum RuBP saturated rate of carboxylation
VPD	vapor pressure deficit
VVP	vertical vent pipes
WUE	water use efficiency
XTH	xyloglucan endotransglycosylase/hydrolase

Part A   Introduction

1 Introduction

J. Nösberger and S.P. Long

1.1 Managed Ecosystems and the Future Supply of Raw Materials

Managed ecosystems provide most of our food, much of our wood and fiber and increasingly are being considered as a source of renewable energy. Forecasting the ability of managed ecosystems to continue these vital roles under global atmospheric change has been the subject of a great deal of modeling effort. Model projections reviewed by the Intergovernmental Panel on Climate Change (IPCC 2001a) suggest that the increased temperature and decreased soil moisture that would otherwise lower crop yields will be offset by the direct fertilization effect of rising carbon dioxide concentration ($[CO_2]$; for a review, see Long et al. 2005). Averaged across the globe, total crop yield may rise; but this would be achieved by generally lower yields in the tropics and increased yields in the temperate zones. The IPCC (2001a) projected that world grain prices, an indicator of the balance between supply and demand, will continue to fall during through this century. However, model projections can only be as good as their parameterization; and establishing the effect of elevated $[CO_2]$ on yield is more challenging than for other abiotic factors. Yet it is critical, since it is CO_2 that provides the security in these projections. Without the "CO_2–fertilization" effect, global climate change would cause large losses in food supply and other products of managed ecosystems. For the world's major crops, vast quantities of data are available which show how yields are affected by inter-annual and geographical variation in temperature, precipitation and soil moisture. This information is used to parameterize and validate models. Approximately 150 years ago, atmospheric $[CO_2]$ was ca. 260 ppm, but in February 2006 it reached 382 ppm, possibly for the first time in several million years. The increase in global atmospheric $[CO_2]$ in 2005 was the largest since records began. At the current accelerating rate of increase, we expect global $[CO_2]$ to reach 700 ppm by the end of the twenty-first century, according to estimations presented in the third assessment report of the IPCC

Ecological Studies, Vol. 187
J. Nösberger, S.P. Long, R.J. Norby, M. Stitt,
G.R. Hendrey, H. Blum (Eds.)
Managed Ecosystems and CO$_2$
Case Studies, Processes, and Perspectives
© Springer-Verlag Berlin Heidelberg 2006

(2001b). But, unlike temperature and precipitation, [CO$_2$] is spatially remarkably uniform across the globe. So, in contrast to temperature and precipitation, there is no consistent spatial variation on which to estimate yield responses to increasing [CO$_2$]. And it is not easy to experimentally alter its concentation within managed ecosystems, except by enclosing them. As a result, most information about crop responses to elevated [CO$_2$] is from greenhouses, laboratory-controlled environment chambers, and transparent field chambers, where released CO$_2$ may be retained and easily controlled. Most of our information about the responses of managed ecosystems to rising [CO$_2$] are from such environments, with the implicit assumption that enclosure does not significantly alter response. Plants grown in protected environments commonly appear very different to those in the field. It has therefore been uncertain whether the response of chamber-grown crop plants to elevated [CO$_2$] will equal that of the crop in the open. FACE (free-air CO$_2$ enrichment), the subject of this book, is the one technique that does allow the impacts of future [CO$_2$] on managed ecosystems be assessed without otherwise altering the environment. Although systematic side-by-side (FACE vs enclosure) trials are lacking, there is now sufficient information from FACE to show by statistical meta-analysis that the effects of elevated [CO$_2$] on managed ecosystems differ significantly from chamber studies. Most notably, the yields of our major C3 grain crops (rice, wheat, soybean) are enhanced by elevated [CO$_2$] by only half the amount observed in enclosures and assumed in the IPCC model projections (IPCC 2001a; for a review, see Long et al. 2005). Elevation of [CO$_2$] to 550 ppm in FACE, the level expected by 2050, resulted in no increase in yield of the C4 cereals sorghum and maize, when a ca. 10 % increase is assumed in model projections (Leakey et al. 2006). By contrast, the yield increases of managed forest systems in FACE at elevated [CO$_2$] are larger than those found in enclosure studies (Ainsworth and Long 2005). Given that production fuels ecosystem processes, these findings show the need to reassess how rising [CO$_2$] will impact managed ecosystems via FACE. The following sections detail these findings.

1.2 Why are [CO$_2$] Enrichment Studies with Managed Ecosystems Important?

Agriculture is one of the most common land uses on Earth and agroecosystems are quite extensive. Globally, 5×10^{12} ha are under agricultural management and some 13×10^6 ha are annually converted to agricultural use, mainly from forests (FAO 2002). The world resources report 2000–2001 World Resources Institute (2002) defined agricultural areas as those where at least 30 %of the land is used as croplands or highly managed pastures. According to this definition, agroecosystems cover approximately 28 % of the total land area excluding Greenland and Antarctica and including some overlap with forest and grassland ecosystems. According to the FAO, 69 % of agroecosystems consist of permanent pastures. However, this global average masks very large differences in regional balances between crops and pastureland. On cropland, annual crops such as wheat, rice, maize, soybeans and tuber crops, which provide us with food, feed and fiber, occupy more than 90 % of the area. Thus, the share of carbon stored in agroecosystems (about 26–28 % of all carbon stored in all terrestrial systems) is about equal to the share of land that is devoted to agroecosystems. Despite the high productivity of global agriculture, much of the world's agricultural land offers less than optimal growing conditions. Soil fertility constraints include low potassium and phosphorus reserves, high sodium concentrations, a low moisture-holding capacity, or limited depth. Hence, a realistic assessment of the effects of $e[CO_2]$ has to consider the interactions between the changing additional constraints.

Between 20 % and 40 % of the world's land surface, depending on the definition used, is covered by grasslands. They are found throughout the world, in both humid and arid zones, but grasslands are particularly important features of the earth's drylands. The current volume contains two chapters on grassland from humid temperate regions only. Vast areas of rangelands, which cover more than double the global cropped area, have until now not received the attention they deserve. Only recently and for the first time have the net ecosystem CO_2 exchanges above a steppe in Mongolia been measured using the eddy covariance technique (Li et al. 2005, 2006). Moreover, grasslands provide a livelihood for 938×10^6 people (White et al. 2000), as well as forage for livestock and habitats for wildlife. Grassland vegetation and soil also store a considerable quantity of carbon. Other grassland ecosystem goods and services include cultural and recreational services, such as tourism and aesthetic gratification, and water regulation and purification.

The third class of ecosystems treated in the current volume are forests, specifically forest plantations. Forests cover about 25 % of the world's land surface, excluding Greenland and Antarctica. Although forest areas have increased slightly in industrial countries since 1980, they have declined by almost 10 % in developing countries. The greatest majority of forests in the industrial countries, except Canada, central Europe and Russia, are reported to be in "semi-natural" conditions or converted to plantations (World Resources Institute 2002). From the range of goods and services provided by forest ecosystems, the World Resources Institute considers the following five as the most important for human development and wellbeing: timber production and consumption, woodfuel production and consumption, biodiversity and watershed protection and carbon storage. Thus, forest FACE experiments seek to answer a critical question for foresters and policy-makers: Can we expect more growth and carbon sequestration in these forests in the future (see Chapters 10–13)?

The brief descriptions of the three classes of ecosystems considered in this volume convincingly highlight their economic and ecological importance. In addition, the focus of the research about the effects of e[CO_2] on managed ecosystems provides two important methodological advantages:

1. Experimental manipulations of the growing conditions, such as irrigation, supply of mineral fertilizers and plants with different functional traits, offer the possibility to detect, at the stand level, interactions with other growth factors.
2. Changes of a few percent in biogeochemical cycles have major implications at the global scale, yet are unlikely to be detected in experiments elevating [CO_2] in natural systems, because of the difficulty of separating [CO_2] effects from the high degree of spatial heterogeneity.

Crops provide genetically uniform monocultures, planted in fields where soil, nutrients and topography are also relatively uniform. As a result, between-plot variation is minimized, allowing a high degree of statistical sensitivity. These systems also therefore serve as model, yet real-world systems, where hypotheses of elevated [CO_2] effects may be tested in a cost-efficient manner, possibly providing guidance to subsequent study in natural ecosystems.

1.3 Free-Air [CO_2] Enrichment

Any attempt to understand the effects of increasing atmospheric [CO_2] concentration on ecosystem function must involve exposing today's ecosystems to expected future [CO_2] concentrations. A solid scale-up of the results from experimental plots to the field scale has to fulfil two minimal requirements:

1. The experimental setup to increase the free-air target gas concentration should not change the microclimate within and above the canopy, including the energy balance of the plant stand.
2. The experimental plots must be large enough to permit the removal of borders with a large enough remainder to provide a reasonable yield sample.

The disregard of this second requirement leads to a seriously flawed base for scaling-up (see also Chapter 14). Free-air carbon dioxide enrichment (FACE) offers a technology which meets the first requirement by minimizing unwanted effects of the system on the plant stand, but the system is not entirely without its own limitations (see Chapter 2). FACE also allows use of the experimental plot sizes needed for a reasonable scale-up. Consequently, FACE experiments offer a distinctive platform for multidisciplinary approaches, vital in addressing the essential features of a plant stand and its soil. The experience gathered with FACE experiments during the past decade has shown that the study of processes in the soil does require an extensive soil sampling (e.g. Van Kessel et al. 2006) and requires large plots.

Briefly, the FACE apparatus consists of a circular or octagonal system of pipes that releases either CO_2 or air enriched with the treatment gas just above the top of the crop canopy. For tall canopies (greater than 1 m), this is released at one or two additional heights below the canopy. Wind direction, wind velocity and $[CO_2]$ are measured at the centre of each plot and the information is used by a computer-controlled system to adjust the gas flow rate, controlled by a massflow control valve, to maintain the target elevated $[CO_2]$ (Long et al. 2005). FACE avoids the changes in micro-climate observed with all types of enclosure, especially warming, altered interception of precipitation and increased relative humidity, that impact evapotranspiration and feed-forward into changes in carbon uptake (Chapter 2).

The case studies presented in Section II of this volume are based on FACE technology. The different systems used include annual and perennial crops with different functional traits. Their systems vary widely in canopy structure, development of the source–sink ratios during the growing cycle and partitioning of photosynthates to the different plant parts. The plant stands were grown under a wide array of evaporative demand, soil fertility, fertilization and availability of water.

The FACE experiments are often designed to investigate fundamental mechanisms that drive ecosystem structure and function, core issues of ecology. Thus the importance of FACE experiments is not only how well they help to predict the impacts of $e[CO_2]$, but also how well they test ecological concepts in plant stands adequately representing the target ecosystem. We contend that the comprehensive studies at the large FACE sites are currently the best method to assess the impact of $e[CO_2]$. Simultaneously, they provide agronomists, foresters and breeders with the best opportunity to test and develop adaptation measures (Chapter 5).

1.4 Spatial and Temporal Scale

The effects of a major environmental variable on plants and ecological systems can be examined at spatial scales ranging from sub-cellular through to geographical regions. An example of this could be CarboEurope, which aims to understand and quantify the present terrestrial carbon balance of Europe and the associated uncertainty at local, regional and continental scales (http://www.carboeurope.org/). The timescales range from parts of seconds for rapid biophysical processes, to centuries for evolutionary changes. It is important to note that most field studies on the effects of $e[CO_2]$ on plants involve a step increase in atmospheric $[CO_2]$. A major assumption of these approaches has rarely been tested – that exposing an ecosystem to a single-step increase in $[CO_2]$ will yield a similar response to those of a gradual increase over several decades. In the real world, $[CO_2]$ is increasing gradually, thus allowing time to adjust processes in the soil within feedback mechanisms (see Chapter 8: Figs. 8.2, 8.3). Here again, managed systems have a distinct advantage over natural systems which include long-lived individuals. Annual crops can and are grown throughout their life cycle in elevated $[CO_2]$, as are short-lived perennial crops. These plants therefore are not subjected to a step increase within their lifetime, although the soils are. The results from the Swiss-FACE project, with a fumigation over 10 years, showed convincingly that the immediate response of an ecosystem to a step increase in $[CO_2]$ at the start of an experiment may overestimate some community responses to increasing $[CO_2]$, because soil biota may be sensitive to ecosystem changes that occur as a result of abrupt increases. The feedback mechanisms in the soil are only revealed during long-term field experiments (Schneider et al. 2004). Therefore, results from short-term experiments may be a critical flaw when scaled-up to longer time periods.

Long time-spans may also be needed to get a deeper insight into the effects of $e[CO_2]$ on soil carbon stabilisation mechanisms (see Chapter 21).

Gifford (1994) suggests that $e[CO_2]$ will result in an increase in C assimilation by plants,and that its subsequent sequestration in the soil could counterbalance CO_2 emissions. However, higher plant growth rates in a CO_2-rich world can only be sustained if the soil supplies plants with additional nutrients. Plant growth may be limited by several external co-limiting factors. The availability of nutrients in the soil is the key to the potential response of a plant to $e[CO_2]$ and is central to correctly predicting the response of terrestrial ecosystems to rising $[CO_2]$ levels. In order for a plant to fully realize the potential of increased $[CO_2]$, it must increase its nutrient uptake for the increased production of biomass. The use of stable isotope N-15 helped to identify the importance of the sources of N from symbiotic fixation and from soil N for plant growth (see Chapter 18). It has been well known since Liebig's time (1803–1873) that plant productivity is not controlled by the total of resources available, but by the scarcest resource (*Law of the minimum*). Enhanced plant growth needs a sustainable supply of N and other minerals. Therefore, the effect of $e[CO_2]$ on soil N availability is of crucial importance when predicting the potential for C storage in terrestrial ecosystems. Most of the studies on C–N interactions in the past two decades were in short-term experiments and conducted primarily from a plantcentric perspective. This perspective focuses on plant growth as regulated by N through short-term N turnover, soil mineral N availability and plant physiological adjustment. However, the plantcentric perspective does not consider the carbon pools and fluxes in the soil controlled by biota (see Chapters 21, 23) and their potential feedback on plant growth.

Temporarily, additional C sequestration into terrestrial ecosystems can be an effective strategy for mitigating the effects of $e[CO_2]$. The meta-analysis presented in Chapter 21 reviews, collates and synthesizes experimental results about the potential and the limitations of C sequestration. Determining the direction and magnitude of soil C sequestration is important because it is used to make decisions regarding the current and future management of ecosystems. At the political level, decisions have indeed been made. The Kyoto Protocol contains a provision of assigning credits for carbon (C) sequestration in forestry and agricultural soils. Thus, enhancement of the verifiable C pool in terrestrial ecosystems (soils and vegetation) can have both economic and environmental benefits. However, soil properties such as soil organic carbon pool are in dynamic equilibrium with the climate, particularly with precipitation and temperature (Lal 2003). Therefore, it has to be considered that the potential of soil organic carbon sequestration is finite.

The consequences of the rise in atmospheric $[CO_2]$ concentration have to be seen over these ranges of both temporal and spatial scales.

1.5 Elevated $[CO_2]$ Affects Plant Growth and Ecosystems via a Multitude of Mechanisms

The mechanisms of the effects of $[CO_2]$ on ecosystem productivity and processes in the soil in the face of other limiting factors are a complex issue, for which data is inadequate. Annual productivity of vegetation is usually constrained by one ore more of the following environmental variables: (1) incident radiation, (2) water supply, (3) temperature, (4) availability of mineral nutrients, (5) adverse soil conditions such as extreme pH, waterlogging, compaction or salinity. Experience has shown that the pacesetting role of the limiting factors and their importance may change during the life cycle of the plants. Functional traits of the species (e.g. a superficial root system that cannot acquire water and mineral nutrients in the deeper soil layers, as can species with a large tap root) may change the response to environmental variables.

Elevated CO_2 can affect plant growth via a multitude of mechanisms which are reviewed and synthesized in Section III of this volume. While many steps in metabolism utilize or respond to $[CO_2]$, the only sites where there is convincing evidence for a response in the concentration range of relevance (240–1000 ppm) are ribulose 1:5 bisphosphate carboxylase/oxygenase (Rubisco) and a yet-undefined metabolic step affecting stomatal aperture that may also involve Rubisco (see Chapter 14). Photosynthesis and stomatal movement are therefore the primary points of response to $e[CO_2]$; and all other changes in the system follow on from the response at this level. The study of Ainsworth et al. (2003) appears the most comprehensive with regards to the effect of the long-term growth at $e[CO_2]$ on leaf photosynthesis in any systems of CO_2 enrichment. It provides no support for the contention that stimulation of leaf photosynthesis will decline with time or that it will be diminished by low nitrogen supply in managed systems. The marked increase of foliar water-soluble carbohydrate content in plants grown at $e[CO_2]$ has consequences on the cross-talk between carbon and nitrogen metabolism. The expression of a number of genes is known to be sensitive to soluble carbohydrates, including glucose and sucrose. FACE experiments provide strong evidence that sink capacity is a key factor in determining the response of foliar carbohydrates to growth at $e[CO_2]$ (see Chapter 16). How is the increased rate of photosynthesis under $e[CO_2]$ reflected in the productivity of the plant stand? An increase of about 30 % is observed when plants grow individually in controlled conditions with ample nutrient supply. However, when plants are grown in the field, where colimiting growth resources other than CO_2 can markedly constrain plant growth response to $e[CO_2]$ (see Chapters 8, 19), the yield response is markedly weaker. The ecosystem's productivity may also be altered by the effect of $e[CO_2]$ on the leaf area, which determines the capacity for radiation interception. For example, in soybean, elevated $[CO_2]$ affects the leaf area primarily by delaying loss to senescence (Dermody et al. 2006). The actual yield increases due to $e[CO_2]$ are reviewed in Section II of this volume. Long et al. (2005) conclude that the evidence for a large response to $e[CO_2]$ is largely based on studies made within chambers at small scales which would be considered unacceptable for agronomic experiments when the results are scaled-up to the field level. The partitioning of carbon and dry matter within plants affects the way in which C enters harvestable products and the soil. The grassland swards at Eschikon, Switzerland, were grown for 10 years at $e[CO_2]$. Growth at $e[CO_2]$ resulted in a 43 % higher rate of light-saturated leaf photosynthesis. However, the yield response of *Lolium perenne* was low, due to increased allocation of biomass to non-harvested plant parts and because of increased night-time respiration (see Chapters 14, 19). FACE studies with the leguminous crop soybean showed that the effect of elevating $[CO_2]$ to 550 ppm on the aboveground net primary production (17–18 %) and yield (15 %) was less than projected from previous chamber experiments (Morgan et al. 2005). The observed decrease in partitioning to seed dry mass indicates that FACE also offers an experimental set-up to gain novel insights into the source–sink relationships of grain crops (see also Chapter 5). The flux of photosynthates and litter in the soil has profound consequences for C sequestration; and these relationships are addressed in Chapter 15. The observation on a planted sweetgum (*Liquidambar styraciflua* L.) site (Chapter 13) showed that this species preferentially allocates additional C to fine roots rather than to woody biomass, resulting in significant implications for the potential of this forest to sequester C. Also, loblolly pines (*Pinus taeda* L.) showed a higher rate of fine root growth at $e[CO_2]$, but this was accompanied by greater root death, resulting in a higher absolute turnover of root tissue (Chapter 11). Elevated CO_2 leads to a partial closure of the stomata and an increase in water-use efficiency. These changes have important positive impacts on plant water relations (Chapter 17). The effects of $e[CO_2]$ differed among plant functional types. However, statistical analysis to compare the effects of $e[CO_2]$ among plant types and experimental sites was exacerbated by the available dataset (Nowak et al. 2004). Because legumes fix N_2 and hence potentially avoid severe N limitations, they have been predicted to have greater responses to $e[CO_2]$ than other plant functional types. Data presented in Chapters 8 and 19 show that symbiotic N_2 fixation plays a key role in the strong stimulation of white clover (*Trifolium repens* L.), which also influenced interspecific interactions. The effect of $e[CO_2]$-induced changes in the population structure of rhizobia was transient in nature and was most likely influenced by the N status of the ecosystem, as well as by the type and concentration of root exudates (see Chapter 18). Processes in the soil are major players for the response of ecosystems to $e[CO_2]$. Thus, the availability of N affects soil C dynamics and plant growth (Chapter 21). The importance of understanding plant growth and soil biota productivity is widely recognized. However, the specific mechanisms that control those connections are poorly understood (Chapters 22, 23). Recent advances in technologies and resources hold the potential to improve our understanding of growth, development and adaptation of ecosystem processes under $e[CO_2]$. Molecular genetics and genomics have to date been largely unused in large-scale ecosystem experiments, despite their potential to elucidate aspects of the genome responsible for adaptive traits (see Chapter 20). The isotopic signature of carbon and nitrogen provides the means of tracking the transformation of carbon and nitrogen as they flow through plant and soil compartments; and it holds the possibility of separately investigating individual soil processes. The concluding Chapter 24 of this volume shows predictive models of biological systems incorporating biochemical and genetic data, an area where advances in technology have spurred realistic experimental platforms for data collection that could provide the information for predictive computational models – needed in order for public policy and management decisions to be underpinned by good critical science.

This volume is focussed on the effect of elevated $[CO_2]$, largely ignoring other environmental factors (with the exception of N availability) that are likely to change as $[CO_2]$ continues to increase. In future, more realistic open-air multifactor experiments will be needed, as the importance of multifactor interactions in the real world is evident. The pursuit of this concept could yield economical and ecological benefits.

1.6 Conclusions

- Managed ecosystems cover the greater part of the global land area and have the greatest potential to alter $[CO_2]$. They facilitate the use of manipulable variables and help to detect new insights into the effects of $e[CO_2]$ on plant growth, development and processes in the soil.
- FACE studies now provide our most realistic estimates of how ecosystems in the open will respond to the atmospheric $[CO_2]$ predicted for the middle or end of this century. They are our best validation data for models predicting the responses of managed ecosystems and natural vegetation to this ongoing change. Production changes and other effects due to elevated $[CO_2]$ in FACE differ significantly from observations made in chamber and greenhouse studies, suggesting that responses are signigicantly modified in protected environments and that models parameterized from such data might be in serious error.
- Elevated $[CO_2]$ affects ecosystem processes via a multitude of mechanisms at different timescales. The relationships between the C and N metabolism in the plants and the N availability in the soil are key factors that influence the direction and magnitude of the response of ecosystems to a $[CO_2]$-enriched atmosphere.
- Projections of ecosystem responses to elevated $[CO_2]$ must incorporate the reality of multiple factor influences. Many suggestions in this volume guide to the development of testable hypotheses for experiments which could provide basic information for predictive computational models. The findings have important implications for public policy decisions and adaptations of management systems.

References

Ainsworth EA, Long SP (2005) What have we learned from 15 years of free-air CO_2 enrichment (FACE)? A meta-analytic review of the responses of photosynthesis, canopy. New Phytol 165:351–371

Ainsworth EA, Davey PA, Hymus GJ, Osborne CP, Rogers A, Blum H, Nösberger J, Long SP (2003) Is stimulation of leaf photosynthesis by elevated carbon dioxide concentration maintained in the long term? A test with *Lolium perenne* grown for 10 years at two nitrogen fertilization levels under free air CO_2 enrichment (FACE). Plant Cell Environ 26: 705–714

Dermody O, Long SP, DeLucia EH (2006) How does elevated CO_2 or ozone affect leaf-area index of soybean when applied independently? New Phytol 169:145–155

FAO (2002) FAO-STAT statistics database. UN Food and Agriculture Organization, Rome

Gifford RM (1994) The global carbon cycle: a viewpoint on the missing sink. Aus J Plant Phys 21:1–15

IPCC (2001a) Climate change 2001: impacts, adaptation and vulnerability. Cambridge University Press, Cambridge

IPCC (2001b) Climate change 2001: the scientific basis. Cambridge University Press, Cambridge

Lal R (2003) Global potential of soil carbon sequestration to mitigate the greenhouse effect. Crit Rev Plant Sci 22:151–184

Leakey ADB, Uribelarrea M, Ainsworth EA, Naidu SL, Rogers A, Ort DR, Long SP (2006) Photosynthesis, productivity, and yield of maize are not affected by open-air elevation of CO2 concentration in the absence of drought. Plant Physiol 140:779–790

Li SG, Asanuma J, Eugster W, Kotani A, Liu JJ, Urano T, Oikawa T, Davaa G, Oyunbaatar D, Sugita M, (2005) Net ecosystem carbon dioxide exchange over grazed steppe in central Mongolia. Global Change Biol 11:1941–1955

Li SG, Eugster W, Asanuma J, Kotani A, Davaa G, Oyunbaatar D, Sugita M (2006) Energy partitioning and its biophysical controls above a grazing steppe in central Mongolia. Agric For Meteorol 137:89–106

Long SP, Ainsworth EA, Leaky ADB, Morgan PB (2005) Global food insecurity. Treatment of major food crops with elevated carbon dioxide or ozone under large-scale fully open-air conditions suggests recent models may have overestimated future yields. Phil Trans R Soc B 360:2011–2020

Morgan PB, Bollero GA, Nelson RL, Dohleman FG, Long SP (2005) Smaller than predicted increase in aboveground net primary production and yield of field-grown soybean under fully open-air [CO_2] elevation. Global Change Biol 11:1856–1865

Nowak RS, Ellsworth DS, Smith SD (2004) Functional responses of plants to elevated atmospheric CO_2 – do photosynthetic productivity data from FACE experiments support early predictions? New Phytol 162:253–280

Schneider MK, Lüscher A, Richter M, Aeschlimann U, Hartwig UA, Blum H, Frossard E, Nösberger J (2004) Ten years of free-air CO_2 enrichment altered the mobilization of N from soil in Lolium perenne L. swards. Global Change Biol 10:377–388

Van Kessel C, Boots B, Graaf M de, Harris D, Blum H, Six J (2006) Soil C and N sequestration in a grassland following 10 years of free air CO_2 enrichment. Global Change Biol (in press)

White R, Murray S, Rohweder M (2000) Pilot analysis of global ecosystems: grassland ecosystems. World Resources Institute, Washington, D.C. World Resources Institute (2002) World resources 2000–2001: people and ecosystems: the fraying web of life. United Nations Development Programme, United Nations Environment Programme, World Bank, World Resources Institute, Washington, D.C.

2 FACE Technology: Past, Present, and Future

G.R. Hendrey and F. Miglietta

> "Plants can only perceive a change in atmospheric concentration through tissues that are exposed to the open air."

2.1 Introduction

FACE (free-air CO_2 enrichment) was developed in order to fill a need for conducting realistic experiments to understand how plants and ecosystems will respond to the increasing ambient concentration of atmospheric carbon dioxide ($c[CO_2]$). The need for such studies is apparent since nearly every life form on Earth is completely dependent on the conversion of CO_2 to plant matter via photosynthesis, so understanding the consequences of changes in $c[CO_2]$ is a critical societal as well as scientific interest. Among the many high-level questions addressed by FACE are:

- How will plants respond to the increases expected in atmospheric $[CO_2]$ and oxidants such as tropospheric ozone (O_3)?
- How are these responses likely to feed forward into ecosystems, become limited or enhanced by ecosystem properties, and feed back to contribute to regulation of $c[CO_2]$ itself, a principal driver of global climate change?
- How will goods and services provided to mankind by forests, crops, and natural ecosystems be altered due to these changes in atmospheric trace gases?

Questions such as these have motivated the development of techniques for conducting controlled experiments in which the concentrations of CO_2 and/or oxidants are manipulated to investigate the responses of plants and their ecosystems. The objectives of this chapter are to provide: (1) background information on the need and development of FACE technology, (2) summary and comparative information on the characteristics of FACE experiments that operate at ecosystem scale (Tables 2.1–2.3), (3) a discussion of

Ecological Studies, Vol. 187
J. Nösberger, S.P. Long, R.J. Norby, M. Stitt,
G.R. Hendrey, H. Blum (Eds.)
Managed Ecosystems and CO_2
Case Studies, Processes, and Perspectives
© Springer-Verlag Berlin Heidelberg 2006

Table 2.1. Physical characteristics of 11 FACE facilities including location, elevation, and both the number and size of the FACE rings. See Tables 2.2, 2.3 for additional information on these FACE facilities

PROJECT	Location	Latitude	Longitude	Elev. (m)	Ring diameter (m)
AZFACE	Maricopa, AZ, USA	33°4' N	111°58' W	361	25
SoyFACE	Champaign, IL, USA	40°02'N	88°14'W	228	20
RiceFACE	Shizukuishi, Japan	39°38'N	140°57'E	200	12
Potato FACE[a]	Rapolano, Italy	43° 17' N	11°36' E	172	8
FAL FACE	Braunschweig, Germany	52° 18' N	10° 26' E	81	20
Swiss FACE	Eschikon, Switzerland	47°27' N	8°42' E	565	18
New Zealand FACE	Bulls, New Zealand	40° 14' S	175°16' E	9	12
FACTS-I	Chapel Hill, NC	35°97' N	79°09' W	178	30
FACTS-II	Rhinelander, WI	89.7° N	45.6° W	490	30
ORNL FACE	Oak Ridge, TN	35°54'N	84°20'W	229	25
PopFACE	Viterbo,Italy	42°22'04" N	11°48' E	150	22

[a] PotatoFACE experiment-1 occurred in 1995, Experiment-2 in 1998–99.
[b] Control rings have equipment similar to CO_2-enriched FACE rings.
[c] Ambient plots have no equipment simulating FACE equipment.
[d] Soybean experiment only. [e]Soybean and Maize experiment.

some problems and limitations of the method, and (4) suggestions for future directions of FACE technology.

Thirty-two FACE facilities of various designs ranging from 1 m to 30 m in diameter were assembled, with 14 in managed ecosystems and 15 in natural or semi-natural ecosystems. A list of these FACE sites and a map of their approximate locations is at http://cdiac.esd.ornl.gov/programs/FACE/face.html. Of these FACE facilities, 12 were in various types of grasslands, six were for crops, five were for bogs, five were for trees, one was for native chaparral, one was for a tree-line ecotone, and one was for native desert. Several of these FACE sites included multiple species or natural communities and addressed interactions of populations in the various systems in which they were set. Subsequent chapters in this book are case studies describing the output of 11 FACE experiments in crops, meadows, and forest ecosystems; and this chapter concentrates on those FACE studies in particular.

Treatmet rings	Control Rings[b]	Ambient plots[c]	References to FACE
4	4	varied, 0 to 4	Lewin et al. 1992; Lipfert et al. 1992; Pinter et al. 2000
12[d] or 16[e]	4[d] or 8[e]	–	Miglietta et al. 2001; Leaky et al. 2003; Morgan et al. 2004
4	4	–	Okada et al. 2001
exp.1 : 3	exp. 1: 1	exp. 1: 0	Miglietta et al. 1997
exp. 2 : 3	exp. 2: 3	exp. 2: 3	
2	2	2	Lewin et al. 1992; Weigel and Dämmgen 2000
3	0	3	Lewin et al. 1992; Hebeisen et al. 1997
3	0	3	Edwards et al. 2001; Lewin et al.1992
4	3	variable	Hendrey et al. 1999; DeLucia et al. 1999.
9	3	–	Hendrey et al. 1999; Karnosky et al 1996; Karnosky et al. 1999
2	2	1	Hendrey et al. 1999; Norby et al. 2001
3	3	–	Miglietta et al. 2001

2.2 Need for Controlled Experiments in the Field: Historical Perspective

Prior to 1985, several investigators at different institutions had been working on various forms of open-air fumigation systems, more or less independently. Some of these included the concept of surrounding a plot with a circular array of emitters for fumigating plants with O_3 or SO_2 and including some form of feedback control system (e.g., McLeod et al. 1985; Mooi and van der Zalm 1985; McLeod 1993). However, many early free-air release systems often experienced concentration excursions equal to three to five times the intended mean elevated (e) $[CO_2]$ concentration for about 10 % of the time over periods of minutes to hours (Shinn and Allen 1985; Allen 1992). Often, a large volumetric buffer was placed between the CO_2 sample-point within the treatment area and the CO_2 analyser to reduce the obvious variability in the treatment $e[CO_2]$. Under such conditions, the reported occurrences of excursions in $e[CO_2]$ (several times the treatment mean) for averaged values implies plants were at times exposed to very high concentrations indeed and for intervals of many minutes duration.

Thus, a reasonable conclusion was reached at that time that poor control of the fumigant and exposure of test plants to variable, high spike concentra-

Table 2.2. Experimental descriptors for 11 FACE studies. See Tables 2.1, 2.3 for more details.

FACE Site	Main experimental variables[a,b]			Intended Treatment level[a,b]	
AZFACE	CO_2	$CO_2 \times H_2O$	$CO_2 \times N$	550	N: 15, 70, 350 kg N ha^{-1}
SoyFACE[c]	CO_2	O_3	$CO_2 \times O_3$	550	O_3: ambient $\times$ 1.2
RiceFACE	CO_2	$CO_2 \times N$	–	Ambient +200	N: 40, 80, 120 kg N ha^{-1}
Potato FACE	CO_2	–	–	Exp. 1: 460, 560, 660	Exp. 2: 560
FAL FACE	CO_2	$CO_2 \times N$	–	550	N: adequate and 50 % of adequate
Swiss FACE	CO_2	N $CO_2 \times N$	$CO_2 \times$ spp $CO_2 \times N \times$ spp	600	140, 560 kg N ha^{-1}
New Zeland FACE	CO_2	CO_2 and grazing	–	475	With or without sheep grazing
PopFACE	CO_2	N[e]	CO2 $\times$ N[e]	550	290 kg N ha^{-1}
FACTS-I	CO_2	–	–	Ambient+200	–
FACTS-II	CO_2	O_3	$CO_2 \times O_3$	560	O_3: 1.5 times ambient
ORNL FACE	CO_2	–	–	565	–

[a] Unless otherwise indicated concentration values are for [CO_2] in ppm. All FACE experiments compare observations of plants grown under [CO_2] enrichment to plants grown in ambient [CO_2].

[b] FACTS-II O_3 treatment aimed for an average of 60 ppb. O_3 fumigation did not occur when leaves were wet due to dew or rain.

[c] The SoyFACE [O_3] target was adjusted to 20 % above the ambient value, the actual seasonal elevation was 21 % and O_3 fumigation did not occur when leaves were wet with dew or rain (D. Ort, personal commuication).

Auxilliary experiment[a]	Exposure period	System	Vegetation	Chapter in this book
Sudangrass	14–24 hour seasonal	Agriculture	Cotton, Wheat, Sorghum	3.1
–	Daylight, seasonal	Agriculture	Soybean, maize	3.2
–	24 hour, seasonal	Agriculture	Rice	3.3
–	Daylight seasonal	Agriculture	Potato	3.4
–	Daylight, multi-seasonal	Agriculture	Winter barley, ryegrass, sugar beet, winter wheat	3.5
12 grassland spp placed in rings	daylight seasonal	Pasture	Perennial ryegrass, white clover and other grassland spp	3.6
–	Daylight 365 days year^{-1}	Pasture	C3 and C4 grasses, forbs and legumes (=20 spp)	3.7
–	Daylight seasonal	Forest plantation	Poplar plantation	3.8
550 × N	24 hour, 365 days year^{-1}	Forest plantation	–	3.9
	Daylight seasonal	Forest plantation	Plantation: aspen, birch, maple	3.10
–	Daylight seasonal	Forest plantation	Sweetgum plantation	3.11

tions made free-air releases unacceptable as an experimental method. Nevertheless, the pressing need to conduct plant and ecosystem fumigation experiments under realistic field conditions was widely perceived, as the limitations of chamber methods became increasingly apparent. For example, open-topped chambers (OTC), developed for fumigation of plants rooted in field soils, introduced many perturbations to the growth environment by altering microclimate variables such as photon flux, the ratio of diffuse to total solar irradiance, temperature, humidity, and wind stress. Typically, OTCs are about 2–4 °C warmer than ambient, light is attenuated by about 10–25 %, and wind speeds are low and steady in contrast to their great variability in the ambient environment, thus protecting the plants from physical stresses. These changes can significantly alter leaf temperature, transpiration, and latent heat flux, and

Table 3. Typical performance and operating costs for 11 FACE experiments

Project	Typical performance: fraction of 1-min average [CO_2][a]		Years of operation	Annual costs[b] (U.S. $ × 1000)				References
	within 20% of setpoint	within 10% of setpoint		Total operation	CO_2	Electricity	Maintenance	
AZFACE	0.99	0.90	10	300	120	5	30	Nagy et al 1994
SoyFACE	0.95	0.85	4	235	81[d]	12	45	Leakey et al. 2004; D. Ort, personal communication
RiceFACE	0.87	0.59	6	–	–	–	–	Okada et al. 2001
Potato FACE	0.96	0.81	exp. 1:1 exp. 2:2	60–90	20–50	5–10	35–30	Miglietta et al. 1997
FAL FACE	0.99	0.99	6	75	70	2	3	Weigel and Dämmgen 2000; H. Weigel, personal communication
Swiss FACE	0.99	0.92	10	150	112	2	3	Zanetti et al. 1996; Lüscher et al. 1998
New Zealand FACE	–	0.83	7	75	37	6	4	Edwards et al. 2001
FACTS-I	0.92	0.69	8	1,311	911	–	194	Hendrey et al. 1999
FACTS-II	0.93	0.73	7	1,000	400–650	15	115	Karnosky et al. 2003
ORNL FACE	0.9		7	650	200	30	50	Norby 2001
PopFACE	0.91	0.75	6	150–300[e]	67–227[e]	1.5	–	Miglietta et al. 2001

[a] Average of 60 1-s "grab samples" except at RiceFACE where the average is for 30 s recorded every 24 min.
[b] Currency conversion: Euro to U.S. $ at 1:1; SFr to U.S. $ at 1:0.666.
[c] For FACTS-II O_3 performance information see accompanying text and Wustman et al. 2001.
[d] Soybean and maize experiment with 12 rings. [e] CO_2 consumption differed year to year.

can alter the apparent response of plants within OTCs, resulting in treatment response artefacts, so-called "chamber effects", equivalent to doubling $[CO_2]$ within the chamber (reviewed by McLeod and Long 1999).

2.3 Advantages of FACE

FACE technology permits $[CO_2]$ and the concentrations of other trace gases, when averaged over minutes or longer, to be maintained stably at levels expected to prevail in the mid- to late twenty-first century. This is accomplished in most cases without biologically significant change in micrometeorological conditions (Hendrey et al. 1993). The size of larger FACE plots, encompassing up to hundreds of square meters, permitted use of a buffer zone that eliminated effects of the plot edge on the rest of the vegetation within the plot, a major problem with small plots and enclosure studies. It also allowed many ecosystem-level questions to be addressed by large teams of investigators sampling above and below ground without damaging much of the plant or soil material or compromising the utility of the plot.

FACE systems can provide a built-in isotopic tracer of carbon uptake and distribution, but the source of CO_2 is an important consideration. If the CO_2 is derived from the combustion of fossil carbon sources, then it is depleted in ^{14}C relative to ambient air and the ^{12}C/^{13}C ratio is substantially different from that of current atmospheric CO_2. This difference can be tracked through biological systems (Leavitt et al. 1994).

FACE experiments make possible the simultaneous application of multiple variables interacting with CO_2, including increased O_3 and multiple plant species (Karnosky et al. 1999), altered nutrient or water regimes (Mauney et al. 1994), and warming of the vegetation by radiant heaters (Nijs et al. 1996).

2.4 Problems and Limitations

While there are numerous and well recognized advantages of FACE, there are also some problems that are less well known. Investigators using FACE experiments should be more aware than they seem to be of some of these inherent limitations. Among the most important of these are: (1) the step increase in $[CO_2]$ when a FACE experiment is initiated within plots of existing vegetation, (2) biologically significant high-frequency variability in e$[CO_2]$, (3) the limited size of the FACE plots, and (4) in some situations, disturbance to microclimate.

2.4.1 CO$_2$ as a Step Treatment

The actual rate of change of c[CO$_2$] due to anthropogenic activities is gradual (2.5 ppm year^{-1} in 2002–2003) but in most FACE studies the experiment is initiated with a step increase on the order of 50 % in e[CO$_2$] relative to c[CO$_2$]. The abrupt increase might result in plants, soil organisms, and ecosystems responding differently to the treatment compared to those exposed to the ongoing, gradual increase in c[CO$_2$] of the atmosphere. The large surge of photosynthate into the soil system with the suddenly increased CO$_2$ may bind nitrogen or via accelerated root uptake place an unrealistically large demand on soil nutrients, causing the stimulation of carbon gain by the system to be transitory (Luo and Reynolds 1999; Pendall et al. 2004). There is some experimental evidence to suggest this may be a problem for FACE experiments. The downward adjustment of net primary production (NPP) due to such soil nutrient limitation was demonstrated for loblolly pine in North Carolina with moderate soil fertility (Oren et al. 2001). An initial large stimulation of NPP with a step increase in e[CO$_2$] from ambient to 550 ppm all but disappeared over a 3-year period. Luo et al. (2003) suggested that nitrogen immobilization in biomass and organic-horizon pools drove the forest plots to a state of acute N limitation. When N enrichment was applied to half of the plot, NPP was restored to near initial stimulation levels but remained lower in the untreated half of the plot (Oren et al. 2001). In the FACTS-I experiment that started 2 years later at the same forest area, the initial CO$_2$-induced stimulation of NPP did not "crash" in the same way or at the same time relative to initiation of the study, suggesting that factors such as local soil variables or even climate conditions in the initial year might influence the outcome and conclusions drawn from the FACE study (Hamilton et al. 2002).

In a simple model ecosystem of a grass (*Bromus inermis*) and its associated arbuscular mycorrhizal symbionts, a stepped e[CO$_2$] treatment of field soils in a greenhouse resulted in a loss of fungal taxa compared to plots receiving a gradual e[CO$_2$] treatment (Kilronomos et al. 2005). A possible explanation for this observation was that the rapid loading of organic carbon to the soil with the step increase in e[CO$_2$] stressing the fungal community more severely than did the slow increase in e[CO$_2$], the latter permitting the rapidly regenerating taxa of the fungal community enough time to adapt so as to take maximal advantage of the new, but slowly increasing resources.

The lesson here is that experimenters need to keep in mind that the inherent artificiality of a manipulation of a natural ecosystem with FACE or any other controlled experimental approach may impact the ecosystem somewhat differently from the century-long, uncontrolled experimental enrichment of the atmosphere. New approaches to modeling carbon acquisition and sequestration within a CO$_2$-enriched ecosystem (e.g. Luo et al. 2003) are likely to add great value to the interpretation of FACE data. Nevertheless, the advantages of FACE make it the best approach available to evaluate whole ecosystem responses to manipulation of atmospheric chemistry and data from FACE studies are required for evaluation of such models.

2.4.2 High-Frequency Variation in [CO_2]

What sort of variability in [CO_2] do plants normally experience? Seasonal oscillations in c[CO_2] are evident in long-term records. Diurnal variability is associated with night-time respiration; and pre-dawn c[CO_2] in mid-summer may exceed 500 ppm within a crop or forest canopy. Diurnal excursions in [CO_2] can persist well after dawn if the atmosphere is stable, but as solar heating introduces turbulence and photosynthesis ramps up, c[CO_2] is soon restored to more typical values. With windy conditions such diurnal variability is greatly reduced to typically ~10 ppm at the canopy top.

For most FACE systems, when e[CO_2] is averaged over long periods, variability appears quite low, but on a second-to-second basis, it is seen to be high. For example, at the Duke Forest FACE site, 1-s e[CO_2] typically ranges from ambient to over 1000 ppm with oscillations sometimes lasting tens of seconds (Fig. 2.1). Although this is among the most variable e[CO_2] of the modern

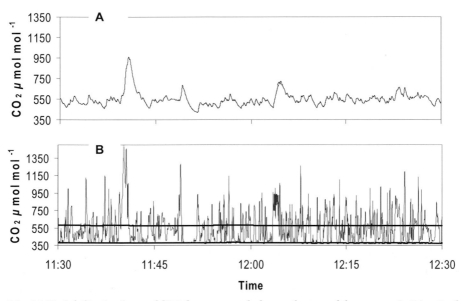

Fig. 2.1 Variability in elevated [CO_2] ppm sampled near the top of the canopy in Ring 2 of the FACTS-I FACE experiment, Duke Forest, 6 November 2004. **A** 1-min average (exponential smoothing) of 1-s observations. **B** 1-s observations. The solid line at ~578 ppm is the target value (ambient plus 200 ppm). The line near 378 ppm is the 1-s observation of ambient [CO_2] ppm. During this period, the average wind speed was 2.6 m s^{-1} (s.d. 1.20)

large FACE systems reported to date (very little of such 1-s data is ever reported), at its worst this high frequency variability is a considerable improvement over the long-term average performance of exposure systems reviewed by Shinn and Allen (1985). What is the significance of this variability?

Wheat leaf photosynthetic fluorescence responded within a few seconds to a step-change of about 50 % above $c[CO_2]$. In leaf cuvettes, symmetrical oscillations of 225 ppm about a constant $e[CO_2]$ set point of 575 ppm or 650 ppm enduring <1 min had no effect on net carbon gain, but oscillations lasting ca. 1 min or longer decreased the net carbon gain observed at the constant $e[CO_2]$ by up to ~17 % (Hendrey et al. 1997). A similar effect was seen with teak and barrigon (*Pseudobombax septenatum*) tree seedlings (Holtum and Winter 2003). Differences of this magnitude are not trivial and there may be a systematic understatement of the effect that globally elevated $c[CO_2]$ has on carbon gain when the assumption is that $e[CO_2]$ is essentially constant in CO_2 enrichment experiments such as FACE.

2.4.3 Limited Plot Size

The largest FACE plots cover ~450 m^2 per ring and may encompass ca. 100 pine trees at canopy closure as well as dozens of secondary plant species, permitting the study of community interactions and nutrient dynamics at the scale of a forest ecosystem. However, a FACE plot is not large enough to capture a meaningful watershed or for understanding effects on large herbivores or predators. A FACE plot resembles an island within its surrounding ecosystem and FACE experiments cannot yet control large-scale feedbacks, for example, between reduced transpiration due to $e[CO_2]$ and relative humidity entering the FACE plot. Small FACE plots, those of a few square meters or less, are essentially all "edge" and both the temporal variability in $e[CO_2]$ and the frequency with which very high concentrations of CO_2 persist for extended periods are not as well documented as with the larger FACE systems.

2.4.4 Blower Effect

A problem arising with the Brookhaven National Laboratory (BNL) FACE design (Hendrey et al. 1992, 1999) is that the CO_2-enriched air blowing out of the vertical vent pipes (VVP) has the potential to cause microclimate perturbations under very stable and calm atmospheric conditions, as on some cold nights with dew or frost formation. Under such conditions in winter, in FACE experiments with winter wheat in Arizona, FACE plots averaged about 1 °C warmer than plots without blowers. The period of time that leaves were wetted with dew was reduced by 30 % and incidence of frost on leaves was reduced. Later in the crop season, on calm nights $c[CO_2]$ within ambient plots (no $e[CO_2]$ treatment, no blowers) could build up to 800 ppm or even 1000 ppm due to respiration. But when blowers were operated 24 h day^{-1}, this respiratory CO_2 was dissipated. Significant biological effects could conceivably occur from even these small perturbations (Pinter et al. 2000). However, a study of the blower effect at the Duke Forest FACE site (He et al. 1996) found no detectable effects on heat and momentum and no biologically significant disturbances to the natural environment. Most FACE experiments avoid this problem and reduce CO_2 use by not fumigating at night.

Whether or not to fumigate with CO_2 at night is a somewhat contentious issue because of the possible inhibition of dark respiration as a consequence of elevation of $[CO_2]$. A recent and thorough analysis by Davey et al. (2004) found neither inhibition of specific rates of respiration, nor a known mechanism to explain such an inhibitory effect. We conclude that no meaningful experimental artefact will be introduced by applying $e[CO_2]$ only during the day.

2.5 FACE Systems Engineering

2.5.1 Historical Perspective

A lengthy review of FACE development is presented elsewhere (Allen 1992), but some additions are made here. Maintenance of stable $e[CO_2]$ in a free-air environment presented a general engineering control issue that can be understood as regulation of fluid dilution into a stream of variable flow. This was handled earlier in an entirely different environment and experiment, the regulation of pH in stream channels with irregular flow rates and with very low chemical conductivity. Effective pH control was achieved with automated predilution of the added acid or base, chemical monitoring downstream with real-time data input to a microprocessor and fast-feedback control of a dosing system using a proportional-integral-differential (PID) controller (Hendrey 1976). Integration of these ideas from a limnological experiment with the concept of a circular array emitting a fumigant into ambient airflow contributed to the success of the FACE prototype system developed in 1986 in the field at BNL (Hendrey et al. 1992). The key innovations (compared to earlier free-air release systems) adopted from the stream channel study included improved and inexpensive control systems, fast-feedback PID control and more effective dilution of the fumigant. In addition, directional control was greatly improved for FACE compared to earlier systems by a microprocessor connected to a wind vane and the ability to alter directional control by 11.4° increments.

Innovative control algorithms were developed based on well-known engineering methods coupled with new ideas for improved mixing of the fumigant with ambient air. A sufficient engineering investment to assure a high degree of system reliability resulted in maintenance of $e[CO_2]$ within FACE plots with much-reduced variability compared to earlier methods (Hendrey et al. 1993; Miglietta et al. 2001). The introduction of directional control, that is the release of CO_2 only from the upwind direction relative to the center of the FACE plot, provided a large increase in the efficiency of CO_2 use (Lipfert et al. 1992), hence a decrease in the cost of FACE experiments, while simultaneously maintaining acceptable control of the gas concentration.

The successful FACE systems used in the experiments described in the case studies of Part B of this book are based, essentially, on one of two types of FACE technology that we will refer to as the BNL and CNR designs.

2.5.2 BNL FACE Design

FACE systems designed by BNL (Hendrey et al. 1992, 1993, 1999; Lewin et al. 1992; Nagy et al. 1994) use VVP arrays connected to a toroidal plenum through which CO_2-enriched air is released (Fig. 2.2). All of the FACE studies described in the case studies are well replicated, with multiple plots for both treatment and ambient conditions. The main features of the BNL design are: (1) pre-dilution of the CO_2, (2) directional control of CO_2 release relative to wind direction, (3) fast feedback regulation with a PID algorithm based on wind speed and direction measured right at the canopy top and $e[CO_2]$ measured within the canopy, and (4) an annular mixing zone (typically 2 m wide) to maintain relatively constant temporal and spatial control of $e[CO_2]$ and to exclude plants exposed to highly variable and excessive $e[CO_2]$ from the sampling programs. Although nine BNL-type FACE facilities were assembled, each differed in ways dictated by requirements of the individual experiments. For example, in the AZFACE, FAL FACE, FACTS-I, and FACTS-II sites (Tables 2.1–2.3), all of the tubing was located on the surface as in Fig. 2.2. At the Swiss FACE site (Fig. 2.3), the plenum was buried to minimize any wind turbulence that might produce an experimental artefact. Ring diameters also differed among, but not within, experiments over a range of 15–30 m. The "performance" of FACE is indicated by the ability to control $e[CO_2]$. BNL-type systems typically maintain 1-min average $e[CO_2]$ within ±20 % of the intended treatment level for 90–99 % of the time (Table 2.3).

The Institute of Biometeorology (formerly IATA, Institute of Agrometeorology) of the National Research Council (CNR) in Italy started to consider the possibility of using FACE systems in the 1990s. A MiniFACE system with a plenum and VVP system was developed and used to fumigate several types of vegetation in plots of one to a few square meters (Miglietta et al. 1996, 1998; Heijmans et al. 2001; Teyssonneyre et al. 2002; Mitchell et al. 2003; Barnard et

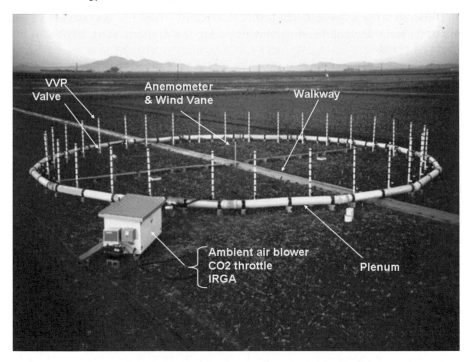

Fig. 2.2 Typical configuration of a BNL FACE design in an agricultural setting (Maracopa, Ariz.) for a FACE ring of 25 m diameter showing arrangement of key elements such as the vertical vent pipes (*VVP*), circular plenum and housing for the blower and gas analyzer (*IRGA*). Photo by K. Lewin

Fig. 2.3 One of the three FACE rings at Eschikon, Switzerland. Note that the plenum is below ground. Photo by G. Hendrey

al. 2004). In other CNR research projects, medium-sized FACE systems of 8 m diameter were designed and used with potatoes (Miglietta et al. 1997; Magliulo et al. 2003).

2.5.3 CNR FACE Design

At the end of the 1990s, a new approach was considered by investigators at the Institute of Biometeorology (IBIMET-CNR), Firenze, Italy, that used pure CO_2 rather than pre-mixed air plus CO_2, thus eliminating the need for blowers and leading to a substantial simplification of the equipment and hardware required to perform FACE experiments. A similar approach was independently chosen by the TNAES (Tohoku National Agricultural Station) scientists in Morioka, Japan (Okada et al. 2001). In this FACE design (Fig. 2.4), two or more layers of 8-m long pipes are arranged in an octagon and pure CO_2 is released to the atmosphere through 350 or 500 small gas jets. Mass flow is controlled by pressure within the pipes. High jet velocity creates rapid dilution with ambient air as described by Miglietta et al. (2001). The same approach has been applied, so far, in the SoyFACE experiment in the United States (Leakey et al. 2004; also see Chapter 4), in the POPFACE and EuroFACE projects in Italy (see Chapter 10), and more recently in the Australian TasFACE

Fig. 2.4 A segment of one of the FACE rings of the RiceFACE experiment at Shizukuishi, Japan. Photo by G. Hendrey

project (Hovenden 2003). CNR-type systems maintain 1-min average $e[CO_2]$ within ±20 % of the intended treatment level for 87–95 % of the time (Table 2.3).

2.5.4 Web-FACE

A different approach to FACE for tall trees, called Web-FACE, has been developed by Pepin and Körner (2002). This does not provide an ecosystem experiment, but simply emits CO_2 into the canopy, so it is not dealt with elsewhere in this book. Nevertheless, the approach deserves some comment. Web-FACE consists of a network of ~8.5 km of surface drip irrigation tubing (4.3 mm i.d.) strung through the canopy of 15 adjacent trees of a variety of species. Installation and biological sampling required a 45-m tall, free-standing tower crane and jib able to reach all of the trees. Pure CO_2 is distributed through the emitter tubes controlled by a simple algorithm based on wind speed and measured $[CO_2]$ within the canopies. Wind speed is not measured at the canopy as in other FACE systems, but at the top of the crane, some 10–15 m above the canopy and control is based on $[CO_2]$ averaged within the enriched zone of the canopy measured at 8-min intervals. Web-FACE performance (47 % and 76 % of 1-min averages within 10 % and 20 %, respectively, of the target $[CO_2]$) is substantially below other FACE systems (Table 2.3). With an intended treatment $[CO_2]$ increment of ~130 ppm (target of 500 ppm), the range of seasonal averages was 120 ppm, so these 15 un-replicated treatments differed among trees by up to 81 % of the intended $[CO_2]$ increment (Fig. 7 in Pepin and Körner 2002). This compares to a seasonal variation from target of <3 % of the intended $[CO_2]$ increment with the BNL forest FACE design (Hendrey et al. 1999). Spatial variation within the treatment zone covers a range >200 ppm and is quite variable within individual trees, as determined from $\delta^{13}C$ of indicator plants grown June–October in small flasks distributed through the treatment volume (Fig. 8 in Pepin and Körner 2002). Pure CO_2 is emitted in very close proximity and even directly onto some leaves (Fig. 1 in Pepin and Körner 2002), so that the actual exposure of individual leaves must cover a concentration range from near ambient to nearly 10^6 ppm, even though monitoring indicates reasonable dilution of the CO_2 some 40 cm away. It seems likely that the short-term variability (minutes) in $e[CO_2]$ is quite large for an un-quantified fraction of these leaves, a biological problem discussed above (Section2.4.2). One may ask just what is the experimental treatment. The principal advantage of Web-FACE is that the initial cost of system hardware may be less than other designs but this is an advantage that must be a trade-off with a problematic experimental design and poorly controlled performance.

2.6 Multiple Variable Experiments

Several of the FACE experiments described in this book (see Part B: Case Studies, for details) have simultaneously interacting variables on plant responses (Table 2.2). From the standpoint of FACE technology currently in use, however, the simultaneous application of O_3 with CO_2 is the most challenging. The FACTS-II (Aspen FACE) study utilizes the BNL FACE system and the SoyFACE project uses the CNR design (Tables 2.1–2.3).

As discussed above, e[CO_2] is subject to frequent oscillations in which the momentary concentration of CO_2 may be double the intended e[CO_2], but this is not acceptable for an experiment with O_3. Andrew McLeod (Fig. 2.5), a pioneer in the early development of FACE technology (e.g., McLeod et al. 1985), conducted engineering studies on the configuration of a system for releasing O_3 in such a way as to assure adequate dilution with minimal likelihood of O_3 spikes that would quickly damage plants. He developed a system in which the emitter ports faced outward from plot center with the jets impinging on a rectangular baffle to facilitate mixing with wind blowing into the plot. We incorporated this concept into the second forest-atmosphere carbon transfer and storage (FACTS-II) FACE facility in Rhinelander, Wis., (AspenFACE; Fig. 2.6). FACTS-II has 12 rings of 30-m diameter in a full factorial randomized com-

Fig. 2.5 Andrew McLeod with an engineering prototype for release of O_3 into FACE plots. Note the emitter jets facing outward toward vertical baffles. Photo by G. Hendrey

plete block experiment with controls, elevated CO_2, elevated O_3, and elevated $CO_2 + O_3$, all in triplicate (Dickson et al. 2000). The SoyFACE project has as many as 16 of the CNR FACE rings, depending on the objectives of annual experiments.

Some FACE experiments intended to maintain a constant, elevated concentration of CO_2 while others applied an increment of circa 50 % to the slightly variable ambient concentration. In contrast, O_3 exposure experiments with FACE do not try to maintain a constant treatment, nor do they apply a controlled increment to the ambient O_3 value. Instead, on any particular day, the elevated $[O_3]$ follows a series of steps from ambient to a maximum daily value and back to ambient in an attempt to mimic what happens in nature, but in a more well defined way. The O_3 treatments in both the SoyFACE and FACTS-II experiments (see Chapters 4 and 12) are applied during daylight hours following the 1.5 times ambient fumigation profile described by Karnosky et al. (1996) for FACTS-II or a 1.2 times ambient for the Soy FACE project. For exam-

Fig. 2.6 FACTS-II facility (AspenFACE experiment) in Rhinelander, Wis. This has three 30-m diameter rings with elevated $[CO_2]$, three with elevated $[O_3]$, three with both $[O_3]$ and $[CO_2]$ elevated and three with ambient air. Photo by D. Karnosky

ple, during the growing season of 1998, $c[O_3]$ at the FACTS-II site was 36 ppb and $e[O_3]$ in the fumigated plots was 56 ppb. On warm sunny days when $c[O_3]$ is normally higher than average, a diurnal exposure curve with a high absolute maximum $e[O_3]$, 100 ppb, was employed. On cool and cloudy days, the diurnal curve had a lower absolute maximum set point, ~55 ppb. Elevated O_3 treatment did not occur when plants were wet with rain or dew due the accelerated rate of O_3 deposition onto wet surfaces. The total O_3 exposure over the growing season was typically near 95 ppb (Dickson et al. 2000; Wustman et al. 2001).

2.7 Future Perspectives

2.7.1 The GradFACE Design

In 2004, the USDA-ARS Rangeland Research Unit in Colorado (USA) in collaboration with the Institute of Biometeorology of CNR (Italy) began testing an innovative FACE design that is intended to create a more-or-less continuous gradient of $e[CO_2]$ over an experimental area of 200 m^2 (GradFACE). Starting from an original unpublished idea of Prof. Kenji Kurata (Department of Biological and Environmental Engineering, University of Tokyo, Japan), this new system was initially developed using FLUENT (www.fluent.com), a computational fluid dynamic (CFD) gas dispersion model. The CFD experiment allowed testing of a series of alternative designs, resulting in the final layout illustrated in Fig. 2.7. In this design, a rectangular array of laser-drilled horizontal pipes is placed over the vegetated surface, a few centimeters above the canopy height. The use of 22 automatic pressure regulators and a CO_2 injection control algorithm allow the release of different amounts of pure CO_2 from each pipe segment, depending on wind direction. In this way, more CO_2 is released at one end of the array and the amount of CO_2, which is injected along the main direction of the array, can be decreased linearly. Model experiments indicated that specific patterns of the static pressure inside the laser-drilled pipes of the array permit maintenance of a relatively constant $e[CO_2]$ gradient over the vegetated surface, irrespective of the wind direction (Fig. 2.8). At present (winter 2004–2005), a GradFACE prototype is under testing in a real field application at the USDA-ARS field station in Cheyenne, Colo. (USA) and the preliminary results indicate that it is possible to obtain a consistent gradient of CO_2 concentrations under different wind directions and wind speed situations as the model anticipated.

There may be a significant advantage in using this type of gradient design in elevated CO_2 studies. Tunnel experiments made in Texas (USA) created a consistent CO_2 concentration gradient (Polley et al. 2003) and demonstrated that non-linearities occur in plant responses as $[CO_2]$ increases. However, the

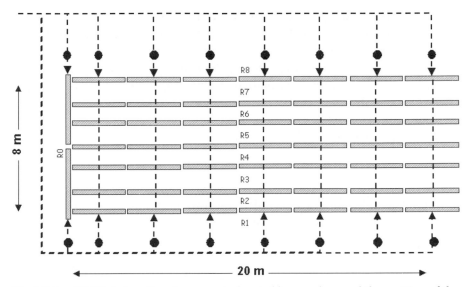

Fig. 2.7 GradFACE design: the pipe array is denoted by *gray bars* and the position of the proportional valves controlling the release of CO_2 from the pipe array is indicated by the *black dots*. Dimensions of the experimental plot used for CFD simulations are shown. As discussed in the text, the direction and intensity of wind controls the pattern of valve aperture and the amount of CO_2 injected over the plot

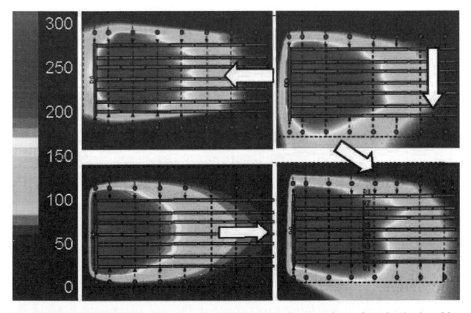

Fig. 2.8 Distribution of $[CO_2]$ in GradFACE (in ppm) above ambient $[CO_2]$ calculated by the FLUENT CFD model at 35 cm above ground in a 15 cm tall short-grass prairie with wind blowing from four different directions (*arrows*). In these simulations, adjustment of valve apertures allows maintenance of a well defined elevated $[CO_2]$

tunnel enclosure itself will impact growth conditions so an open-air approach would be advantageous. Such observations provide very valuable information for modellers. In addition, gradient experiments may enhance the statistical power of elevated CO_2 experiments, as they are amenable to regression analysis rather than the standard analysis-of-variance approach typically used for block experimental designs by most FACE studies.

2.7.2 HotFACE

Plant responses to $e[CO_2]$ are strongly affected by increasing ambient air temperature (T_a) as the kinetics of Rubisco saturation, stomatal conductance, and many other biochemical processes are strongly temperature-dependent (Long 1991; Idso et al. 1992; Tjoelker et al. 1998; Bernacchi et al. 2002). So too are the inhibitory effects of elevated O_3. However, acclimatory responses of respiration to changes in T_a have been observed (Arnone and Körner 1997; Frantz et al. 2004). Accordingly, if T_a increases with $[CO_2]$ due to global warming, one may expect a shift in the ratio of respiration to photosynthesis, with a consequent change in net ecosystem production. But while many experimental data are accumulating on the response of plants, soils, and ecosystems to rising $[CO_2]$ and to experimentally elevated temperature (Rustad et al. 2001), the interactive effects of $e[CO_2]$ and temperature are insufficiently explored in a natural field setting. Thus, there is high interest in exposing ecosystems to elevated CO_2 and/or O_3 simultaneously with elevated T_a, but the combination of free air temperature enhancement (FATE) and FACE has not yet been successfully demonstrated.

Two approaches to experiments simulating effects of global warming on ecosystems without enclosure have been implemented. For example (among many), Melillo et al. (2002) began a 10-year soil warming experiment in 1991 in a mid-latitude hardwood forest, using electrical heating cables at 20 cm spacing buried at 10 cm to provide a 5 °C elevation of soil temperature. Not surprisingly, this resulted in a respiratory loss of 11 % of soil carbon in the top 60 cm of soil. But what about the potential effect of warming above ground on carbon supply to the soil? From the standpoint of understanding carbon flux at ecosystem-scale, soil warming disconnects the impact of warming among ecosystem components. The effect of warming on translocation of photosynthate from leaf to soil storage is largely unknown, while accelerated respiration due to soil warming depletes labile carbon stocks, which surely impacts the warmed microbial community with ramifications for nutrient cycling and long-term carbon balance.

Another type of warming experiment is the use of radiant heating on aboveground vegetation. Nijs et al. (1996) placed infrared (IR) radiant heaters at a 40° angle from the north above *Lolium perenne* pasture plots in the Swiss FACE experiment for an intended increase in leaf temperature (T_L) of 2.5 °C.

Because of a large standard deviation relative to the heating increment, the heating treatment did not produce a significant (at $P=0.05$) spatial T_L gradient and the standard deviation of T_L within the profiles (four thermocouples each, 10.075 measurements over 3 weeks) was 0.71 °C and 1.00 °C for the $c[CO_2]$ and $e[CO_2]$ plots, respectively. However, when the canopy reached 20 cm in height, the mean vertical T_L profile within the canopy was convex, with a maximum increment of ~3.0 °C at $z = 15$ cm and a minimum of ~1.7 °C at $z = 0.5$ cm. One might expect to observe the maximum heating at the top of the canopy near the IR source; and the convex pattern is suggestive of wind cooling. IR radiation penetration into the canopy should be limited by interception by the uppermost leaves just as visible light is, following something like a Beer's Law relationship with canopy depth and density. Perhaps this absorption of IR energy would be negligible in FACE experiments with very short canopies, particularly if they are not very dense, but for other types of ecosystems such as wheat and cotton or in forests, radiant heating seems an impractical approach simply because of the absorption with canopy depth. Furthermore, there are very significant physiological drawbacks to this approach to manipulating leaf temperature within a FACE experiment. IR heating of leaves will increase transpiration, introducing a potential for increased water stress at different levels within the canopy, with all the consequences this has for plant growth.

As with the soil heating studies, IR heating of vegetation (above some as yet undefined minimal height) will de-couple ecosystem components, as it is unlikely to provide heating of the soil that is analogous to that produced by atmospheric warming. But a bigger problem is that IR heaters and, to a lesser extent soil warming experiments, differentially increase the temperature of plant organs, creating a temperature gradient between the air and the plant as well as within the plant itself, which is an experimental artifact and unrealistic relative to the concept that leaves will be bathed in warmer air in the future.

A third type of warming experiment employs transparent covers over the plants at night, a form of passive heating that prevents long-wave radiation losses from the ecosystem during night (Beier et al. 2004). Global warming simulations suggest that night-time temperatures will increase more than during the day so night-time passive heating is not inappropriate. This approach to plot heating may be appropriate with FACE since, for most FACE experiments, CO_2 fumigation is not applied at night. This will not be entirely realistic, but it may be a good compromise.

Perhaps the best option to make realistic investigations on the potential effects of warming on open-air ecosystems is to elevate T_a rather than soil or plant tissue temperatures independently. Could warming air in a FACE system provide heat? Simple calculations indicate that continuous heating of air for injection into a FACE system would be costly. Consider a FACE plot 10 m wide with vegetation 0.5 m tall and an average wind speed of 1 m s^{-1}. A volume of

18.000 m³ h⁻¹ would have to be heated. If the heat treatment were set at 4 °C above ambient, this would require continuous application of 1448 kWh. For 24 h operation and at U.S.$ 0.10 (kWh)⁻¹, this would cost U.S.$ 108.000 per month for each FACE ring, perhaps an impractical cost for an experiment running in triplicate for months. Just think of heating the FACTS-I experiment's four rings, 30 m in diameter by 20 m tall!

Another drawback of these approaches to heating is that all are likely to alter relative humidity (RH) and vapor pressure deficit (VPD). A warming experiment would be confounded by the effect of warming on RH and feedback on stomatal conductance and, thus, CO_2 uptake. Nijs et al. (1996) estimate that, at 20 °C and 75 % RH, there would be an 8 % reduction in RH with their radiant warming treatment. This can be expected to reduce stomatal conductance and carbon uptake (Ellsworth 2000; Katul et al. 2000). We suggest that, in a FACE experiment using blowers as in the BNL-FACE design, RH might be adjusted rather easily by adding water injection (perhaps by mist sprayers into the plenum) and controlled by means analogous to the control of CO_2 or O_3 to counteract the effects of warming on RH.

However, if heated air (with controlled addition of water vapor) were injected during periods of strong atmospheric stability with low wind speed, then relatively little energy would suffice to produce heating of a few degrees. Furthermore, this will be particularly effective at night with atmospheric stability with radiative cooling of the ground. Then, providing blowing air may induce the same sort of turbulent mixing and warming (with a heating increment) as produced by the "blower effect" observed in the Arizona FACE studies (Pinter et al. 2000). Blowing air near mid-day might work opposite to the intended heating.

The variability of temperature might be problematic in a manipulation experiment. But under normal conditions, plant leaves may be subjected to highly variable temperature conditions over short periods of time due to natural variability in cloud cover, leaf shading and wind speed; and rapid changes in T_L by 10 °C or more above air temperature are observed (Gates et al. 1968). But since boundary layer conductance to heat transfer is often assumed to be very large, leaf temperature is expected to be very close to air temperature (e.g., Angell and Miller 1994). However, at low wind speed, leaf temperatures may be quite a bit above T_a, even in conifers with quite small leaves, e.g., Douglas fir may be up to 6 °C above air temperature in direct sunlight (Martin et al. 1999), whereas when the boundary layer is reduced by wind, leaf temperatures approach air temperature. This suggests an acceptable strategy for FACE warming would be to provide episodic heating, even up to 10 °C above ambient at times judged to be appropriate (as yet undefined), but only when the energy requirement is low, aiming for an increment over time of 2–4 °C.

In a recent exploratory study, IBIMET-CNR considered using spatially distributed heat sources, a series of VVPs equally distributed over the experimental plot at canopy height to release warm air at an adjustable rate. CFD simulations suggest that using a FATE approach, an appropriate T_a increment could be achieved with a sufficient degree of efficiency. This is supported by similar FLUENT CFD studies at BNL, showing that the buoyancy of warmed air is limited by the rate of invasion of cooler ambient air from around the edges of the warm air mass. Thus, with suitable engineering of an emitter system, the warm air might remain at ground level at wind speeds <0.5 m s^{-1} over a distance of 30 m if emitted from a FACE-like system or from VVPs distributed over the plot. Much work remains to be done before our studies lead to a prototype. However, our opinion is that exploratory studies of FATE and both design and field tests of prototypes should be given high priority if there is to be any sort of realistic heating experiment in an open-air setting.

2.8 Conclusions

FACE is a mature technology for ecological experiments that is flexible and readily adapted to a wide range of ecosystems. It offers many advantages over growth chambers or field enclosures and avoids nearly all of the artefacts associated with enclosures, canopy covers, and wind breaks. Hundreds of scientists use FACE in experiments persisting for over a decade.

FACE is now applied to ecosystems ranging from the Mojave Desert to crops, bogs, conifer and deciduous forest plantations. Yet, FACE technology is constantly changing and improving. FACE did not spring de novo from a single concept but was built upon a series of developments, principally improved and inexpensive micro-computer control systems, fast-feedback control algorithms, more effective dilution of the fumigant in a plenum or by jet-induced mixing, and directional control. The diversity of interests of the experimenters who contributed to FACE development led to an engineering synthesis of these technical advances with an understanding of the needs of biological field experiments and a commitment to rigorous testing from concept, to prototype, to facility. Experiments with FACE avoid changes in micro-climate observed in any type of enclosure, especially warming and altered relative humidity that impact evapotranspiration and feed forward into changes in carbon uptake.

FACE technology, despite its obvious success and contributions to science, has limitations that are often overlooked or insufficiently appreciated by users of the technology, including: the unnatural step increase in e[CO_2], unnatural short-term variability in e[CO_2], and plot size that cannot capture some large-scale ecological processes.

A first limitation is that FACE creates a step increase in e[CO_2] that is somewhat unrealistic and may distort early impressions of the effects of e[CO_2] on plants and ecosystem processes. For a crop or plantation experiment within

FACE, it is obvious that the aboveground vegetation has been exposed through its entire life cycle to $e[CO_2]$, but the living soil has not. Soil ecosystems acclimated to current $[CO_2]$ might take a long time to fully adjust to the increased delivery of plant products. This is one reason why it is important to continue such experiments for a long time. Is there a way around problems that might be associated with the step increase in $e[CO_2]$, yet maintain a realistic manipulation experiment in the field? While one might think of ramping up $e[CO_2]$ over a period of years in order to ameliorate artefacts potentially associated with a step change, that would not eliminate them unless you increment $[CO_2]$ at the same rate as our global enrichment experiment.

A second limitation is that, although variability of $e[CO_2]$ in FACE is as good as in many chambered systems, it is still highly variable, with amplitudes sometimes twice the intended treatment $e[CO_2]$. For well run FACE systems, the duration of such excursions are generally short, <60 s, and they do not significantly alter net carbon gain. Nevertheless, it is important for each type of FACE system to demonstrate convincingly the extent of the inherent variability in its ability to control $e[CO_2]$. To assure continuous and stable operation with a high level of operational integrity, adequate staffing is needed.

The third limitation is the plot size. While the large diameter FACE plots may encompass hundreds of square meters and be sufficiently large to capture most critical ecosystem processes relating to plants and soils, they are still like an island within the surrounding ecosystem. The annular mixing zone between the points of CO_2 emission and the biological material to be sampled is an essential part of FACE. Experimenters are often tempted to use plants grown in this zone, but that is inappropriate as the treatment increment and condition in this zone is very irregular, some leaves receiving unacceptably high $e[CO_2]$ exposures. Edge effects likely impact the entire surface of very small FACE plots.

It may be possible to provide elevated temperature within FACE and preliminary studies are underway.

FACE as a technology continues to advance. New FACE concepts for producing $e[CO_2]$ gradient experiments are under development with promising progress, and concepts for heated air experiments are being explored. There is a critical need to investigate the simultaneous and interacting effects of increased temperature with elevated CO_2 and O_3 concentrations. We suggest that, with further engineering development and testing, plus an appropriate operational strategy, free-air temperature elevation or HotFACE may become realistic and cost-effective method for elevated temperature experiments in open-air settings.

Despite its known disadvantages, FACE is the best available approach for conducting realistic trace-gas enrichment experiments at ecosystem-scale.

FACE experiments present us with a window into the likely future of ecosystem function in a CO_2-enriched and warmer world, but are not themselves a facsimile of the future. Despite these limitations, FACE, in several different configurations, is by far the best approach to realistic, open-air field experiments yet conceived.

Acknowledgements. The Carbon Dioxide Research Program of the U.S. Department of Energy, Office of Science, Environmental Sciences Division, Office of Biological and Environmental Research Contract No. DE-AC02-98CH10886 to Brookhaven National Laboratory has supported this research. Dozens of collaborators contributed to FACE development and we thank particularly Dr. Lance Evans, Manhattan College, Riverdale, N.Y. (USA) and Dr. Bruce Kimball, USDA-ARS Water Research Laboratory, Phoenix, Ariz. (USA). The POPFACE and EuroFACE projects were initially funded by the Commission of the European Communities, Contract No. ENV4-CT97-0657. This project contributed to the GCTE (Global Change and Terrestrial Ecosystem), which is a Core Project of International Geosphere-Biosphere Programme. The USDA-ARS Rangeland Research Unit, Ft. Collins, Colo. (USA) is acknowledged for support during the development phase of the GradFACE idea. Authors also acknowledge numerous colleagues and assistants who provided invaluable help and useful suggestions for the development and implementation of FACE systems, worldwide.

References

Allen LH (1992) Free-air CO_2 enrichment field experiments: an historical overview. Crit Rev Plant Sci 11:121–134

Angell RF, Miller RF (1994) Simulation of leaf conductance and transpiration in *Juniperus occidentalis*. For Sci 40:5–17

Arnone JA III, Körner Ch (1997) Temperature adaptation and acclimation potential of leaf dark respiration in two species of *Ranunculus* from warm and cold habitats. Arctic Alpine Res 29:122–125

Barnard R, Barthes L, Le Roux X, Harmens H, Raschi A, Soussana JF, Winkler B, Leadley PW (2004) Atmospheric CO_2 elevation has little effect on nitrifying and denitrifying enzyme activity in four European grasslands. Global Change Biol 10:488–497

Beier C, Emmett B, Gundersen P, Tietema A, Penuelas J, Estiarte M, Gordon C, Gorissen A, Llorens L, Roda F, Williams D (2004) Novel approaches to study climate change effects on terrestrial ecosystems in the field. Ecosystems 7:583–597

Bernacchi CJ, Portis AR, Nakano H, Caemmerer S von, Long SP (2002) Temperature responses of mesophyll conductance. Implications for the determination of Rubisco enzyme kinetics and for limitations to photosyntheses in vivo. Plant Physiol 130:1992–1998

Davey PA, Hunt S, Hymus GJ, DeLucia EH, Drake BG, Karnosky DF, Long SP (2004) Respiratory oxygen uptake is not decreased by an instantaneous elevation of $[CO_2]$, but is increased with long-term growth in the field at elevated $[CO_2]$. Plant Physiol 134:520–527

Dickson RE, Lewin KF, Isebrands JG, Coleman MD, Heilman WE, Riemenschneider DE, Sôber J, Host GE, Zak DR, Hendrey GR, Pregitzer KS, Karnosky DF (2000) Forest atmosphere carbon transfer storage-II (FACTS II) – The aspen free-air CO_2 and O_3 enrichment (FACE) project: an overview (USDA Forest Service General Technical Report NC-214). USDA Forest Service, St. Paul, Minn.

DeLucia EH, Hamilton JG Naidu SL, Thomas RB, Andrews JA, Finzi A, Lavine M, Matamala R, Mohan JE, Hendrey GR, Schlesinger WH (1999) Net primary production of a forest ecosystem with experimental CO_2 enrichment. Science 284:1177–1179

Edwards GR, Newton PCD, Tilbrook JC, Clark H (2001) Seedling performance of pasture species under elevated CO_2. New Phytol 150:359–369

Ellsworth DS (2000) Seasonal CO_2 assimilation and stomatal limitations in a *Pinus taeda* canopy with varying climate. Tree Physiol 20:435–445

Frantz JM, Cometti NN, Bugbee B (2004) Night temperature has a minimal effect on respiration and growth in rapidly growing plants. Ann Bot 94:155–166

Gates D M; Alderfer R, Taylor E (1968) Leaf temperatures of desert plants. Science 159:994–995

Hamilton JS, DeLucia EH, George K, Naidu S., Finzi AC, Schlesinger WH (2002) Forest carbon balance under elevated CO_2. Oecologia 131:250–260

He Y, Yang X, Miller DR, Hendrey GR, Lewin KF, Nagy J (1996) Effects of FACE system operation on the micrometeorology of a loblolly pine stand. Trans Am Soc Atmos Environ 39:1551–1556

Hebeisen T, Lüscher A, Zanetti S, Fischer BU, Hartwig UA, Frehner M, Hendrey GR, Blum H, Nösberger J (1997) Growth response of *Trifolium repens* L. and *Lolium perenne* L. as monocultures and bi-species mixture to free-air CO_2 enrichment and management. Global Change Biol 3:149–160

Heijmans MMPD, Berendse F, Arp WJ, Masselink AK, Klees H, Visser W de, Breemen N van (2001) Effects of elevated carbon dioxide and increased nitrogen deposition on bog vegetation in the Netherlands. J Ecol 89:268–279

Hendrey GR (1976) Effects of pH on the growth of periphytic algae in artificial stream channels (IR 25/76 SNSF project). SNSF, Ås-NLH, 50 pp

Hendrey GR, Lewin KF, Kolber Z, Evans L (1992) Controlled enrichment system for experimental fumigation of plants in the field with sulphur dioxide. J Air Waste Manage Assoc 42:1324–1327

Hendrey GR, Lewin KF, Nagy J (1993) Free air carbon dioxide enrichment: development, progress, results. Vegetatio 104/105:17–31

Hendrey GR, Long SP, McKee IF, Baker NR (1997) Can photosynthesis respond to short-term fluctuations in atmospheric carbon dioxide? Photosynth Res 51:179–184

Hendrey GR, Ellsworth DS, Lewin KF, Nagy J (1999) A free-air CO_2 enrichment system for exposing tall forest vegetation to elevated atmospheric CO_2. Global Change Biol 5:293–309

Holtum JAM, Winter K (2003) Photosynthetic CO_2 uptake in seedlings of two tropical tree species exposed to oscillating elevated concentrations of CO_2. Planta 218:152–158

Hovenden MJ (2003) Growth and photosynthetic responses to elevated $[CO_2]$ in grasses from Tasmanian native pasture. Pap Proc R S Tasmania 137:81–86

Idso SB, Kimball BA, Hendrix DL (1992) Air temperature modifies the size-enhancing effects of atmospheric CO_2 enrichment on sour orange tree leaves. Environ Exp Bot 33:293–299

Karnosky DF, Gagnon ZE, Dickson RE, Coleman MD, Lee EH, Isebrands JG (1996) Changes in growth, leaf abscission, and biomass associated with seasonal tropospheric ozone exposures for *Populus tremuloides* clones and seedlings. Can J For Res 26:23–37

Karnosky DF, Mankovska B, Percy K, Dickson RE, Podila GK, Sober J, Noormets A, Hendrey G, Coleman MD, Kubiske M, Pregitzer KS, Isebrands JG (1999) Effects of tropospheric O_3 on trembling aspen and interaction with CO_2: Results from an O_3-gradient and a FACE experiment. J Water Air Soil Pollut 116:311–322

Karnosky DF, Zak DR, Pregitzer KS, Awmack CS, Bockheim JG, Dickson, RE, Hendrey GR, Host GE, King JS, Kopper BJ, Kruger EL, Kubiske ME, Lindroth RL, Mattson WJ, McDonald EP, Noormets A, Oksanen E, Parsons WFJ, Percy KE, Podila GK, Riemenschneider DE, Sharma P, Thakur R, Sôber A, Sôber J, Jones WS, Anttonen S, Vapaavuori E, Mankovska B, Heilman W, Isebrands JG (2003) Tropospheric O_3 moderates responses of temperate hardwood forests to elevated CO_2: a synthesis of molecular to ecosystem results from the Aspen FACE project. Funct Ecol 17:289–304

Katul GG, Ellsworth DS, Lai C-T (2000) Modelling assimilation and intercellular CO_2 from measured conductance: a synthesis of approaches. Plant Cell Environ 23:1313–1338

Kilronomos JN, Allen MF, Rillig MC, Piotrowski J, Makvandi-Nejad S, Wolfe BE, Powell JR (2005) Abrupt rise in atmospheric CO_2 overestimates community response in a model plant–soil system. Nature 433:621–624

Leakey ADB, Bernacchi CJ, Dohleman FG, Ort DR, Long SP (2004) Will photosynthesis of maize (*Zea mays*) in the US Corn Belt increase in future [CO_2] rich atmospheres? An analysis of diurnal courses of CO_2 uptake under free-air concentration enrichment (FACE). Global Change Biol 10:951–962

Leavitt SW, Paul E, BKimball A, Hendrey GR, Mauney JR, Rauschkolb R, Rogers H, Lewin KF, Nagy J, Pinter PJ Jr, Johnson HB (1994) Carbon isotope dynamics of free-air CO_2 enriched cotton and soils. Agric For Meteor 70:87–101

Lewin KF, Hendrey GR, Kolber Z (1992) Brookhaven National Laboratory free-air carbon dioxide enrichment facility. Crit Rev Plant Sci 11:135–141

Lipfert FW, Alexander Y, Hendrey GR, Lewin KF, Nagy J (1992) Performance of the BNL FACE gas injection system. Crit Rev Plant Sci 11:143–163

Long SP (1991) Modification of the response of photosynthetic productivity to rising temperature by atmospheric CO_2 concentrations – has its importance been underestimated. Plant Cell Environ 14:729–739

Luo Y, Reynolds JF (1999) Validity of extrapolating field CO_2 experiments to predict carbon sequestration in natural ecosystems. Ecology 80:1568–1583

Luo Y, White LW, Canadell JG, DeLucia EH, Ellsworth DS, Finzi A, LichterJ, Schlesinger WM (2003) Sustainability of terrestrial carbon sequestration: a case study in Duke Forest with inversion approach. Global Biogeochem Cycles 17:1021–1034

Lüscher A, Hendrey GR, Nösberger J (1998) Long-term responsiveness to free air CO_2 enrichment of functional types, species and genotypes of plants from fertile permanent grassland. Oecologia 113:37–45

Magliulo V, Bindi M, Rana G (2003) Water use of irrigated potato (*Solanum tuberosum* L) grown under free air carbon dioxide enrichment in central Italy. Agric Ecosyst Environ 97:65–80

Martin TA, Hinkley TM, Meinzer FC, Sprugel DG (1999) Boundary layer conductance, leaf temperature and transpiration of *Abies amabilis* branches. Tree Physiol 19:435–443

Mauney JR, Kimball BA, Pinter PJ Jr, LaMorte RL, Lewin KF, Nagy J, Hendrey GR (1994) Growth and yield of cotton in response to a free-air carbon dioxide enrichment (FACE) environment. Agric For Meteor 70:49–67

McLeod AR (1993) Open-air exposure systems for air pollutants studies – their potential and limitations. In: Schulze ED, Mooney HA (eds) Design and execution of experiments on CO_2 enrichment. (Proceedings of a workshop held at Weidenberg, Germany, 26–30 October 1992. Ecosystems Research Report 6), Commission of the European Communities/ Guyot, Brussels, pp 355–365.

McLeod AR, Long SP (1999) Free-air carbon dioxide enrichment (FACE) in global change research: a review. Adv Ecol Res 28:1–56

McLeod AR, Alexander K, Hatcher P (1985) Open-air fumigation of field crops: criteria and design for a new experimental system. Atmos Environ 19:1639-1649

Melillo JM, Steudler PA, Aber JD, Newkirk K, Lux H, Bowles FP, Catricala C, Magill A, Ahrens T, Morrissau S (2002) Soil warming and carbon-cycle feedbacks to the climate system. Science 298:2173-2176

Miglietta F, Giuntoli A, Bindi M (1996) The effect of free air carbon dioxide enrichment (FACE) and soil nitrogen availability on the photosynthetic capacity of wheat. Photosynth Res 47:281-290

Miglietta F, Lanini M, Bindi M, Magliulo V (1997) Free air CO_2 enrichment of potato (*Solanum tuberosum* L.): design and performance of the CO_2-fumigation system. Global Change Biol 3:417-427

Miglietta F, Magliulo V, Bindi M, Cerio L, Vaccari FP, Loduca V, Peressotti A (1998) Free air CO_2 enrichment of potato (*Solanum tuberosum* L.): development, growth and yield. Global Change Biol 4:163-172

Miglietta F, Peressotti A, Vaccari FP, Zaldei A, deAngelis P, Scarascia-Mugnozza G (2001) Free-air CO_2 enrichment (FACE) of a poplar plantation: the POPFACE fumigation system. New Phytol 150:465-476

Mitchell EAD, Gilbert D, Buttler A, Amblard C, Grosvernier P, Gobat JM (2003) Structure of microbial communities in *Sphagnum* peatlands and effect of atmospheric carbon dioxide enrichment. Microb Ecol 46:187-199

Mooi J, Zalm AJA van der (1985) Research on the effects of higher than ambient concentrations of SO_2 and NO_2 on vegetation under semi-natural conditions: the development and testing of a field fumigation system: execution (Second interim report). Instituut voor Plantenziektenkundig Onderzoek, Wageningen, 19 pp

Morgan PB, Bernacchi CJ, Ort DR, Long SP (2004) An in vivo analysis of the effect of season-long open-air elevation of ozone to anticipated 2050 levels on photosynthesis in soybean. Plant Physiol 135:2348-2357

Nagy J, Lewin KF, Hendrey GR, Hassinger E, LaMorte R (1994) FACE facility CO_2 concentration control and CO_2 use in 1990 and 1991. Agric For Meteorol 70:31-48

Nijs I, Kockelbergh F, Teughels H, Blum H, Hendrey GR, Impens I (1996) Free air temperature increase (FATI): a new tool to study global warming effects on plants in the field. Plant Cell Environ 19:495-502

Norby RJ, Todd DE, Fults J, Johnson DW (2001) Allometric determination of tree growth in a CO_2-enriched sweetgum stand. New Phytol 150:477-487

Okada M, Lieffering M, Nakamura H, Yoshimoto M, Kim HY, Kobaayashi K (2001) Free-air CO_2 enrichment (FACE) using pure CO_2 injection: system Description. New Phytol 150:251-260

Oren R, Ellsworth DE, Johnsen KH, Phillips N, Ewers BE, Maier C, Schäfer KVR, McCarthy H, Hendrey G, McNulty SG, Katul GG (2001) Soil fertility limits carbon sequestration by forest ecosystems in a CO_2-enriched atmosphere. Nature 411:469-472

Pendall E, Bridgham S, Hanson PJ, Hungat, B, Kicklighter DW, Johnson DW, Law BE, LuoY, Megonigal JP, Olsrud M, RyaN MG, Wan S (2004) Below-ground process responses to elevated CO_2 and temperature: a discussion of observations, measurement methods, and models. New Phytol. 162:311-322

Pepin S, Körner C (2002) Web-FACE: a new canopy free-air CO_2 enrichment system for tall trees in mature forests. Oecologia 133:1-9

Pinter PJ Jr, Kimball BA, Wall GW, LaMorte RL, Hunsaker DJ, Adamsen FJ, Frumau KFA, Vugts HF, Hendrey GR, Lewin KF, Nagy J, Johnson HB, Wechsung F, Leavitt SW, Thompson TL, Matthias AD, Brooks TJ (2000) Free-air CO_2 enrichment (FACE): blower effects on wheat canopy microclimate and plant development. Agric For Meteorol 103:319-333

Polley HW, Johnson HB, Derner JD (2003) Increasing CO_2 from subambient to superambient concentrations alters species composition and increases above-ground biomass in a C3/C4 grassland. New Phytol 160:319–327

Rustad LE, CampbellJL, Marion GM, Norby RJ, Mitchel MJ, Hartley AE, Cornelissen JHC, Gurvich J (2001) A meta-analysis of the response of soil respiration, net nitrogen mineralization, and aboveground plant growth to experimental ecosystem warming. Oecologia 126:543–562

Shinn JH, Allen LH Jr (1985) An evaluation of free-air carbon dioxide enrichment (FACE) as a field method for investigation of direct effects of carbon dioxide on plants (UCRL-93635 Report for the US Dept of Energy, Carbon Dioxide Research Division). Lawrence Livermore National laboratory, Livermore, Calif., 24 pp

Teyssonneyre F, Picon-Cochard C, Falcimagne R, Soussana J-F (2002) Effects of elevated CO_2 and cutting frequency on plant community structure in a temperate grassland. Global Change Biol 8:1034–1046

Tjoelker MG, Oleksyn J, Reich PB (1998) Temperature and ontogeny mediate growth response to elevated CO_2 in seedlings of five boreal tree species. New Phytol 140:197–210

Weigel HJ, Dämmgen U (2000) The Braunschweig carbon project: atmospheric flux monitoring and free air carbon dioxide enrichment (FACE). J Appl Bot 74:55–60

Wustman BA, Oksanen E, Karnosky DF, Sober J, Isebrands JG, Hendrey GR, Pregitzer KS, Podila GK (2001) Effects of elevated CO_2 and O_3 on aspen clones varying in O_3 sensitivity: can CO_2 ameliorate the harmful effects of O_3? Environ Pollut 115:473–481

Zanetti S, Hartwig UA, Lüscher A, Hebeisen T, Frehner M, Fischer BU, Hendrey GR, Blum H, Nösberger J (1996) Stimulation of symbiotic N_2 fixation in *Trifolium repens* L. under elevated atmospheric pCO_2 in a grassland ecosystem. Plant Physiol 112:575–583

Part B: Case Studies

3 The Effects of Free-Air [CO₂] Enrichment of Cotton, Wheat, and Sorghum

B.A. KIMBALL

3.1 Introduction

Free-air CO_2 enrichment (FACE) experiments were conducted on cotton (*Gossypium hirsutum* L., cv. Deltapine 77), wheat (*Triticum aestivum* L. cv. Yecora Rojo), and sorghum (*Sorghum bicolor* (L.) Möench cv. Dekalb DK54) at the University of Arizona Maricopa Agricultural Center (MAC), Maricopa, Ariz., USA (Table 3.1). The site is about 50 km south of Phoenix within an irrigated agricultural area extending several kilometers in every direction, which in turn is surrounded by desert. The land has been leveled, and flooding is the usual irrigation method. The soil in the experimental field is classified as a reclaimed Trix clay loam [fine-loamy, mixed (calcareous), hyperthermic Typic Torrifluvents]; additional details about the soil properties are given by Post et al. (1988) and Kimball et al. (1992a).

3.2 Description of the FACE System and Experimental Methodology

The first experiment in 1989 was on cotton, with only elevated (e)[CO_2] and current (c)[CO_2] treatments (Table 3.1). Then in 1990 and 1991, two additional FACE experiments were conducted on cotton, this time with the main CO_2 plots being split into halves with ample and limiting water supplies. Two similar CO_2 × water experiments were conducted on wheat during 1992–1993 and 1993–1994. Another two experiments were conducted on wheat during 1995–1996 and 1996–1997, but for these experiments, the main CO_2 plots were split instead into ample and low nitrogen supplies. For all of the cotton and wheat experiments, the water was supplied via a drip irrigation system. Then in 1998 and 1999, sorghum became the experimental crop, with the main CO_2

Ecological Studies, Vol. 187
J. Nösberger, S.P. Long, R.J. Norby, M. Stitt,
G.R. Hendrey, H. Blum (Eds.)
Managed Ecosystems and CO₂
Case Studies, Processes, and Perspectives
© Springer-Verlag Berlin Heidelberg 2006

plots split into ample and limiting water supplies; and the water was applied by flooding. Plot plans are shown in Wall and Kimball (1993) for the cotton and 1992–1994 wheat, in Kimball et al. (1999) for the 1995–1997 wheat, and in Ottman et al. (2001) for the sorghum. After the 1991 cotton experiment and after the 1994 and 1997 wheat experiments, the plots were moved to new locations in the same or an adjacent field with the same soil type. For all the experiments, there were four replicates of each treatment.

The FACE apparatus was of the Brookhaven design (Hendrey 1993; see Chapter 2). Briefly, replicate 25-m diam. toroidal plenum rings constructed from 0.305-m diam. pipe were placed in the field shortly after planting. The rings had 2.5-m high vertical stand pipes with individual valves spaced about every 2.4 m around the periphery. Air enriched with CO_2 was blown into the rings; and it exited near the canopy top through tri-directional jets in the vertical pipes. Wind direction, wind speed, and $[CO_2]$ were measured at the center of each ring. A computer-control system used the wind speed and $[CO_2]$ information to adjust the CO_2 flow rates to maintain the desired $[CO_2]$ at the centers of the FACE rings. The system used the wind direction information to turn on only those stand pipes upwind of the plots, so that CO_2-enriched air flowed across the plots no matter which way the wind blew. When wind speeds were low (<0.4 m s^{-1}) and it was difficult to detect direction, the CO_2-enriched air was released from every other vertical pipe around the rings.

The strategies used for FACE and for the c$[CO_2]$ control plots changed somewhat from one experiment to another (Table 3.1). For the first experiment in 1989, enrichment was done only during daytime at a fixed set-point of 550 ppm, and the control plots were demarked by some stakes and tape. For the next two experiments, the same enrichment strategy was used, but pipes were added to the control plots which were identical to those of the FACE plots but without blowers like those in the FACE plots. The pipes greatly helped personnel use identical sampling protocols in all plots. Reports began appearing in the literature about this time suggesting there were effects of e$[CO_2]$ on dark respiration, so starting with the 1992–1993 wheat experiment, enrichment was done day and night. Unfortunately, however, we discovered that the blowers disturbed the microclimate at night, thereby affecting plant growth and development somewhat (Pinter et al. 2000). Therefore, starting with the 1995–1996 season, air blowers were installed in the non-CO_2-enriched c$[CO_2]$ control plots to provide air movement similar to that of the FACE plots. However, there were no valves on the vertical pipes, so the air flowed all the time and it was not changed in response to changing wind speed or direction. This strategy was justified because we believe the air flow in these control plots was important only under calm conditions (wind <0.4 m s^{-1}) when the FACE plots were operated in the mode of releasing CO_2-enriched air from every other vertical pipe. Also starting in 1995–1996 (unlike the prior experiments, which had a constant set-point of 550 ppm), in order to mimic the natural diurnal variations in $[CO_2]$, the FACE plots were enriched

Table 3.1 Experimental protocols and site characteristics for nine free-air CO$_2$ enrichment (FACE) experiments conducted at Maricopa, Ariz. (33°4'N, 111°59'W, 358 m asl), on cotton, wheat, and sorghum (adapted from Kimball et al. 2002)

	FACE experiment			
Experiment ID[a]	MCCot89-91	MCWht93-4	MCWht96-7	MCSor98-99
Species or ecosystem	Cotton (*Gossypium hirsutum* L.)	Wheat (*Triticum aestivum* L.)	Wheat (*T. aestivum* L.)	Sorghum [*Sorghum bicolor* (L.) Möench]
Experiment start date[b]	17-04-1989 23-04-1990 16-04-1991	15-12-1992 08-12-1993	15-12-1995 15-12-1996	16-07-1998 15-06-1999
Experiment end date[b]	17-09-1989 17-09-1990 16-09-1991	24-05-1993 01-06-1994	29-05-1996 28-05-1997	21-12-1998 26-10-1999
FACE start[c]	19-05-1989 04-05-1990 26-04-1991	01-01-1993 28-12-1993	01-01-1996 03-01-1997	31-07-1998 01-07-1999
FACE end[c]	17-09-1989 17-09-1990 16-09-1991	16-05-1993 18-05-1994	15-05-1996 12-05-1997	21-12-1998 26-10-1999
Solar Rad. (MJ m^{-2} day^{-1})[d]	Ave. 25.1	Ave. 18.7	Ave. 19.9	Ave. 21.6
Max. air temp. (° C)[e]	Ave. 46.3	Ave. 38.5	Ave. 40.9	Ave. 44.2
Min. air temp. (° C)[e]	Ave. 7.6	Ave. −2.8	Ave. −3.7	Ave. 2.0
Plot diameter (m)[f]	18	20	20	21
No. of replicates[g]	4	4	4	4
No. of [CO$_2$] levels[h]	2	2	2 + ambient	2
Pre-dilution of [CO$_2$]?[i]	Yes	Yes	Yes	Yes
Set-point or increment?[j]	Set-point	Set-point	Increment	Increment
FACE [CO$_2$][k]	550	550	+200	+200
Daily enrichment time[l]	Daylight	24 h day^{-1}	24 h day^{-1}	24 h day^{-1}
"No-enrichment" criteria[m]	None	None	None	None

Table 3.1. (*Continued*)

Experiment ID[a]	FACE experiment			
	MCCot89-91	MCWht93-4	MCWht96-7	MCSor98-99
Species or ecosystem	Cotton (*Gossypium hirsutum* L.)	Wheat (*Triticum aestivum* L.)	Wheat (*T. aestivum* L.)	Sorghum [*Sorghum bicolor* (L.) Möench]
Additional treatments in strip-split plot design[n]	Water in 1990 and 1991	Water	Nitrogen	Water
Level 1 (low water or low N)[p]	Ave. 1009 mm	Ave. 335 mm	70 kg N ha^{-1} in 1996 15 kg N ha^{-1} in 1997	Ave. 483 mm
Level 2 (ample water or ample N)[p]	Ave. 1202 mm	Ave. 679 mm	350 kg N ha^{-1}	Ave.1133 mm
References[q]	Hendrey (1993), Wall and Kimball (1993), Dugas and Pinter (1994), Lewin et al. (1994), Mauney et al. (1994), Pinter et al. (1994)	Wall and Kimball (1993), Hunsaker et al. (1996), Kimball et al. (1999)	Kimball et al. (1999), Hunsaker et al. (2000)	Ottman et al. (2001)

[a] Experiment identification names.
[b] Dates for the start and end of the experiments and of the growing season.
[c] Start and stop dates of the FACE treatments.
[d] Mean daily solar radiant exposure obtained during growing seasons.
[e] Maximum and minimum air temperatures during growing seasons.
[f] Diameter of useable plot area.
[g] Number of replicate [CO_2] and other treatment plots.
[h] Number of treatment [CO_2] levels. A "2" means experiments with FACE and control treatments or with FACE and current treatments. By "control" is meant plots with air flow near-identical to that of the FACE plots but without added CO_2, and by "current" is meant plots at today's current atmospheric [CO_2] levels with no forced air flow and no added CO_2. A "2 + current" means experiments with FACE, control, and current treatments.
[i] "yes" means a blower system in the FACE apparatus pre-diluted CO_2 with air.
[j] "set-point" or "increment" indicates whether a constant target [CO_2] set-point was used or whether a target increment in concentration above current air CO_2 levels was used.

[k] The set-point $[CO_2]$ or the increment in $[CO_2]$ preceded by a "+". Units of ppm = ppm by volume or μmol CO_2 mol^{-1} air.

[l] Portion of day when CO_2 enrichment was done.

[m] Constraints that were put on the enrichment, such high-wind cut-off or low-temperature cut-off.

[n] Additional factorial treatments in the experiment (low water or low nitrogen) in a strip-split-plot design.

[p] Levels of the additional treatment factors.

[q] References that best describe the experimental conditions.

by 200 ppm above $c[CO_2]$ (~360 ppm). The enrichment and control plot strategies for the sorghum experiments were the same as those for the latter wheat experiments.

3.3 Cotton

Extensive details about the FACE cotton experiments are available from Hendrey (1993), Wall and Kimball (1993), and Dugas and Pinter (1994; Table 3.1). Extracted values for many aspects of the results have been tabulated by Kimball et al. (2002).

3.3.1 Resource Availability

During the 1989 FACE cotton experiment, irrigation water was supplied in amounts judged to be non-limiting (Mauney et al. 1992), but for 1990 and 1991, the plots were split into halves with each side receiving either an ample or a low supply of water (Table 3.1; Hunsaker et al. 1994; Mauney et al. 1994). Irrigation water requirements for the ample treatment were estimated as Class A pan evaporation (measured adjacent to one of the plots) in 1990 multiplied by measured leaf area index (LAI) divided by three. Above an LAI of 3.0, they were taken as pan evaporation. In 1991, a grass-reference crop evapotranspiration (ET_o) was used in place of pan evaporation; and it was calculated from a modified Penman equation using data from a weather station 2 km from the site. Irrigations were done approximately weekly in the spring and half-weekly in the summer. The low-water plots were irrigated on the same days as the ample plots, and they were given 75 % and 67 % of the amounts of water applied to the ample plots in 1990 and 1991, respectively. Excluding initial irrigations for plant establishment, the amounts of water received by the ample plots in 1990 were 890 mm and 125 mm from irrigation and rain, respectively. In 1991, they were 780 mm and 41 mm, respectively. The low-water plots received 760 mm and 520 mm from irrigation in 1990 and 1991, respectively.

The irrigation water was applied via a sub-surface drip irrigation system (Hunsaker et al. 1994). The tubes were buried 0.18–0.24 m below the soil surface, with emitters spaced 0.40 m along each tube. The tubes were placed between the cotton rows, which were on raised beds, with 1.02 m row spacing.

For all 3 years, the cotton was supplied ample nutrients (Mauney et al. 1992, 1994). Approximately 130 kg ha^{-1} of N from Uran 32 dissolved in the irrigation water was applied each year. Two or three times per year, foliar applications of a commercial nutrient solution containing all major and minor elements were done. With one exception, leaf tissue analyses indicated all nutrients were adequate. The exception was zinc in the latter part of the 1990 season (22–25 ppm compared to 30 ppm considered adequate), so $ZnSO_4$ was added to the irrigation water to correct the deficiency.

3.3.2 Resource Acquisition and Transformation

3.3.2.1 CO_2 and Carbon

The above-ground growth response of the cotton to e[CO_2] at 550 ppm in the ample irrigation treatment was large compared to that of the c [CO_2] control plots (Table 3.2), amounting to 32, 34, and 37 % for 1989, 1990, and 1991, respectively. Below-ground fine and tap roots were stimulated even more. Boll (seed + lint) yields were increased 22, 51, and 43 %, respectively. The smaller boll yield response in 1989 was due in part to an apparent e[CO_2]-caused shift in development such that the bolls in the FACE plots matured sooner; and then the FACE-grown plants put on a second flush of growth resulting in more green immature bolls near the top of the crop. However, the 1989 crop was planted somewhat later than normal using a short season variety and a growth regulator was applied. More normal planting dates, a medium-length season variety, and no growth regulator were used in 1990 and 1991; and the development shift was not observed. Although variability was high, lint yield increases were even higher than those of biomass or boll yield, amounting to 21, 73, and 81 %.

In the low-water treatment, the above-ground biomass was stimulated 18 % and 35 % by e[CO_2] in 1990 and 1991, respectively (Table 3.2), while boll yields went up by 43 % and 42 %. Lint yields were increased about 52 % both years. Thus, the cotton growth and yield were stimulated by comparable amounts under both the ample- and low-water irrigation treatments.

After the three seasons, Wood et al. (1994) reported that soil C in the FACE plots tended to increase by 14.3 % at 0–5 cm depth, by 4.6 % at 5–10 cm, and by 9.8 % at 10–20 cm under the ample irrigation regime (Table 3.2). Leavitt et

Table 3.2 Percent increases in several plant response parameters to e[CO$_2$] of cotton (*Gossypium hirsutum* L.), wheat (*Triticum aestivum* L.), and sorghum [*Sorghum bicolor* (L.) Möench] relative to their responses at c[CO$_2$]. The elevated [CO$_2$] was provided by free-air CO$_2$ enrichment (FACE) at 550 ppm by volume. The "+SE" values in columns were calculated whenever possible (in several cases the authors did not supply SE). The SE of the ratios of FACE/(control or current) were calculated using the equation: $\Delta r = (|D\Delta N| + |N\Delta D|)D^{-2}$, where Δ indicates the SE, D is the current or control mean value in the denominator, N is the FACE mean value in the numerator, and the vertical bars denote absolute values. The ratios and SEs were then transformed to percent changes and scaled to 550 ppm. In some cases, the authors supplied probability levels of a difference being significant; and these are indicated by an asterisk (*) in the table. (Partially adapted from Kimball et al. 2002)

		Percentage increases due to e[CO$_2$]						
		Ample water				Low water		
		Ample N		Low N		Ample N		
Item	Year	%	+SE	%	+SE	%	+SE	Reference
Cotton (Gossypium hirsutum L.)								
Above-ground biomass	1989	32.3	11.8					Mauney et al. (1992)
	1990	34.2				17.8		Mauney et al. (1994)
	1991	36.8				35.4		Mauney et al. (1994)
Below-ground biomass								
Taproot	1989	156.9	63.8					Rogers et al. (1992)
	1990	37.8	10.1					Prior et al. (1994)
	1991	60.4	16.6					Prior et al. (1994)
Fine roots	1989	100.0	31.3					Rogers et al. (1992)
	1990	29.3						Prior et al. (1994)
	1991	34.4						Prior et al. (1994)
Boll yield (seed + lint)	1989	22.0				43.1		Mauney et al. (1992)
	1990	50.9				42.0		Mauney et al. (1994)
	1991	42.6						Mauney et al. (1994)

B.A. Kimball

Table 3.2 (*Continued*)

Item	Year	Percentage increases due to e[CO_2]						Reference
		Ample water				Low water		
		Ample N		Low N		Ample N		
		%	+SE	%	+SE	%	+SE	
Lint yield	1989	20.7	11.4					Pinter et al. (1996)
	1990	73.4	45.8			51.7	27.3	Pinter et al. (1996)
	1991	81.1	39.3			51.5	20.4	Pinter et al. (1996)
Soil carbon sequestration								
0–5 cm depth	1989–1991	14.3	18.2		−6.6	11.3		Wood et al. (1994)
5–10 cm depth	1989–1991	4.6	7.9		1.5	3.0		Wood et al. (1994)
10–20 cm depth	1989–1991	9.8	8.4		3.8	5.9		Wood et al. (1994)
0–30 cm depth	1989–1991	14.3	42.3					Leavitt et al. (1994)
fAPAR		0						
Early season (day of year, DOY 193)	1990	8.1	5.8			4.9	5.8	Pinter et al. (1994)
Early season (DOY 183)	1991	16.1	7.8			18.0	7.8	Pinter et al. (1994)
After canopy closure (DOY 241)	1990	−1.0	2.6			−2.8	2.6	Pinter et al. (1994)
After canopy closure (DOY 240)	1991	2.3	2.7			6.8	2.7	Pinter et al. (1994)
Light use efficiency (mass basis)	1989	32.5	9.2					Pinter et al. (1994)
	1990	19.0	8.6			22.2	8.6	Pinter et al. (1994)
	1991	23.4	12.8			23.3	12.8	Pinter et al. (1994)
Water use (evapotranspiration)								
From soil water balance	1990	−1.1				−1.6		Hunsaker et al. (1994)

	Year							Reference
From soil water balance	1991	-1.9				-1.6		Hunsaker et al. (1994)
From stem flow	1991	0.0						Dugas et al. (1994)
Nitrogen concentration								
Leaves	1989	1.3	ns					Prior et al. (1998)
	1990	-15.4	0.1*					Prior et al. (1998)
Seed	1989	-7.6	0.1*					Prior et al. (1998)
	1990	-7.2	ns					Prior et al. (1998)
Whole plant	1989	-10.0	0.1*					Prior et al. (1998)
	1990	-12.9	0.1*					Prior et al. (1998)
Nitrogen yield (whole plant)	1989	24.0	0.2*					Prior et al. (1998)
	1990	17.1	ns					Prior et al. (1998)
Wheat (*Triticum aestivum* L.)								
Above-ground biomass	1993	9.3[a]	12.2			12.3[a]	9.1	Pinter et al. (1996, 2005)
	1994	7.7[a]	14.4			16.6[a]	20.6	Pinter et al. (1996, 2005)
	1996	4.8	5.8	8.1	13.8			Pinter et al. (2006)
	1997	11.7	15.8	2.8	15.4			Pinter et al. (2006)
Below-ground biomass	1993	27.9	51.3			22.7		Wechsung et al. (1999)
	1997	14.4	ns (0.5)*					Pendall et al. (2001)
Grain yield	1993	8.0[a]	7.4			21.0[a]	7.2	Pinter et al. (1996, 1997, 2006)
	1994	12.0[a]	8.6			25.0[a]	18.6	Pinter et al. (1996, 1997, 2006)
	1996	15.0	8.8	12.0	13.1			Pinter et al. (1997, 2006)
	1997	17.0	12.7	5.0	16.5			Pinter et al. (1997, 2006)

Table 3.2 (*Continued*)

Item	Year	Percentage increases due to e[CO_2]								Reference
		Ample water				Low water				
		Ample N		Low N		Ample N				
		%	+SE	%	+SE	%	+SE			
Soil carbon sequestration										
0–15 cm depth	1992–1994	6.0	10.6							Leavitt et al. (1994)
15–30 cm depth	1992–1994	21.4	26.4							Leavitt et al. (1994)
0–5 cm depth	1992–1994	14.5	<0.1							Prior et al. (1997)
5–10 cm depth	1992–1994	13.4	<0.03							Prior et al. (1997)
10–20 cm depth	1992–1994	10.9	<0.01							Prior et al. (1997)
Water use										
Soil water balance	1993	−3.6				4.5				Hunsaker et al. (1996)
	1994	−3.3				−2.2				Hunsaker et al. (1996)
	1996	−3.5		−0.7						Hunsaker et al. (2000)
	1997	−3.9		−1.3						Hunsaker et al. (2000)
Energy balance	1993–1997	−6.7								Kimball et al. (1999)
	1996–1997			−19.5						Kimball et al. (1999)
Nitrogen concentration										
Leaves	1993	−2.6	6.2			−5.5	5.9			Sinclair et al. (2000)
	1994	−8.5	3.9			−12.0	3.8			Sinclair et al. (2000)
	1996	−2.9	10.7	−24.6	10.6					Sinclair et al. (2000)
	1997	2.9	8.1	−18.7	20.2					Sinclair et al. (2000)
Grain	1993–1994	−5.8	1.2			−4.0	1.2			Kimball et al. (2001)
	1996–1997	0.0	2.1	−8.5	3.1					Kimball et al. (2001)

Measurement	Year							Reference
Total above-ground biomass at harvest	1993	2.7	ns			−11.4	ns	Adamsen et al. (2005)
	1994	−19.6	ns			33.3		Adamsen et al. (2005)
Nitrogen yield								
Total above-ground at harvest	1993	−5.3	ns			−15.1	ns	Adamsen et al. (2005)
	1994	−20.1	ns			16.8	ns	Adamsen et al. (2005)
Grain	1993–1994	3.5	5.3			17.6	8.2	Kimball et al. (2001)
	1996–1997	16.8	8.8	−0.9	15.0			Kimball et al. (2001)
Total soil inorganic N in top 0.3 m at harvest	1993	−3.2	ns			−32.0	ns	Adamsen et al. (2005)
	1994	−10.9	ns			−14.0	ns	Adamsen et al. (2005)
Sorghum [*Sorghum bicolor* (L.) Möench]								
Above-ground biomass	1998	6.7	3.1			13.3	13.9	Ottman et al. (2001)
	1999	−1.0	4.5			18.0	16.8	Ottman et al. (2001)
Grain yield	1998	0.9	4.9			17.2	20.8	Ottman et al. (2001)
	1999	−10.7	6.9			34.0	42.6	Ottman et al. (2001)
Soil carbon sequestration	1998–1999	5.8				7.7		Leavitt et al. (2006)
Water use								
Soil water balance	1998	−11.1	3.0			0.0	6.2	Conley et al. (2001)
	1999	−8.7	2.6			−6.5	4.0	Conley et al. (2001)
Energy balance	1998	−13.8	1.8			−8.5	3.7	Triggs et al. (2004)
	1999	−11.8	1.9			+10.5	5.1	Triggs et al. (2004)
Nitrogen concentration[b]	1998	−22.5	41.2					Cousins et al. (2003)

[a] Blower effect may have reduced response in 1993 and 1994 (Pinter et al. 2000)
[b] Values are actually for percent change in soluble protein per area of fifth leaf for section close to leaf tip.

al. (1994) similarly found a trend of a 14.3 % increase over 0–30 cm depth (Table 3.2). Yet, taken individually, none of these values were statistically significant. No significant changes due to FACE were detected under the limited-water regime either. It should be noted that Wood et al. (1994), Leavitt et al. (1994), and other authors measuring soil C who are cited in later sections all removed visible pieces of organic matter such as plant and insect parts before doing their soil C analyses. Therefore, the changes in soil C also imply similar changes (or lack thereof) in soil organic matter.

3.3.2.2 Light

The fractions of absorbed photosynthetically active radiation (fAPAR) were higher in the FACE plots compared to those of the control plots early in the 1990 and 1991 growing seasons (Table 3.2; Pinter et al. 1994; n.b. a quantum sensor was used, so to be precise, photosynthetic photon irradiance rather than photosynthetically active radiation was detected). However, about midway through the seasons when canopy closure was achieved, there were no significant differences in fAPAR due to $e[CO_2]$ in either irrigation treatment. The low-water plots had slightly lower fAPAR than the amply irrigated plots, however. When fAPAR was plotted against a normalized difference vegetation index (NDVI) for data obtained from all four treatments and from all observation dates throughout both growing seasons, a single curve emerged. In other words, the magnitude of fAPAR was dependent solely on the size of the plants (as manifested in NDVI), and there were no effects of $[CO_2]$ or irrigation on the character of the relationship.

Light use efficiency (LUE; season-long biomass produced per number of photosynthetic photons absorbed) was also stimulated by $e[CO_2]$, about 22 % for both ample and low irrigation regimes in 1990 and 1991 (Table 3.2; Pinter et al. 1994).

3.3.2.3 Water

Water use per unit of land area was not significantly changed by $e[CO_2]$ under either the ample or low irrigation regimes (Table 3.2; Dugas et al. 1994; Hunsaker et al. 1994). This result was surprising because stomatal conductance was reduced about 20 % due to the FACE treatment (Hileman et al. 1994), so water use per unit of leaf area was decreased and canopy temperatures were increased about 0.8 °C (Kimball et al. 1992b). However, as already discussed, cotton growth was greatly stimulated by the $e[CO_2]$. Therefore, the lack of detectable change in cotton water use per unit of land area suggests that the increase in effective leaf area must have been exactly compensated by reduced transpiration per unit of leaf area.

Because there was no significant change in water use due to e[CO$_2$], water use efficiency (WUE; biomass produced per unit of water used) was increased exactly as much as the increase in growth, as already presented (Table 3.2).

3.3.2.4 Nutrients

Cotton tissues (leaves, burs, seeds, stems, roots) from the plots with ample irrigation were analyzed for concentrations of N, P, K, Ca, Mg, Cu, Fe, Mn, and Zn (Prior et al. 1998). For several cases, concentrations were significantly reduced by e[CO$_2$] (See Table 3.2 for N data); and for several other cases, there were no significant changes. However, K, Cu, Fe, and Zn were significantly increased by in the leaves (but not other organs) from the FACE treatment only in 1990, but the concentrations were significantly decreased or no significant difference in 1991. Thus, while there were many cases of significant nutrient concentration decreases, there were no cases of consistent significant increases due to e[CO$_2$].

Some differences in nutrient concentrations among organs were observed, with leaves being affected more than seeds (Table 3.2). Whole plant N concentrations were reduced about 10 %.

However, because of the large increases in cotton growth due to e[CO$_2$] (Table 3.2), the whole plant uptakes of nutrients were significantly increased in most cases (Prior et al. 1998). Averaging over both years, the increases were: N by 20.9 %, P by 33.1 %, K by 26.2 %, Ca by 17.1 %, Mg by 27.5 %, Cu by 24.1 %, Fe by 26.3 %, Mn by 24.3 %, and Zn by 31.4 %.

3.3.3 Consequences for Management

The large yield increases due to e[CO$_2$] should be encouraging to growers. A slightly reduced planting density may be advantageous, but further study is required before recommendations can be made. Growers often use a growth regulator to induce cotton to become more reproductive than vegetative; and this aspect also needs further study. The greater growth under e[CO$_2$] resulted in greater uptake of all nutrients. Therefore, to sustain productivity over the long term, it is likely that fertilizer application rates will need to increase and growers will need to be more vigilant in order to detect incipient nutrient deficiencies.

It appears that irrigation requirements for cotton will not change significantly due to higher future[CO$_2$] (although if global warming occurs, cotton irrigation requirements likely would increase somewhat; Kimball 2006). However, a promising management tool for scheduling irrigations, the crop water stress index (CWSI), may need some adjustments. The CSWI is based on measurements of canopy temperature, using infrared thermometers. When the canopy temperature starts to rise above that of a well watered crop, it is an indication that the crop is starting to exhaust the soil moisture supply and it is time to irrigate. A key concept is the establishment of "non-water-stressed baselines" for each crop, which are plots of canopy minus air temperature differences against the vapor pressure deficit of the air (Idso 1982). However, when e[CO_2] causes canopy temperatures to increase, this rise in temperature could masquerade as an increase in water stress (Idso et al. 1987). Therefore, the "non-water-stressed baselines" will have to be adjusted for the increasing atmospheric [CO_2] in order to use the CWSI for scheduling irrigations in the future.

3.3.4 Consequences for Plant Breeding

Only one cultivar per year was grown in the FACE cotton experiments, in order to make the canopy as uniform and grower-field-like as possible. As already mentioned, differences in response were observed between the cultivar used in 1989 and that used in 1990 and 1991, but the cultivar change was confounded with changes in year and in application of a growth hormone. Thus, it is difficult to conclude any specific consequences for plant breeding. Rather, as they do now, the breeders need to continue to select for high yield and lint quality, for disease and insect resistance, and for tolerance and/or avoidance of heat stress.

One interesting study by Radin et al. (1994) has shown that, for irrigated cotton in an arid environment, selection for yield has lead to an inadvertent selection for high stomatal conductance and greater transpiration with consequent cooler canopy temperatures. e[CO_2] causes partial stomatal closure and an elevation in canopy temperatures (e.g., Kimball et al. 2002). Therefore, if cultivars differ in their stomatal conductance response to e[CO_2], it is possible that the continued selection for yield by irrigated-cotton breeders will similarly in turn lead to an inadvertent selection for stomatal insensitivity to e[CO_2].

3.4 Wheat

3.4.1 Resource Availability

During the 1993 and 1994 FACE wheat experiments, the plots were split into halves, with each side receiving either an ample or a low supply of water (Table 3.1; Hunsaker et al. 1994). Irrigation water requirements for the ample treatment were based on a computer-based irrigation scheduling program

(AZSCHED; Fox et al. 1992). After about 30 % of the available water in the rooted zone was depleted, they were irrigated with an amount calculated to replace 100 % of the potential evapotranspiration since the last irrigation (adjusted for rainfall). The low-water treatment plots received 50 % of the amounts of the ample treatment for each irrigation in 1993, whereas in 1994 they were irrigated only every other time the ample plots were irrigated. Cumulative irrigation totals between crop emergence and harvest were 600 mm and 620 mm for ample-treatment plots in 1993 and 1994, respectively, and 275 mm and 257 mm for the low-water plots. Corresponding rainfall amounts were 76 mm and 61 mm, respectively.

Except for the use of a drip irrigation system and imposition of the low-water treatment on half the plots, all plots were fertilized and managed according to recommended practice for the region for both the 1993 and 1994 wheat experiments (Hunsaker et al. 1996). Totals of 271 kg N ha^{-1} and 44 kg P ha^{-1} were supplied to the 1993 crop, and for 1994 the amounts were 261 kg N ha^{-1} and 29 kg P ha^{-1}.

After the 1994 wheat harvest, a summer N-removal barley (*Hordeum vulgare*) crop was grown, cut green, and removed from the field in preparation for a nitrogen limitation experiment. In order to minimize residual effects of the prior FACE and water treatments, the FACE and blower plots were positioned into new locations within the same field and then split into semicircular halves, with each half receiving either an ample (High-N) or a limiting (Low-N) level of nitrogen fertilizer (Kimball et al. 1999). The High-N plots received a total of 350 kg N/ha^{-1} from ammonium nitrate in four applications during both seasons. The Low-N plots received 70 kg N ha^{-1} and 15 kg N ha^{-1} during 1995–1996 and 1996–1997, respectively. An additional 33 kg ha^{-1} and 30 kg ha^{-1} of N were added to the high- and low-N plots, respectively, from the irrigation water itself.

Similar to the 1993 and 1994 ample irrigation treatment, all plots in 1996 and 1997 were irrigated as determined by the AZSCHED program (Fox et al. 1992). In 1996, the cumulative seasonal irrigation amounts were 692 mm and 631 mm in the high- and low-N treatments, respectively, and in 1997 they were 621 mm and 548 mm (Kimball et al. 1999). The amounts of high- and low-N would have been nearly identical each year except that the last irrigations in the low-N treatments were curtailed due to the earlier maturity of the N-stressed plants. The seasonal rainfall amounted to 39 mm and 29 mm for the 2 years, respectively.

3.4.2 Resource Acquisition and Transformation

3.4.2.1 CO_2 and Carbon

The average above-ground biomass increases of the wheat in the $e[CO_2]$ plots amounted to about 8.5 % under the ample water regime and about 14.5 % under the low-water regime in 1993 and 1994 (Table 3.2). However, these increases may have been lower than they ought to have been because there was an effect of the blowers in these experiments (Pinter et al. 2000), as discussed earlier. In 1996 and 1997, the above-ground biomass increases were about 8 % for the ample N, ample water case but only about 5 % for the Low-N plots.

Below-ground root biomass was increased relatively more due to $e[CO_2]$ than above-ground (Table 3.2), amounting to about 28 % in 1993 and 14 % in 1997 for the ample-water, ample-N case. When water was limiting, root growth was stimulated about 23 % in 1993.

Average grain yield increases of the wheat due to $e[CO_2]$ amounted to about 10 % under the ample water regime and about 23 % under the low-water regime in 1993 and 1994 (Table 3.2). As with the biomass values, however, these increases may have been lower than they ought to have been because of the blower effect (Pinter et al. 2000). In 1996 and 1997, the yield increases were about 16 % for the ample-N, ample-water case, but only about 8.5 % for the Low-N plots.

Soil carbon concentrations appeared to increase, on average, about 13 % over the two growing seasons of the first (1992–1994) FACE wheat experiment (Table 3.2), although variability was high for some of the measurements. However, Leavitt et al. (2001) using C isotope techniques did not detect any significant increase in new C added to the soil in FACE plots compared to the control plots during the second wheat experiment.

3.4.2.2 Water

Water use per unit of land area declined modestly for wheat when water and nitrogen were ample, about 3.6 % based on water balance measurements or 6.7 % based on the energy balance approach (Table 3.2). When nitrogen was limiting, soil water balance measurements indicated only about a 1 % decrease, whereas energy balance measurements indicated about a 20 % reduction in water use. However, simulations with the *ecosys* model by Grant et al. (2001) predicted a reduction in ET of 16 % at low nitrogen, caused by reductions in rubisco activity and concentration, which forced greater reductions in stomatal conductance in order to maintain a constant Ci:Ca ratio (ratio of internal leaf $[CO_2]$ concentration to that of outside air). Thus, the energy balance result of a 20 % reduction in ET of wheat at low N seems reasonable. When seasonal water supply is severely growth-limiting, one would expect plants to utilize all the water available to them, so that effects of e[CO_2] on seasonal ET would be minimal, unless perhaps CO_2-enriched plants with more robust root systems might extract more water from the soil and actually use more water. The latter phenomenon might have occurred for wheat in 1993 (Table 3.2), but generally the observed effects of FACE on the ET of wheat, as well as sorghum, under limited water have been inconsistent and small (Table 3.2).

3.4.2.3 Nutrients

Nitrogen concentrations in the leaves of wheat declined an average of about 4 % under e[CO_2] when soil water and N were ample (Table 3.2). When water was limited, the decline was somewhat larger, about 9 %. However, as expected, when soil N was low, N concentrations in the leaves declined markedly, about 22 %. The N concentrations of the wheat grain also declined due to e[CO_2], but not as much as those of the leaves (Table 3.2), amounting to about 3 % when water and N were ample, about 4 % when water was low, and about 9 % when soil N was low. The N concentrations of the total aboveground biomass of the crop declined through the season like those of the leaves, with the FACE-treated plants being about 10 % lower in N than the controls for both the ample and low-water treatments (Adamsen et al. 2005). However, by harvest time, the N differences due to e[CO_2] were no longer detectable (Table 3.2).

Soil inorganic N concentrations in the top 0.3 m declined rapidly during the 1993 growing season, with the concentrations in the FACE plots below those of the controls (Adamsen et al. 2005). During 1994, they were more steady with time, but again those in the FACE plots tended to be below those in the control plots. Although not statistically significant at $P=0.05$, there was a tendency at harvest time for the soil in the FACE plots to have lower inorganic N concentrations that those of the corresponding control plots for both the ample and low-water treatments (Table 3.2).

3.4.3 Consequences for Management

The results of the FACE wheat experiments are generally encouraging for growers, especially in areas where water is limited. Wheat yields increased under both ample and limited supplies of water and nitrogen. Water requirements declined slightly under the ample regime, yet when water was in short supply, the relative yield increase due to e[CO_2] was larger. Both cases represent an increase in water use efficiency.

In 2003, a severe drought occurred in Europe which caused drastic reductions in wheat and maize yields, as documented by Ciais et al. (2005). The results from the FACE wheat experiments suggest that the yield reductions could have been even worse if, for example, the atmospheric $[CO_2]$ had been at pre-industrial levels of about 280 ppm, rather than current levels of about 370 ppm.

The N concentrations of the leaves decreased markedly under $e[CO_2]$ when soil N was low, which represents a large decline in the nutritional value of the crop for forage. Similarly, grain N concentrations declined due to $e[CO_2]$ and low soil N, although not as much as the leaves, and such decreases represent declines in both baking quality and in nutritional value. However, under an ample soil N regime, the decreases in N concentration of both leaves and grain were minimal, which means that future management practices must include application of the ample levels of N used in the FACE experiments and possibly even higher.

3.4.4 Consequences for Plant Breeding

Only one cultivar was grown in the FACE wheat experiments in order to make the canopy uniform and as much like that of a commercial grower's field as possible. Therefore, no conclusions can be drawn with regard to the variability among cultivars with respect to their responses to $e[CO_2]$. However, in general, we can advise breeders to continue with efforts to improve wheat's ability to maintain productivity under water and heat stress conditions, which are likely to come with possible global warming.

3.5 Sorghum

3.5.1 Resource Availability

During the 1998 and 1999 FACE sorghum experiments, the plots were again split into halves with each side receiving either an ample or a low supply of water (Table 3.1; Ottman et al., 2001). Unlike the previous cotton and wheat experiments, the irrigation system was changed from drip to flood. But similar to the wheat experiments, irrigation water requirements for the ample treatment were based the AZSCHED program (Fox et al. 1992). After about 30 % of the available water in the rooted zone was depleted, the plots were irrigated with an amount calculated to replace 100 % of the potential evapotranspiration since the last irrigation (adjusted for rainfall). The low-water treatment plots were irrigated shortly after planting and only once more during the growing season. Cumulative irrigation totals between crop emergence and harvest were 1198 mm and 894 mm for ample-treatment plots in 1998 and 1999, respectively, and 454 mm and 338 mm for the low-water plots, respectively. Corresponding rainfall amounts were 20 mm and 153 mm, respectively.

All plots were fertilized and managed according to recommended practice for the region (Ottman et al., 2001). Totals of 279 kg N ha^{-1} and 266 kg N ha^{-1} were supplied to the 1998 and 1999 crops, respectively.

3.5.2 Resource Acquisition and Transformation

3.5.2.1 CO_2 and Carbon

There was a small mean above-ground biomass increase of only about 3 % to e[CO_2] under ample water (Table 3.2). In contrast, at low water, the sorghum grew about 16 % more under FACE compared to the controls. There was a small average decrease in grain yield 5 % under ample water, but variability was high (Table 3.2). In contrast, at low water, the sorghum yielded about 25 % more grain under FACE compared to the controls. The biomass and grain yield increases under FACE and low water were probably due to high-CO_2-induced partial stomatal closure, which enabled the FACE-grown plants to maintain photosynthesis and grow longer into each drought cycle.

3.5.2.2 Water

At ample water and nitrogen, reductions in water use of about 10 % were observed, based on soil water balance, or about 13 %, based on energy balance (Table 3.2). When water supply was low, changes in water use were inconsistent, as observed also for cotton and wheat. As already mentioned, when seasonal water supply is severely growth-limiting, one would expect plants to utilize all the water available to them, so that effects of e[CO_2] on seasonal water use would be minimal.

3.5.2.3 Nutrients

Cousins et al. (2003) reported on the utilization of nutrients by the sorghum (Table 3.2). Nitrogen concentration (actually soluble protein) near the tips of the fifth leaves was reduced about 23 % due to e[CO_2], which is at the larger end of the range of reductions generally observed (Kimball et al. 2002).

3.5.3 Consequences for Management

As expected, based on prior chamber-based experiments, the FACE C_4 sorghum was less responsive to $e[CO_2]$ than the C_3 cotton and wheat under ample water and N. Thus, sorghum yield increases due to the increasing atmospheric $[CO_2]$ are likely to be small or none. However, this same yield was obtained with significantly less water consumption. In contrast, when water was low, the relative yield increase due to $e[CO_2]$ was substantial. Both cases represent significant increases in water use efficiency.

As mentioned previously, in 2003 a severe drought occurred in Europe which caused drastic reductions in wheat and maize yields (Ciais et al. 2005). Because maize is very similar to sorghum (both C4 tropical grasses), the results from the FACE sorghum experiments suggest that the yield reductions could have been even worse if, for example, the atmospheric CO_2 concentrations had been at pre-industrial levels of about 280 ppm, rather than current levels of about 370 ppm.

3.5.4. Consequences for Plant Breeding

Like the cotton and wheat experiments, only one cultivar was grown in the FACE sorghum experiments in order to make the canopy as uniform and grower-field-like as possible. Therefore, no conclusions can be drawn with regard to the variability among cultivars with respect to their responses to $e[CO_2]$. However, in general, we can advise breeders to continue with efforts to improve sorghum's ability to maintain productivity under water and heat stress conditions, which are likely to come with possible global warming.

3.6 Conclusions

FACE experiments were conducted on cotton, wheat, and sorghum at Maricopa, Ariz., which has a hot, arid climate and is located within a large irrigated area surrounded by desert. The FACE treatments ($e[CO_2]$) initially were 500 ppm (by vol.) for cotton; and later they were 200 ppm above $c[CO_2]$ levels. Important results included the following:
- At ample water and nutrients, cotton, a C_3 woody plant grown in the summer, was highly responsive to $e[CO_2]$, with above-ground biomass and boll yields increased about 40%. In contrast, sorghum, a C_4 herbaceous plant also grown in the summer with little photosynthetic response to $e[CO_2]$, had little biomass and yield response. Wheat, a C_3 herbaceous plant grown in the winter had about a 9% increase in above-ground biomass and about a 13% increase in grain yield.

- When water was limiting, the growth and yield stimulations of cotton were about the same as under ample water (about 40 %), whereas those for wheat increased to about 15 % for above-ground biomass and 22 % for grain yield. In contrast to the negligible response at ample water, sorghum's above-ground biomass increased about 16 % and grain yield about 26 % due to e[CO$_2$] under limited water, which was attributed to water conservation from reduced stomatal conductance at e[CO$_2$].
- When nitrogen was limiting, the increases in above-ground biomass and yield of wheat due to e[CO$_2$] were only about 5 % and 9 %, respectively.
- Root growth responses of cotton and wheat due to e[CO$_2$] were generally larger than those of above-ground biomass.
- Soil carbon concentrations tended to increase (average of about 11 %), but variability was too high for statistical significance in individual studies.
- Leaf nitrogen concentrations at ample soil N supply and water were reduced by about 7 % for cotton and about 4 % for wheat. When soil N was limited, wheat leaf N concentrations decreased about 21 %.

References

Adamsen FJ, Wechsung G, Wechsung F, Wall GW, Kimball BA, Pinter PJ Jr, LaMorte RL, Garcia RL (2005) Temporal changes in soil and biomass nitrogen for irrigated wheat grown under free air carbon dioxide enrichment (FACE). Agron J (in press)

Ciais Ph, Reichstein M, Viovy N, Granier A, Ogee J, Allard V, Aubinet M, Buchmann N, Bernhofer C, Carrara A, Chevallier F, De Noblet N, Friend AD, Friedlingstein P, Grunwald T, Heinesch B, Keronen P, Knohl A, Krinner G, Loustau D, Manca G, Matteucci G, Miglietta F, Ourcival JM, Papale D, Pilegaard K, Rambal S, Seufert G, Soussana JF, Sanz MJ, Schulze ED, Vesala T, Valentini R (2005) Europe-wide reduction in primary productivity caused by heat and drought in 2003. Nature 437:529–533

Conley MM, Kimball BA, Brooks TJ, Pinter PJ Jr, Hunsaker DJ, Wall GW, Adam NR, LaMorte RL, Matthias AD, Thompson TL, Leavitt SW, Ottman MJ, Cousins AB, Triggs JM (2001) Free-air carbon dioxide enrichment (FACE) effects on sorghum evapotranspiration in well-watered and water-stressed irrigation treatments. New Phytol 151:407–412

Cousins AB, Adam NR, Wall GW, Kimball BA, Pinter PJ Jr, Ottman MJ, Leavitt SW, Webber AN (2003) Development of C$_4$ photosynthesis in Sorghum leaves grown under free-air CO$_2$ enrichment (FACE). J Exp Bot 54:1969–1975

Dugas WA, Heuer ML, Hunsaker DJ, Kimball BA, Lewin KF, Nagy J, Johnson M (1994) Sap flow measurements of transpiration from cotton grown under ambient and enriched CO$_2$ concentrations. Agric For Meteorol 70:231–245

Dugas WA, Pinter PJ Jr (1994) The free-air carbon dioxide enrichment (FACE) cotton project: a new field approach to assess the biological consequences of global change. Agric For Meteorol 70:1–342

Fox FA Jr, Scherer T, Slack DC, Clark LJ (1992) AriZona irrigation SCHEDuling user's manual. University of Arizona, Tucson

Grant RF, Kimball BA, Brooks TJ, Wall GW, Pinter PJ Jr, Hunsaker DJ, Adamsen FJ, LaMorte RL, Leavitt SW, Thompson TL, Matthias AD (2001) Interactions among CO$_2$, N, and climate on energy exchange of wheat: model theory and testing with a free air CO_2 enrichment (FACE) experiment. Agron J 93:638–649

Hendrey GR (1993) Free-air carbon dioxide enrichment for plant research in the field. Smoley, Boca Raton

Hileman DR, Huluka G, Kenjige PK, Sinha N, Bhattacharya NC, Biswas PK, Lewin KF, Nagy J, Hendrey GR (1994) Canopy photosynthesis and transpiration of field-grown cotton exposed to free-air CO_2 enrichment (FACE) and differential irrigation. Agric For Meteorol 70:189–207

Hunsaker DJ, Hendrey GR, Kimball BA, Lewin KF, Mauney JR, Nagy J (1994) Cotton evapotranspiration under field conditions with CO_2 enrichment and variable soil moisture regimes. Agric For Meteorol 70:247–258

Hunsaker DJ, Kimball BA, Pinter PJ Jr, LaMorte RL, Wall GW (1996) Carbon dioxide enrichment and irrigation effects on wheat evapotranspiration and water use efficiency. Trans ASAE 39:1345–1355

Hunsaker DJ, Kimball BA, Pinter PJ Jr, Wall GW, LaMorte RL, Adamsen FJ, Leavitt SW, Thompson TL, Matthias AD, Brooks TJ (2000) CO_2 enrichment and soil nitrogen effects on wheat evapotranspiration and water use efficiency. Agric For Meteorol 104:85–105

Idso SB (1982) "Non-water-stressed baselines:" a key to measuring and interpreting plant water stress. Agric Meteorol 27:59–70

Idso SB, Kimball BA, Mauney JR (1987) Atmospheric carbon dioxide enrichment effects on cotton midday foliage temperature: implications for plant water use and crop yield. Agron J 79:667–672

Kimball BA (2006) Global changes and water resources. In: Lascano RJ, Sojka RE (eds) Irrigation of agricultural crops monograph. American Society of Agronomy, Madison (in press)

Kimball BA, LaMorte RL, Peresta GJ, Mauney JR, Lewin KF, Hendrey GR (1992a) Appendices: weather, soils, cultural practices, and cotton growth data from the 1989 FACE experiment. Crit Rev Plant Sci 11:271–308

Kimball BA, Pinter PJ Jr, Mauney JR (1992b) Cotton leaf and boll temperatures in the 1989 FACE experiment. Crit Rev Plant Sci 11:233–240

Kimball BA, LaMorte RL, Pinter PJ Jr, Wall GW, Hunsaker DJ, Adamsen FJ, Leavitt SW, Thompson TL, Matthias AD, Brooks TJ (1999) Free-air CO_2 enrichment and soil nitrogen effects on energy balance and evapotranspiration of wheat. Water Resour Res 35:1179–1190

Kimball BA, Morris CF, Pinter PJ Jr, Wall GW, Hunsaker DJ, Adamsen FJ, LaMorte RL, Leavitt SW, Thompson TL, Matthias AD, Brooks TJ (2001) Wheat grain quality as affected by elevated CO_2, drought, and soil nitrogen. New Phytol 150:295–303

Kimball BA, Kobayashi K, Bindi M (2002) Responses of agricultural crops to free-air CO_2 enrichment. Adv Agron 77:293–368

Leavitt SW, Paul EA, Kimball BA, Hendrey GR, Mauney GR, Rauschkolb R, Rogers H, Lewin KF, Nagy J, Pinter PJ Jr, Johnson HB (1994) Carbon isotope dynamics of free-air CO_2-enriched cotton and soils. Agric For Meteorol 70:87–101

Leavitt SW, Pendall E, Paul EA, Brooks T, Kimball BA, Pinter PJ Jr, Johnson HB, Matthias A, Wall GW, LaMorte RL (2001) Stable-carbon isotopes and soil organic carbon in wheat under CO_2 enrichment. New Phytol 150:305–314

Leavitt SW, Cheng L, Williams DG, Brooks T, Kimball BA, Pinter PJ Jr, Wall GW, Ottman MJ, Matthias A, Paul EA, Thompson TL, Adam NR (2006) Soil organic carbon dynamics in sorghum under ambient CO_2 and free-air CO_2 enrichment (FACE). Global Change Biol (in revision)

Lewin KF, Hendrey GR, Nagy J (1994) Design and application of a free-air carbon dioxide enrichment facility. Agric For Meteorol 70:15–30

Mauney JR, Lewin KF, Hendrey GR, Kimball BA (1992) Growth and yield of cotton exposed to free-air CO$_2$ enrichment (FACE). Crit Rev Plant Sci 11:213–222

Mauney JR, Kimball BA, Pinter PJ Jr, LaMorte RL, Lewin KF, Nagy J, Hendrey GR (1994) Growth and yield of cotton in response to a free-air carbon dioxide enrichment (FACE) environment. Agric For Meteorol 70:49–68

Ottman MJ, Kimball BA, Pinter PJ Jr, Wall GW, Vanderlip RL, Leavitt SW, LaMorte RL, Matthias AD, Brooks TJ (2001) Elevated CO$_2$ effects on sorghum growth and yield at high and low soil water content. New Phytol 150:261–273

Pendall E, Leavitt SW, Brooks T, Kimball BA, Pinter PJ Jr, Wall GW, LaMorte RL, Wechsung G, Wechsung F, Adamsen F, Matthias AD, Thompson TL (2001) Elevated CO$_2$ stimulates soil respiration and decomposition in a FACE wheat field. Basic Appl Ecol 2:193–201

Pinter PJ Jr, Kimball BA, Mauney JR, Hendrey GR, Lewin KF, Nagy J (1994) Effects of free-air CO$_2$ enrichment on PAR absorption and conversion efficiency by cotton. Agric For Meteorol 70:209–230

Pinter PJ Jr, Kimball BA, Garcia RL, Wall GW, Hunsaker DJ, LaMorte RL (1996) Free-air CO$_2$ enrichment: responses of cotton and wheat crops. In: Koch GW, Mooney HA (eds) Carbon dioxide and terrestrial ecosystems. Academic, San Diego, pp 215–248

Pinter PJ Jr, Kimball BA, Wall GW, LaMorte RL, Adamsen FJ, Hunsaker DJ (1997) Effects of elevated CO$_2$ and soil nitrogen fertilizer on final grain yields of spring wheat (Annual research report, US Water Conservation Laboratory). USDA, Agricultural Research Service, Phoenix, pp 71–74

Pinter PJ Jr, Kimball BA, Wall GW, LaMorte RL, Hunsaker DJ, Adamsen FJ, Frumau KFA, Vugts HF, Hendrey GR, Lewin KF, Nagy J, Johnson HB, Wechsung F, Leavitt SW, Thompson TL, Matthias AD, Brooks TJ (2000) Free-air CO$_2$ enrichment (FACE): blower effects on wheat canopy microclimate and plant development. Agric For Meteorol 103:319–333

Pinter PJ Jr, et al (2006) Elevated CO$_2$ effects on wheat at ample and limited water and nitrogen. Agron J (in preparation)

Post DF, Mack C, Camp, PD, Suliman AS (1988) Mapping and characterization of the soils on the University of Arizona Maricopa Agricultural Center. Proc Hydrol Water Resour Southwest, Arizona–Nevada Acad Sci 18:49–60

Prior SA, Rogers HH, Runion GB, Mauney JR (1994) Effects of free-air CO$_2$ enrichment on cotton root growth. Agric For Meteorol 70:69–86

Prior SA, Torbert HA, Runion GB, Mullins GL, Rogers HH, Mauney JR (1998) Effects of carbon dioxide enrichment on cotton nutrient dynamics. J Plant Nutr 21:1407–1426

Radin JW, Zhenmin L, Percy RG, Zeiger E (1994) Genetic variability for stomatal conductance in Pima cotton and its relation to improvements of heat dissipation. Proc Natl Acad Sci USA 91:7217–7221

Rogers HH, Prior SA, O'Neill EG (1992) Cotton root and rhizosphere responses to free-air CO$_2$ enrichment. Crit Rev Plant Sci 11:251–264

Sinclair TR, Pinter PJ Jr, Kimball BA, Adamsen FJ, LaMorte RL, Wall GW, Hunsaker DJ, Adam NR, Brooks TJ, Garcia RL, Thompson TL, Leavitt SW, Matthias AD (2000) Leaf nitrogen concentration of wheat subjected to elevated [CO$_2$] and either water or nitrogen deficits. Agricult Ecosyst Environ 79:53–60

Triggs JM, Kimball BA, Pinter PJ Jr, Wall GW, Conley MM, Brooks TJ, LaMorte RL, Adam NR, Ottman MJ, Matthias AD, Leavitt SW, Cerveny RS (2004) Free-air carbon dioxide enrichment (FACE) effects on energy balance and evapotranspiration of sorghum. Agric For Meteorol 124:63–79

Wall GW, Kimball BA (1993) Biological databases derived from free-air carbon dioxide enrichment experiments. In: Schulze ED, Mooney HA (eds) Design and execution of experiments on CO_2 enrichment (Ecosystems Report 6, Environmental Research Programme). Commission of the European Communities, Brussels, pp 329–348

Wechsung G, Wechsung F, Wall GW, Adamsen FJ, Kimball BA, Pinter PJ Jr, LaMorte RL, Garcia RL, Kartschall T (1999) The effects of free-air CO_2 enrichment and soil water availability on spatial and seasonal patterns of wheat root growth. Global Change Biol 5:519–529

Wood CW, Torbert HA, Rogers HH, Runion GB, Prior SA (1994) Free-air CO_2 enrichment effects on soil carbon and nitrogen. Agric For Meteorol 70:103–116

4 SoyFACE: the Effects and Interactions of Elevated [CO_2] and [O_3] on Soybean

D.R. Ort, E.A. Ainsworth, M. Aldea, D.J. Allen, C.J. Bernacchi,
M.R. Berenbaum, G.A. Bollero, G. Cornic, P.A. Davey, O. Dermody,
F.G. Dohleman, J.G. Hamilton, E.A. Heaton, A.D.B. Leakey,
J. Mahoney, T.A. Mies, P.B. Morgan, R.L. Nelson, B. O'Neil,
A. Rogers, A.R. Zangerl, X.-G. Zhu, E.H. DeLucia, and S.P. Long

4.1 Introduction

SoyFACE is the first FACE experiment to focus on a seed legume and on corn and the first to explore the interactions of both elevated (e)[CO_2] and e[O_3] on the growth and development of an arable crop. The intent of the SoyFACE experiment is to orchestrate a coordinated and comprehensive investigation of the impact of atmospheric change on this expansive agroecosystem, addressing questions ranging from rhizosphere biology through to seed composition and employing techniques from genomics to micrometeorology. Soy-FACE completed its fourth season of operation in 2004. This chapter provides a description of the SoyFACE facility and its operation and overviews the published results from the 2001–2003 growing seasons.

4.2 Site Description

The SoyFACE facility is located in Champaign, IL, USA (40°02' N, 88°14' W, 228 m above sea level; http://www.soyface.uiuc.edu) situated on 32 ha of farmland within the Experimental Research Station of the University of Illinois. The soil is a deep (up to 1.25 m), organically rich Flanagan/Drummer series typical of northern and central Illinois "prairie soils" (fine-silty, mixed, mesic Typic Endoaquoll). Highly detailed information on the physical and chemical characteristics of Champaign County Illinois soils from the USDA Natural

Ecological Studies, Vol. 187
J. Nösberger, S.P. Long, R.J. Norby, M. Stitt,
G.R. Hendrey, H. Blum (Eds.)
Managed Ecosystems and CO_2
Case Studies, Processes, and Perspectives
© Springer-Verlag Berlin Heidelberg 2006

Resources Conservation Service can be found at: http://soils.usda.gov/survey/online_surveys/illinois/. The field is tile-drained and has been in continuous cultivation to arable crops for over 100 years. Agronomic practices in use at the site are typical for this region of Illinois. No nitrogen fertilizer is added to the soybean crop, whereas the corn (*Zea mays*) crop receives 202 kg N ha^{-1} (157 kg ha^{-1} as 28% 1:1 urea:ammonium nitrate liquid pre-plant and 45 kg ha^{-1} credit from previous soybean N$_2$ fixation). Soybean (*Glycine max* L. Merr. cv Pana in 2001; cv Pioneer 93B15 in 2002 and thereafter) and corn (Pioneer cv 34B43) each occupy one-half of the field and follow an annual rotation. In 2001, it was found that cv. Pana grew about 1.5 m at this site and was vulnerable to lodging, leading to its replacement by the shorter but related cv 93B15 for subsequent years. Both soybean cultivars were similar indeterminate lines of maturity group III which formed 20 000–30 000 nodules m^{-2}, amounting to 25 g m^{-2} in mass in the control and 32 g m^{-2} in the e[CO$_2$] treatment plots. Soybean was seeded using a mechanical seed planter to a field density of about 200,000 plants ha^{-1}at row spacing of 0.38 m (15 in); and the corn row spacing was 0.76 m (30 in at a seed density of 74 100 ha^{-1}). The experimental plots were oversown by hand on the day of planting and thinned after emergence to ensure uniform plant density.

An on-site weather station (MetData 1-type; Cambell Scientific, Logan, Utah) measured air temperature (T_{air}; for an explanation of abbreviations, see end of chapter) and relative humidity at a height of 3 m. A quantum sensor (model QSO; Apogee Instruments, Logan, Utah) measured incident photosynthetic photon irradiance (PPI) at a height of 3 m. Data were averaged and logged at 10 min intervals throughout the growing season. Tipping bucket rain gauges (model 52202; R.M. Young, Traverse City, Mich.) were distributed throughout the field and recorded rainfall events in 0.0001 m increments throughout the season. Weather data is posted on the SoyFACE website (http://www.soyface.uiuc.edu/weather.htm). The Illinois State Water Survey weather station (http://www.sws.uiuc.edu/data/climatedb/) in Urbana, Ill. (40°05'N, 88°14'W) is situated 3 km north of the SoyFACE site and at the same altitude.

4.3 Experimental Treatments

4.3.1 Field Layout and Blocking

To control for topographic (<1 m) and soil variation, each 16-ha half of the field was divided into 16 blocks, each able to accommodate two 20-m diameter octagonal plots. One plot in each block was untreated but otherwise outfitted with treatment equipment. The atmosphere in the second plot was amended during daylight hours, from crop emergence until harvest, using a FACE system (Miglietta et al. 2001; see Chapter 2).

4.3.2 CO$_2$ Treatment

The target e[CO$_2$] was 550 ppm, as projected for the year 2050 by the Intergovernmental Panel on Climate Change (Prentice et al. 2001). The average [CO$_2$] over the 2001–2003 seasons was 372 ppm in four ambient and 549 ppm in four elevated soybean plots. One-minute-average CO$_2$ concentrations were within ±10 % of the 550 mmol mol^{-1} target for more than 85 % of the time. On those rare instances when wind speeds dropped below 0.2 m s^{-1}, CO$_2$ fumigation cycled around the octagon to maintain the [CO$_2$] within the plot as close to the 550 mmol mol^{-1} set point as possible. Air temperature, PPI, and precipitation were recorded at 15-min intervals throughout the growing season. In alternate years (i.e., 2002, 2004) there were four e[CO$_2$] treatment plots and four control plots in corn.

4.3.3 O$_3$ Treatment

The seasonal target of 1.2 times the current ozone concentration was based on projected future mean global tropospheric concentrations, which suggest a 20 % increase by 2050 (Prather et al. 2001). Because of the reactivity of O$_3$ with water, fumigation was held in abeyance during periods when leaf surfaces were damp (e.g., most mornings). Thus to achieve the target concentration over the entire growing season, the set point was 1.5 times the continuously monitored ambient level. Elevation of [O$_3$] was based on the method of Miglietta et al. (2001; see also Chapter 2), but in this instance using compressed air enriched in ozone instead of compressed CO$_2$. As described previously for CO$_2$ by Miglietta et al. (2001), the quantity and duration of the ozone release was controlled by a proportional integral derivative algorithm for computer feedback that compares achieved [O$_3$] to the target [O$_3$] of 1.5 % current with a gas concentration monitor (model 49 O$_3$ analyzer; Thermo Environmental Instruments, Franklin, MA; calibration by US EPA Equivalent Method EQQA-0880-047; ranges 0–0.05–1.0 ppm), anemometer, and wind direction vane mounted in the center of each ring. Ozone was generated by passing pure oxygen through a high-voltage electrical field generating a composite gas consisting of approximately 10 % ozone and 90 % oxygen (GSO-40; Wedeco Environmental Technologies, Herford, Germany). Using a mass flow controller, the ozone/oxygen mixture was added to a compressed air stream through a venturi bypass system. Ozone fumigation began 20 days after seeding and continued for the remainder of the growing season until harvest. Fumigation operated during daylight hours and stopped to prevent damage to leaves when the crop was wet with dew or rain or when wind speeds dropped below 0.2 m s^{-1}, when control would be inadequate to avoid accumulation of high $[O_3]$ (>300 ppb) near the edges of the plots (similar protocols were followed at Aspen FACE; see Chapter 12). In 2002 and 2003, the 8-h average $[O_3]$ was 62 ppb and 50 ppb, respectively, under current conditions (control) and 75 ppb and 63 ppb in the $e[O_3]$ treatment. The effective treatment over the season was 121 % in 2002 and 125 % in 2003 of ambient $[O_3]$ and fumigation control was maintained at ±10 % of the set point concentration for 77 % of the time, and at ±20 % for 93 % of the time. Sampling plots were located at a minimum of 2 m internal to the octagon of horizontal pipes, to minimize any residual effect of the injection system.

4.3.4 $CO_2 \times O_3$ Treatment

The combined $CO_2 \times O_3$ treatment, begun in the 2003 growing season, was achieved by combining the treatments described above for the individual gases within the same experimental plots. The technology and the performance are as described above. No results have yet been published on the combined treatment.

4.4 Resource Acquisition

4.4.1 Effects of $[CO_2]$ Treatment on Photosynthesis

In the short term, an increase in $[CO_2]$ stimulates net photosynthesis in C_3 plants because the current $[CO_2]$ is insufficient to saturate Rubisco activity and because CO_2 inhibits the competing process of photorespiration. Therefore, an increase in net photosynthesis in $e[CO_2]$ is anticipated, regardless of whether Rubisco activity or regeneration of ribulose-1,5-bisphosphate (RuBP) is limiting assimilation, and regardless of whether light is saturating or not (Long and Bernacchi 2003). The seasonal profile of stimulation of soybean leaf photosynthesis by an increase in $[CO_2]$ to 552 mmol mol^{-1} was examined under open-field conditions (Rogers et al. 2004). Diurnal measurements of net leaf CO_2 uptake (A) were supported by simultaneous measurements of leaf carbohydrate dynamics, water vapor flux, modulated chlorophyll fluorescence, and microclimate conditions. Measurements were made from pre-dawn to post-dusk on 7 days, covering different developmental stages from the first node formation through complete seed fill. Across the 2001 growing season, the daily integrals of leaf photosynthetic CO_2 uptake (A') increased by nearly 25 % in $e[CO_2]$ even as the average mid-day stomatal conductance (g_s) decreased by 22 % (Table 4.1). However, while theory

Table 4.1 The effects of growth at elevated carbon dioxide on resource allocation in FACE-grown soybean. Values *in italics* indicate significance at α = 0.1 or better. *Dashes* indicate missing data. See list of abbreviations for parameter definitions

Parameter	Percentage change e[CO$_2$]		Reference
	2001	2002	
A'	*24.6*	–	Rogers et al. (2004)
g_s (mid-day)	*–29.9*	–	Rogers et al. (2004)
g_s (diurnal)	*–10*	*–10*	Bernacchi et al. (2005)
l	*–8*	*–4.5*	Bernacchi et al. (2005)
J_{PSII}'	4.5	–	Rogers et al. (2004)
A_{sat} (season-long)	*20*	*16*	Bernacchi et al. (2005)
$V_{c,max}$ (season-long)	*–4*	*–6*	Bernacchi et al. (2005)
g_m	–	No change	Bernacchi et al. (2005)
J_{max} (season-long)	No change	No change	Bernacchi et al. (2005)
$V_{c,max}/J_{max}$ (season-long)	*–5*	*–5*	Bernacchi et al. (2005)

predicts that stimulation of A should be seen at all light levels, [CO$_2$] enhancement of A was apparent only when PPI was above 1000 µmol m^{-2} s^{-1}. The greatest stimulation of A was observed during the early- to mid-grain filling stages (Bernacchi et al. 2005). There was no evidence of any loss of stimulation toward the end of the growing season; in fact the largest stimulation of A occurred during late-seed filling (Rogers et al. 2004; Bernacchi et al. 2005). In contrast to A', daily integrals of PSII electron transport (J_{PSII}'), measured by modulated chlorophyll fluorescence, were not significantly increased by e[CO$_2$] (Table 4.1). Although the results show sustained increase in A by soybean in response to growth in e[CO$_2$], it is only approximately half of the maximum stimulation predicted from theory (Rogers et al. 2004).

Down-regulation of light-saturated photosynthesis (A_{sat}) at e[CO$_2$], which typically involves a decrease in the amount and/or activity of Rubisco, has been demonstrated for many C$_3$ species (Long et al. 2004; Chapter 14, however see Chapter 6). But, did down-regulation occur in the SoyFACE experiment and can it account for the smaller than predicted stimulation of A that was observed? Potential Rubisco carboxylation ($V_{c,max}$) and electron transport through photosystem II (J_{max}) were determined from the responses of A_{sat} to intercellular [CO$_2$] (C_i) at biweekly intervals over the 2001 and 2002 growing seasons (Bernacchi et al. 2005). Measurements were made under controlled conditions on leaves harvested at predawn to ensure that determination of A_{sat} was not obscured by factors such as transient water stress or mid-day photoinhibition. Elevated [CO$_2$] increased A_{sat} by 15–20 %, even though stomatal conductance was reduced (Table 4.1). There was a small, yet statistically significant decrease in $V_{c,max}$ which in turn drove a decrease in $V_{c,max}/J_{max}$ (Table 4.1), inferring a shift in resource investment away from Rubisco. The decrease in $V_{c,max}/J_{max}$ was not an illusion caused by a decrease in mesophyll conductance (g_m), which was unchanged by e[CO_2]. While g_s was significantly decreased across both growing seasons, the limitation imposed on photosynthesis by the stomata (l) was lower (Table 4.1), implying that stomata represented less of a limitation to photosynthesis for plants growing at e[CO_2] (see also Chapter 14). Although there was no progressive decline in A_{sat} during either season, analyses of the A versus C_i responses showed that, even in an N-fixing species grown without rooting restriction and under open-field conditions, down-regulation of photosynthesis occurred. This down-regulation, small yet statistically significant, is in effect an optimization of photosynthesis to e[CO_2] in that the decrease in $V_{c,max}$ would result in lower rates of A only when measured at lower [CO_2] with little or no loss at e[CO_2].

Experiments were also conducted at SoyFACE to test the "source-sink" hypothesis of down-regulation by examining acclimation of photosynthesis in lines of soybean differing by single genes that altered sink capacity either by an ability to nodulate or by switching between determinate and indeterminate growth habits (Ainsworth et al. 2003). By restricting vegetative growth after flowering, the stem termination associated with determinate growth would be expected to limit the size of the carbon sink. Because the respiratory rate of a nodulated root system can be an order of magnitude greater than its non-nodulated counterpart (Vessey et al. 1988), root nodules are strong sinks for carbohydrates (see Chapter 18). Soybean isolines, in which single-locus gene substitutions changed indeterminate growth to determinate and nodulated roots to non-nodulated, resulted in enhanced down-regulation of photosynthesis at e[CO_2]. Whereas down-regulation in the nodulating indeterminate varieties Pana and Pioneer 93B15 discussed above (Bernacchi et al. 2005) was driven solely by decreases in $V_{c,max}$, both $V_{c,max}$ and J_{max} decreased when sink strength was reduced by genetically limiting nodulation and vegetative stem growth (Ainsworth et al. 2003). Increases in the total non-structural carbohydrate (i.e., starch plus ethanol-extractable carbohydrates), which frequently portend photosynthetic down-regulation in response to e[CO_2] (see Chapter 16), were twice as great when sink capacity was reduced by genetically controlled stem termination. These sink-manipulation experiments strongly support the premise that genetic capacity for the utilization of photosynthate is critical for the ability of plants to sustain enhanced photosynthesis when grown at e[CO_2].

Corn is the third most important food crop globally in terms of production; and demand is predicted to increase by 45 % from 1997 to 2020 (Pingali 2001). Although our FACE experiment has focused primarily on soybean, corn has also been grown within the experiment, such that the ecosystem in this agricultural rotation is continuously treated with e[CO_2]. Previous laboratory studies suggest that, under favorable growing conditions, C_4 photosynthesis is not typically enhanced by e[CO_2], yet stomatal conductance and transpiration are decreased, which can indirectly increase photosynthesis in dry climates. Given the deep soils and relatively high rainfall of the United States

Corn Belt, it was predicted that photosynthesis would not be enhanced by e[CO_2]. The diurnal course of gas exchange of upper canopy leaves was measured in situ across the growing season of 2002 (Leakey et al. 2004). Contrary to the prediction, growth at e[CO_2] significantly increased A by up to 41 % and by 10 % on average. Greater A was associated with greater intercellular [CO_2], lower stomatal conductance, and lower transpiration. In two of four cultivars grown, significant increases in production were observed. Summer rainfall during 2002 was very close to the 50-year average for this site, indicating that the year was neither atypical nor a drought year. However, stimulation of photosynthesis was limited to periods of mild drought, as following rainfall there was no effect of e[CO_2] on photosynthesis (Leakey et al. 2004). The results suggest that, even in the wetter areas of the Corn Belt, photosynthesis and yield increase if there are any periods of water stress (see Chapter 3, concerning related results with sorghum).

4.4.2 Effects of [O_3] Treatment on Photosynthesis

A number of prior enclosure studies with soybean (Mulchi et al. 1992) and other plants (McKee et al. 1995, 2000; Zheng et al. 2002) suggest that the effects of e[O_3] on photosynthesis accumulate with leaf age, reflecting the cumulative uptake of ozone. In the 2002 SoyFACE experiment, two cohorts of leaves were followed from completion of expansion through senescence, a period of approximately 35 days (Morgan et al. 2004b). The first cohort of leaves was formed during the vegetative stage of growth and remained green until about halfway through the flowering phase. At complete leaf expansion, both control and e[O_3]-treated leaves showed a high A_{sat} which declined over their lifetime, but there was no evidence of any e[O_3] effect relative to the controls. The later leaf cohort completed expansion near the beginning of pod-filling and persisted throughout pod-filling. At full leaf expansion, A_{sat} was high but, in contrast to the earlier cohort, there was a significant e[O_3] treatment effect from the accumulation of damage, resulting in the treated leaves reaching an average A_{sat} of 0 (leaf senescence) while control leaves still maintained >30 % of the original A_{sat}. This accelerated decline of A_{sat} in the e[O_3] treated leaves was accompanied by accelerated losses in $V_{c,max}$ and a lesser but significant loss in J_{max}. Unlike the first cohort, which moved deeper into the canopy as more nodes and leaves were added above, the second cohort developed near the completion of node formation and remained close to the top of the canopy throughout its life. This difference in canopy position likely explains the response differences as the two cohorts aged, given that under open-air conditions ozone only reaches the lower canopy by diffusion down through the upper canopy, resulting in a rapid decline of [O_3] with canopy depth. It should be noted that the significant effect of e[O_3] on the second cohort is of particular agronomic importance, since these are the leaves most responsible for providing photoassimilate during seed filling.

4.4.3 Effects of [CO₂] and [O₃] on Canopy Development

Any changes in canopy structure and duration caused by growth in $e[CO_2]$ and $e[O_3]$ would be expected to impact the productivity of agro-ecosystems (Drake et al. 1997; Long et al. 2004). By improving carbon assimilation and efficiency of water use, $e[CO_2]$ may increase the leaf area index (LAI). Additionally, by raising the maximum quantum yield of photosynthesis ($\phi_{CO2,max}$), $e[CO_2]$ may decrease the light compensation point, increasing the carbon gain in deeper-shaded leaves, which may in turn maintain a positive carbon balance thereby driving a further increase in LAI by delaying leaf abscission. In contrast, $e[O_3]$ accelerates senescence and may thereby reduce LAI. These predictions were tested at SoyFACE during the 2001 and 2002 growing seasons (Dermody et al. 2005). Maximum LAI increased from 6.7±0.2 in ambient air to 7.4±0.2 in $e[CO_2]$. The $\phi_{CO2,\ max}$ of shade leaves in $e[CO_2]$ increased from 0.059±0.002 in ambient air to 0.067±0.002. Elevated [CO₂] also extended the growing season: for example, the average LAI of soybeans growing in $e[CO_2]$ on 23 September (1.6±0.1) was ~33 % greater than in ambient air (1.1±0.1). This was not delayed senescence, as there was no enhanced retention of shade leaves but rather a sustained addition of new nodes and leaves at the top of canopy later in the growing season. Although $e[O_3]$ did not affect the maximum LAI, it shortened the growing season by as much as 40 %, reducing LAI from 3.5±0.2 in ambient air to 2.1±0.2 near the end of the growing season. No effect of $e[O_3]$ on $\phi_{CO2,max}$ was detected.

4.4.4 Effects of [CO₂] and [O₃] on Insect Herbivory

A common feature of growth at $e[CO_2]$ and $e[O_3]$ is an alteration of leaf chemical composition that can influence the palatability and nutritional quality of foliage for herbivorous insects. For example, plants grown at $e[CO_2]$ and $e[O_3]$ often produce leaves with a lower nitrogen and soluble protein content (Mulchi et al. 1992; Cotrufo et al. 1998) and plants grown at $e[CO_2]$ commonly accumulate sugars and starch in their foliage, also affecting palatability by altering C:N (Cotrufo et al. 1998; Long et al. 2004). To meet their nutritional requirements, some herbivores exhibit "compensatory feeding" by increasing their consumption of foliage with lower N content (Bezemer and Jones 1998; Whittaker 1999). To test these predictions, levels of herbivory were measured in soybean grown in ambient air and air enriched with CO₂ or O₃ at SoyFACE (Hamilton et al. 2005). FACE is unique among facilities for elevating either [CO₂] or [O₃], in allowing the free movement of insect pests and predators. Exposure to $e[O_3]$ appeared to have no effect on insect herbivory. Growth at e[CO_2] significantly increased the susceptibility of soybeans to herbivory early in the season, with the amount of leaf area removed increasing from 5 % in controls to more than 11 %. There was no evidence of compensatory feeding in that leaf nitrogen content and C:N ratio were unaltered in those leaves experiencing increased herbivory. Rather than feeding in an effort to compensate for poor nutritional value, it appears that elevated sugar concentrations stimulated Japanese beetles (*Popillia japonica* Newman) to increase consumption of leaves grown at e[CO_2]. Levels of soluble leaf sugars were increased by >30 % at e[CO_2] (Chapter 16) and coincided with a significant increase in the density of Japanese beetle. In two-choice feeding trials, Japanese beetles and Mexican bean beetles (*Epilachna varivestis* Mulsant.) preferred foliage grown at e[CO_2] to foliage grown at ambient [CO_2] (see Chapter 6 for potato herbivory). These results imply that growth at e[CO_2] has the potential to increase crop susceptibility to pests, particularly those insects stimulated to feed by sugar availability, and thus possibly increasing the need for insect pest management.

4.5 Resource Transformation

4.5.1 Effects of e[CO_2] Treatment on Crop Production and Yield

To date, only two large-scale and fully replicated FACE experiments have examined effects of e[CO_2] on yields of C_3 grain crops: wheat and rice. Over 3 years of growth, rice seed yield was increased by 7–15 % (Kim et al. 2003) and wheat yield increased by 8 % in two growing seasons (Kimball et al. 1995; Chapter 3) at e[CO_2]. Modern soybean cultivars grown in the mid-west United States include many indeterminate cultivars that fix nitrogen, thereby creating and sustaining additional carbon sinks. Indeterminate nodulated soybeans provide a good test of the maximum response to the e[CO_2] of the future atmosphere that can be anticipated under actual field conditions. The effect of growth in e[CO_2] on above-ground net primary production (ANPP) and yield was investigated at SoyFACE over three growing seasons. Additionally, via sequential harvests at 2-week intervals, a study investigated how the patterns of production and partitioning were differentially affected with time and developmental stage across the growing season (Morgan et al. 2005a). Although a different cultivar was used in 2001 and a hailstorm defoliated the crop mid-season in 2003, the relative enhancement of seed yield due to e[CO_2] was remarkably similar (~15 %) across the 3 years (Table 4.2). For cv Pana grown in 2001, e[CO_2] increased seed yield by greater number of seeds per pod. The increased seed yield in e[CO_2] for cv Pioneer 93B15 grown during 2002 and 2003 was due to an increase in the number of pods per plant, with no

Table 4.2 The effects of growth at $e[CO_2]$ or $[O_3]$ on resource transformation in FACE-grown soybean. Values *in italics* indicate significance at $\alpha = 0.1$ or better. *Dashes* indicate missing data

Parameter	Treatment	Percentage change at $e[CO_2]$ or $[O_3]$			Reference
		2001	2002	2003	
Seed yield	$[CO_2]$	*16*	*15*	*15*	Morgan et al. (2005a)
	$[O_3]$	–	*–15*	*–25*	Morgan et al. (2005b)
Harvest index	$[CO_2]$	*–3*	*–2.0*	*–2*	Morgan et al. (2005a)
	$[O_3]$	–	–2.5	–3	Morgan et al. (2005b)
Litterfall	$[CO_2]$	–	38	16	Morgan et al. (2005a)
	$[O_3]$	–	No change	*–39*	Morgan et al. (2005b)
ANPP	$[CO_2]$	–	*15*	*17*	Morgan et al. (2005a)
	$[O_3]$	–	*–11*	*–23*	Morgan et al. (2005b)
Nodes	$[CO_2]$	*21*	*22*	8	Morgan et al. (2005a)
	$[O_3]$	–	–	–	

increase in the number of seeds per pod. There was a consistent and significant, albeit small, decline in harvest index in all 3 years (Table 4.2). During much of the season, the portion of the assimilated carbon that accumulated in leaves as non-structural carbohydrate was small (<10%) for both cultivars and was independent of growth $[CO_2]$, implying enhanced export at $e[CO_2]$. However, towards the end of the season, export of photosynthate slowed and there was a significant, $[CO_2]$-dependent daytime accumulation of non-structural carbohydrates in source leaves (Rogers et al. 2004, Chapter 16). Biweekly litter production was not measured in 2001, but was significantly increased by $e[CO_2]$ in 2002, although not in 2003 (Table 4.2), probably due to the removal of much of the canopy by the July hailstorm. Above-ground net primary production, the sum of the dry mass and cumulative litter production, was significantly increased by 2002 and 2003 (Table 4.2), but the difference was only evident in the later part of the growing season and was the result of the prolonged growing season under $e[CO_2]$. Continued addition of nodes in $e[CO_2]$ likely explains the increased stem dry mass and height; and the additional leaves associated with these nodes may explain the significant extension of the growing season by 2–7 days across the 3 years.

4.5.2 Effects of O$_3$ Treatment on Crop Production and Yield

Among the major crops, soybean is one of the most susceptible to ozone, with adverse effects apparent at concentrations as low as 40 ppb (Ashmore 2002; Fuhrer et al. 1997). Nearly one-quarter of the earth's surface is currently at risk from tropospheric ozone in excess of 60 ppb during mid-summer, with even greater concentrations occurring in isolated regions (Fowler et al. 1999a, b), with Western Europe, the mid-west and eastern United States, and eastern China being exposed to some of the highest background levels (Prather et al. 2001). The SoyFACE ozone experiment is the first on soybean or any other row crop using free air fumigation. While the effects of ozone on soybean photosynthesis at SoyFACE are subtle (see Section 4.4.2; Morgan et al. 2004b), the effects on biomass and seed yield are robust (Morgan et al. 2005b). Seed yield decreased by 15 % in 2002 and 25 % in 2003 for soybean grown in e[O$_3$] (Table 4.2). The larger yield losses in 2003 likely are the consequence of the July hailstorm that severely damaged the crop and from which the e[O$_3$] plants recovered more slowly. In 2002, yield reduction was entirely due to a 14 % decrease in individual seed weight. While yield losses in 2003 resulted from both 7 % decrease in seed weight and a 16 % decrease in pods per plant (i.e., four fewer pods per plant). Elevated [O$_3$] had similar impacts on both yield and shoot dry mass at maturity but there was a trend, albeit not statistically significant, to a slightly reduced harvest index (Table 4.2). Decreases in the shoot dry mass of e[O$_3$] grown plants relative to controls appeared late in the 2002 growing season; and leaf, stem, and pod dry mass all reflected this late season difference. However, in 2003 decreases were apparent earlier, notably in stem dry mass, possibly reflecting a weakened ability to recover following the mid-July hailstorm. The decreased production in e[O$_3$] grown plants in 2003 following the hail may also explain the lower biweekly litterfall (Table 4.2). The depression in cumulative ANPP caused by e[O$_3$] increased as the growing season progressed. In 2002, significant differences developed late in the growing season and persisted throughout the remainder of the soybean lifecycle. The cumulative effect of e[O$_3$] over the 2002 season decreased ANPP by 11 % compared to controls (Table 4.2). In 2003, ANPP of the control was 50 % lower than in 2002 and the impact of e[O$_3$] was greater, decreasing ANPP by 23 % relative to controls.

4.6 Consequences for Future Soybean Crop Management and Plant Breeding

The soybean/corn rotation occupied about 62 × 10^6 ha in the United States during 2003, more than that of any other crop system, making it among the largest ecosystems in the contiguous 48 States (USDA 2004). Soybean is globally the most important dicotyledonous seed crop in terms of area planted and is also a major source of food protein worldwide (FAO–UN 2002). When grown in the field under FACE fumigation, stimulation of ANPP and yield by e[CO_2] were smaller than predicted from open-top chamber studies (reviewed by Ainsworth et al. 2002). Elevated [CO_2] increased the ANPP and seed yield of field-grown soybeans by about 15%, even while the harvest index decreased by about 3%. From the statistical summary of published reports of the e[CO_2] response of soybean in enclosure studies, the yield was 24% greater in plants grown in 450–550 ppm [CO_2] than in current [CO_2], with a 9% decrease in harvest index (Ainsworth et al. 2002). Although the response found at SoyFACE for soybean is only 60% of that predicted from the meta-analysis of soybean enclosure studies, the overall yield stimulation by e[CO_2] is greater than in FACE grown rice (7–15%; Kim et al. 2003) and spring wheat (8%; Kimball et al. 1995). This difference among FACE experiments with C_3 crops may reflect the indeterminate nature of the soybean cultivars grown and/or the nitrogen-fixing capacity of soybean. The effects of e[CO_2] are apparent in soybean during early vegetative growth in enclosure experiments and are sustained through the duration of the experiment (Ziska and Bunce 1995; Miller et al. 1998). The greatest stimulation of dry mass due to e[CO_2] was found to occur during flowering, declining through pod-filling (Ainsworth et al. 2002). In contrast to these findings, no significant increase in any growth parameter was observed until pod-filling in SoyFACE in any of the 3 years. Furthermore, the subsequent relative stimulation by e[CO_2] remained constant throughout pod-filling to maturity. The extended growing season of the SoyFACE soybean crop was also counter to the previous reports for soybean from enclosure studies where there was either a lack of effect or a shortened growing cycle. Taken together, the comparison of results between FACE and enclosure experiments suggest that our current projections of future food supply, based largely on chamber studies, are overly optimistic. Resolving this potential overestimation of global food supply will require more studies with major food crops and comparisons of the different technologies for examining the effects of e[CO_2] on crops (see Chapter 24).

In contrast to e[CO_2], the effect of ozone on soybean yield at SoyFACE was close to that predicted from chamber studies, although this was perhaps by happenstance since the cause for the decrease in yield appears to be different. The meta-analysis of prior chamber studies suggests that decreased net photosynthesis alone was responsible for decreased production at the moderate elevations of [O_3] used in the SoyFACE experiment (Morgan et al. 2004a). However, photosynthetic analyses of soybean grown under FACE fumigation showed decreases in leaf photosynthesis only as leaves entered senescence, which was accelerated in e[O_3] (Morgan et al. 2004b). Accelerated senescence induced by e[O_3] would limit season-long canopy photosynthesis and account for the dry mass decreases and yield losses, despite no response in leaf photosynthesis. Thus projection of crop yield should focus on the relationship of ozone deposition during the reproductive developmental stage. In addition to the effects of ozone on the normal developmental program of soybean, the defoliation resulting from the 2003 hailstorm provided a unique, albeit unplanned, illustration of the effect of e[O$_3$] on a field crop's compromised ability to recover from an extreme event. This might include outbreaks of defoliating insects, high winds, as well as hail; all of which could increase with global climate change and further exacerbate the impact on soybean production. Based on calculations from published linear responses, every 1 ppb increase in tropospheric [O$_3$] potentially results in up to a 0.6 % yield reduction calculated from soybean yield response in the unstressed year, based on a 40-ppb threshold for damage and an assumed linear response (Mills et al. 2000; Ashmore 2002). With global [O$_3$] increasing by 10 ppb over the next half century (Prather et al. 2001), this potentially will have significant impact on global agriculture, especially in two of the major soybean growing areas of the globe, which are projected to see large [O$_3$] increases, i.e., China and the United States mid-west (Fowler et al. 1999a; Prather et al. 2003).

4.7 Conclusions

The SoyFACE experiment is the first to focus on the affects of e[CO$_2$] and e[O$_3$] on a seed legume under fully open-air conditions. The experiment mimicked e[CO$_2$] and e[O$_3$] predicted for the middle of this century and was conducted in one of the world's major production areas for corn and soybean under cultivation and management techniques standard for the industry in the United States corn-belt region. Growth of soybean at e[CO$_2$] resulted in an approximately 25 % increase in the daily integral of net leaf CO$_2$ uptake, a 20 % increase in the rate of light saturated CO$_2$ uptake, a 15 % increase in seed yield, a 15 % increase in above ground primary productivity, and a 20 % increase in node number. Growth of soybean at e[CO$_2$] also resulted in approximately a 30 % decrease in mid-day stomatal conductance, a 10 % decrease in stomatal conductance averaged over the day, an 8 % decrease in the limitation of photosynthesis by stomatal conductance, and a 2–3 % decrease in harvest index.

Growth of soybean at e[CO$_2$] caused about a 5 % decrease in the ratio of maximum carboxylation capacity compared to maximum electron transport capacity, indicative of acclimation to optimize photosynthetic performance to the higher [CO$_2$] conditions. Growth of soybean at e[CO$_2$] extended the growing season and resulted in increased herbivory by Japanese beetles.

Growth of soybean at e[O$_3$] was largely deleterious to soybean although the effects developed slowly over the course of the growing season. e[O$_3$] resulted in decreases in seed yield (15–25 %), above-ground primary productivity (11–23 %), and harvest index (2–3 %). Growth at e[O$_3$] caused accelerated senescence of the crop.

Abbreviations

A = Net leaf CO_2 uptake
A' = Daily integral of net leaf CO_2 uptake
A_{sat} = Light-saturated CO_2 uptake
ANPP = Above-ground primary productivity
C_i = Intercellular CO_2 concentration
g_m = Leaf mesophyll conductance
g_s = Stomatal conductance
J_{max} = Maximum rate of electron transport
J_{PSII}' = Daily integral of photosystem II electron transport
l = Stomatal limitation to photosynthesis
LAI = Leaf area index
PPI = Photosynthetic photon irradiance
Rubisco = Ribulose-1,5-bisphosphate carboxylase/oxygenase
RuBP = Ribulose-1,5-bisphosphate
T_{air} = Air temperature
$V_{c,max}$ = Maximum RuBP saturated rate of carboxylation
$\phi_{CO2,max}$ = Maximum apparent quantum efficiency of CO_2 uptake

References

Ainsworth EA, Davey PA, Bernacchi CJ, Dermody OC, Heaton EA, Moore DJ, Morgan PB, Naidu SL, Ra HY, Zhu X, Curtis PS, Long SP (2002) A meta-analysis of elevated [CO_2] effects on soybean (*Glycine max*) physiology, growth and yield. Global Change Biol 8:1–15

Ainsworth EA, Rogers A, Nelson R, Long SP (2003) Testing the "source-sink" hypothesis of down-regulation of photosynthesis in elevated [CO_2] in the field with single gene substitutions in *Glycine max*. Agric For Meteorol 122:85–94

Ashmore MR (2002) Effects of oxidants at the whole plant and community level. In: Bell JNB, Treshow M (eds) Air pollution and plants. Wiley, London

Bernacchi CJ, Morgan PB, Ort DR, Long SP (2005) The growth of soybean under free air [CO_2] enrichment (FACE) stimulates photosynthesis while decreasing in vivo Rubisco activity. Planta 220:434–446

Bezemer TM, Jones TH (1998) Plant–insect herbivore interactions in elevated atmospheric CO_2: quantitative analyses and guild effects. Oikos 82:212–222

Cotrufo M, Ineson P, Scott A (1998) Elevated CO_2 reduces the nitrogen concentration of plant tissues. Global Change Biol 4:43–54

Dermody O, Long SP, DeLucia EH (2005) How does elevated CO_2 or ozone affect the leaf-area index of soybean when applied independently? New Phytol doi:10.1111/j.1469-8137.2005.01565.x

Drake BG, Gonzalez-Meler MA, Long SP (1997) More efficient plants: a consequence of rising CO_2. Annu Rev Plant Physiol 48:609–639

FAO–UN (2002) FAO trade yearbook (vol 165). FAO, Rome

Fowler D, Cape JN, Coyle M, Flechard C, Kuylenstierna J, Hicks K, Derwent D, Johnson C, Stevenson D (1999a) The global exposure of forests to air pollutants. Water Air Soil Pollut 116:5–32

Fowler D, Cape JN, Coyle M, Smith RI, Hjellbrekke AG, Simpson D, Derwent RG, Johnson CE (1999b) Modeling photochemical oxidant formation, transport, deposition and exposure of terrestrial ecosystems. Environ Pollut 100:43–55

Fuhrer J, Skarby L, Ashmore MR (1997) Critical levels for ozone effects on vegetation. Eur Environ Pollut 97:91–106

Hamilton JG, Dermody O, Aldea M, Zangerl AR, Rogers A, Berenbaum MR, DeLucia EH (2005) Anthropogenic changes in tropospheric composition increase susceptibility of soybean to insect herbivory. Environ Entomol 34:479–485

Kim HY, Lieffering M, Kobayashi K, Okada M, Miura S (2003) Seasonal changes in the effects of elevated CO$_2$ on rice at three levels of nitrogen supply: A free air CO$_2$ enrichment (FACE) experiment. Global Change Biol 9:826–837

Kimball BA, Pinter PJ, Garcia RL, Lamorte RL, Wall GW, Hunsaker DJ, Wechsung G, Wechsung F, Kartschall T (1995) Productivity and water use of wheat under free-air CO$_2$ enrichment. Global Change Biol 1:429–442

Leakey ADB, Bernacchi CJ, Ort DR, Long SP (2004) Will photosynthesis of maize (*Zea maize*) in the US cornbelt increase in future [CO$_2$] rich atmospheres? An analysis of diurnal courses of CO$_2$ uptake under free-air enrichment. Global Change Biol 10:951–962

Long SP, Bernacchi CJ (2003) Gas exchange measurements, what can they tell us about the underlying limitations of photosynthesis? Procedures and sources of error. J Exp Bot 54:2393–2401

Long SP, Ainsworth EA, Rogers A, Ort DR (2004) Rising atmospheric carbon dioxide: Plants FACE the future. Annu Rev Plant Biol 55:591–628

McKee IF, Farage PK, Long SP (1995) The interactive effects of elevated CO$_2$ and O$_3$ concentration on photosynthesis in spring wheat. Photosynth Res 45:111–119

McKee IF, Mulholland BJ, Craigon J, Black CR, Long SP (2000) Elevated concentrations of atmospheric CO$_2$ protect against and compensate for O$_3$ damage to photosynthetic tissues of field-grown wheat. New Phytol 146:427–435

Miglietta F, Peressotti A, Vaccari FP, Zaldei A, deAngelis P, Scarascia-Mugnozza G (2001) Free-air CO$_2$ enrichment (FACE) of a poplar plantation: the POPFACE fumigation system. New Phytol 150:465–476

Miller JE, Heagle AS, Pursley WA (1998) Influence of ozone stress on soybean response to carbon dioxide enrichment II. Biomass and development. Crop Sci 38:122–128

Mills G. Hayes F, Buse A, Reynolds B (2000) Air pollution and vegetation. In: UN/ECE IPC (ed) Annual report 1999/2000 of UN/ECE IPC vegetation. Center for Ecology and Hydrology, Bangor

Morgan PB, Ainsworth EA, Long SP (2004a) Elevated O$_3$ impact on soybeans, a meta-analysis of photosynthetic, biomass, and yield responses. Plant Cell Environ 26:1317–1328

Morgan PB, Bernacchi CJ, Ort DR, Long SP (2004b) An in vivo analysis of the effect of season-long open-air elevation of ozone to anticipated 2050 levels on photosynthesis in soybean. Plant Physiol 135:2348–2357

Morgan PB, Bollero GA, Nelson RL, Dohleman FG, Long SP (2005a) Smaller than predicted increase in aboveground net primary production and yield of field-grown soybean under fully open-air [CO$_2$] elevation. Global Change Biol 11:1856–1865

Morgan PB, Bollero GA, Nelson RL, Long SP (2005b) Season-long elevation of ozone concentration by 20 % under fully open-air conditions decreases the growth and production of Midwest soybean crops by ca. 20 %. Environ Pollut (in press)

Mulchi CL, Slaughter L, Saleem M, Lee EH, Pausch R, Rowland R (1992) Growth and physiological-characteristics of soybean in open-top chambers in response to ozone and increased atmospheric CO_2. Agric Ecosyst Environ 38:107–118

Pingali PL (2001) Meeting world maize needs: technological opportunities and priorities for the public sector. In: CIMMYT (ed) 1999–2000 World maize facts and trends. CIMMYT, Mexico City

Prather M, Ehhalt D, Dentener F, Derwent R, Dlugokencky E, Holland E, Isaksen I, Katima J, Kirchhoff V, Matson P, Midgley P, Wang M (2001) Atmospheric chemistry and greenhouse gases. In: Houghton JT, Ding Y, Griggs DJ, Noguer M, Linder PJ van der, Dai X, Maskell K, Johnson CA (eds) Climate change 2001: the scientific basis. Contribution of working group I to the third assessment report of the intergovernmental panel on climate change. Cambridge University Press, Cambridge, pp 239–280

Prather M, Gauss M, Berntsen T, Isaksen I, Sundet J, Bey I, Brasseur G, Dentener F, Derwent R, Stevenson D, Grenfell L, Hauglustaine D, Horowitz L, Jacob D, Mickley L, Lawrence M, Von Kuhlmann R, Muller J-F, Pitari G, Rogers H, Johnson M, Pyle J, Law K, Van Weele M, Wild O (2003) Fresh air in the 21st century? Geophys Res Let 30:1100, doi: 10.1029/2002GL016285

Prentice IC, Farquahar GD, Fasham MJR, Goulden ML, Heimann M, Jaramillo VJ, Kheshgi HS, Le Quere C, Scholes RJ, Wallace DWR (2001) The carbon cycle and atmospheric carbon dioxide. In: Houghton JT, Ding Y, Griggs DJ, Noguer M, van der Linder PJ, Dai X, Maskell K, Johnson CA (eds) Climate change 2001: the scientific basis. Contribution of working group I to the third assessment report of the intergovernmental panel on climate change. Cambridge University Press, Cambridge, pp 183–230

Rogers A, Allen DJ, Davey PA, Morgan PB, Ainsworth EA, Bernacchi CJ, Cornic G, Dermody O, Dohleman FG, Heaton EA, Mahoney J, Zhu X-G, DeLucia EH, Ort DR, Long SP (2004) Leaf photosynthesis and carbohydrate dynamics of soybeans grown throughout their life-cycle under free-air carbon dioxide enrichment. Plant Cell Environ 27:449–458

USDA (2004) Crop production 2003 summary. Agricultural Statistics Board, National Agricultural Statistics Service, Washington, D.C.

Vessey JK, Walsh KB, Layzell DB (1988) Oxygen limitation of N_2 fixation in stem-girdled and nitrate-treated soybean. Physiol Plant 73:113–121

Whittaker JB (1999) Impacts and responses at population level of herbivorous insects to elevated CO_2. Eur J Entomol 96:149–156

Zheng Y, Shimizu H, Barnes JD (2002) Limitations to CO_2 assimilation in ozone-exposed leaves of *Plantago major*. New Phytol 155:67–78

Ziska LH, Bunce JA (1995) Growth and photosynthetic response of 3 soybean cultivars to simultaneous increases in growth temperature and CO_2. Physiol Plant 94:575–584

5 Paddy Rice Responses to Free-Air [CO$_2$] Enrichment

K. Kobayashi, M. Okada, H.Y. Kim, M. Lieffering, S. Miura, and T. Hasegawa

5.1 Introduction to Rice

Rice (*Oryza sativa* L.) is one of the world's three major crops. It differs from wheat and maize, the world's top two crops, in the distribution of its production areas: a predominant proportion of the global rice harvest comes from regions at latitudes between 30° N and 30° S, mostly in Asia (FAOSTAT 2004). In contrast, the majority of the wheat and maize crops are produced at higher latitudes. Rice is also quite unique in that the majority (ca. 90 %) of the world's harvest comes from flooded fields. As such, rice is grown under natural and socioeconomic environments that are different from those for the other major crops. This would further imply that the impacts of global change on rice may differ from those on other crops due to these differences in the growing environment.

The importance of rice as the most important food crop in Asia justified the commencement of the Rice FACE project in Japan in 1996. The primary objective of the project was to improve our capability to predict responses of rice plants and paddy ecosystems to increasing atmospheric CO$_2$ concentration ([CO$_2$]). The FACE experiment was conducted for 3 years (1998–2000, Phase 1), and after a 2-year hiatus, for an additional 2 years (2003–2004, Phase 2). This chapter summarizes rice plant responses to elevated [CO$_2$] in the Rice FACE experiment during Phase 1.

Ecological Studies, Vol. 187
J. Nösberger, S.P. Long, R.J. Norby, M. Stitt,
G.R. Hendrey, H. Blum (Eds.)
Managed Ecosystems and CO$_2$
Case Studies, Processes, and Perspectives
© Springer-Verlag Berlin Heidelberg 2006

5.2 The Rice FACE Experiment: Phase 1

5.2.1 Site Description, Plot Layout, and Crop Management

The Rice FACE experiment was conducted in Shizukuishi township, Iwate prefecture in northern Honshu island, Japan (39° 38'N, 140° 57'E, elevation 200 m a.s.l.).

The area is typical of the agro-climatic region that produces a large proportion of Japan's rice crop. Average air temperature and incident shortwave radiation across the rice season ranged from 19.7 °C (1998) to 21.4 °C (2000) and from 12.5 MJ day^{-1} m^{-2} (1998) to 16.0 MJ day^{-1} m^{-2} (2000), respectively. The temperature was close to normal in 1998, but much warmer than normal in 1999 and 2000. Shortwave radiation input was lower than normal in 1998 and higher in 1999 and 2000 (Kim et al. 2003a).

We conducted the FACE experiment in farmers' paddy fields. On the basis of crop and soil surveys prior to the experiment, we chose eight fields and blocked them into four pairs, each of which had fields with similar agronomic history and soil properties. The fields within a block were randomly designated as either elevated [CO_2] (e[CO_2]) plots or current [CO_2] (c[CO_2]) plots. The e[CO_2] and c[CO_2] plots were identical with respect to the plot configuration and management practices, except for [CO_2].

In each of the four e[CO_2] plots, we installed a [CO_2] enrichment apparatus (hereafter referred to as a FACE ring). Each FACE ring was made of eight 5-m long flexible plastic tubes arranged in an octagon and serving as the CO_2 emitters. Pure CO_2 was sprayed without blowers via tiny holes in the tubes which were suspended about 0.5 m above the plant canopy (Okada et al. 2001). The c[CO_2] plots were situated at least 90 m (center to center) away from any e[CO_2] plots to minimize contamination by CO_2 released in the FACE rings.

We used the rice cultivar Akitakomachi, which is popular in this region. The seedlings were raised in plastic greenhouses under either current or elevated (current plus 200 ppm) [CO_2]. Seedlings were hand-transplanted into the corresponding c[CO_2] or e[CO_2] plots in groups of three (referred to as a hill) on 21 May 1998, 20 May 1999, and 22 May 2000. Hills were spaced at 17.5 cm and rows were 30 cm apart (equivalent to 19.05 hills m^{-2}).

The soil was an Andosol paddy soil typical of northern Japan. On average, the plow layer depth was 12.3 cm, and the soil contained 82.5 g kg^{-1} organic C and 5.1 g kg^{-1} total N with an average soil pH of 5.6. The fields were plowed and then flooded and puddled to establish the flat paddies. The fields were flooded throughout the rice season, except for a period of about 5 days for drainage in mid-July and for about 10 days prior to harvest. Rice plants were sampled for destructive measurements to determine crop growth parameters (Kim et al. 2003b). Final harvest varied by year and to a lesser extent by N fertilizer application rate. In the medium-N subplots (see below), harvests were done on 29 September 1998, 20 September 1999, and 19 September 2000.

5.2.2 Experimental Treatments

5.2.2.1 [CO$_2$] Enrichment

Twenty-four-hour [CO$_2$] enrichment was carried out from transplanting through to harvest. The target [CO$_2$] at the center of the FACE ring was 200 ppm above the [CO$_2$] in c[CO$_2$] plots. Actual daytime [CO$_2$] at the FACE ring center at canopy height was 599, 568, and 559 ppm for 1998, 1999, and 2000, respectively, on average across the e[CO$_2$] plots throughout the season, whereas the corresponding values for the c[CO$_2$] plots were 368, 369, and 365 ppm. The actual [CO$_2$] was thus substantially higher than the target in 1998, but was close to the target in 1999 and 2000 due to improvements of the [CO$_2$] control algorithms.

5.2.2.2 N Fertilizer Application

Within each of the c[CO$_2$] and e[CO$_2$] plots, N fertilizer was supplied as ammonium sulfate at three rates: Low-N (40 kg N ha^{-1} in 1998, 1999, 2000), medium-N (80 kg N ha^{-1} in 1998; 90 kg N ha^{-1} in 1999, 2000), and high-N (120 kg N ha^{-1} in 1998; 150 kg N ha^{-1} in 1999, 2000). The medium-N rate is representative of the local farmers' practice. In all years we split-applied N to mimic the local farmers' practice: 63 % of the total N applied as a basal dressing 4 days prior to transplanting, 20 % as a top-dressing at mid-tillering, and 17 % as a top-dressing at panicle initiation (PI). Phosphate and potash were supplied as basal fertilizers at optimum levels: 300 kg P$_2$O$_5$ ha^{-1} (1998, 2000) or 480 kg P$_2$O$_5$ ha^{-1} (1999), and 150 kg K$_2$O ha^{-1} (1998–2000).

5.3 Effects of e[CO$_2$] on Paddy Rice

5.3.1 Effects on Resource Acquisition

5.3.1.1 Phenology

Under the normal weather conditions in 1998, the rice plants reached the PI stage around 60 days after transplanting (DAT), heading stage 84 DAT, and maturity 133 DAT averaged across the N fertilization levels. The warmer-than-normal weather in 1999 and 2000 resulted in quicker phenological development than in 1998: the PI stage was reached 52 DAT (1999) and 47 DAT (2000), the heading stage 76 DAT (1999) and 71 DAT (2000), and maturity was reached 124 DAT (1999) and 120 DAT (2000).

The [CO_2] enrichment slightly accelerated phenological development with heading accelerated by ca. 2 days and maturity by 2–3 days earlier (Kim et al. 2001). Although the accelerated development reduced the opportunity for e[CO_2] plants to capture light and to fix carbon, the difference was less than 2 % of the whole growth duration (120–133 days) and hence the direct effect of the reduced growth duration would not have a large impact on the rice biomass and grain yield.

5.3.1.2 Light Capture by Leaves

We determined green leaf lamina area by destructively sampling and calculated the leaf area index (LAI). We also estimated LAI non-destructively with a light interceptometer (LAI-2000; LiCor) under sky conditions without direct sun light. The estimated LAI was used to supplement the direct measurement before heading, after which the interceptometer measurements tended to overestimate LAI and thus were not used.

The response of LAI to e[CO_2] differed among the growth stages across the 3 years (Table 5.1). Elevated [CO_2] increased LAI at tillering ($P=0.002$) and PI ($P=0.0005$) stages, but only weakly so at heading ($P=0.09$). At maturity, in contrast, e[CO_2] *reduced* LAI ($P=0.06$) due to the larger loss of LAI in e[CO_2] for the period from heading to maturity ($P=0.007$). It is noteworthy that the enhancement of LAI by e[CO_2] was N-dependent. The[CO_2] × N interaction was statistically significant at PI ($P=0.005$), when LAI was changed little at the low-N level but was increased by e[CO_2] at the medium-N and high-N levels (Table 5.1).

The above effects of e[CO_2] on LAI can be converted to effects on fractional interception of PAR (photosynthetically active radiation) by the rice canopy, assuming an exponential reduction of incident light within the rice canopy, viz:

$$f_{PAR} = 1 - \exp(-k \, LAI),$$

where f_{PAR} is the fraction of PAR intercepted by the canopy, and k is the light extinction coefficient. We estimated the values of k as 0.5 for tillering stage, and 0.55 for PI and heading stages with diurnal measurements of PAR incidence above and below the plant canopies (Okada M., unpublished data). At the tillering stage, the LAI increase of 12–21 % due to e[CO_2] (Table 5.1) would have increased f_{PAR} by 7–11 %, whereas at PI the f_{PAR} increase was 1–4 % and at heading only 0–1 %. Thus it appears that the [CO_2] enrichment increased light

Table 5.1 Responses of LAI, root biomass, and tillers number to [CO$_2$] enrichment at three rates of N fertilizer application[a]

| Growth stage | LAI (m^2 m^{-2})[b] | | | | | |
| | Low-N | | Medium-N | | High-N | |
	c[CO$_2$][c]	e[CO$_2$][d]	c[CO$_2$]	e[CO$_2$]	c[CO$_2$]	e[CO$_2$]
Tillering	1.52	1.76 (16%)	1.89	2.11 (12%)	2.07	2.49 (21%)
PI[e]	2.71	2.78 (3%)	3.38	3.71 (10%)	3.91	4.54 (16%)
Heading	3.52	3.53 (0%)	4.15	4.38 (6%)	4.85	5.11 (5%)
Maturity	1.64	1.49 (−9%)	1.91	1.70 (−11%)	2.26	2.16 (−4%)

| Growth stage | Root biomass dry weight (g m^{-2})[f] | | | | | |
| | Low-N | | Medium-N | | High-N | |
	c[CO$_2$]	e[CO$_2$]	c[CO$_2$]	e[CO$_2$]	c[CO$_2$]	e[CO$_2$]
PI	54.4	53.0 (−3%)	50.2	60.5 (21%)	51.9	63.5 (22%)
Maturity	41.9	62.2 (48%)	41.9	52.1 (24%)	40.1	52.9 (32%)

| Growth stage | Number of tillers (m^{-2})[g] | | | | | |
| | Low-N | | Medium-N | | High-N | |
	c[CO$_2$]	e[CO$_2$]	c[CO$_2$]	e[CO$_2$]	c[CO$_2$]	e[CO$_2$]
PI	508	545 (7%)	565	660 (17%)	609	701 (15%)
Maturity	377	396 (5%)	480	506 (5%)	512	551 (8%)

[a] N fertilizer application rates: low-N (40 kg N ha^{-1} in 1998, 1999, 2000), medium-N (80 kg N ha^{-1} in 1998; 90 kg N ha^{-1} in 1999, 2000), high-N (120 kg N ha^{-1} in 1998; 150 kg N ha^{-1} in 1999, 2000)
[b] Averages of four replicates across 3 years (1998–2000)
[c] Values for current atmospheric [CO$_2$]
[d] Values for elevated [CO$_2$]. Percent increase over c[CO$_2$] shown in parentheses
[e] Panicle initiation
[f] Averages of four replicates across 2 years (1999–2000)
[g] Averages of four replicates across 2 years (1999–2000)

capture through to PI but not after. A similar finding has been made with cotton (*Gossypium hirsutum*, L.; see Chapter 3).

5.3.1.3 Leaf Photosynthesis

We measured leaf photosynthesis with a portable gas exchange system (LI-6400; LiCor). According to Farquhar et al. (1980), light-saturated photosynthesis (A_{sat}) is limited by carboxylation or by RuBP regeneration and these processes are well represented by two parameters obtained from leaf gas exchange measurements: maximum carboxylation rate ($V_{c,max}$) and maximum rate of electron transport (J_{max}).

In the Rice FACE, Seneweera et al. (2002) found that $e[CO_2]$ decreased both $V_{c,max}$ and J_{max}, with the larger reduction observed in the former, which is commonly observed in C_3 species (Long et al. 2004). The $e[CO_2]$ effects were larger in the flag leaf than in the eighth leaf at the active tillering stage. Nitrogen content per unit leaf area (N_{area}) was also reduced with $e[CO_2]$ by 6 %, which is similar to the reduction reported in the meta-analysis (4 %) by Long et al. (2004).

Data reported by Seneweera et al. (2002) indicated that both $V_{c,max}$ and J_{max} are highly sensitive to decreases in N_{area}; and these relationships largely explain the variation in both parameters with $e[CO_2]$ and leaf age. According to the relationships among $V_{c,max}$, J_{max} and N_{area}, the enhancement of A_{sat} by $e[CO_2]$ is greater where N_{area} is large, and becomes progressively smaller as N_{area} declines. In fact, A_{sat} measured on the eighth leaf at the active tillering stage showed a large enhancement of about 40 % (Seneweera et al. 2002), and measurements thereafter by Anten et al. (2003) and those on the flag leaf by Seneweera et al. (2002) resulted in lower enhancement (about 20 % and 4 %, respectively); and these changes related well to the changes in N_{area}.

Elevated $[CO_2]$ also increases the maximum apparent quantum yield of CO_2 uptake (QY). Anten et al. (2003) reported ca. 20 % increase in QY at around the heading stage with FACE; this was slightly larger than the enhancement in the meta-analysis of C_3 species (Long et al. 2004). Consequently, both A_{sat} and QY contribute to the carbon gain enhancement of the canopy. The enhancement of canopy photosynthesis estimated from the leaf photosynthetic parameters was 24 % for rice (Anten et al. 2003), which was slightly lower but comparable to the daily integral of leaf CO_2 uptake (A') in the meta-analysis (Long et al. 2004).

5.3.1.4 Root Development

Root biomass was used as a measure of root development, since it is closely related to N uptake during the early growth stages (Kim et al. 2001). At PI, root biomass was increased by $e[CO_2]$ ($P=0.014$) at medium-N and high-N fertilizer rates, but was unchanged at low-N (Table 5.1) with a significant $[CO_2] \times$ N interaction ($P=0.04$). At maturity, the enhancement of root biomass by $e[CO_2]$ was even more evident ($P=0.0004$), with a weak tendency for a $[CO_2] \times$ N interaction ($P=0.09$). The direction of the interaction was, however, opposite to that at PI: root biomass was increased by elevated $[CO_2]$ more in low-N than the higher N levels (Table 5.1).

5.3.1.5 Tillering

During vegetative growth in rice, tillers are formed at the nodes of the main stem, and as the tillers grow, they serve as additional plants, each of which develops leaves and roots during vegetative growth and a panicle at the apex during reproductive growth. The numbers of leaves and roots are thus strongly affected by tillering, hence the resource capture above- and below-ground is also dependent on the number of tillers.

Elevated [CO$_2$] enhanced tillering by 7–15 % at PI (P=0.0002), but had no significant effect on it at maturity (P=0.13; Table 5.1) without a [CO$_2$] × N interaction at either PI or maturity. Considering the role of tillers as 'additional plants', it is not surprising to see that this response to e[CO$_2$] was parallel to those of root biomass and LAI (Table 5.1) at the PI stage. At maturity, in contrast, these responses diverged: LAI was reduced and root biomass was increased, while the number of tillers was little changed by e[CO$_2$].

5.3.1.6 Accumulation of Plant Biomass and Nitrogen

Total plant biomass was increased by e[CO$_2$] at the PI (P<0.0001) and maturity (P=0.001) stages, but the extent of the increase was much higher at PI (21–34 %) than at maturity (11–12 %; Fig. 5.1A). It has been shown that, at the medium-N fertilizer level, the biomass response to e[CO$_2$] declined almost linearly from the tillering to maturity stages (Kim et al. 2003b).

The distinction between the PI and maturity stages was clear also in the plant biomass [CO$_2$] × N interaction, which was significant at PI (P=0.02) but not significant at maturity (P=0.8). At PI, the response to e[CO$_2$] was less in low-N than the higher fertilization levels. This interaction was mostly due to the response of green leaf laminae biomass to [CO$_2$] and N: it was increased by only 9 % in low-N, but up to 25 % in the higher N levels (P=0.005 for [CO$_2$] × N interaction). The response of green leaf biomass was, however, smaller than that of stems and leaf sheaths, which showed an increase of ca. 40 % in response to e[CO$_2$] irrespective of the N levels (Fig. 5.1).

At maturity, the responses of the leaf laminae and stems to e[CO$_2$] diverged: leaf laminae biomass was reduced by 5–10 % (P=0.03) whereas stem biomass increased by ca. 18 % across the N levels (P<0.0005). The loss of leaf laminae biomass (P=0.03) was caused by the increase in senescent leaves (P=0.10). At maturity, no [CO$_2$] × N interaction was found in the biomass responses to e[CO$_2$], except for the root mass response (see Section 5.3.1.4).

Nitrogen accumulation was significantly increased by e[CO$_2$] at PI (P=0.008) with the interacting effect of N fertilization rate (P=0.0009; Fig. 5.1B). The [CO$_2$] × N interaction at PI was clearly seen in the amount of

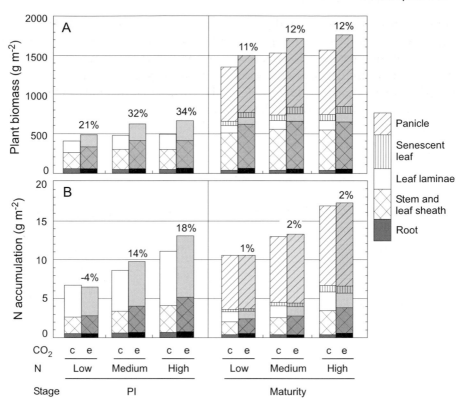

Fig. 5.1 Accumulation of plant biomass (**A**) and nitrogen (**B**) within different organs at panicle initiation (*PI*) and grain maturity averaged across 2 years (1999, 2000). The experimental treatments are: two levels of [CO_2] (*c* current, *e* elevated), three rates of N fertilizer application (*Low, Medium, High*). The *figures* over the e[CO_2] bars are the percent increase over the corresponding c[CO_2] bars

N accumulated in the leaf laminae (*P*=0.002) and stem (*P*=0.0006) with responses being smaller at low-N versus higher N rates.

At maturity, N accumulation in stems and roots was increased by e[CO_2] (*P*=0.002 for roots, *P*=0.05 for stems), but that in leaf laminae was reduced (*P*=0.002). Because of the cancellation between these opposite responses, total N accumulation in plants was unchanged (*P*=0.36). It follows that *less* N was accumulated in e[CO_2] than c[CO_2] for the period between PI and maturity at medium-N and high-N fertilization rates (Kim et al. 2003b). At the low-N rate, in contrast, the amount of N accumulation for the same period was not influenced by [CO_2].

It is noteworthy that the seasonal pattern of N accumulation differed from that of plant biomass accumulation. Nitrogen accumulation through to PI represented more than 60 % of total N at maturity, whereas the plant biomass accumulation through to PI was only 30–38 % of the final biomass. The discrepancy between the seasonal patterns of accumulation of N and plant biomass indicates that the accumulation of plant biomass, mostly grain biomass, in the reproductive growth period was performed with progressively more limited N than that during the period of vegetative growth. Plants were subjected to N shortage more under e[CO$_2$] than c[CO$_2$] at medium-N and high-N fertilization rates, whereas at low-N there was no such difference between [CO$_2$] treatments with respect to N availability.

5.3.2 Effects on Resource Transformation

5.3.2.1 Distribution of Plant Biomass and N During Reproductive Growth

As shown above, accumulation of plant biomass at maturity exhibited contrasting responses to e[CO$_2$] among the plant parts. This is due to the difference between the plant parts in the change in biomass during reproductive growth in response to e[CO$_2$]. Interestingly, the rate of increase of total plant

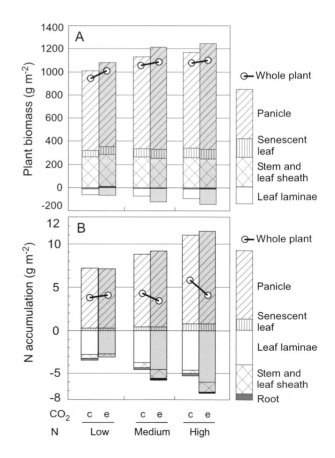

Fig. 5.2 Distribution of plant biomass (**A**) and nitrogen (**B**) among different organs for the period between panicle initiation and maturity averaged across 2 years (1999, 2000). The change of root mass for this period was very small (ca. 10 g m^{-2}), hence its omission in **A**. See Fig. 5.1 for a description of the treatment combinations

mass for this period was *not* enhanced by $e[CO_2]$ ($P=0.3$), but panicle biomass was stimulated compared to $c[CO_2]$ ($P=0.01$). The higher gain of panicle biomass was counterbalanced by the higher loss of plant mass in leaf laminae under elevated $[CO_2]$ ($P<0.0005$; Fig. 5.2A). Whereas it appears that the plant mass in the stems did not respond to $e[CO_2]$ for the period between PI and maturity (Fig. 5.2A), this was in fact the result of opposite responses to $e[CO_2]$: an enhancement of stem biomass increase from PI to heading and the greater loss of stem weight from heading to maturity. The former response represented a greater increase of culm biomass and carbohydrates; and the latter was presumed to represent a greater translocation of carbohydrates in culms and leaf sheaths to grains in $e[CO_2]$.

The distribution of N during reproductive growth was even more pronounced than that of plant biomass (Fig. 5.2B). The total amount of N accumulation for this period was reduced by $e[CO_2]$ at medium-N and high-N fertilization rates, but unchanged at the low-N rate ($P=0.06$ for the $[CO_2] \times$ N interaction). The interaction was clearer ($P=0.003$) in the amount of N lost from leaf laminae: more N was lost under $e[CO_2]$ than $c[CO_2]$ at medium-N and high-N fertilization rates, whereas this was not the case at the low-N rate (Fig. 5.2B). The $[CO_2] \times$ N interaction was also significant in the loss of N from stems ($P=0.02$) but to a lesser extent than in leaf laminae. Most, if not all, N lost from leaves and stems should have been redistributed to the panicles, which accumulated about the same amount of N for both $[CO_2]$ levels (Lieffering et al. 2004) despite the decline of total N accumulation in $e[CO_2]$ for the reproductive period (Fig. 5.2B).

5.3.2.2 Grain Yield, Yield Components, and Harvest Index

Grain yield was significantly increased by $e[CO_2]$ across the 3 years ($P<0.01$; Fig. 5.3A; Kim et al. 2003a). Yield increase due to $e[CO_2]$ was 15 % at medium-N and high-N fertilization levels, whereas that at the low-N level was only 7 %. The $[CO_2] \times$ N interaction was significant ($P<0.05$), but the $[CO_2] \times$ year interaction was not. The effect of $e[CO_2]$ was therefore not different among the 3 years, despite the significant difference in the yield among years ($P<0.01$; Kim et al. 2003a).

Among the yield components, the number of fertile spikelet per unit land area dominated the yield response to $e[CO_2]$ (Fig. 5.3B). This component was also responsible for yield gain due to the increased N fertilizer rate, but in this case, individual grain mass declined at the higher N rates, which partly cancelled out the effect of increased grain number (Fig. 5.3C). Spikelet fertility was least affected by $[CO_2]$ and N among the yield components (Fig. 5.3D).

The yield increase by $e[CO_2]$ was due mostly to an increase in the number of panicles, whereas yield increase by higher N fertilization was due to increases in the number of panicles and the number of spikelets per panicle

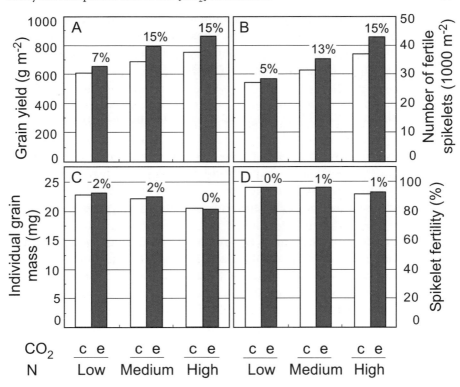

Fig. 5.3 Fertile grain yield (**A**), number of spikelets per unit land area (**B**), mass of individual grains (**C**), and fertility percentage (**D**) averaged across 3 years (1998–2000). The grain yield is expressed on an oven-dried mass basis. See Fig. 5.1 for a description of the treatment combinations and the *figures* above the e[CO$_2$] bars.

(Kim et al. 2003a). The harvest index was reduced ca. 2 % by e[CO$_2$] ($P<0.05$) on average across the N fertilization rates (Kim et al. 2003a).

5.3.2.3 Grain Quality

Lieffering et al. (2004) analyzed the elemental composition of rice grains from plants grown in the Rice FACE experiment. Among the elements analyzed (N, P, S, Mg, Ca, Na, K, Mn, Zn, Cu, Fe, B, Co, Mo, Se), only N showed a statistically significant concentration response to e[CO$_2$] consistently across the 2 years of 1999 and 2000 ($P<0.05$). The decline in grain N concentration was in line with the enhanced panicle mass accumulation (Fig. 5.2A) compared with little change in N accumulation in panicles (Fig. 5.2B) under e[CO$_2$].

We also found that milled grains from e[CO$_2$] plots had lower protein content ($P=0.0001$) with whiter ($P=0.002$) and softer ($P=0.009$) surfaces than those in c[CO$_2$] plots (Terao et al. 2005). While amylose content in grains was unchanged (P=0.76), rice starch had higher maximum viscosity in e[CO_2] (P=0.0015) due to lower protein content. These changes suggest better palatability of cooked rice in e[CO_2], but sensory evaluation tests indicated no detectable changes (P=0.31; Terao et al. 2005).

5.3.3 Synthesis of Rice Plant Responses to e[CO_2] and N Fertilization

It is clear that photosynthetic enhancement is the primary cause of the greater biomass accumulation and grain yield under e[CO_2]. There are, nonetheless, questions yet to be answered about how the extent of the yield increase was determined, and how the N fertilization rate altered the yield response to e[CO_2].

With respect to plant biomass accumulation, an obvious determinant of [CO_2] response is the developmental stage. In the medium-N subplots, plant biomass was increased more than 30 % by e[CO_2] at early stages, but the enhancement had declined to ca. 10 % at maturity (Kim et al. 2003b). The seasonal decline in biomass response to e[CO_2] could be explained by declining N content in leaves via changes in leaf photosynthetic parameters (see Section 5.3.1.3). Enhancement of leaf area growth at early stages (Table 5.1) was evidently a result of enhanced N accumulation due to e[CO_2] (Fig. 5.1B) and also should have contributed to greater biomass response to e[CO_2] via increased light capture early in the season. Likewise, enhancement of root development by e[CO_2] at PI should have enabled plants to capture more N at early stages (see Section 5.3.1.4). The enhancement of N accumulation due to e[CO_2] clearly depends on N availability in soil and hence accounts for the low response of plant biomass to e[CO_2] at the low-N treatment during the PI stage.

The response of the grain yield to elevated [CO_2] and N was determined by the number of spikelets per unit land area (Fig. 5.3; Kim et al. 2003a). In other experiments with chambers (Baker and Allen 1993; Kim 1996) and FACE (Huang et al. 2004) also, the number of grains per unit land area dominated rice yield response to e[CO_2].

It has been reported in rice that the number of grains is closely related to the amount of N accumulated through to the early reproductive stage; and the relationship holds for a cultivar across locations of widely varied climate and edaphic conditions (Horie et al. 1997). In the Rice FACE, a single relationship holds across [CO_2] and N rates between the number of spikelets per unit land area and N accumulation through to the PI stage (Fig. 5.4), although it appears that the cultivar we used and that in Horie et al. (1997) differ in response to N accumulation.

Linking N accumulation and grain number, we have hypothesized a sink-mediated mechanism for interactive responses of rice yield to elevated [CO_2] under varying N-availability as follows.

Fig. 5.4 Relationships between the number of fertile spikelets at harvest and N accumulated through to PI. Observations in the Rice FACE experiment for c[CO$_2$] and e[CO$_2$] for 3 years (1998–2000) are shown, along with the exponential curve fitted to the FACE results (*solid line*) and the relationship (*broken line*) reported by Horie et al. (1997)

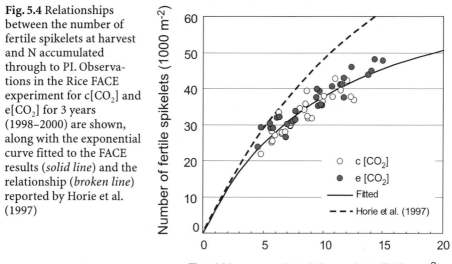

The [CO$_2$] enrichment increased N accumulation up to the PI stage consistent with positive feedback between carbon assimilation and N uptake via an increased capability to capture above- and below-ground resources. With greater N accumulation at the PI stage, plants developed a greater number of spikelets via increased numbers of panicles, and to a lesser extent, increased number of spikelets per panicle at medium-N and high-N fertilization rates. At low-N rate, in contrast, N uptake through to PI had already been constrained by N availability and was unchanged by [CO$_2$] enrichment. The number of grains was hence increased much less under low-N than higher N rates.

In the grain-filling stage, the larger number of grains under e[CO$_2$] demanded more carbohydrates and N than those under c[CO$_2$] at the medium-N and high-N rates. The higher demand of grains for carbohydrates and N was met by increased re-translocation of stored resources, which resulted in greater loss of biomass and N in leaf laminae and stems during the grain-filling period. Leaf senescence was then accelerated by e[CO$_2$]; and the plant's capabilities to capture light (LAI) and fix CO$_2$ (photosynthesis) were lost progressively. This did not happen in the low-N plots, where the number of grains was increased much less by e[CO$_2$].

The nitrogen-spikelet number relationship (Fig. 5.4) thus plays a key role in the sink-mediated mechanism. This hypothesis is supported by findings in a [CO$_2$] enrichment of 300 ppm with a chamber facility (Sakai et al. 2001), where the initial enhancement of canopy photosynthesis by e[CO$_2$] disappeared around the heading stage, whereas grain yield was increased by 22 %. It was also found that loss of leaf area was stimulated by e[CO$_2$] from the heading stage onwards. Using ^{13}C in the same chamber experiment, Sasaki et al. (2005) found accelerated transport of current and stored photosynthates to filling grains under $e[CO_2]$. Involvement of sink capacity in photosynthetic responses has also been reported for soybean (*Glycine max* L. Merr.; see Chapter 4).

5.4 Implications for Rice Production in $e[CO_2]$

5.4.1 Prediction of Global Change Impacts

These Rice FACE results are being used to test models of rice growth and ecosystem processes against observations. The FACE experiment was done in the field with few artifacts; and hence systematic deviations of model estimates from observations would present us with an opportunity for model improvements, with which we can improve our understandings of the real system responses to $e[CO_2]$. By comparison, in experiments with enclosures such as greenhouses or open-top chambers, the deviation of simulations from observations could be due to the artifacts (e.g., un-natural microclimate, small plot size) that have not been accounted for by the model (Ewert et al. 2002). Modeling of the artifacts does not improve our scientific understandings; and it is arguably harder than the modeling of plant responses to $e[CO_2]$.

We found that one of the widely used models of rice growth, ORYZA2000 (Bouman et al. 2001), was able to simulate the responses of plant biomass and grain yield to $e[CO_2]$ reasonably well for the medium-N and high-N application rates, but it overestimated yield response at the low-N rate (Bannayan et al. 2005). It follows that model predictions of climate change impacts on rice production (e.g. Matthews et al. 1997) may have overestimated the positive effect of $e[CO_2]$ for low-N input production systems, which are commonly found in developing countries. If negative impacts of increased temperature have been correctly estimated, the combined effects of climate change on rice production should be less positive or more negative than widely accepted (e.g. Gitay et al. 2001). Moreover, the negative impact of increased temperature on the pollination process (Matsui et al. 1997) could be exacerbated by increased canopy temperature due to stomatal closure in response to $e[CO_2]$ (Yoshimoto et al. 2005). This negative effect of $e[CO_2]$ would be more pronounced in low-N input systems (see Chapter 17) in developing countries particularly at low latitude, where temperature is already high.

5.4.2 Adaptations to e[CO$_2$]

In adaptation, the main concern is in the change of manipulable variables, e.g. fertilizer rate or crop variety, in response to increased [CO$_2$]. If there is no change due to increased [CO$_2$] in the optimal combination of other variables, then no adaptation would occur. Therefore, it is important to see if there are interacting effects of other variables on crop responses to e[CO$_2$].

In the Rice FACE, we found a [CO$_2$] × N interaction in the yield response. The interaction is, however, significant only for increased N from low to medium rates, while N application above the medium rate did not enhance yield response to e[CO$_2$] (Fig. 5.3A). Therefore, no adaptation to e[CO$_2$] would be attained by applying more N, unless there are changes in factors that determine the current N rate. In Japan, one of the major considerations in determining the N rate is lodging. Higher grain yield at high-N than at medium-N rates (Fig. 5.3A) was made possible by artificially supporting the plants in the high-N subplots to prevent lodging damage later in the season. If [CO$_2$] enrichment reduces the risk of lodging, as suggested by a preliminary finding (H. Shimono, personal communication), then the optimal rate of N application may increase, which constitutes an adaptation.

Another constraint is plant diseases which tend to become more severe with higher N applications. As we found in the Rice FACE (Kobayashi et al. 2006), occurrences of rice blast and sheath blight were enhanced by e[CO$_2$] and hence increasing N rate would be counterproductive. Adaptation may be possible by altering the timing and proportion of the top dressings of N relative to the basal application. Since the plants in e[CO$_2$] were more N-stressed than those in c[CO$_2$] at later growth stages (see Section 5.3.3), applying more N at the later stages may ameliorate N-deficiency in plant mass accumulation. It is unclear, however, if it is effective in grain mass accumulation as well.

Varieties may also offer an opportunity for adaptation. We may have seen a larger response of grain number to e[CO$_2$] if we had used a cultivar that is more responsive to enhanced N accumulation in producing grains (see Fig. 5.4). Varietal difference has indeed been reported in rice yield response to e[CO$_2$] (Ziska et al 1996; Moya et al. 1998) with the major difference being in the responsiveness of tillering to e[CO$_2$]. It must be noted, however, that the major issue in adaptation is the shift of optimal cultivars in response to e[CO$_2$] rather than responsiveness of individual cultivars. It may not help adaptation to use a cultivar that is high in responsiveness to e[CO$_2$] but low in productivity. Rather, it will be more relevant to determine whether the varieties that are presently optimal will be optimal or not under conditions of higher [CO$_2$] and temperature. Optimality here is not limited to grain yield but includes grain quality and resistances against biological and environmental stresses.

With rice paddies being among the major anthropogenic sources of methane (CH_4), the second most important greenhouse gas, the adaptation of rice production to the future atmosphere may entail reduction of CH_4 emission. While CH_4 emission from rice paddies will be stimulated by e[CO_2] (Inubushi et al. 2003), it could be reduced by changes in agronomic practices for water and organic matter management (Yan et al. 2005).

FACE experiments provide agronomists and breeders with the best opportunity to test adaptation measures. By taking agronomic adaptations into account, we will have a better capability to predict future climate change impacts on crop production.

5.5 Conclusions

Rice plants were grown from the seedling to maturity stages under [CO_2] ca. 200 ppm higher than the ambient level in farmers' fields for three seasons. The results showed the responses of plant growth and grain yield to e[CO_2] with some implications for future rice production.

At standard rate of N fertilizer application (80–90 kg N ha^{-1}), plant biomass at maturity and grain yield were increased by 12% and 15%, respectively. An increase in N fertilizer application by up to 66% did not change the plant responses, whereas halving the N fertilizer rate diminished the yield response to 7%.

The grain yield increase was mostly due to the increase in the number of fertile grains, which was closely related with plant N accumulation through to early reproductive stage. Under standard and higher N fertilizer rates, N accumulation was enhanced by greater root growth early in the season, with a resultant increase in the number of fertile grains. Under lower N fertilizer rate, limited N availability constrained the N accumulation response to e[CO_2], and hence, the increase in grain number.

The linkage between grain yield and N accumulation in plant responses to elevated [CO_2] implies the possibilities of adapting rice production to higher [CO_2] by adjusting fertilizer application and variety. In designing the adaptation, however, we should consider changes caused by elevated [CO_2] and temperature in other aspects of rice production as well. Such aspects include lodging, pest damage, temperature stresses, and methane emission.

Process-based models of plant growth and agricultural ecosystems could be a powerful tool to design the adaptation of crop production to future environment, but they have to be tested against observations before being used reliably. FACE would provide a very good opportunity for the model testing.

Acknowledgements. The Rice FACE project was supported by the Core Research for Evolutional Science and Technology (CREST) program of the Japan Science and Technology Corporation (JST). The first author (K.K.) also received a financial support from the Japan Society for the Promotion of Science.

References

Anten NPR, Hirose T, Onoda Y, Kinugasa T, Kim H-Y, Okada M, Kobayashi K (2003) Elevated [CO$_2$] and nitrogen availability have interactive effects on canopy carbon gain in rice. New Phytol 161:459–471

Baker JT, Allen LH Jr (1993) Contrasting crop species responses to CO$_2$ and temperature: rice, soybean and citrus. Vegetatio 104/105:239–260

Bannayan M, Kobayashi K, Kim H-Y, Lieffering M, Okada M, Miura S (2005) Modeling the interactive effects of atmospheric CO$_2$ and N on rice growth and yield. Field Crops Res 93:237–251

Bouman BAM, Kropff MJ, Tuong TP, Wopereis MCS, Ten Berge HFM, Laar HH van (2001) Oryza2000: modeling lowland rice. IRRI, Los Banos, 235 pp

Ewert F, Rodriguez D, Jamieson P, Semenov MA, Mitchell RAC, Goudriaan J, Porter JR, Kimball BA, Pinter PJ Jr, Manderscheid R, Weigel HJ, Fangmeier A, Fereres AE, Villalobos F (2002) Effects of elevated CO$_2$ and drought on wheat: testing crop simulation models for different experimental and climatic conditions. Agric Ecosys Environ 93:249–266

FAOSTAT (2004) http://faostat.fao.org/

Farquhar GD, Caemmerer S von, Berry JA (1980) A biochemical model of photosynthetic CO$_2$ assimilation in leaves of C$_3$ species. Planta 149:78–90

Gitay H, Brown S, Easterling W, Jallow B, Antle J, Apps M, Beamish R, Chapin T, Cramer W, Frangi J, Laine J, Lin E, Magnuson J, Noble I, Price J, Prowse T, Root T, Schulze E, Sirotenko O, Sohngen B, Soussana J (2001) Contribution of working group II to the Third Assessment Report of the Intergovernmental Panel on Climate Change. In: McCarthy JJ, Canziani OF, Leary NA, Dokken DJ, White KS (eds) Climate change 2001: impacts, adaptation, and vulnerability. Cambridge University Press, Cambridge, pp 235–342

Horie T, Ohnishi M, Angus JF, Lewin LG, Tsukaguchi T, Matano T (1997) Physiological characteristics of high-yielding rice inferred from cross-location experiments. Field Crops Res 52:55–57

Huang JY, Yang HJ, Yang LX, Liu HJ, Dong GC, Zhu JG, Wang YL (2004) Effects of free-air CO$_2$ enrichment (FACE) on yield formation of rice (*Oryza sativa* L.) and its interaction with Nitrogen. Sci Agric Sin 37:1824–1830

Inubushi K, Cheng W, Aonuma S, Hoque MM, Kobayashi K, Miura S, Kim H-Y, Okada M (2003) Effects of free-air CO$_2$ enrichment (FACE) on CH$_4$ emission from a rice paddy field. Global Change Biol 9:1458–1464

Kim H-Y (1996) Effects of elevated CO$_2$ concentration and high temperature on growth and yield of rice. PhD thesis. Kyoto University, Kyoto

Kim H-Y, Lieffering M, Miura S, Kobayashi K, Okada M (2001) Growth and nitrogen uptake of CO$_2$-enriched rice under field conditions. New Phytol 150:223–229

Kim H-Y, Lieffering M, Kobayashi K, Okada M, Mitchell MW, Gumpertz M (2003a) Effects of free-air CO$_2$ enrichment and nitrogen supply on the yield of temperate paddy rice crops. Field Crops Res 83:261–270

Kim H-Y, Lieffering M, Kobayashi K, Okada M, Miura S (2003b) Seasonal changes in the effects of elevated CO_2 on rice at three levels of nitrogen supply: a free air CO_2 enrichment (FACE) experiment. Global Change Biol 9:826–837

Kobayashi T, Ishiguro K, Nakajima T, Kobayashi K (2006) Effects of elevated atmospheric CO_2 concentration on rice blast and sheath blight epidemics. Phytopathology (in press)

Lieffering M, Kim H-Y, Kobayashi K, Okada M (2004) The impact of elevated CO_2 on the elemental concentrations of field-grown rice grains. Field Crops Res 88:279–286

Long SP, Ainsworth EA, Rogers A, Ort DR (2004) Rising atmospheric carbon dioxide: plants FACE the future. Annu Rev Plant Biol 55:557–594

Matsui T, Namuco OS, Ziska LH, Horie T (1997) Effects of high temperature and CO_2 concentration on spikelet sterility in indica rice. Field Crops Res 51:213–219

Matthews RB, Kropff MJ, Horie T, Bachelet D (1997) Simulating the impact of climate change on rice production in Asia and evaluating options for adaptation. Agric Sys 54:399–425

Moya TB, Ziska LH, Namuco OS, Olszyk D (1998) Growth dynamics and genotypic variation in tropical, field-grown paddy rice (*Oryza sativa* L.) in response to increasing carbon dioxide and temperature. Global Change Biol 4:645–656

Okada M, Lieffering M, Nakamura H, Yoshimoto M, Kim H-Y, Kobayashi K (2001) Free-air CO_2 enrichment (FACE) using pure CO_2 injection: system description. New Phytol 150:251–260

Sakai H, Yagi K, Kobayashi K, Kawashima S (2001) Rice carbon balance under elevated CO_2. New Phytol 150:241–249

Sasaki H, Aoki N, Sakai H, Hara T, Uehara N, Ishimaru K, Kobayashi K (2005) Effect of CO_2 enrichment on the translocation and partitioning of carbon at the early grain-filling stage in rice (*Oryza sativa* L.). Plant Prod Sci 8:8–15

Seneweera SP, Conroy JP, Ishimaru K, Ghannoum O, Okada M, Lieffering M, Kim H-Y, Kobayashi K (2002) Changes in source-sink relations during development influence photosynthetic acclimation of rice to free air CO_2 enrichment (FACE). Funct Plant Biol 29:945–953

Terao T, Miura S, Yanagihara T, Hirose T, Nagata N, Tabuchi H, Kim H-Y, Lieffering M, Okada M, Kobayashi K (2005) Influence of free-air CO_2 enrichment (FACE) on the eating quality of rice. J Sci Food Agric 85:1861–1868

Yan X, Yagi K, Akiyama H, Akimoto H (2005) Statistical analysis of the major variables controlling methane emission from rice fields. Global Change Biol 11:1131–1141

Yoshimoto M, Oue H, Kobayashi K (2005) Energy balance and water use efficiency of rice canopy under free-air CO_2 enrichment. Agric For Meteorol 133:226–246

Ziska LH, Manalo PA, Ordonez RA (1996) Intraspecific variation in the response of rice (*Oryza sativa* L.) to increased CO_2 and temperature: growth and yield response of 17 cultivars. J Exp Bot 47:1353–1359

6 Growth and Quality Responses of Potato to Elevated [CO₂]

M. Bindi, F. Miglietta, F. Vaccari, E. Magliulo and A. Giuntoli

6.1 Introdution

Root and tuber crops are highly important food resources. They comprise several genera and supply the main part of the daily carbohydrate intake of large populations. These carbohydrates are mostly starches found in storage organs, which may be enlarged roots, corms, rhizomes, or tubers. Food and Agriculture Organization (FAO) statistics indicate that in 2004 root and tuber crops were cultivated over more than 53 Mha and the total production was greater than 710 Mt (FAOSTAT 2004). Among these, potato (*Solanum tuberosum* L.) is the most cultivated species in the world, as its current cultivated area accounts for 36% of the total harvested areas of root and tuber crops (FAOSTAT 2004). Potato also makes the largest contribution to the total production of tuber and root crops, representing 46% of the global production of root and tuber crops (FAOSTAT 2004). Moreover, the yield per unit of potato steadily increased during the last 40 years passing from 12 t ha^{-1} in 1961 to 17 t ha^{-1} in 2004 (FAOSTAT 2004). Among other factors (e.g. improved crop varieties, crop management, increased use of fertilisers, reduced losses from pest and disease infestations, improved harvesting and conservation methods, extended irrigation), the continuous increase in atmospheric carbon dioxide may also have contributed to these yield increases. The characteristics of potato source and sink organs and the transport capacity of assimilates seem to be important pre-requisites for a large CO₂ response.

However, despite its economic and global importance and its expected strong response to increasing [CO₂], little research has been done on the potential effects of elevated [CO₂] on potato, especially if this is compared with that made on cereals and if we look at experiments performed under field conditions (i.e. free air CO₂ enrichment; FACE). In order to improve pre-

Ecological Studies, Vol. 187
J. Nösberger, S.P. Long, R.J. Norby, M. Stitt, G.R. Hendrey, H. Blum (Eds.)
Managed Ecosystems and CO₂
Case Studies, Processes, and Perspectives
© Springer-Verlag Berlin Heidelberg 2006

dictions of how elevated [CO_2] might affect the growth and quality responses of potato, a series of FACE experiments have been carried out, starting from 1995 at Rapolano Terme, Italy.

6.2 Site Description

6.2.1 Physical: Location, Size, Elevation, Layout of Experiment and Blocking

The field site of Rapolano Terme (43.2856° N, 11.6048° E; 172 m a.s.l.) is located in a region of Central Italy that is very rich in carbon dioxide springs. Most of these springs are localised in undisturbed areas, thus in recent years they were extensively used to study in detail the [CO_2] long-term effects on natural vegetation. Some CO_2 springs have been exploited for industrial CO_2 production, offering an unique opportunity to get CO_2 at a low price, in this way making FACE affordable (Miglietta et al. 1993). Starting from 1993, a series of FACE experiments have been carried out on several agro-forest crops (wheat, potato, grapevine, olive, forage crops, grassland plants, poplar, castor bean, etc.) in the fields of Poggio Santa Cecilia farm where CO_2 was supplied, after purification, from the CO_2 storage equipment of the Geogas Company.

In particular, as regards potato crops, two series of FACEs have been organised in a field of the Poggio Santa Cecilia farm. The first FACE experimental campaign was carried out in 1995 and the second, for two consecutive years, in 1998–1999 (Table 6.1). In both experiments, the size of the FACE ring was 50.24 m² (8 m diameter) and all crop measurements were made on plants located in the "sweet spot" (28.3 m²), sensu Lewin et al. (1992), in which [CO_2] was minimally affected by wind speed and direction. Further details on the CO_2 fumigation system used in these experiments are given elsewhere (Miglietta et al. 1998; Chapter 2).

6.2.2 Soil Types, Tillage Practices, Fertilisation, Crop Samplings and Measurements

Both experiments were carried out in a field of 1.8 ha with a sandy-clay-loam soil (Table 6.2). In 1995, the experiment was made from June to September. Tubers of cv. Primura were planted in rows with North–South orientation, at 0.03 m depth. Rows were 0.8 m apart with plant spacing of 0.2 m. In 1998 and 1999, the experiments were made from May to August and tubers of cv. Bintje were planted in nine square plots with North–South orientation and a plant-

Table 6.1 Experimental protocols and site characteristics of the potato FACE experiments at Rapolano Terme (43.2856° N, 11.6048° E; 172 m a.s.l.)

Parameter	FACE experiment	
	1995	1998–1999
Potato cultivar	Primura	Bintje
Experiment start dates	27-05-1995	20-05-1998
		05-05-1999
Experiment end dates	05-09-1995	18-08-1998
		17-08-1999
Growing season start dates	10-06-1995	28-05-1998
		26-05-1999
Growing season end dates	05-09-1995	18-08-1998
		17-08-1999
Plot diameter (m)	8	8
No. of replicates	1	3
No. of CO$_2$ levels	4	2
FACE CO$_2$ concentrations (ppm)	Current, 460, 560, 660	Current, 550

ing density of 5.7 plants m^{-2} (0.73 m between rows and 0.24 m within rows; Table 6.2). Planting depth was 0.10 m. In the rest of the field, ray-grass was sown to reduce weed problems and to ensure an almost uniform spatial distribution of [CO$_2$] within the rings. The height of the ray-grass canopy was controlled not to exceed that of the potato plants.

The crops were abundantly fertilised in all three FACE experimental campaigns (Table 6.2). Irrigation was provided throughout the growing season, using sprinklers with complete restitution after cumulative pan evaporation had attained a threshold level of 15 mm.

Cultural practices included pre- and post-sowing weed control treatments (glyphosate 2 l ha^{-1}, Sencor 1 kg ha^{-1}) and weekly sprays to control downy mildew (copper sulfate 4 kg ha^{-1}) and Colorado beetle infestations (*Bacillus thuringiensis* var. Kurstaki 5 kg ha^{-1}). In 1995, to allow specific examination of the effect of elevated [CO$_2$] on Colorado feeding behaviour, chemical pest control was suspended in the last part of the growing season.

In the FACE experimental campaigns, detailed analyses of phenology, growth, yield and gas exchanges of the potato crops were made. In particular, in all the three experiments: (a) crop phenology was monitored and leaf number and plant height were recorded non-destructively at weekly intervals, (b) leaf area index (LAI) was determined using an optical system (LAI-2000; Licor, Lincoln, Neb.) (c) soil moisture was monitored just prior every irrigation using a TDR cable tester (model 1502B; Tektronix, Beaverton, Ore.), (d) chlorophyll concentration was measured at weekly intervals on new and labelled leaves using a SPAD-Meter, (e) specific leaf area (SLA) was deter-

Table 6.2 Physical and chemical characteristics of soil of the experimental field. Dates of fertilisation, type and amount added fertilisers are also given

Soil texture (%)		Chemical analysis		Added fertiliser (kg ha^{-1})					
	Mean (SE)	N (ppm)	Traces	1995		1998		1999	
Clay	11.65 (0.42)	P$_2$O$_5$ (ppm)	32	26 May	N (170) P (100) K (200)	18 May	N (200) P (80) K (150)	5 May	N (200) P (80) K (150)
Silt	18.70 (0.40)	K$_2$O (ppm)	120						
Sand	69.65 (0.65)	pH	8	3 July	N (83)	26 June	N (50)	16 June	N (50)
		Organic matter (%)	0.91						
		Conductivity (mS cm^{-1})	0.0054						
		Total lime	0						

Table 6.3 Climatic conditions recorded during the three FACE experiments

Growth Season	Temperature (°C, daily average)			Solar radiation (MJ m^{-2} day^{-1})	Wind speed (m s^{-1})	Total precipitation (mm)
	Average	Minimum	Maximum			
1995	17.4	10.5	26.6	20.1	1.8	130
1998	19.9	11.3	29.2	21.3	2.1	81
1999	19.5	11.0	27.7	21.4	2.3	102

mined at several occasions during the growing seasons by cutting leaf disks from the centre of the ultimate fully expanded leaflet, (f) plant growth and yield were determined by destructive samplings during growing seasons and at the maturity. Additionally, in 1995: (g) Colorado potato beetle feeding behaviour (i.e. larval growth rate, consumption rate) was investigated by collecting a large number of young larvae, (h) photosynthetic capacity of potato leaves was investigated (i.e. A/C_i curves, leaf conductance), (i) leaf reflectance was monitored using a laboratory spectroradiometer, (l) concentration of total non-structural carbohydrate (TNC) was determined on the same leaves sampled for the determination of SLA. In 1998–1999: (m) actual crop evapotranspiration (ET) was monitored throughout the crop cycle, on an hourly basis, using the residual energy balance approach (Jackson et al. 1987; Kimball et al. 1994), (n) physical (malformations, occurrence of common scab, glassy and green tubers, specific gravity) and chemical (nitrate, Kjeldahl nitrogen, starch, sugar and organic acids, glycoalkaloids) yield quality were determined at the final harvest.

6.2.3 Meteorological Description

Weather data were collected by an automatic weather station (ACME, Firenze) located 50 m from the centre of the experimental plots. Main seasonal characteristics of climate conditions during the FACE experiments are reported in Table 6.3. Average minimum daily temperatures were higher than 10 °C, whilst daily average maximum temperatures were rather close 30 °C, especially in 1998. Global radiation was, on average, above 20 MJ m^{-2} day^{-1}. There were no evident seasonal variation in average daily wind speed that ranged from about 1.8 m s^{-1} to 2.3 m s^{-1}. Wind speed generally increased during the day from an average of about 1 m s^{-1} in the morning to 3 m s^{-1} in the afternoon and predominantly came from the SE quadrant (28 % frequency). High-wind episodes from the NW were observed. Precipitation was rather low in all the three seasons (from 81 mm to 130 mm); and most of the precipitation events were concentrated in the first part of the growing seasons (May–June), whilst in the rest of the seasons only a few rain events were recorded. This type of weather explains the large amount of irrigation water that was used in all the experiments (from 360 mm to 460 mm).

6.3 Experimental Treatments

6.3.1 Elevated [CO_2]

In 1995, the field was divided into four main blocks and five FACE rings were installed immediately after planting. Three of five rings were enriched with CO_2 and were kept at 660, 560 and 460 ppm, respectively, while the two remaining rings were left at current [CO_2] (360 ppm). In the 1998 and 1999 FACE experiments, six FACE rings were installed immediately after planting and fumigation started at crop emergence. Three rings were kept at 550 ppm and three were left at current [CO_2]; and the experimental setup of FACE and control rings were identical. Moreover, square plots having the same areas of the ringed plots were also used as non-ringed controls.

6.4 Resource Acquisition

6.4.1 Effect of Treatments

6.4.1.1 Photosynthesis

Measurements of gas exchange made in July showed that photosynthetic capacity of the leaves (A/C_i curve) was not affected by long term [CO_2] exposure. Both RuBP saturated ($V_{c,max}$) and RuBP regeneration-limited (J_{max}) carboxylation rates were unaffected by the CO_2 treatment (Table 6.4). This allowed the conclusion that there was no photosynthetic acclimation under elevated [CO_2].

TNC content was higher in leaves sampled from plants exposed to elevated [CO_2] than in plants grown in current [CO_2] (Table 6.4), reflecting a substantial imbalance between CO_2 uptake and transport out of the leaves, although potato crops have a very large sink organs for carbohydrates (Farrar and Williams 1991) and use an apoplastic mechanism for phloem loading, based on a sucrose transporter (Riesmeier el al. 1994). This is in agreement with the results of a recent meta-analyses performed by Ainsworth and Long (2005) in which, despite an unrestricted rooting volume, plants grown at elevated [CO_2] in FACE facilities accumulated significantly more sugars and starch than those plants grown at current [CO_2] (Chapter 16).

SLA was decreased in plants exposed to elevated [CO_2], as well as leaf N content, expressed on a dry weight basis (Table 6.4). Such a decrease was only evident when leaf N was expressed on a dry weight basis (Table 6.4); and it disappeared when leaf N was expressed on a leaf area basis (data not shown).

Table 6.4 Effects of elevated CO$_2$ on potato resource acquisition. The response ratios were calculated using the equation $r = (p_{e[CO_2]}) / (p_{a[CO_2]})$, where $(p_{e[CO_2]})$ = parameter under elevated [CO$_2$], $(p_{a[CO_2]})$ = parameter under current [CO$_2$]; the standard errors of the response ratios (reported within brackets) were calculated using the equation $\Delta r = (|p_{e[CO_2]} \times \Delta p_{a[CO_2]}| + |p_{a[CO_2]} \times \Delta p_{e[CO_2]}|) / p_{a[CO_2]}^2$, where Δ indicates the standard error and the vertical bars denote absolute values. The response ratios and the standard errors were then transformed to percentage changes

Crop parameter	Percentage changes due to Elevated CO$_2$ 1995			1998	1999
	460	560	660	550	550
Photosynthesis (Vaccari et al. 2001)					
A/C$_i$ curves					
J_{max}	19.3 (12.9)	−7.4 (8.4)	−2.9 (7.7)		
$V_{c,max}$	14.3 (18.3)	0.2 (17.5)	−1.3 (13.8)		
Total non-structural carbohydrate content					
	33.0 (15.1)	41.8 (20.2)	48.4 (12.2)		
Assimilation					
July	17.2 (15.9)	53.6 (20.5)	81.9 (22.9)		
August	9.8 (23.3)	43.8 (29.9)	46.8 (28.0)		
Specific leaf area					
	−15.2 (9.6)	−19.2 (11.5)	−22.2 (6.4)		
Leaf N					
July	−10.2 (3.0)	−15.3 (2.6)	−17.3 (2.7)		
August	−6.6 (0.6)	−44.8 (0.8)	−51.4 (0.4)		
Leaf reflectance					
at 550 nm	12.5 (8.3)	31.3 (10.3)	50.0 (10.5)		
Canopy temperature and energy balance (Magliulo et al. 2003)					
Mean seasonal daytime canopy temperature				2.3 (1.6)	3.7 (0.8)
Energy balance terms					
R_n				1.4 (2.5)	1.7 (1.8)
G				8.4 (5.7)	11.3 (3.1)
H				−76.7 (8.7)	−214.1 (1.0)
L_e				−5 (2.1)	−11.4 (1.5)
Water consumption (Magliulo et al. 2003)					
Consumptive water use				−8.8 (4.4)	−15.9 (7.5)
Water use efficiency				69.6 (16.8)	66.3 (26.0)
Crop phenology and development (Miglietta et al. 1998; Bindi et al. 1999)					
Phenological stages					
Anthesis	−4.6 (9.0)	−4.6 (10.0)	−6.9 (12.0)	−3.7 (9.0)	−1.2 (7.0)
Peak leaf area index					
	−5.3 (17.9)	5.3 (22.9)	−2.6 (15.5)	−10.3 (27.5)	−16.0(3.8)
Maximum plant height					
	−1.1 (5.9)	−1.1 (5.8)	−9.1 (5.3)	−14.4 (6.5)	−15.4 (4.9)
Leaf greenness (SPAD)					
Early season				−7.7 (3.2)	−2.3 (1.9)
Late season				−16.7 (6.0)	−5.1 (2.4)
Herbivory (Vaccari et al. 2000)					
Larval dry mass	−23.3 (8.9)	−10.6 (4.6)	−9.9 (2.4)		
Leaf ingested	−7.2 (19.9)	−10.9 (20.0)	−7.6 (13.8)		

This supports the hypothesis that leaf N decrease under elevated $[CO_2]$ was just a N-dilution effect caused by increased carbohydrate accumulation in the leaves (Evans 1989) and that the amount of nitrogen invested by the plant in the light harvesting complex and in the carboxylation enzyme per unit of intercepted light was not reduced under elevated $[CO_2]$. Nitrogen use efficiency was therefore increased under elevated $[CO_2]$, as previously reported for other species (Gifford 1992; Sage 1994; Drake et al. 1997).

Gas-exchange measurements made in August showed that the stimulation of elevated $[CO_2]$ on photosynthetic rates were lower than in July (Table 6.4). Accordingly, leaf reflectance measured in August showed that the reduction in the concentration of the leaf pigments, that occurred in response to leaf senescence, was significantly correlated to the $[CO_2]$. This suggested that increasing atmospheric $[CO_2]$ did promote leaf senescence (Table 6.4).

6.4.1.2 Canopy Temperature and Energy Balance

The canopy temperature was always higher in elevated $[CO_2]$ plots, both in 1998 and 1999 (Table 6.4). In 1998, the mean seasonal daytime canopy temperature was 25.9±0.2 °C and 26.5±0.2 °C under current $[CO_2]$ and elevated $[CO_2]$, respectively. In 1999, the average values were 24.0±0.1 °C and 24.9±0.1 °C (Table 6.4). The higher values for the first year should be ascribed to the higher air temperature and the shorter monitoring window, with the exclusion from the seasonal mean of some of the cooler days of late spring. Peak differences in canopy radiative temperature were recorded in the early afternoon, between 1530 hours and 1600 hours (data not shown). Differences between treatments of 0.73 °C and 1.15 °C in maximum canopy temperature, averaged across the entire growing season, were detected in 1998 and 1999, respectively.

The seasonal mean values of the CO_2 response of the different terms of the energy balance are reported in Table 6.4. The patterns of energy partitioning among the different components were similar for the two treatments. In both cases Net Radiation (R_n) represents the main term of the energy budget ($\approx 50\,\%$). When averaged across the entire season, R_n was higher in 1998, due to the delayed planting date, and resulted slightly lower in the current $[CO_2]$ treatment, with an average difference of 1.4 % in 1998 and 1.7 % in 1999 that, however, was not statistically significant. In both years, soil heat flux (G) was increased in the elevated $[CO_2]$ plot, as a result of a faster crop development and a consequent increased light penetration in the final part of the growing season. Sensible heat flux (H) was higher (more negative) in 1998, since canopy was cooler than air for most of the time and was decreased at elevated $[CO_2]$. Latent heat flux (Le), the residual term of the budget, was significantly decreased at elevated $[CO_2]$ (Table 6.4).

6.4.1.3 Water Consumption

From the beginning of the measurements, the actual crop ET of the elevated [CO$_2$] plots was consistently lower than that of control, in either growing season. Overall mean daily water use in 1998 was 6.58 mm day^{-1} and 7.22 mm day^{-1} for enriched and current [CO$_2$] plots, respectively. On a seasonal basis, elevated [CO$_2$] plots used 342 mm of water compared with 375 mm used by the crop under current [CO$_2$] conditions, with a water saving of 8.9 % (Table 6.4). During the second year of the study, the effect of fumigation on water use by the crop was more evident. Daily mean water use in elevated [CO$_2$] was 4.98 mm, while under current [CO$_2$] conditions it was 5.90 mm; and total seasonal water consumption was 297 mm and 353 mm, which corresponds to a reduction of 15.6 % (Table 6.4).

The mean seasonal values of the water use efficiency (WUE) showed a substantial effect of the CO$_2$ enrichment on the WUE of the crop, with an overall increase of 70 % in 1998 and 67 % in 1999 (Table 6.4).

6.4.1.4 Crop Phenology and Development

Crop phenology was not substantially affected by elevated [CO$_2$] (Table 6.4), even if flower anthesis (corresponding to tuber initiation) under elevated [CO$_2$] started slightly early. The LAI only evidenced minor variations among the treatments, although some differences were detectable in the late part of the growing cycle (Table 6.4), whilst significant differences between the current and elevated [CO$_2$] plots were observed in plant height in 1998 and 1999 (Table 6.4). Moreover, in 1998 and 1999, the measurement of greenness of leaves made with SPAD showed, as expected, lower readings in CO$_2$ treated plants both for new and labelled leaves (Table 6.4), confirming a lower leaf N concentration under elevated [CO$_2$] plots that has been observed to occur also in the 1995 FACE experiment (see Section 6.3.1.1). This CO$_2$ effect increased gradually during the growing season, suggesting (together with that observed in growth measurements; see Section 6.4.1.1) that plant senescence was accelerated.

6.4.1.5 Herbivory

The analysis of larval growth of Colorado potato beetle indicated that growth was sensitive to changes in leaf composition (N concentration, see Table 6.4). Larvae grew faster when feeding on leaves grown in current [CO$_2$] than on elevated [CO$_2$] leaves, but differences between mean daily growth rates of larvae fed with leaves of plants exposed to 460, 560 and 660 ppm [CO$_2$] were not appreciable. Larval size at the end of the experiment was affected by the quality of the foliage ingested, with larvae from the ambient treatments having a larger dry mass than those fed from the elevated $[CO_2]$ grown leaves (Table 6.4). However, the total amount of food ingested by the larvae and the leaf consumption rates were almost identical for all the CO_2 treatments (Table 6.4). Reduced growth of the larvae may be translated into a decreased dry mass and a reduced amount of reserves of the larvae at the time of pupation. This may have consequences on the ability of the insect to survive winter conditions while diapausing in the soil.

6.5 Resource Transformation

6.5.1 Effect of Treatments on Biomass Growth

6.5.1.1 Aboveground

The effects of elevated $[CO_2]$ on aboveground biomass components (leaf and shoot dry weight, leaf number) were in general small and not consistent during the whole growing periods. More specifically, during the early part of the growing season, some CO_2 stimulation was observed (Table 6.5), while this was not the case when the plants approached maturity (Table 6.5). At this development stage, due also to the overall accelerated senescence of plants exposed to elevated $[CO_2]$, total aboveground biomass produced by these plants was lower than that of plants under current $[CO_2]$.

6.5.1.2 Belowground

Tuber growth was strongly affected by increasing levels of $[CO_2]$. Both the number and dry mass of tubers were stimulated in FACE plots; and large differences with current $[CO_2]$ plots were found (Table 6.5). In particular, the increase in the number of tubers was particularly high from the beginning of the growing seasons (+55 % in 1998, +52 % in 1999); and then it remained still apparent, showing a positive CO_2 effect of about 26–27 % in the late part of the growing seasons (Table 6.5). The dry mass of tubers showed that plants under elevated $[CO_2]$ accumulated carbon in the tubers at a significantly higher rate than those in current $[CO_2]$. At final harvest, tuber dry mass was 59 % greater under elevated $[CO_2]$ than under ambient conditions (Table 6.5).

Table 6.5 Effects of elevated CO$_2$ on potato resource transformation. The response ratios were calculated using the equation $r = (p_{e[CO_2]}) / (p_{a[CO_2]})$, where $(p_{e[CO_2]})$ = parameter under elevated [CO$_2$], $(p_{a[CO_2]})$ = parameter under current [CO$_2$]; the standard errors of the response ratios (reported within brackets) were calculated using the equation $\Delta r = (|p_{e[CO_2]} \times \Delta p_{a[CO_2]}| + |p_{a[CO_2]} \times \Delta p_{e[CO_2]}|) / p_{a[CO_2]})^2$, where Δ indicates the standard error and the vertical bars denote absolute values. The response ratios and the standard errors were then transformed to percentage changes

| Crop parameter | Percentage changes due to elevated CO$_2$ | | | | |
| | 1995 | | | 1998 | 1999 |
	460	560	660	550	550
Biomass growth (Miglietta et al. 1998; Bindi et al. 1999)					
Above ground biomass					
Leaf dry mass					
Early season				25.5 (18.9)	16.6 (12.5)
Mid season				18.0 (15.1)	2.9 (1.9)
Late season				20.8 (17.6)	−7.8 (9.6)
Leaf number					
Early season				17.2 (14.6)	10.9 (8.1)
Mid season				14.3 (12.8)	−6.5 (5.5)
Late season				26.5 (24.3)	−16.8 (14.3)
Shoot dry mass					
Early season				19.0 (16.3)	5.6 (3.6)
Mid season				15.7 (13.0)	−6.6 (6.7)
Late season				15.9 (13.1)	−16.1 (14.2)
Below ground biomass					
Tuber dry mass					
Early season	15.4 (8.5)	30.9 (16.7)	46.3 (24.1)	64.8 (28.6)	118.7 (39.1)
Mid season	31.1 (14.3)	62.2 (21.8)	93.3 (25.9)	56.7 (28.5)	78.0 (27.2)
Late season				53.8 (26.2)	59.5 (22.6)
Tuber number					
Early season				50.4 (20.9)	52.2 (20.4)
Mid season				43.6 (20.1)	33.8 (17.4)
Late season				35.5 (19.9)	27.7 (15.8)
Final yield (Miglietta et al. 1998; Bindi et al. 1999)					
Above and below ground biomass					
Leaf dry mass				−12.7 (16.9)	−30.1 (25.3)
Shoot dry mass				10.2 (9.5)	−19.3 (18.0)
Tuber dry mass	13.8 (8.1)	27.7 (15.7)	41.5 (25.7)	35.6 (16.2)	50.0 (18.3)
Tuber number	23.8 (12.3)	25.7 (15.6)	41.9 (24.0)	23.2 (12.9)	35.1 (17.6)
Tuber physical quality					
Dry matter				5.1 (2.4)	6.7 (2.7)
Weight under water				4.0 (4.5)	7.4 (5.0)
Number of tuber per size class (<35 mm)				56.6 (20.8)	22.9 (20.1)
Number of tuber per size class (35–50 mm)				45.5 (7.8)	42.5 (9.3)
Number of tuber per size class (>50 mm)				−2.0 (22.9)	37.7 (46.6)

Table 6.5 (*Continued*)

Crop parameter	Percentage changes due to elevated CO_2				
	1995			1998	1999
	460	560	660	550	550
Dry weight of tuber per size class (<35 mm)				31.1 (28.8)	41.3 (177.7)
Dry weight of tuber per size class (35–50 mm)				65.5 (12.1)	53.2 (23.9)
Dry weight of tuber per size class (>50 mm)				15.5 (26.6)	58.6 (42.2)
Number of glassy tuber				−35.3 (9.4)	−6.8 (14.4)
Number of malformed tuber				9.0 (27.9)	17.9 (11.4)
Tuber chemical quality					
Starch				6.4 (3.1)	9.3 (1.5)
Glucose				−3.6 (2.5)	−2.2 (1.2)
Fructose				−8.0 (4.7)	−3.6 (1.8)
Sucrose				−17.9 (14.3)	−22.9 (12.9)
Citric acid				−8.9 (5.7)	−19.7 (5.0)
Malic acid				−21.6 (10.9)	−10.7 (8.5)
Glycoalkaloids: α-solanine				1.7 (3.3)	−1.2 (3.8)
Glycoalkaloids: α-chaconine				1.7 (3.9)	−0.4 (4.2)
Nitrate				−50.3 (9.6)	−42.2 (5.0)

6.5.2 Effect of Treatments on Yield Quantity and Quality

6.5.2.1 Above- and Belowground Biomass

According to the data collected during the growing seasons, the final below-ground biomass showed significant differences among the CO_2 treatments (Table 6.5), with increases under elevated [CO_2] that ranged from the 13% (460 ppm) to 50% (550 ppm in 1999) for tuber dry matter and from the 23% (460 ppm) to 35% (550 ppm in 1999) for tuber number. Moreover, the final aboveground biomass also seemed to be affected by [CO_2] (Table 6.5), show-ing lower dry matter for both components (stem, leaf) under current [CO_2].

6.5.2.2 Tuber Physical Quality

Since the cv. Bintje selected in the 1998 and 1999 FACE experiments was com-mercially used mainly for producing fried products (e.g. french fries), several physical qualities of the tubers are fundamental in the industrial processing. In particular, the dimension of the tubers and the presence of glossy or mal-formed tubers contribute to the quality of the tubers and thus their commer-cial price. Thus, in 1998 and in 1999, these physical quality parameters were analysed. The physical quality analyses showed clearly that the effect of [CO_2]

on tuber production is mainly due to the increase in the number of tubers in commercial classes (from 35 mm to 50 mm; Table 6.5), whilst the number of smaller or larger tubers (<35 mm, >50 mm) were less affected, as well as the weight of these classes. With the exception of glassy tubers in 1998, the fraction of glassy tubers or malformed tubers was not significantly affected by elevated [CO$_2$] (Table 6.5). The dry matter of tubers grown under elevated [CO$_2$] was increased (Table 4). Both treatments, however, reached under water weight (UWW) values higher than 360 g m^{-2} (minimum score for normal tuber quality), but under elevated [CO$_2$] the UWW was over the bonus limit (UWW >400 g m^{-2}; data not shown).

6.5.2.3 Tuber Chemical Quality

Potato components such as starch, reducing sugars and organic acids are important for potato quality and nutritive value. Starch is the major component of potato tubers, accounting for as much as 65–80 % of the tuber dry weight. The dry matter content of potatoes is very often equated with its starch content and, because it is easily measured, it is widely used by the processing industry to assess quality for a particular use. The dry matter concentration needs to be high to avoid excessive fat absorption in frying, whereas the reducing sugar concentration (i.e. glucose and fructose) should be low because a high concentration causes the fried product to turn dark brown. Organic acids influence flavour directly by their tartness and affect colour by inhibiting aftercooking darkening and enzymatic browning. Citric acid is present in the largest amount, followed by malic acid. Another quality aspect concerns the toxic compounds, glycoalkaloids, which are naturally occurring toxins found in all parts of the potato plant. α-Solanine and α-chaconine make up about 95 % of the total glycoalkaloid concentration in cultivated potatoes. A low concentration of glycoalkaloids enhances potato flavour, whereas a high concentration may cause a bitter taste and gastroenteritis. Nitrates are other naturally occurring toxins in the potato tuber. High nitrate values in foodstuffs constitute a potential health hazard because of the precursor role of this compound in the formation of nitrites.

In the 1998 and 1999 FACE experiments, the dry matter and starch concentration of the potato tubers was increased under elevated [CO$_2$], thus improving the processing quality of potatoes (Table 6.5). The decrease in reducing sugars under elevated [CO$_2$] reduces the risk that fried product turns dark brown (Table 6.5), but the concomitant decrease in organic acids content (Table 6.5) determines a reduction in potato quality, as there is a higher risk of discoloration. Glycoalkaloids were not significantly affected by elevated [CO$_2$] (Table 6.5); however, the strong reduction in nitrate under elevated [CO$_2$] (Table 6.5) determines a nutritional improvement in the quality of potato tubers.

6.6 Conclusions

The results of the FACE experiments made with potato indicate that, although the responses of the two selected cultivars are in some cases rather variable, under future climatic conditions with increased atmospheric [CO_2] there will be significant changes in the acquisition and transformation of potato crop resources, and these will determine an important improvement not only in the potato yield but also in the quality of tubers (physical and chemical qualities):

- Measurements of gas exchange showed that the photosynthetic capacity of the leaves was not affected by long-term [CO_2] exposure. This allowed the conclusion that there was no photosynthetic acclimation under elevated [CO_2].
- Mean seasonal values of the WUE showed a substantial effect of the CO_2 enrichment, with an overall increase that allowed a decrease in water use.
- Crop phenology and development were not substantially affected by elevated [CO_2].
- Analysis of larval growth of Colorado potato beetle indicated that growth was sensitive to changes in leaf composition (N concentration). Larval size at the end of the experiment was affected by the quality of the foliage ingested, with larvae from the ambient treatments having larger dry mass than those fed from the elevated [CO_2]-grown leaves.
- Effects of elevated [CO_2] on aboveground biomass components (leaf and shoot dry weight, leaf number) were in general small and not consistent during the whole growing periods. More specifically, during the early part of the growing season some CO_2 stimulation was observed, while this was not the case when the plants approached maturity.
- Tuber growth and yield were strongly affected by increasing levels of [CO_2]; and both the number and dry mass of tubers were stimulated in FACE plots.
- Physical quality analyses showed clearly that the effect of [CO_2] on tuber production is mainly due to the increase in the number of tubers and dry matter of tubers.
- Chemical quality analyses showed that the dry matter and starch concentration of the potato tubers was increased under elevated [CO_2], thus improving the processing quality of potatoes. Moreover, the strong reduction in nitrate under elevated [CO_2] determines a nutritional improvement in the quality of potato tubers.

References

Ainsworth EA, Long SP (2005) What have we learned from 15 years of free-air CO$_2$ enrichment (FACE)? A meta-analytic review of responses to rising CO$_2$ in photosynthesis, canopy properties and plant production. New Phytol 165:351–372

Bindi M, Fibbi L, Frabotta A, Chiesi M, Selvaggi G, Magliulo V (1999) Free air CO$_2$ enrichment of potato (*Solanum tuberosum* L.). In: CHIP (ed) Changing climate and potential impacts on potato yield and quality: CHIP project, final report; contract ENV4-CT97-0489. Commission of the European Union, Brussels, pp 160–196

Drake BG, Gonzales-Meler M, Long SP (1997) More efficient plants: a consequence of rising atmospheric CO$_2$? Annu Rev Plant Physiol Plant Mol Biol 48:609–639

Evans JR (1989) Photosynthesis and nitrogen relationship in leaves of C3 plants. Oecologia 78:9–19

FAOSTAT (2004) On line and multilingual database. FAO, Rome

Farrar JF, Williams ML (1991) The effects of increased atmospheric carbon dioxide and temperature on carbon partitioning, source-sink relations and respiration. Plant Cell Environ 14:819–830

Gifford RM (1992) Interaction of carbon dioxide with growth-limiting environmental factors: implications for the global carbon cycle. Adv Bioclimatol 1:24–58

Jackson RD, Moran MS, Gay LW, Raymond LH (1987) Evaluating evaporation from field crops using airborne radiometry and ground-based meteorological data. Irrig Sci 8:81–90

Kimball BA, LaMorte RL, Seay RS, Pinter PJ, Rokey RR, Hunsaker DJ, Dugas WA, Heuer ML, Mauney JR, Hendrey GR, Lewin KF, Nagy J (1994) Effects of free-air CO$_2$ enrichment on energy balance and evapotranspiration of cotton. Agric For Meteorol 70:259–278

Lewin KF, Hendrey G, Kolber Z (1992) Brookhaven national laboratory free-air carbon dioxide enrichment facility. Crit Rev Plant Sci 11:135–141

Magliulo V, Bindi M, Rana G (2003) Water use of irrigated potato (*Solanum tuberosum* L.) grown under free air carbon dioxide enrichment in central Italy. Agr Ecosyst Environ 97:65–80

Miglietta F, Raschi A, Bettarini I, Resti R, Selvi F (1993) Natural CO$_2$ springs in Italy – a resource for examining long-term response of vegetation to rising atmospheric CO$_2$ concentrations. Plant Cell Environ 16:873–878

Miglietta F, Magliulo V, Bindi M, Cerio L, Vaccari F, Peressotti A (1998) Free air CO$_2$ enrichment of potato (*Solanum tuberosum* L.): development, growth and yield. Global Change Biol 4:163–172

Riesmeier JW, Willmitzer L, Frommer WB (1994) Evidence for an essencial role of the sucrose transporter in phloem loading and assimilate partitioning. EMBO J 13:1–17

Sage RF (1994) Acclimation of photosynthesis to increasing atmospheric CO$_2$. The gas exchange perspective. Photosynth Res 39:351–368

Vaccari F, Miglietta F, Magliulo V, Raschi A, Bindi M (2000) Free air CO$_2$ enrichment of potato: leaf consumption and larval growth of Colorado potato bettle (*Leptinotarsa decemlineata* Say). Ital J Agron 4:37–41

Vaccari F, Miglietta F, Magliulo E, Giuntoli A, Cerio L, Bindi M (2001) Free air CO$_2$ enrichment of potato (*Solanum tuberosum* L.): photosynthetic capacity of leaves. Ital J Agron 5:3–10

7 Responses of an Arable Crop Rotation System to Elevated [CO_2]

H.J. Weigel, R. Manderscheid, S. Burkart, A. Pacholski,
K. Waloszczyk, C. Frühauf, and O. Heinemeyer

7.1 Introduction

During the past three decades, research on the effects of elevated atmospheric CO_2 concentrations [CO_2] on agricultural ecosystems has mainly been driven by two concerns. First, in order to assess the potential impacts of future worldwide climatic changes on global food supply, studies have focussed on how elevated [CO_2], either singly or in combination with other growth variables, affects crop growth and yields (e.g. Rosenzweig and Hillel 1998; Reddy and Hodges 2000; Kimball et al. 2002). The second and more recent concern resulted from the need to better understand ecosystem feedbacks to climate change and to changes in atmospheric chemistry (i.e. elevated [CO_2] levels; e[CO_2]), particularly in terms of atmospheric–biospheric exchange of carbon (C), nitrogen (N) and water (H_2O; e.g. Mosier 1998; Polley 2002).

Existing studies of e[CO_2] effects on arable crops in Europe are dominated by assessments for wheat (e.g. Mitchell et al. 1993; Batts et al. 1997; Bender et al. 1999) and potato (e.g. Miglietta et al. 1998; Schapendonk et al. 2000; Craigon et al. 2002). Much less or even hardly any information is available for barley (e.g. Weigel et al. 1994; Martin-Olmedo et al. 2002), sugar beet (Demmers-Derk et al. 1998) and maize (Bethenod et al. 2001).

Large biomass and yield enhancements by e[CO_2] have been found in these studies. For example, wheat growth was stimulated by e[CO_2] (+ ca. 250–350 ppm above current levels) between + ca. 20–30%, which is in line with earlier information for other plants (e.g. Cure and Acock 1986; Kimball et al. 1993). Realization of such large growth enhancements for major crops would be of significant importance with respect to future European and global food supply. However, most of the studies cited above were performed on isolated plants or under conditions of repeated monocultures, or crops

Ecological Studies, Vol. 187
J. Nösberger, S.P. Long, R.J. Norby, M. Stitt,
G.R. Hendrey, H. Blum (Eds.)
Managed Ecosystems and CO_2
Case Studies, Processes, and Perspectives
© Springer-Verlag Berlin Heidelberg 2006

were grown under optimized conditions with respect to nutrient and water supply. Moreover, nearly all studies were carried out in chambers or enclosures (open-top chambers, field tunnels), i.e. under conditions where the different microclimates might have resulted in an overestimation of the $[CO_2]$ effect (e.g. Kimball et al. 1997; Van Oijen et al. 1999). The question remains open whether similar growth stimulations by $e[CO_2]$ can be realized in fully open air under field conditions (e.g. McLeod and Long 1999; Parry et al. 2004). The only two recent field studies with cereals worldwide which have used the free air carbon dioxide enrichment (FACE) technology (ca. 200 ppm CO_2 above current levels) resulted in yield enhancements of only ca. 8 % for wheat in Arizona (Kimball et al.1995; see Chapter 3) and ca. 7–15 % for rice in Japan (Kim et al. 2003; see Chapter 5).

Arable crops are not only vital for the supply of food, fibre and energy, but they also represent a significant type of land use in Europe, as roughly 25 % of the total land area is covered by arable land (FAOSTAT 2002). Thus, arable agro-ecosystems significantly contribute to the biogeochemical cycling of elements and water on a continental scale and provide important habitats for fauna and flora. Potential impacts of $e[CO_2]$ on this role of arable land are of equal importance as the growth and yield responses of the crops. For example, it still remains open how fluxes of CO_2 and H_2O (and of other trace gases like N_2O, CH_4 or tropospheric O_3) are related to agricultural management practices and particularly how these processes may respond to future atmospheric $[CO_2]$ (e.g. Amthor 1995; Lal et al. 1998; Canadell et al. 2000).

In order to improve predictions of how climate change might affect these different functions of agro-ecosystems, more realistic field data of possible effects of $e[CO_2]$, e.g. as obtained with FACE experiments (Hendrey 1992; see Chapter 2) are required. Arable crop rotations are still a dominant form of agricultural land use in Europe and in Germany, respectively. Due to their inherent growth conditions, rotations exhibit different ecosystem properties in comparison to e.g. repeated monocultures; and the responses of a particular crop species or cultivar to environmental (i.e. climatic) or management changes may thus differ from the response of that crop in monocultures. Assessments of potential effects of $e[CO_2]$ on regional crop production have to consider these particular growth conditions. Up to now, CO_2 enrichment studies have not been carried out in Europe under conditions of crop rotations.

A FACE experiment has been installed in an arable crop rotation at Braunschweig, Germany, in order to investigate long-term effects of $e[CO_2]$ on field-grown crops and related agro-ecosystem properties. The experiment is combined with micrometeorological and chamber measurements of atmospheric–biospheric fluxes of CO_2 and water and of other air constituents which are carried under the same site conditions, i.e. on a uniform arable field with sufficient fetch and under otherwise identical atmospheric, soil, vegetation and management conditions (Weigel and Dämmgen 2000). The objectives of the FACE experiment at Braunschweig are: (i) to validate existing information on [CO$_2$] effects on European arable crops obtained from chamber experiments under real agricultural growth conditions and to extend the database of potential [CO$_2$] effects to crop species, which have received little attention so far, and (ii) to assess possible feedback effects of e[CO$_2$] levels on the ecosystem properties (CO$_2$ and H$_2$O fluxes, soil biology) of arable land. In the following sections, the design of the ongoing experiment is briefly described and some preliminary results obtained during the first crop rotation cycle are shown.

7.2 Site Description

7.2.1 Location, Climate, Meteorological and Soil Conditions

The 22-ha experimental field plot is located at the Federal Agricultural Research Centre (FAL) in Braunschweig, south-east Lower Saxony, Germany (52°18' N, 10°26' E, 79 m a.s.l.). The long-term average climate is characterized by an annual mean temperature of 8.8 °C, a mean July temperature of 17 °C and total precipitation of 618 mm year^{-1} (half of the precipitation is deposited during May to September), sunshine for 1514 h year^{-1} and a solar radiation of approximately 350 kJ cm^{-2} year^{-1}. A description of the meteorological conditions during the first crop rotation cycle (1999–2002) is given in Table 7.1.

The soil is a luvisol of a loamy sand texture (69 % sand, 24 % silt, 7 % clay) in the plough horizon. The profile has a depth of about 60 cm (–30 cm Ap, –15 cm Al, –15 cm Bt, >60–70 cm CII). The lower layers, in particular the parent material (>70 cm), are characterized by a coarser soil texture (almost pure sand) and are structured by the succession of thin silt/clay layers. The plough layer has a pH of 6.5 and a mean organic matter (C$_{org}$) content of 1.4 %. In line with the soil texture, the soil has a volumetric plant available water content (PAWC) of ca. 18 % in the plough layer, which decreases slightly with increasing profile depth. Overall, the soil of the study site is of low to intermediate fertility and provides a comparatively shallow rooting zone.

7.2.2 Crop Rotation and Agricultural Management

A typical crop rotation in the northern part of Germany comprises a winter grain crop, followed by a cover crop, a (spring-sown) row crop and a winter grain crop again. The FACE experiment is applied to a locally important rotation, consisting first of winter barley (*Hordeum vulgare*), then a ryegrass mixture (a mixture of different cultivars of *Lolium multiflorum*) as a cover crop,

Table 7.1 Meteorological conditions at the FACE site at Braunschweig during the first crop rotation 1999–2002 and deviations from the long-term mean (1961–1990; source: German Weather Service, FAL-Braunschweig)

Variable	Units	Winter barley	Italian rye grass mixture[a]	Sugar beet	Winter wheat
Vegetation period	Months	Mar–Jun 2000	Aug–Sept 2000	May–Sep 2001	Mar–Jul 2002
Air temperature (mean; 2 m)	°C/deviation	12.2/±2.1	16.2/±0.7	15.9/±0.54	12.7/±1.16
Precipitation (sum)	mm/deviation	195/±29	76/±37	376/±74	468/±186
Evaporation (sum)	mm/deviation	313/±78	152/±9	437/±51	308/±11
Global radiation (sum)	$J\ cm^{-2}$/deviation	184 929/±13 934	80 301/±3814	252 326/±15 100	215 750/±7850

[a] Mixture of different cultivars of *Lolium multiflorum*.

Table 7.2 Crop cultivars, treatment conditions and agricultural management measures during the first crop rotation cycle, 1999–2002, at the FACE site at Braunschweig

Management	Units	Winter barley cv. "Theresa"	Italian rye grass mixture[a]	Sugar beet cv. "Wiebke"	Winter wheat "Batis"
Seeding	Date	23 Sept 1999	26 Jul 2000	11 Apr 2001	6 Nov 2001
Seeding density	Seeds m^{-2}	280	40 kg ha^{-1}	11	360
Start of CO_2 enrichment	Date	4 Oct 1999	5 Aug 2000	14 May 2001	22 Jan 2002
N fertilisation; total amount: 100 %/50 %	(No.) kg ha^{-1}	(4) 264/105	(1) 197/99	(2) 126/63	(3) 181/91
Irrigation	No./mm	3/69	2/44	5/107	3/60
CO_2 enrichment duration	Days	260	70	138	183
CO_2 concentration (current/elevated)	ppm	373/549	373/550	371/550	377/548
Final harvest	Date	22 Jun 2000	12 Oct 2000	24 Sept 2001	31 Jul 2002

[a] Mixture of different cultivars of *Lolium multiflorum*

then sugar beet (*Beta vulgaris*) and then winter wheat (*Triticum aestivum*). The rotation cycle is repeated once, resulting in a total duration of the CO$_2$ exposure experiment of 6 years. Agricultural management measures of the field (total 22 ha) are carried out according to local farm practices, using the same technologies for the FACE ring plots. Plough tillage is applied to the soil immediately after the harvest of each crop. In order to avoid interacting effects with drought stress in the FACE experiment, the field is irrigated using a linear irrigation system to keep the soil water content above 50% PAWC. Field irrigation to avoid drought stress conditions is a common practice in the area. Crop details and management measures as well as [CO$_2$] treatment details are summarized in Table 7.2.

7.2.3 Treatment Design

A FACE system consisting of rings with 20 m diameter engineered by Brookhaven National Laboratory (New York, USA; see Chapter 2) is operated. The system is described in detail by Hendrey (1992) and Lewin et al. (1992). During each growing season, the FACE rings are assembled immediately after seeding of the crop (before emergence) and removed from the field plot ca. 3–5 days prior to the final harvest of the total 22-ha field. Based on GPS-supported information, the rings are placed at exactly the same location (± 5 cm) during each growing season. Treatments include two rings equipped with blowers and enriched with CO$_2$ (i.e. e[CO$_2$]) and two rings operated with blowers and ambient air only (i.e. current [CO$_2$]; c[CO$_2$]). Circular plots equipped with vertical vent pipes only and without blowers serve as control/reference areas. The total area enriched with CO$_2$ is approximately 510 m^2. The target CO$_2$ concentration in the enriched rings is set to 550 μmol mol^{-1} (ppm) during daylight hours (i.e. daylight solar altitude $\theta > -0.833$). No-enrichment criteria for CO$_2$ are wind speeds >6.0–6.5 m s^{-1} and air temperatures <5 °C. The CO$_2$ supplied to the FACE rings is derived from natural gas and is depleted in ^{13}C (δ^{13}C $= -47$‰). In order to simulate alternative nutrient management scenarios in agriculture and to study interactions between C and N turnover in the plant–soil system, respectively, N supply is restricted to 50% (N50) of adequate N (N100) in half of each of the FACE rings (Table 7.2), resulting in a CO$_2$ × N split-plot design. Assessment of the treatment effects under the conditions of the present experimental design requires the inclusion of the replication over time. Statistical analysis of the crop growth and soil data (as shown here) is done on four replicate samples per treatment by regarding one-quarter of each ring as one replicate plot.

7.3 Results

The following paragraphs present selected examples of canopy gas exchange and the crop and soil responses to e[CO_2] treatment during the first rotation cycle with winter barley, sugar beet and winter wheat.

7.3.1 Resource Acquisition

7.3.1.1 [CO_2] Effects on Photosynthesis (Canopy CO_2 Exchange Rates)

Changes in CO_2 fluxes at the canopy level are more relevant than leaf-level changes for assessing the effects of climate change on ecosystem functioning and agricultural productivity. They can provide first information on altered C fluxes into the plant–soil system. Due to the high resolution in time, such measurements also allow a quantification of daily and seasonal crop growth dynamics. As limited information is available how e[CO_2] affects canopy photosynthesis (Lawlor and Mitchell 1991; Dijkstra et al. 1993), which is especially true for FACE conditions (Brooks et al. 2000), we measured day-time canopy CO_2 exchange rates (CCER) as an estimate of canopy photosynthesis for all crop species under investigation (Burkart et al. 2000). As an example, Fig. 7.1a shows a typical weekly time-course of CCER of sugar beet. Variation of CCER was strongly dependent on photosynthetic photon irradiance (PPI) and vapour pressure deficit (VPD), which was also true for the other crop species.

Figure 7.2a, b shows the relationship between CCER and absorbed PPI flux density for sugar beet and wheat obtained using a weekly measuring interval. In accordance with other studies, the canopy CO_2 fluxes (photosynthesis) were related to radiation in a non-linear way (Ruimy et al. 1995). e[CO_2] significantly enhanced CO_2 fluxes into the canopy, especially at higher radiation levels. It is hardly evident from Fig. 7.2 that, at lower light intensities, the slope of the relationship between CCER and light intensity – which may be regarded as the apparent canopy quantum yield – was also enhanced by e[CO_2]. However, analysing these initial slopes (absorbed PPI <500 μmol m^{-2} s^{-1}) revealed that e[CO_2] increased apparent canopy quantum yield significantly (c[CO_2]/e[CO_2] sugar beet = 0.044/0.052 mol mol^{-1}, $P<0.05$; c[CO_2]/e[CO_2] wheat = 0.041/0.049 mol mol^{-1}, $P<0.01$). Averaged across the time-period shown in Fig. 7.2 (green LAI >2.0), at saturating PPI levels and under adequate N supply (N100), the stimulation of CCER by e[CO_2] reached ca. +45 % and +37 % for sugar beet and wheat, respectively. With respect to FACE conditions Brooks et al. (2000), using a similar canopy chamber device as in the present case, found a 19 % stimulation of wheat canopy CO_2 fluxes (photosynthesis) due to a 550 ppm [CO_2] treatment under growth conditions in Arizona. Overall, the pre-

Fig. 7.1 Typical weekly time-course of (**a**) canopy CO$_2$ exchange rate (CCER) and (**b**) canopy evapotranspiration ET (canopy H$_2$O flux rate; CO$_2$- and H$_2$O- exchange rates per unit ground area) of sugar beet (N100 treatment only) measured under c[CO$_2$] and e[CO$_2$]. Flux rates were measured continuously during most days of the growing seasons (LAI >2) using four open-system canopy chambers (1.2 m^2 area, 1.5 m^3 volume), as described in detail by Burkart et al. (2000)

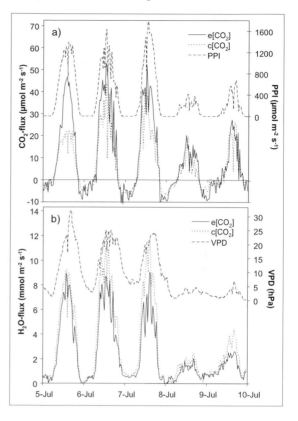

sent data show that a significant amount of additional C from the atmosphere was fixed by the crop canopies.

7.3.1.2 [CO$_2$] Effects on Canopy Evapotranspiration ET (Canopy H$_2$O Exchange Rate)

Responses of canopy evapotranspiration (ET) of crop plants to e[CO$_2$] have been found to vary from enhanced canopy H$_2$O fluxes (e.g. Kimball et al. 1994; Hui et al. 2001), unchanged fluxes (e.g. Hilemann et al. 1994) to decreased H$_2$O fluxes (e.g. Drake et al. 1997; Kimball et al. 1997). As shown for sugar beet in Fig. 7.1b, ET (daily canopy H$_2$O fluxes) showed a similar pattern to canopy CO$_2$ fluxes. It is evident that e[CO$_2$] reduced ET at different radiation levels. Figure 7.2 c, d depicts the linear relationships between ET and global radiation of sugar beet and wheat. During the measuring periods presented in Figs. 7.1 and 7.2, mean daily ET (mm H$_2$O day^{-1}) was reduced by e[CO$_2$] by –21 % and –6 % for sugar beet and wheat, respectively. Overall, ET for barley,

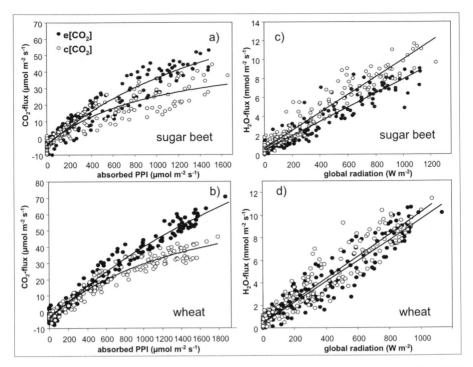

Fig. 7.2 CCER as a function of absorbed photosynthetic photon irradiance (**a, b**) and ET related to global radiation (**c, d**) of sugar beet (**a, c**) and winter wheat (**b, d**) measured by a CO_2/H_2O gas exchange chamber technique under $c[CO_2]$ and $e[CO_2]$ (Burkart et al. 2000). Data are based on 10-min average values measured during one week in August 2001 (sugar beet; see Fig. 7.1) and in May 2002 (wheat; N100 treatment only). Canopy radiation absorption was measured with a SunScan line sensor (Delta-T Devices, Cambridge, UK). Fig. 7.2c. $c[CO_2]$: $y = 0.0094x + 0.6345$; $r^2 = 0.94$; $e[CO_2]$: $y = 0.0074x + 0.4881$; $r^2 = 0.92$. Fig. 7.2d. $c[CO_2]$: $y = 0.0097x + 0.5116$; $r^2 = 0.92$; $e[CO_2]$: $y = 0.0092x + 0.4459$; $r^2 = 0.93$

sugar beet and wheat was generally lower under $e[CO_2]$ in comparison to $c[CO_2]$, although the relative reduction differed between the crops. Averaged across the different measurements campaigns which were carried out during the growing season of the particular crop, daily mean ET (mm H_2O day^{-1}) was reduced by $e[CO_2]$ by -7.5%, -19.8% and -2.6% for barley, sugar beet and wheat, respectively (data not shown). While there is no comparable information for barley and sugar beet, the effect of $e[CO_2]$ on ET of wheat was smaller, as compared to the wheat experiments in the Arizona FACE under well watered conditions, where seasonal evapotranspiration was reduced by $e[CO_2]$ by $4.5–11.0\%$ (Kimball et al. 1995, 1997; see Chapters 3, 17).

7.3.1.3 [CO$_2$] Effects on Leaf Area Index

Large differences in the response of leaf area index (LAI) of crops to e[CO$_2$] concentrations have been found, ranging from negative effects to strong enhancements (Kimball et al. 2002). A recent meta-analysis of the percent change in LAI in previous FACE experiments yielded a non-significant increase of 7 % averaged over all published FACE studies (Long et al. 2004). From this analysis, it was concluded that LAI is obviously not changed by CO$_2$ enrichment under such exposure conditions. Available information on the effect of [CO$_2$] elevation on LAI of the four species in the present crop rotation experiment is summarized in Table 7.3. Numbers of LAI are presented for those periods when LAI was at its maximum or – as in the case of sugar beet – shortly after canopy closure. As expected, reduced N supply resulted in smaller LAI numbers. Averaged across both N levels e[CO$_2$] increased LAI of sugar beet by ca. 13 %, although this effect was only slightly significant ($P=0.08$). Due to the large variation, the effects of e[CO$_2$] on barley were not consistent. Hence, LAI of all crop species was hardly affected by elevated CO$_2$ concentrations, which confirmed the conclusion drawn by Long et al. (2004).

Table 7.3 Green leaf area index (LAI)[a] of species in a crop rotation system exposed to c[CO$_2$] and e[CO$_2$] and to two N levels (N50/N100; see Table 7.1)[b]

Nitrogen	CO$_2$	Crop			
		Winter barley	Italian rye grass mixture[c]	Sugar beet	Winter wheat
		LAI (m^2 m^{-2})			
N50	c[CO$_2$]	4.6±0.6	4.4±0.2	2.8±0.1	3.0±0.5
N50	e[CO$_2$]	4.0±0.3	4.5±0.1	3.0±0.4	3.1±0.1
Relative [CO$_2$] effect (%)		−13.0	+2.0	+7.2	+3.0
N100	c[CO$_2$]	6.5±0.7	5.6±0.2	3.2±0.1	3.8±0.6
N100	e[CO$_2$]	7.2±1.3	5.5±0.2	3.7±0.4	3.9±0.1
Relative [CO$_2$] effect (%)		+11.7	−1.8	+16.2	+2.1

[a] LAI was measured near anthesis (barley, wheat), in the middle of September (ryegrass) or after canopy closure at the end of June (sugar beet)
[b] Numbers represent mean±SE, $n=4$
[c] Mixture of different cultivars of *Lolium multiflorum*

7.3.2 Resource Transformation

7.3.2.1 [CO₂] Effects on Above-ground Biomass Production

Shoot biomass responses to $e[CO_2]$ levels have repeatedly been found to be smaller when nutrients – especially N – are limiting, as compared to ample nutrient supply (e.g. Kimball 1993; Wand et al. 1999; see Chapter 3), although the differences between the N nutrition may be small (e.g. Idso and Idso 1994). In the first crop rotation cycle of the present experiment, stimulation of above-ground biomass production by $e[CO_2]$ ranged from ca. 7% to ca. 13% when the results were averaged across both N supply levels (Table 7.4). While for winter barley, the relative CO_2 enhancement was higher under N limitation (ca. 13%) than under ample N supply (ca. 8%), sugar beet and wheat showed comparable responses under both N supply levels with only a slightly higher (ca. 3%) growth stimulation under ample N supply. Therefore, these results obtained so far do not support the observations that suboptimal N supply levels limit the $[CO_2]$ response. As summarized by Kimball et al. (2002), $e[CO_2]$ levels in FACE studies with other C_3 grasses (wheat, rice, ryegrass) stimulated above-ground biomass production under unrestricted supply of other

Table 7.4 Above-ground biomass production (g m⁻²) of species in a crop rotation system at final harvest exposed to $c[CO_2]$ and $e[CO_2]$ and to two N levels (N50/N100; see Table 7.1)[a]

Nitrogen	CO₂	Crop			
		Winter barley	Italian rye grass mixture[b]	Sugar beet	Winter wheat
		g m⁻²			
N50	$c[CO_2]$	1360±44	484±19	1919±44	1163±33
N50	$e[CO_2]$	1546±26	531±24	2036±29	1292±37
Relative [CO₂] effect (%)		+13.7*	+9.6	+6.1	+11.1*
N100	$c[CO_2]$	1679±19	484±16	2295±42	1272±27
N100	$e[CO_2]$	1815±3	543±27	2481±23	1456±43
Relative [CO₂] effect (%)		+8.1*	+12.1	+8.1*	+14.4*

[a] Numbers represent means ±SE, $n = 4$
[b] Mixture of different cultivars of *Lolium multiflorum*
* Significant at least at $P < 0.05$

resources by ca. +11.5 %. This is rather similar to the relative CO_2 enhancement (+10.7 %) found in the present FACE experiment with cereal species under ample N supply. The storage crop sugar beet showed a comparably smaller biomass enhancement by e[CO_2]. Such low or even negative effects of e[CO_2] on above-ground biomass production have also been observed with potato (e.g. Miglietta et al. 1998). Effects of e[CO_2] on the economic yield of barley, sugar beet and wheat in the present FACE experiment were in the same order of magnitude as for total above-ground biomass (data not shown). From the present data, it can also be assumed that the strong stimulation of canopy photosynthesis by e[CO_2] (CCER; see Section 4.1) – albeit the CCER measurements did not cover the total growing season – was not fully translated into enhanced above-ground biomass production.

7.3.2.2 [CO_2] Effects on Below-ground Biomass Production

There is still little information on root growth of crops under [CO_2] elevation (Rogers et al. 1997). Moreover, most previous studies have been done under conditions which restrict root growth (e.g. pot studies). In the present study, fine root biomass production was measured repeatedly during the growing season at different developmental stages of the different crop species (Fig. 7.3). While root growth of barley, sugar beet and wheat was enhanced by high [CO_2] levels, this effect varied during the growing season. For example, fine root biomass production of barley was highest during flowering and was enhanced by e[CO_2] prior to this growth stage. This stimulation was higher under reduced N supply (ca. 50 %) than under ample N supply (ca. 30 %). At the end of the growing season, a significant [CO_2] effect could no longer be detected. Roots of sugar beet showed a tendency for enhanced growth under e[CO_2], but this effect was significant only towards the end of the growing season. Again, the relative [CO_2] effect was higher (ca. 90 %) under reduced N supply. For wheat, the slight stimulation of fine root biomass production by e[CO_2], which was evident under full N supply, could also not be proved statistically. In previous FACE experiments with wheat, rice and ryegrass an average of 47 % growth stimulation by e[CO_2] under non-restricted nutrient and water supply was found (Kimball et al. 2002; see Chapter 3). The present data only partly support this information. Moreover, as already outlined by Pritchard and Rogers (2000), our results point to the fact that root growth responses to e[CO_2] might be a transient response which changes throughout plant development.

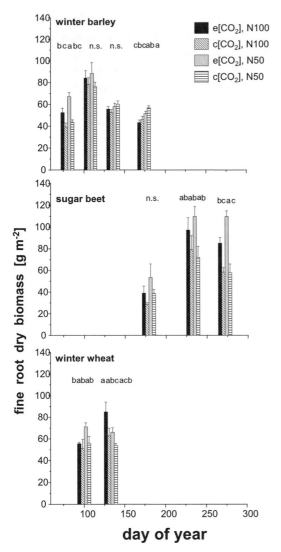

Fig. 7.3 Fine root production of winter barley, sugar beet and winter wheat at different times of the growing season in a crop rotation under c[CO$_2$] and e[CO$_2$] and at different N supply levels (N100/N50; see Table 7.1). Soil cores (barley, diam. 3 cm; sugar beet, wheat, diam. 8 cm) were taken up to 30 cm depths. Root biomass was determined by extracting visible roots and washing and sieving the soil (>0.63 mm). Numbers represent means ± SD; $n = 4$; significant differences between the different treatments are indicated by *different letters* above the columns

7.3.2.3 [CO$_2$] Effects on Soil Microbial Biomass

Soil microbial biomass (C$_{mic}$) is a key variable of organic carbon (C$_{org}$) turnover in soil, e.g. as the turnover of "new" carbon entering the soil is processed by soil microorganisms. C$_{mic}$ can thus be used as an early and susceptible indicator of changes of C$_{org}$. During the first rotation cycle, C$_{mic}$ was measured biweekly or monthly by the substrate-induced-respiration (SIR) method (Heinemeyer et al. 1989). C$_{mic}$, which varied over 150–300 µg C g^{-1} soil, depending on the crop species and the time of the year, was not affected by e[CO$_2$] treatment (data not shown). Averaged across the whole time period

1999–2002, there was a small but significant positive effect of the N treatments ($P<0.05$). Available information on the effects of e[CO₂] on soil biological variables obtained from very different experimental approaches in different ecosystem types points to an increase in C_{mic} due to e[CO₂] treatments (Zak et al. 2000; see Chapter 21). However, there is considerable scatter in these data, ranging from negative to positive [CO₂] effects on C_{mic}. In the present FACE experiment, where soil carbon turnover is also strongly affected by repeated soil management measures throughout the year, there is no evidence even after 3 years of CO₂ enrichment that an effect of the extra C input into the soil can be observed at the level of soil microbial biomass.

7.3.2.4 [CO₂] Effects on In Situ Soil CO₂ Efflux

Interest in the effects of elevated atmospheric CO₂ concentrations on CO₂ efflux from soils (in situ soil respiration; R_S) is related to the question whether the additional C-input into the plant–soil system results in an enhanced C-turnover rate or a net C-sequestration. R_S mainly originates from decomposition of organic matter (heterotrophic R_{SH}) and respiration by living roots (autotrophic R_{SA}). However, the contribution of each component to the total flux is difficult to estimate. In the present case, R_S could only be measured for sugar beet and wheat (Fig. 7.4). While R_S of sugar beet and wheat showed typ-

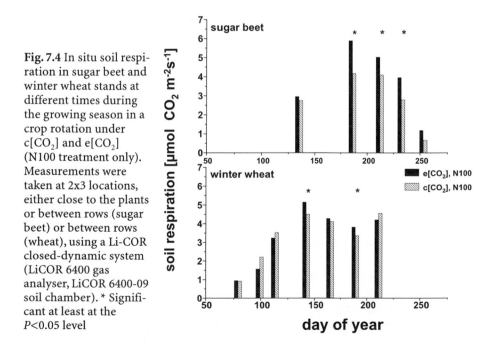

Fig. 7.4 In situ soil respiration in sugar beet and winter wheat stands at different times during the growing season in a crop rotation under c[CO₂] and e[CO₂] (N100 treatment only). Measurements were taken at 2x3 locations, either close to the plants or between rows (sugar beet) or between rows (wheat), using a Li-COR closed-dynamic system (LiCOR 6400 gas analyser, LiCOR 6400-09 soil chamber). * Significant at least at the $P<0.05$ level

ical seasonal courses, the effects of $e[CO_2]$ on R_S differed between these two crop species. Under sugar beet canopies, R_S was significantly stimulated by $e[CO_2]$ during most of the growing season with a maximum stimulation of ca. 30%, while for wheat only a small stimulation could be observed at the time of the maximum of the seasonal respiration rate. In a compilation of available soil respiration data across different CO_2 exposure techniques and different plant species, Zak et al. (2000) concluded that despite a considerable variability of the data, there is an indication of a significant stimulation of in situ soil respiration by $e[CO_2]$. In previous FACE experiments, elevated CO_2 concentrations caused an increase of R_S under cotton by 16–23% (Nakayama et al. 1994), 40–70% under wheat (Pendall et al. 2001) and 13% in a grassland (Craine et al. 2001).

7.4 Conclusions

The specific objectives and the design of the FACE experiment at Braunschweig required repetition of the CO_2 exposure of the different plants at least once during the time-course of crop rotation. Consequently, overall assessments of the different FACE effects on the plant–soil system, including conclusions for management options, will be more reasonable after the end of the 6-year experiment. This is particularly true for the soil compartment, as soil processes often show delayed responses to environmental perturbations (i.e. $e[CO_2]$).

Considering that, for sugar beet and barley, no comparable field data are available, the following crop responses to $e[CO_2]$ have been observed so far:

- Canopy photosynthesis of all crops was stimulated by ca. +35% to +45%.
- Canopy evapotranspiration was reduced to a greater extent for sugar beet (ca. –20%) than for cereals (ca. –10%).
- There were no effects of $e[CO_2]$ on LAI of cereals, but a stimulation of LAI of sugar beet by ca. +13%.
- A comparable stimulation of final biomass production was observed between ample (+10.7%) and restricted (+10.1%) nitrogen supply treatment, which is in contrast to many previous studies.
- The stimulation of final biomass production of sugar beet (ca. +7%) was smaller in comparison to cereals (ca. +12%), which is in contrast to theoretical expectations with regard to a root crop.

Below-ground processes of carbon turnover responded inconsistently to $e[CO_2]$:

- Root growth was stimulated by $e[CO_2]$, from ca. +30% to +50%. However, this response was transient during the growing seasons.

- Soil microbial biomass as a key variable of soil carbon turnover did not respond to e[CO$_2$].
- There were transient stimulations (max. ca. +30 %) of soil respiration by e[CO$_2$] during the growing seasons.

Overall, it remains open whether the differences observed between the crops and soil responses are differences related to the species or to the year-to-year variability in growth conditions.

Acknowledgements. The FACE system used at Braunschweig was provided by Brookhaven National Laboratory. We are especially grateful to George Hendrey, Keith Lewin and John Nagy for their help in establishing the system and their continuous advice during its operation.

References

Amthor JS (1995) Terrestrial higher-plant response to increasing atmospheric [CO$_2$] in relation to the global carbon cycle. Glob Change Biol 1:243–274

Batts GR, Morison JIL, Ellis RH, Hadley P, Wheeler TR (1997) Effects of CO$_2$ and temperature on growth and yield of crops of winter wheat over four seasons. Eur J Agron 7:43–52

Bender J, Hertstein U, Black CR (1999) Growth and yield responses of spring wheat to increasing carbon dioxide, ozone and physiological stresses: a statistical analysis of 'ESPACE-Wheat' results. Eur J Agron 10:185–195

Bethenod O, Ruget F, Katerji N, Combe L, Renard D (2001) Impact of atmospheric CO$_2$ concentration on water use efficiency of maize. Maydica 46:75–80

Brooks TJ, Wall GW, Pinter PJ, Kimball BA, LaMorte RL, Leavitt S, Matthias AD, Adamsen FJ, Hunsacker DJ, Webber AN (2000) Acclimation response of spring wheat in a free air CO$_2$ enrichment (FACE) atmosphere with variable soil nitrogen regimes. 3. Canopy architecture and gas exchange. Photosynth Res 66:97–108

Burkart S, Manderscheid R, Weigel HJ (2000) Interacting effects of photosynthetic flux density and temperature on canopy photosynthesis of spring wheat under different CO$_2$ concentrations. J Plant Physiol 157:31–39

Canadell JG, Mooney HA, Baldocchi DD (2000) Carbon metabolism of the terrestrial biosphere: a multitechnique approach for improved understanding. Ecosystems 3:115–130

Craigon J, Fangmeier A, Jones M, Donnelly A, Bindi M, De Temmerman L, Persson K, Ojanpera K (2002) Growth and marketable-yield responses of potato to increased CO$_2$ and ozone. Eur J Agron 17:273–289

Craine JM, Wedin DA, Reich PB (2001) The response of soil CO$_2$ flux to changes in atmospheric CO$_2$, nitrogen supply and plant diversity. Global Change Biol 7:947–1053

Cure JD, Acock B (1986) Crop responses to carbon dioxide doubling: a literature survey. Agric For Meteorol 38:127–145

Demmers-Derk H, Mitchell RAC, Mitchel VJ, Lawlor DW (1998) Response of sugar beet (*Beta vulgaris L.*) yield and biochemical composition to elevated CO$_2$ and temperature at two nitrogen applications. Plant Cell Environ 21:829–836

Dijkstra P, Schapendonk AHCM, Groenwold J (1993) Effects of CO_2 enrichment on canopy photosynthesis, carbon economy and productivity of wheat and faba beans under field conditions. In: DeGeijn SC van, et al (eds) Climate change: crops and terrestrial ecosystems (Agrobiol.Themas 9). AB-DLO, Wageningen, pp 23–41

Drake BG, Gonzalezmeler MA, Long SP (1997) More efficient plants: a consequence of rising atmospheric CO_2? Annu Rev Plant Physiol 48:609–639

FAOSTAT (2002) Food and Agriculture Organization of the United Nations, Statistical Databases. FAO, Rome

Heinemeyer O, Insam H, Kaiser EA, Walenzik G (1989) Soil microbial biomass and respiration measurements: an automated technique based on infra-red gas analysis. Plant Soil 116:191–195

Hendrey G (1992) Global greenhouse studies: need for a new approach to ecosystem manipulation. Crit Rev Plant Sci 11:61–74

Hileman DR, Huluka G, Kenjige PK, Sinha N, Bhattacharya NC, Biswas PK, Lewin, KF, Nagy J, Hendrey GR (1994) Canopy photosynthesis and transpiration of field-grown cotton exposed to free-air CO_2 enrichment (FACE) and differential irrigation. Agric For Meteorol 70:189–207

Hui D, Luo Y, Cheng W, Coleman S, Johnson DW, Sims DA (2001) Canopy radiation- and water-use efficiencies as affected by elevated CO_2. Global Change Biol 7:75–91

Idso KE, Idso SB (1994) Plant responses to atmospheric CO_2 enrichment in the face of environmental constraints: a review of the past 10 years' research. Agric For Meteorol 69:153–203

Kim HY, Lieffering M, Kobayashi K, Okada M, Miura S (2003) Seasonal changes in the effects of elevated CO_2 on rice at three levels of nitrogen supply: a free air CO_2 enrichment (FACE) experiment. Global Change Biol 9:826–837

Kimball BA, Mauney JR, Nakayama FS, Idso SB (1993) Effects of increasing atmospheric CO_2 on vegetation. Vegetatio 104/105:65–75

Kimball BA, Lamorte RL, Seay RS, Pinter PJ, Rokey PJ, Hunsaker DJ, Dugas A, Heuer ML, Mauney JR, Hendrey GR, Lewin KF, Nagy J (1994) Effects of free-air CO_2 enrichment on energy-balance and evapotranspiration of cotton. Agric For Meteorol 70:259–278

Kimball BA, Pinter PJ, Garcia RL, Lamorte RL, Wall GW, Hunsacker DJ, Wechsung G, Wechsung F, Kartschall T (1995) Productivity and water use of wheat under free-air CO_2 enrichment. Global Change Biol 1:429–442

Kimball BA, Pinter PK, Wall GW, Garcia RL, LaMorte RL, Jak PMC, Frumau KFA, Vugts HF (1997) Comparison of responses of vegetation to elevated carbon dioxide in free-air and open-top chamber facilities. In: Allen LH, Kirkham MB, Olszyk DM, Whitman CE (eds) Advances in carbon dioxide effects research (ASA, CSSA and SSSA special publication 61) ASA/CSSA/SSSA, Madison, pp 113–130

Kimball BA, Kobayashi K, Bindi M (2002) Responses of agricultural crops to free air CO_2 enrichment. Adv Agron 77:293–368

Lal R, Kimble JM, Follet KF, Cole CV (1998) The potential of U.S. cropland to sequester carbon and mitigate the greenhouse effect. Sleeping Bear Press, Chelsea, 128 pp

Lawlor D, Mitchell RAC (1991) The effects of increasing CO_2 on crop photosynthesis and productivity: a review of field studies. Plant Cell Environ 14:807–818

Lewin KF, Hendrey G, Kolber Z (1992) Brookhaven National Laboratory free-air carbon dioxide enrichment facility. Crit Rev Plant Sci 11:135–141

Long SP, Ainsworth EA, Rogers A, Ort DR (2004) Rising atmospheric carbon dioxide: plants FACE the future. Annu Rev Plant Biol 55:591–628

Martin-Olmedo P, Rees RM, Grace J (2002) The influence of plants grown under elevated CO_2 and N fertilization on soil nitrogen dynamics. Global Change Biol 8:643–657

McLeod A, Long SP (1999) Free air carbon dioxide enrichment (FACE) in global change research: a review. Adv Ecol Res 28:1–55

Miglietta F, Magliulo V, Bindi M, Cerio L, Vaccari F, LoDuca V, Peressoti A (1998) Free air CO$_2$ enrichment of potato (*Solanum tuberosum* L.): development, growth and yield. Global Change Biol 4:163–172

Mitchell RAC, Mitchell VJ, Driscoll SP, Franklin J, Lawlor DW (1993) Effects of increased CO$_2$ concentration and temperature on growth and yield of winter wheat at 2 levels of nitrogen application. Plant Cell Environ 16:521–529

Mosier AR (1998) Soil processes and global change. Biol Fertil Soils 27:221–229

Nakayama FS, Huluka G, Kimball BA, Lewin KF, Nagy J, Hendrey GR (1994) Soil carbon dioxide fluxes in natural and CO$_2$-enriched systems. Agric For Meteorol 70:131–140

Parry ML, Rosenzweig C, Iglesias A, Livermore M, Fischer G (2004) Effects of climate change on global food production under SRES emissions and socio-economic scenarios. Global Environ Change 14:53–67

Pendall E, Leavitt SW, Brooks T, Kimball BA, Pinter PJ, Wall GW, LaMorte RL, Wechsung G, Wechsung F, Adamsen F, Matthias AD, Thompson TL (2001) Elevated CO$_2$ stimulates soil respiration in a FACE wheat field. Basic Appl Ecol 2:193–201

Polley HW (2002) Implications of atmospheric and climatic change for crop yield. Crop Sci 42:131–140

Pritchard SG, Rogers HH (2000) Spatial and temporal development of crop roots in CO$_2$-enrichment environments. New Phytol 147:55–71

Reddy KR, Hodges, HF (2000) Climate change and global crop productivity. CAB International, Wallingford, 472 pp

Rogers HH, Runion GB, Krupa SV, Prior SA (1997) Plant responses to atmospheric CO$_2$ enrichment: implications in root–soil–microbe interactions. In: Allen LH, et al (eds) Advances in carbon dioxide effects research (ASA special publications 61). ASA, Madison, pp 1–34

Rosenzweig C, Hillel D (1998) Climate change and the global harvest. Oxford University Press, Oxford, 324 pp

Ruimy A, Jarvis PG, Baldocchi DD, Saugier B (1995) CO$_2$ fluxes over plant canopies and solar radiation: a review. Adv Ecol Res 25:2–68

Schapendonk AHCM, Van Oijen M, Dijkstra P, Pot CS, Jordi WJRM, Stoopen GM (2000) Effects of elevated CO$_2$ concentration on photosynthetic acclimation and productivity of two potato cultivars grown in open-top chambers. Aust J Plant Physiol 27:1119–1130

Van Oijen M, Schapendonk AHCM, Jansen MJH, Pot CS, Maciorowski R (1999) Do open-top chambers overestimate the effects of rising CO$_2$ on plants? An analysis using spring wheat. Global Change Biol 5:411–421

Wand SJE, Midgley GF, Jones MH, Curtis PS (1999) Responses of wild C$_4$ and C$_3$ grasses (Poaceae) species to elevated atmospheric CO$_2$ concentrations: a meta-analytic test of current theories and perceptions. Global Change Biol 5:723–741

Weigel HJ, Manderscheid R, Jäger HJ, Mejer GJ (1994) Effects of season-long CO$_2$ enrichment on cereals. I. Growth performance and yield. Agric Ecosyst Environ 48:231–240

Weigel HJ, Dämmgen U (2000) The Braunschweig carbon project: atmospheric flux monitoring and free air carbon dioxide enrichment (FACE). J Appl Bot 74:55–60

Zak DR, Pregnitzer KS, King JS, Holmes WE (2000) Elevated atmospheric CO$_2$, fine roots and the response of soil microorganisms: a review and hypothesis. New Phytol 147:201–222

8 Short- and Long-Term Responses of Fertile Grassland to Elevated [CO$_2$]

A. LÜSCHER, U. AESCHLIMANN, M.K. SCHNEIDER, and H. BLUM

8.1 Introduction

Grassland covers about 70% of the world's agricultural area. Rising atmospheric CO$_2$ concentrations will likely affect several aspects of importance for grassland, such as: the quantity and quality of the forage produced, species composition, soil fertility and the potential to sequester carbon (C) in the soil in order to mitigate the rise of CO$_2$.

The effects of elevated CO$_2$ (e[CO$_2$]) on individual plants growing under controlled conditions and with restricted rooting volume have been studied extensively. However, these studies do not fully reflect the much more complex situation in ecosystems. Single plants in pots differ from ecosystems in a lack of interactions between plants (e.g. competition for light and nutrients) and interactions between plants and the soil (e.g. nutrient cycling). Moreover, most studies are not conducted over a time-period which is long enough that adaptation of feedback mechanisms in the soil–plant relationship can be observed.

The Swiss FACE experiment offered a unique opportunity to study the effects of e[CO$_2$] on model grassland ecosystems for 10 years. This was the longest-lasting FACE experiment on grassland so far and one of the longest continuous CO$_2$ experiments on fertile soil. Most studies in this experiment concentrated on the CO$_2$ and N fertilizer effects on *Lolium perenne*, a grass with a high demand of mineral nitrogen (N), and *Trifolium repens*, a legume which has access to atmospheric N$_2$. Both species are typical components of agricultural grassland in temperate regions.

Here we discuss the results for biomass, its components and allocation of the two species grown in pure swards with special emphasis on *L. perenne*. We demonstrate how the plant response to e[CO$_2$] was affected by N avail-

Ecological Studies, Vol. 187
J. Nösberger, S.P. Long, R.J. Norby, M. Stitt,
G.R. Hendrey, H. Blum (Eds.)
Managed Ecosystems and CO$_2$
Case Studies, Processes, and Perspectives
© Springer-Verlag Berlin Heidelberg 2006

ability in the soil and how these effects changed in the course of the 10 years caused by feedback mechanisms in the ecosystem. Beyond this, we discuss consequences of e[CO_2] for C and N cycling in grassland ecosystems. Effects of e[CO_2] on interspecific interactions in mixed swards are presented in Chapter 19.

8.2 Site Description

The site of the Swiss FACE experiment was at Eschikon (8°41'E, 47°27'N, 550 m a.s.l.), 20 km north-east of Zurich. The experiment consisted of three FACE rings of 18 m diameter with the BNL type of installations (see Chapter 2) and three control rings without any installations. The six rings were separated at least 100 m from each other to prevent interaction of the CO_2 released. The fields used were on farmland which had been cultivated in a crop rotation system before the start of the experiment. Pairs of one FACE and one control ring were placed on fields with the same crop history, thus forming three blocks. The crops grown in the three blocks in the year preceding the FACE set up were winter wheat (two blocks) and a grass/white clover mixture.

The soil was a fertile, eutric cambisol with pH between 6.5 and 7.6. The clay loam (US classification) consisting of 28 % clay, 32 % silt and 36 % sand. The organic matter was between 2.8 % and 5.1 % of dry soil. Available phosphorus and potassium concentrations in the top 20 cm soil were measured before the start of the experiment in 1993. Phosphorus weight concentrations ranged between 0.12 % and 0.6 % and potassium between 1.8 % and 4.7 % (Table 8.1).

Monthly average air temperature at 2 m above ground and monthly sum of precipitation at 1.5 m above ground were calculated from daily values mea-

Table 8.1 Chemical and physical properties of topsoil (0–20 cm) in the six areas of the Swiss FACE in Eschikon near Zurich (from Lüscher et al. 1998). pH in water, P and K in CO_2-saturated water extract

[CO_2] treatment (ppm)	Block	pH	P (mg kg⁻¹ soil)	K (mg kg⁻¹ soil)	Clay (%)	Silt (%)	Sand (%)	Organic material (%)
360	I	7.1	6.0	35.0	29.7	31.5	34.0	4.8
600	I	7.6	5.3	46.7	26.8	28.5	40.8	3.9
360	II	7.6	5.1	27.5	27.8	33.2	34.6	4.4
600	II	7.5	3.6	29.2	33.2	30.4	31.3	5.1
360	III	6.9	1.3	17.9	23.7	36.3	37.2	2.8
600	III	6.5	2.9	25.8	24.9	34.5	37.7	2.9

sured at the Institute's meteorological station, at approximately 300 m distance from the FACE experiment (see Schneider et al. 2004).

8.3 Experimental Treatments

CO$_2$ enrichment started at the end of May 1993. In each year until 2002, elevated CO$_2$ concentration (e[CO$_2$]) was controlled to 600 ppm during daylight hours in the FACE rings and was at current CO$_2$ concentration (c[CO$_2$]) in the control rings. No CO$_2$ was released during winter, i.e. before 20 March and after 20 November in each year. [CO$_2$] was measured approximately 25 cm above ground in the centre of each ring. In spring and autumn, CO$_2$ release was switched off automatically when the air temperature at 2 m was below 5 °C. Current [CO$_2$] during the day varied largely, depending on the season, wind speed and solar irradiance. At sunrise, the c[CO$_2$] dropped rapidly from the much higher values of 450–1000 ppm observed during the night to 340–380 ppm at about noon. With full sunlight at noon, c[CO$_2$] varied between approximately 340 ppm in summer and 360 ppm in spring and autumn. The CO$_2$ released was of fossil origin with depleted ^{13}C concentration. This resulted in a measurable ^{13}C signature in the plant material and after about 1 year also in the soil.

In August 1992, in each ring, four plots of 5.3 m^2 each were sown with *Lolium perenne* cv. Bastion and four plots with *Trifolium repens* cv. Milkanova. From spring 1993 onwards, the four plots were managed differently, thus constituting four treatments: two cutting frequencies (four and eight cuts per year) combined with two N fertilizer supplies (14 g m^{-2} year^{-1} and 56 g m^{-2} year^{-1}, respectively; 10 g m^{-2} year^{-1} and 42 g m^{-2} year^{-1} only in 1993). After 1995, all plots were cut five times per year. Unsown species were removed manually throughout the experiment. Further experimental details are given by Zanetti et al. (1996) and Schneider et al. (2004).

Each year, all plots received 5.5 g m^{-2} phosphorus (P) and 24.1 g m^{-2} potassium (K). In the first 3 years, the plots in block three received 35 % increased amounts of P and K to compensate for a lower initial availability of these nutrients (Table 8.1). These amounts are considered non-limiting for plant growth under the experimental conditions. The annual amounts of N fertilizer were split into portions according to the expected yield of the respective regrowth period. These portions were applied at the beginning of each regrowth period. The 1-m^2 sampling area in each plot was fertilized with ^{15}N-enriched NH$_4$NO$_3$ (Zanetti et al. 1996).

The harvesting area was 1 m^2 surrounded by a 40-cm border area and was not disturbed by soil sampling. The remaining area of the plots was used for soil sampling and other, partly destructive, studies, but received the same cutting and fertilizer treatments as the harvesting area. During the first 3 years of the experiment, the intervals between the harvests were constant, 4 weeks and 8 weeks, in the eight and four cut treatments, respectively. After 1995, the intervals between the five cuts varied between 5 weeks and 8 weeks, corresponding to the different growth rates during the season. Cutting height was approximately 5 cm above ground level.

8.4 Nutrient Availability: A Key Factor for the Plant's Response to e[CO$_2$]

8.4.1 Above-Ground Yield

Under e[CO$_2$], an increased dry matter production of about 30 % (Newton 1991; Suter et al. 2002) is observed when plants grow as individuals in controlled conditions at ample nutrient supply. However, when plants are grown in communities in the field, competition (see Chapter 19) and limiting growth resources other than CO$_2$ can restrict the plant's response to e[CO$_2$]. Under e[CO$_2$], there was only an 8 % average increase in the yield of *L. perenne* swards in the Swiss FACE, varying in a range from –11 % to +32 %, depending

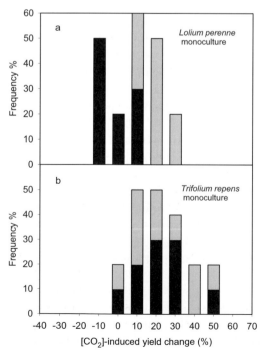

Fig. 8.1 Frequency distribution (*n*=10; 10 years) of the effect of e[CO$_2$] on annual mean harvested biomass of: (**a**) *L. perenne* and (**b**) *T. repens* grown in monoculture. *Black bars* Low N, *grey bars* high N

on the year and N fertilization treatment (Fig. 8.1). This weak yield response is in line with other field experiments. For the New Zealand FACE, a yield increase of 8% under e[CO$_2$] is reported for the whole sward (see Chapter 9) and the results from the grassland ecosystems of the 'Global change and terrestrial ecosystems (GCTE) pastures and rangelands core research project network' showed a stimulatory effect of e[CO$_2$] on yield of +17% (Campbell et al. 2000).

In the Swiss FACE experiment, the sward's yield response strongly depended on the availability of N. The *L. perenne* yield response to e[CO$_2$] was –2% (range –11% to + 10%) under the low N fertilization rate of 14 g N m^{-2} year^{-1}, but it was +17% (range +7 to +32%) under the high N fertilization rate of 56 g N m^{-2} year^{-1} (Fig. 8.1). The Swiss FACE was a highly fertile system on an eutric cambisol with high availability of P and K in the soil (Table 8.1) and with P and K fertilization appropriate for highly productive grassland. Thus, N availability was the main growth limiting factor for the highly productive grass *L. perenne*, which produced annual yields of up to 1800 g m^{-2} year^{-1} (Schneider et al. 2004; Chapter 19). The experiment of Daepp et al. (2001) impressively demonstrated the dominant role of N availability as the primary growth limiting factor in the Swiss FACE system. Even an increase in N fertilization from high (56 g N m^{-2} year^{-1}) to excessive (112 g m^{-2} year^{-1}) further increased the yield at c[CO$_2$] and, more importantly, the yield response to e[CO$_2$] of *L. perenne*.

Depending on the system, not only N but also other nutrients may act as the main limiting factor restricting growth and yield response to e[CO$_2$]. The effect of e[CO$_2$] on leaf lamina mass of *T. repens* strongly depended on the concentration of P in the nutrient solution in a growth room experiment (Almeida et al. 1999, 2000). The effect of e[CO$_2$] varied from –14% at the lowest, severely limiting P supply to +50% at the highest, non-limiting P supply. A similar P-dependent response to e[CO$_2$] was also observed in calcareous grassland (Stöcklin et al. 1998). In the non-limiting P treatment, the total concentration of P in *T. repens* leaves ranged from 4.4 mg g^{-1} to 6.6 mg g^{-1} DM (Almeida et al. 1999, 2000). In the Swiss FACE experiment, the average concentration of P in the herbage of *T. repens* was 4.3±0.3 mg g^{-1} DM, demonstrating that P nutrition was indeed non-limiting.

Significant differences in the yield response to e[CO$_2$] were observed between functional types of plant species. While it ranged from –11% to +32% for pure *L. perenne* swards, it was between +10% and +49% for pure *T. repens* swards (Hebeisen et al. 1997; Zanetti et al. 1996; 1997; Fig. 8.1). These differences were confirmed for other legume and grass species in the Swiss FACE experiment (Lüscher et al. 1998, 2000) and have to be seen in the light of resource availability. While the grasses depend solely on the strongly limited source of mineral N from the soil, the legumes have access to the unlimited N source of the atmosphere. The stronger yield increase under e[CO$_2$] of legumes as compared to grasses seems to be a general phenomenon. It was found also in the New Zealand FACE (see Chapter 9), a grazed system with no fertilizer N application. In a wide range of field experiments, the advantage of the legumes under $e[CO_2]$ led to an increased proportion of legumes in the sward (Lüscher et al. 2005; Chapter 19).

8.4.2 Resource Acquisition and Resource Allocation

The yield responses to $e[CO_2]$ of nutrient limited and non-limited systems as described above are the result of resource acquisition in and resource allocation between the root and shoot zones. Results of these two processes are discussed here in relation to the effects of nutrient limitation.

The effects of $e[CO_2]$ on photosynthesis are well known. Long et al. (2004) found across many FACE experiments an average increase in leaf photosynthesis at $e[CO_2]$ of more than 30 %. In the Swiss FACE experiment, light-saturated leaf photosynthesis at $e[CO_2]$ was increased by 37 % in *T. repens* and by 43 % in *L. perenne*, resulting in an increase of the daily integral of leaf CO_2 uptake by 36 % in the latter (Ainsworth et al. 2003a, b). In *L. perenne*, the C uptake was increased at $e[CO_2]$ by 44 % in the first 20 days of the regrowth, independent of the N treatment, compared to only 23 % in the later part of regrowth. This reduction in the $e[CO_2]$-induced increase in leaf photosynthesis was stronger in low N swards than in high N swards (Rogers et al. 1998). The increased leaf photosynthesis resulted in increased net ecosystem CO_2 uptake (Aeschlimann et al. 2005). Net ecosystem CO_2 uptake during midday was increased at $e[CO_2]$ by up to 32 % in both species and N treatments, depending on canopy size and radiation. It is crucial to realize that there were no major differences in the $e[CO_2]$-induced effects on net C-uptake between the non-N-limited system (*T. repens*) and the different levels of N limitation in the *L. perenne* swards (high N and low N fertilization).

An experiment with N_2-fixing and non-fixing Lucerne (*M. sativa*) showed that in contrast to C uptake, the uptake of mineral N from the soil did not increase under $e[CO_2]$ (Lüscher et al. 2000). All the additionally harvested N under $e[CO_2]$ derived solely from increased activity of symbiotic N_2 fixation. This is in line with the results of *T. repens* in the Swiss FACE experiment (Zanetti et al. 1996, 1997) and from an experiment conducted with microswards (Soussana and Hartwig 1996). Under $e[CO_2]$, *L. perenne* showed a significant reduction in the shoot N concentration. Even though at $e[CO_2]$ the relative N requirement for the maximal growth of *L. perenne* was reduced, an analysis of the reduced N concentration revealed that the restricted availability of mineral N in the soil was a major factor (Soussana et al. 1996; Zanetti et al. 1997) for the limitation of growth and of yield response to $e[CO_2]$. The ratio between the measured leaf N concentration and the corresponding N concentration enabling maximum growth is referred to as the 'N nutrition index' (NNI) for a given sward and does not depend on yield parameters (Lemaire et al. 1989). The NNI of frequently defoliated *L. perenne* monocultures was reduced by 36 % in the low N treatment and by 17 % in the high N treatment during the first 3 years of the Swiss FACE experiment (Zanetti et al. 1997). This occurred even when the lower critical N concentration for the e[CO$_2$] conditions was used for the calculations (Soussana et al. 1996). Thus, this decline cannot be explained by the lower N requirement under e[CO$_2$]; it is the result of the limited availability of mineral N in the soil.

These strong differences occurring in *L. perenne* swards in the response to e[CO$_2$] between C uptake in the shoot zone and N uptake from the soil led to a changed C/N balance, i.e. C supply was increased at e[CO$_2$] whilst N supply was not. This strongly affected resource allocation within the plant. To counteract the C/N imbalance, the plants invested more N in the root system (Stulen and Den Hertog 1993; Luo et al. 1994), as evident from the relationship between the concentration of N in the above-ground plant material and the proportion in the root fraction (Daepp et al. 2001). This relationship was similar to that found in a short-term experiment (Schenk et al. 1995) and was unaffected by e[CO$_2$]. Therefore, the greater investment of assimilates in the root (Jongen et al. 1995; Hebeisen et al. 1997) was not a result of [CO$_2$] per se but a response to a limited supply of N, and was not observed when *L. perenne* was grown on nutrient solution (Suter et al. 2002) or in the FACE under excessive N fertilization (112 g N m^{-2} year^{-1}) (Daepp et al. 2001). Besides the proportion of roots also the proportion of stubble increased in N-limited *L. perenne* under e[CO$_2$] (Suter et al. 2001; Daepp et al. 2001). In conclusion, the lack of yield response to e[CO$_2$] of N limited *L. perenne* swards despite a strong increase in C assimilation is due to an increased biomass allocation to non-harvested plant parts.

8.5 Changes over 10 Years in the e[CO$_2$] Response of Pure *L. perenne* Swards

The following section mostly considers the temporal dynamics of the relative changes of parameters by e[CO$_2$] compared to c[CO$_2$] (Figs. 8.2b, 8.3b) in order to distinguish effective [CO$_2$] effects from the dynamics over time occurring independently of [CO$_2$] (Figs. 8.2a, 8.3a).

Over the 10 years of fumigation, the most important change was the increasing response to e[CO$_2$] of the annual N yield (from –13 % to +29 %; Fig. 8.2) and DM yield (from +7 % to +32 %) of *L. perenne* monocultures in the high N treatment but not in the low N treatment (Fig. 8.2; Daepp et al. 2000; Schneider et al. 2004). These results demonstrate that the immediate response of an ecosystem to a step increase in [CO$_2$] at the start of the experiment may not represent an appropriate base to predict the response of the ecosystem to the ongoing slow increase in [CO$_2$] in the atmosphere.

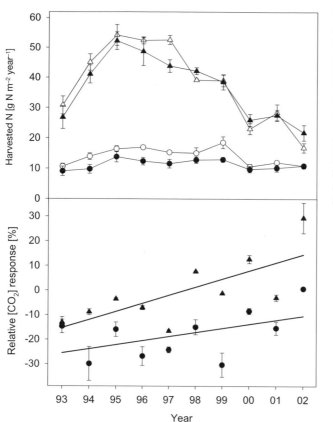

Fig. 8.2 (a) Annual amounts of harvested N in *L. perenne* at current and elevated [CO$_2$] in combination with two levels of N supply (○, △: ~360 ppm; ●, ▲: 600 ppm; ○, ● 14 g m^{-2} year^{-1}; △, ▲: 56 g m^{-2} year^{-1}). **(b)** Relative e[CO$_2$] response for the data shown in **a** (from Schneider et al. 2004)

In the third, fourth, sixth, seventh, eighth and tenth years of fumigation, DM yield in *L. perenne* monocultures in the high N treatment was stimulated by e[CO$_2$] to the same extent as in *T. repens* in the first two years of the same experiment (Hebeisen et al. 1997). This high yield response of *L. perenne* to e[CO$_2$] may indicate that, in the high N treatment, the response to e[CO$_2$] was not as limited by N as it was during the first 2 years. This hypothesis is clearly supported by the unchanged weak response of *L. perenne* to e[CO$_2$] in the low N treatment, where the amount of plant available N severely limited the growth of *L. perenne* over the whole duration of the experiment, even under c[CO$_2$]. This is obvious from the low yields of DM and N, and the low concentration of N (Fig. 8.2; Schneider et al. 2004).

The temporal dynamics of specific leaf area (SLA) confirms that N limitation to plant growth decreased with time in the high N treatment. The SLA of *L. perenne* was reduced under e[CO$_2$] in the beginning of CO$_2$ enrichment (Daepp et al. 2000), indicating that the size of the C sink limited the utilization of the additionally fixed C at e[CO$_2$], due to a lack of available mineral N (Fis-

Fig. 8.3 (a) Proportion of harvested N derived from soil organic matter (SOM) in *L. perenne* at current and elevated CO_2 (O, △: ~360 ppm; ●, ▲: 600 ppm) and at two levels of N supply (O, ●: 14 g m^{-2} year^{-1}; △, ▲: 56 g m^{-2} year^{-1}). **(b)** Relative e[CO₂] response. Data is derived from source separation in biomass N using ^{15}N in labeled fertilizer and SOM (from Schneider et al. 2004)

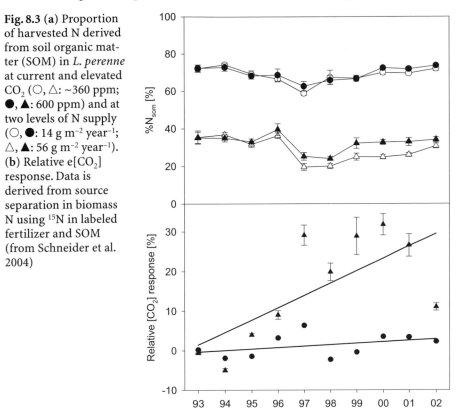

cher et al. 1997; Rogers et al. 1998; Isopp et al. 2000). From the third year on, however, elevated [CO₂] did not cause a decrease in SLA in the high N treatment. An analysis of several hundred measurements found no decline of the photosynthetic response to e[CO₂] over the 10 years of the FACE experiment (Ainsworth et al. 2003a; see Chapter 14), contradicting previous speculations that the increase of photosynthesis was a transient response. Thus, the lack of the e[CO₂] effect on SLA from year 3 on has to be related to an increased sink for assimilates and not to a reduced source activity under e[CO₂]. In contrast, at low N, the SLA remained low in both [CO₂] treatments and was reduced by e[CO₂] during the entire experimental period. Furthermore, the reducing effect of e[CO₂] on N concentration tended to diminish over the 10 years, suggesting an increased availability of N to the plants. The strong response of *L. perenne* swards to e[CO₂] was not related to the age of the plants, the cutting frequency or climatic conditions (Daepp et al. 2000). Thus, it is evident that the key factor for the increased responsiveness of *L. perenne* to e[CO₂] is the decreasing N limitation of growth.

Independent of [CO_2], the absolute amounts of biomass and N yield increased in the first 3 years and decreased after year 5 (Fig. 8.2). Until year 7, the biomass and N yields were equal or above those of year 2. The declining yield towards the end of the experiment was attributed to a thinning of the sward, probably due to mutual shading of tillers and frost damage (Schneider et al. 2004). Despite being a perennial species and despite all management efforts, *L. perenne* was not able to form a stable, productive monoculture after year 7. The lower yields towards the end of the experiment may have further reduced N limitation of *L. perenne*, due to a reduced N demand, which may additionally have favored a strong response to e[CO_2]. Nevertheless, the results suggest that the main factor to overcome N limitation was an increased N availability in the soil, which allowed already a high response to e[CO_2] in years 3, 4, 6 and 7, when yields were very high.

8.6 N Availability in Soil

The question remains which processes were able to affect under e[CO_2] the availability of mineral N in the soil over the years. It is generally assumed that e[CO_2] does not directly affect processes or organisms in the soil but these can be changed by plants through biomass inputs and altered litter quality, i.e. different (mostly increased) C/N ratio or lignin/N ratio (Norby et al. 2001). These changes in the quantity and quality of organic matter input to the soil affect soil organisms and their diversity (see Chapters 22, 23). However, biomass inputs and their effects are methodologically difficult to simulate in experiments, since we do not know the exact amounts and quality of these inputs (e.g. at which stage a root is dead and being decomposed). Also, the biomass used in such experiments usually differs in quality from the litter which enters the soil in the field. These difficulties and the temporal fluctuations of fast-turning pools explain why, in the Swiss FACE experiment in some studies, microbial biomass had increased (e.g. Sowerby et al. 2000) but activity had not changed, whereas other studies did not report changes in microbial biomass (e.g. De Graaff et al. 2004) but instead found that microbial activity had increased in some cases. Moreover, there may be complex species effects on the composition of the microbial community. Drissner et al. (2006) identified classes of bacteria, protozoa and fungi which increased and other bacteria which decreased at e[CO_2] in the soil under *T. repens*. Rezácová et al. (2005) showed that the communities of soil saprophytic micro-fungi, organisms important for the degradation of stable humic substances, differed significantly under *L. perenne* and *T. repens*. Due to the limited mechanistic knowledge available today, the effects on N availability of all these e[CO_2]-induced changes in biomass inputs, biomass quality, micro-organism community structure and activity cannot be quantified.

However, the plant itself may provide valuable long-term information on the sources from which it derived its N. This comes about because the N fertilizer used in the FACE experiment throughout the years was labeled with ^{15}N, allowing determination of the proportion of labeled N in the plant originating from applied fertilizer and the unlabeled N derived from "old" (i.e. generated before the experiment started) soil organic matter (SOM). Over 10 years, plants at high N derived increasingly more N from unlabeled SOM under e[CO$_2$] than c[CO$_2$] (Fig. 8.3). Transplantation experiments were carried out to examine the response of L. perenne to e[CO$_2$] when grown on e[CO$_2$]-adapted soil and non-adapted soil. These experiments clearly showed that the increased N availability on e[CO$_2$]-adapted soil was a true effect triggered by high N fertilization (Schneider et al. 2004). In a decomposition study using intact soil cores, Sowerby et al. (2005) observed that at e[CO$_2$], despite an increased ^{15}N release from litter, not more ^{15}N was found in the growing plants than at c[CO$_2$]. This supports the observation that plants at e[CO$_2$] depend to a larger extent on N mineralized from unlabeled SOM which diluted the ^{15}N released from the labeled litter. The decomposition of unlabeled, old SOM and hence an increased N mineralization (Ebersberger et al. 2003) may be one of these sources and contribute to the increased N availability at e[CO$_2$]. However, with a mobilized amount of around 5 g m^{-2} year^{-1}, the effects are small relative to the applied amounts of fertilizer; and other factors such as temporal and spatial changes in availability and in plant acquisition have to be considered.

The complex of changes observed in the highly N-fertilized swards of L. perenne may be summarized as decreasing N limitation in reference to the concept of progressive N limitation in natural systems (Luo et al. 2004). One principal trigger for either phenomenon to occur may be the presence of a significant external input of N. In the case of decreasing N limitation, the input of N in the form of mineral fertilizer led to the observations described above. The higher C assimilation under e[CO$_2$] was met by high N inputs to the soil and resulted in an increased mobilization of N from SOM which, consequently, enabled a stronger growth response to e[CO$_2$]. In contrast, when N inputs were small, the stimulated C sequestration into plant biomass and SOM at e[CO$_2$] may have increasingly limited available N in the soil of natural ecosystems (Luo et al. 2004).

A further indication of decreasing N limitation can be found in altered C fluxes at e[CO$_2$]. Since C and N fluxes in soils are tightly coupled (Van Groenigen et al. 2003), an increased mobilization of N from old SOM may result in an increased release of C from this pool and thus counteract C sequestration. Indeed, comparing the evolution and the ^{13}C signature of the CO$_2$ emission from soils with and without roots, Xie et al. (2005) concluded that higher new C input at e[CO$_2$] stimulated decomposition of older SOM. They suggested that the stimulation of the decomposition of old carbon was driven by the quantity of biomass inputs in L. perenne and by the quality of old SOM in T.

repens which may affect the potential of C sequestration. Also, Glaser et al. (2006) found that in *L. perenne* 100 % of the aminosugars, a residue of microbial decomposition, consisted of new C after 7 years, indicating that the old C in this component of SOM had been completely replaced in a relatively short time.

8.7 C and N Sequestration

Assimilation at the leaf level (Ainsworth et al. 2003a) and at the ecosystem level (Aeschlimann et al. 2005) at e[CO_2] were increased in *L. perenne* and e[CO_2] led to preferential allocation to residual above- and below-ground biomass (Daepp et al. 2001; Suter et al. 2002; Schneider et al. 2006). These e[CO_2]-induced changes could potentially increase C sequestration.

In fact, the net ecosystem C input in swards of *L. perenne* tended to be higher at e[CO_2] compared to c[CO_2], but this difference was not statistically significant (Aeschlimann et al. 2005; Xie et al. 2005). Xie et al. (2005) showed that the C input at e[CO_2] was mainly into the less protected soil particles >53 µm in size. The increased net CO_2 uptake during day-time at e[CO_2] was mostly compensated for by a higher night-time respiration (Aeschlimann et al. 2005). As a consequence, the relatively small effect of e[CO_2] on the net ecosystem C input did not result in a significant increase in total soil C (Van Groenigen et al. 2002). A small (3.3 %) increase in total soil C at e[CO_2] is a consistent trend among a wide range of terrestrial ecosystems and plant species (see Chapter 21) but, due to a large variability of soil C, significant differences are very hard to detect in individual studies.

Since there is a strong indication that sequestration of C and fertilizer N are strongly linked, it is not surprising that e[CO_2] did not affect the total soil N in the Swiss FACE experiment (Van Groenigen et al. 2003). Net ecosystem C input in the Swiss FACE experiment was markedly higher at low N than at high N supply (Aeschlimann et al. 2005; Xie et al. 2005). This indicates that, at least in the last years of the experiment, more C was sequestered at low than at high N supply, due to decreasing N limitation at the high N treatment (see Section 8.5).

The ratio of residual to harvestable biomass (Daepp et al. 2001) was higher at low N than at high N supply, leading to smaller C exports by harvest at low N supply. In addition, a large proportion of the non-structural carbohydrates which accumulated in plant leaves at low N supply were exported to stubble and roots (Suter et al. 2002). Therefore, decomposition of the residual plant biomass constitutes a C flux to the soil and a potential for C sequestration which is especially great at low N supply (Schneider et al. 2006).

However, these results do not imply that natural, unfertilized ecosystems represent a greater C sink than N-rich agricultural grasslands. On the one hand, a severe N limitation in natural grassland may inhibit any [CO$_2$] response of assimilation and consequently of the C balance (see Chapter 21; Luo et al. 2004). On the other hand, excessive N supply may increase C exports disproportionately, by promoting both shoot growth and respiration, resulting in a reduced net ecosystem C input compared to moderate N supply. The net ecosystem C input in the Swiss FACE experiment was higher in *L. perenne* than in *T. repens* (Aeschlimann et al. 2005; Xie et al. 2005). This may be due to the higher root biomass of *L. perenne*, which builds the potential for a C flux to the soil. Additionally, C losses by night-time ecosystem respiration were smaller in *L. perenne* than in *T. repens* (Aeschlimann et al. 2005). The species-dependency of C sequestration may not be a general finding (see Chapter 21) but is restricted to species which differ in their way of nitrogen acquisition, such as legumes and grasses. However, as a consequence of the species-specific C balance in the Swiss FACE experiment, CO$_2$ and N-related changes in the species composition of grasslands (Hebeisen, Lüscher, & Nösberger 1997; Lüscher, Hendrey, & Nösberger 1998; Navas et al. 1999) are likely to affect the net ecosystem C input. For the Swiss FACE experiment in the growing season 2001, a net C input of 210–631 g C m^{-2} was calculated (Aeschlimann et al. 2005). This indicates that, according to the history of a grassland soil, the annual net C input may be considerable. This finding is in line with modeling studies that suggest the effects of land-use change on C sequestration are larger than the effects of [CO$_2$] (Schimel et al. 2000). After a change in land use from arable crops to grassland, such as the Swiss FACE site has experienced, the soil C content can increase for a period of 50–70 years (Sauerbeck 2001). Because the land-use change in our system was relatively recent (in 1993), the soil C was most probably not yet in equilibrium, which enabled the observed net C input. However, the net C input calculated on a growing season basis may overestimate the annual net C input, because respiration during winter is neglected. In a calcareous grassland, respiration rates during winter were about 150 g C m^{-2} (Volk and Niklaus 2002).

The results from the Swiss FACE experiment clearly show that the sequestration of C to grassland is importantly affected by the nutrient status (especially N availability), the plant species composition and the history of land use.

8.8 Conclusions

The following summarizes the information about the long-term response of *L. perenne* and *T. repens* model ecosystems to e[CO$_2$] and N fertilization in the Swiss FACE experiment.

- The average 8 % yield increase of *L. perenne* at e[CO$_2$] under field conditions was clearly weaker than expected from short-term laboratory experiments.

- Photosynthesis of *L. perenne* was stimulated by 36 % under e[CO_2]. The weak yield response of *L. perenne* to e[CO_2] was due to increased allocation (up to 108 % increase) of biomass to non-harvested plant parts, and due to increased night-time respiration.
- Under low N fertilization, the limited availability of mineral N in the soil strongly limited the yield response of *L. perenne* to e[CO_2] to an average of –2 % (range –11 % to +10 %). Consequently, unlimited access to atmospheric N_2 (symbiotic N_2 fixation) played a key role in the strong stimulation of yield production of *T. repens* by e[CO_2] (10–49 %).
- The yield response of *L. perenne* to e[CO_2] increased from 7 % to 32 % over the years under high N fertilization. This increase was probably due to a decreased N limitation of plant growth.
- Changes in micro-organism community structure and activity as well as increased mineralization of N from SOM under e[CO_2] combined with high N fertilization indicate that processes in the soil adapted to the new environmental conditions. These processes led to decreasing N limitation at high N fertilization and the increased availability of mineral N stimulated the response of plant growth to e[CO_2] in the long term.
- The sequestration of C was more importantly affected by N supply, species composition and land-use change than by [CO_2], resulting in a greater net ecosystem C input at low N than at high N fertilization. At high N fertilization, decreasing N limitation reduced the potential for C sequestration into the ecosystem.

These results demonstrate that, for a realistic prediction of the effects of e[CO_2], long-term experiments under real field conditions are needed, where the availability of growth resources is varied and where processes of competition and feed-back mechanisms are active.

Acknowledgements. Our studies were supported by the Swiss National Energy Fund, ETH Zurich, Swiss Department of Science and Education (COST), Federal Office of Agriculture and Federal Office of Energy.

References

Aeschlimann U, Nösberger J, Edwards PJ, Schneider MK, Richter M, Blum H (2005) Responses of net ecosystem CO_2 exchange in managed grassland to long-term CO_2 enrichment, N fertilization and plant species. Plant Cell Environ 28:823–833

Ainsworth EA, Davey PA, Hymus GJ, Osborne CP, Rogers A, Blum H, Nösberger J, Long SP (2003a) Is stimulation of leaf photosynthesis by elevated carbon dioxide concentration maintained in the long term? A test with *Lolium perenne* grown for ten years at two nitrogen fertilization levels under free air CO_2 enrichment (FACE). Plant Cell Environ 26:705–714

Ainsworth EA, Rogers A, Blum H, Nösberger J, Long SP (2003b) Variation in acclimation of photosynthesis in *Trifolium repens* after eight years of exposure to free air CO_2 enrichment (FACE). J Exp Bot 54:2769–2774

Almeida JPF, Lüscher A, Frehner M, Oberson A, Nösberger J (1999) Partitioning of P and the activity of root acid phosphatase in white clover (*Trifolium repens* L.) are modified by increased atmospheric CO_2 and P fertilization. Plant Soil 210:159–166

Almeida JPF, Hartwig UA, Frehner M, Nösberger J, Lüscher A (2000) Evidence that P deficiency induces N feedback regulation of symbiotic N_2 fixation in white clover (*Trifolium repens* L.). J Exp Bot 51:1289–1297

Campbell BD, Stafford Smith DM, Ash AJ, Fuhrer J, Gifford RM, Hiernaux P, Howden SM, Jones MB, Ludwig JA, Manderscheid R, Morgan JA, Newton PCD, Nösberger J, Owensby CE, Soussana JF, Tuba Z, ZuoZhong C (2000) A synthesis of recent global change research on pasture and rangeland production: reduced uncertainties and their management implications. Agric Ecosyst Environ 82:39–55

Daepp M, Suter D, Almeida JPF, Isopp H, Hartwig UA, Frehner M, Blum H, Nösberger J, Lüscher A (2000) Yield response of *Lolium perenne* swards to free air CO_2 enrichment increased over six years in a high-N-input system on fertile soil. Global Change Biol 6:805–816

Daepp M, Nösberger J, Lüscher A (2001) Nitrogen fertilization and developmental stage alter the response of *Lolium perenne* to elevated CO_2. New Phytol 150:347–358

De Graaff MA, Six J, Harris D, Blum H, van Kessel C (2004) Decomposition of soil and plant carbon from pasture systems after 9 years of exposure to elevated CO_2: impact on C cycling and modeling. Global Change Biol 10:1922–1935

Drissner D, Blum H, Kandeler E (2006) Nine years of elevated CO_2 shifts function and structural diversity of soil microorganisms in a grassland. Eur J Soil Sci (in press)

Ebersberger D, Niklaus PA, Kandeler E (2003) Long term CO_2 enrichment stimulates N-mineralisation and enzyme activities in calcareous grassland. Soil Biol Biochem 35:965–972

Fischer BU, Frehner M, Hebeisen T, Zanetti S, Stadelmann F, Lüscher A, Hartwig UA, Hendrey GR, Blum H, Nösberger J (1997) Source-sink relations in *Lolium perenne* L. as reflected by carbohydrate concentrations in leaves and pseudo-stems during regrowth in a free air carbon dioxide enrichment (FACE) experiment. Plant Cell Envir 20:945–952

Glaser B, Millar N, Blum H, Zech W (2006) Sequestration and turnover of microbial-derived carbon into a temperate grassland soil under elevated atmospheric pCO_2. Global Change Biol (submitted)

Hebeisen T, Lüscher A, Zanetti S, Fischer BU, Hartwig UA, Frehner M, Hendrey GR, Blum H, Nösberger J (1997) Growth response of *Trifolium repens* L. and *Lolium perenne* L. as monocultures and bi-species mixture to free air CO_2 enrichment and management. Global Change Biol 3:149–160

Isopp H, Frehner M, Almeida JPF, Blum H, Daepp M, Hartwig UA, Lüscher A, Suter D, Nösberger J (2000) Nitrogen plays a major role in leaves when source-sink relations change: C and N metabolism in *Lolium perenne* growing under free air CO_2 enrichment. Aust J Plant Physiol 27:851–858

Jongen M, Jones MB, Hebeisen T, Blum H, Hendrey G (1995) The effects of elevated CO_2 concentrations on the root-growth of *Lolium perenne* and *Trifolium repens* grown in a FACE system. Global Change Biol 1:361–371

Lemaire G, Gastal F, Salette J (1989) Analysis of the effect of N nutrition on dry matter yield and optimum N content. Proc Int Grassland Congr 16:179–180

Long SP, Ainsworth EA, Rogers A, Ort D (2004) Rising atmospheric carbon dioxide: plants FACE the future. Annu Rev Plant Biol 55:591–628

Luo Y, Field CB, Mooney HA (1994) Predicting responses of photosynthesis and root fraction to elevated [CO_2] (a) interactions among carbon, nitrogen, and growth. Plant Cell Environ 17:1195–1204

Luo Y, Su B, Currie WS, Dukes JS, Finzi A, Hartwig U, Hunate B, McMurtrie RE, Oren R, Parton WJ, Pataki DE, Shaw MR, Zak DR, Field CB (2004) Progressive N limitation of ecosystem responses to rising atmospheric carbon dioxide. Biosci 54: 731–739

Lüscher A, Hendrey GR, Nösberger J (1998) Long-term responsiveness to free air CO_2 enrichment of functional types, species and genotypes of plants from fertile permanent grassland. Oecologia 113:37–45

Lüscher A, Hartwig UA, Suter D, Nösberger J (2000) Direct evidence that symbiotic N_2 fixation in fertile grassland is an important trait for a strong response of plants to elevated atmospheric CO_2. Global Change Biol 6:655–662

Lüscher A, Daepp M, Blum H, Hartwig UA, Nösberger J (2004) Fertile temperate grassland under elevated atmospheric CO_2 – role of feed-back mechanisms and availability of growth resources. Eur J Agron 21:379–398

Lüscher A, Fuhrer J, Newton PCD (2005) Global atmospheric change and its effect on managed grassland systems. In: McGilloway DA (ed) Grassland: a global resource. Wageningen Academic, Wageningen, pp 251–264

Navas ML, Garnier E, Austin MP, Gifford RM (1999) Effect of competition on the responses of grasses and legumes to elevated atmospheric CO_2 along a nitrogen gradient: differences between isolated plants, monocultures and multi-species mixtures. New Phytol 143:323–331

Newton PCD (1991) Direct effects of increasing carbon dioxide on pasture plants and communities. NZ J Agric Res 34:1–24

Norby RJ, Cotrufo MF, Ineson P, O'Neill EG, Canadell JG (2001) Elevated CO_2, litter chemistry, and decomposition: a synthesis. Oecologia 127:153–165

Řezáčová V, Blum H, Hršelová H, Gamper H, Gryndler M (2005) Saprobic microfungi under *Lolium perenne* and *Trifolium repens* at different fertilization intensities and elevated atmospheric CO_2 concentration. Global Change Biol 11:224–230

Rogers A, Fischer BU, Bryant J, Frehner M, Blum H, Raines CA, Long SP (1998) Acclimation of photosynthesis to elevated CO_2 under low-nitrogen is affected by the capacity for assimilate utilization. Perennial ryegrass under free-air CO_2 enrichment. Plant Physiol 118:683–689

Sauerbeck DR (2001) CO_2 emissions and C sequestration by agriculture – perspectives and limitations. Nutrient Cycling Agroecosyst 60:253–266

Schenk U, Manderscheid R, Hugen J, Weigel HJ (1995) Effects of CO_2 enrichment and intraspecific competition on biomass partitioning, nitrogen-content and microbial biomass carbon in soil of perennial ryegrass and white clover. J Exp Bot 46:987–993

Schimel D, Melillo J, Tian HQ, Mcguire AD, Kicklighter D, Kittel T, Rosenbloom N, Running S, Thornton P, Ojima D, Parton W, Kelly R, Sykes M, Neilson R, Rizzo B (2000) Contribution of increasing CO_2 and climate to carbon storage by ecosystems in the United States. Science 287:2004–2006

Schneider MK, Lüscher A, Richter M, Aeschlimann U, Hartwig UA, Blum H, Frossard E, Nösberger J (2004) Ten years of free air CO_2 enrichment altered the mobilization of N from soil in *Lolium perenne* L. swards. Global Change Biol 10:377–388

Schneider MK, Lüscher A, Frossard E, Nösberger J (2006) An overlooked carbon source to grasdsland soils: Loss of structural C from stubble in response to pCO_2 and N supply. New Phytologist, in press

Soussana JF, Hartwig UA (1996) The effects of elevated CO_2 on symbiotic N_2 fixation: a link between the carbon and nitrogen cycles in grassland ecosystems. Plant Soil 187:321–332

Soussana JF, Casella E, Loiseau P (1996) Long-term effects of CO$_2$ enrichment and temperature increase on a temperate grass sward. II. Plant nitrogen budgets and root fraction. Plant Soil 182:101–114

Sowerby A, Blum H, Gray TRG, Ball AS (2000) The decomposition of Lolium perenne in soils exposed to elevated CO$_2$: comparisons of mass loss of litter with soil respiration and soil microbial biomass. Soil Biol Biochem 32:1359–1366

Sowerby A, Blum H, Ball AS (2005) Elevated atmospheric CO$_2$ affects the turnover of nitrogen in a European grassland. Appl Soil Ecol 28:37–46

Stöcklin J, Schweizer K, Körner C (1998) Effects of elevated CO$_2$ and phosphorus addition on productivity and community composition of intact monoliths from calcareous grassland. Oecologia 116:50–56

Stulen I, Den Hertog J (1993) Root growth and functioning under atmospheric CO$_2$ enrichment. Vegetatio 104:99–111

Suter D, Nösberger J, Lüscher A (2001) Response of perennial ryegrass to free-air CO$_2$ enrichment (FACE) is related to the dynamics of sward structure during regrowth. Crop Sci 41:810–817

Suter D, Frehner M, Fischer BU, Nösberger J, Lüscher A (2002) Elevated CO$_2$ increases carbon allocation to the roots of Lolium perenne under Free-Air CO$_2$ Enrichment but not in a controlled environment. New Phytol 154:65–75

Van Groenigen KJ, Six J, Harris D, Blum H, Van Kessel C (2003) Soil [13]C-[15]Ndynamics in an N$_2$-fixing clover system under long-term exposure to elevated atmospheric CO$_2$. Global Change Biol 9:1751–1762

Volk M, Niklaus PA (2002) Respiratory carbon loss of calcareous grasslands in winter shows no effects of 4 years CO$_2$ enrichment. Funct Ecol 16:162–166

Xie Z, Cadisch G, Edwards G, Baggs EM, Blum H (2005) Carbon dynamics in a temperate grassland soil after 9 years exposure to elevated CO$_2$ (Swiss FACE). Soil Biol Biochem 37:1387–1395

Zanetti S, Hartwig UA, Lüscher A, Hebeisen T, Frehner M, Fischer BU, Hendrey GR, Blum H, Nösberger J (1996) Stimulation of symbiotic N$_2$ fixation in Trifolium repens L under elevated atmospheric pCO$_2$ in a grassland ecosystem. Plant Physiol 112:575–583

Zanetti S, Hartwig UA, Van Kessel C, Lüscher A, Hebeisen T, Frehner M, Fischer BU, Hendrey GR, Blum H, Nösberger J (1997) Does nitrogen nutrition restrict the CO$_2$ response of fertile grassland lacking legumes? Oecologia 112:17–25

9 Impacts of Elevated CO$_2$ on a Grassland Grazed by Sheep: the New Zealand FACE Experiment

P.C.D. NEWTON, V. ALLARD, R.A. CARRAN, and M. LIEFFERING

9.1 Introduction

Grasslands cover 20 % of the Earth's land area and their response to global change is important because they contain 10 % of global carbon (C) stores and they sustain significant agricultural activity, ranging from high-input, intensive animal production systems to extensive nomadic pastoralism. Grasslands are arguably the most intensively investigated ecosystem in relation to global change, with experiments conducted on a range of grassland types and employing a variety of experimental approaches. These include open-top chamber experiments on tallgrass prairie, shortgrass steppe and calcareous grassland (see Morgan et al. 2004 for references) and free air carbon dioxide enrichment (FACE) experiments on native Australian grassland (Hovenden et al. 2005), temperate grassland in Switzerland (Chapter 8) and tropical savannah (A. Ash, personal communication). Although these systems would in practice be grazed by livestock, with the exception of a brief period of grazing by sheep in the tallgrass prairie open-top chamber experiment (Owensby et al. 1996), these experiments have not included animals and have simulated grazing by cutting. Because grazing exerts such a strong influence over the dynamics of plant and soil processes, there is a strong possibility of interactions between grazing and e[CO$_2$] (Newton et al. 2001). Consequently, a FACE experiment was established in New Zealand to study grassland responses to e[CO$_2$] under grazing.

Ecological Studies, Vol. 187
J. Nösberger, S.P. Long, R.J. Norby, M. Stitt,
G.R. Hendrey, H. Blum (Eds.)
Managed Ecosystems and CO$_2$
Case Studies, Processes, and Perspectives
© Springer-Verlag Berlin Heidelberg 2006

9.2 Site Description

The New Zealand FACE experiment is situated in the Rangitikei region of the North Island (40°14' S, 175°16' E, 9 m a.s.l.). The FACE rings are 12 m in diameter and CO_2 is delivered using pulse width modulation controlled by solenoid valves (Newton et al. 2001) in a variation of the BNL system (Lewin et al. 1994). Enrichment started in October 1997. The rings are in a 2.5-ha field and contained within permanent fenced areas approximately 25 × 25 m; the perimeter of the ring is fenced with electric wires to contain stock during grazing events. There are three enriched and three control rings that are blocked according to their initial botanical composition.

The soil at the site is a Pukepuke black sand (Mollic Psammaquent) with a 0.25-m black loamy fine-sand topsoil (Cowie and Hall 1965) that is hydrophobic (Newton et al. 2003). At the start of the experiment, soil tests showed soil $pH_{(water)}$= 5.8, exchangeable K = 0.15 cM(+) kg^{-1} soil, Olsen P = 20 $\mu g\ ml^{-1}$ soil and sulfate-S = 7 $\mu g\ ml^{-1}$ soil. The field has been in permanent grass (natural re-seeding only) for at least 40 years and contains 15–25 plant species depending on the season, including C3 and C4 grasses, legumes and forbs. Grazing has been by sheep, cattle and goats. Prior to the experiment, fertiliser additions were sporadic, but from the start of the experiment fertiliser was applied as shown in Table 9.1. Note that the fertiliser regime reflects a management system that relies on legumes to provide the nitrogen (N) input.

There are strong seasonal patterns in mean daily temperature and soil moisture (Fig. 9.1) with frequent soil moisture deficits in summer (December–February), some of which are severe, as seen in 2003.

Table 9.1 Fertiliser applied to e[CO_2] and c[CO_2] rings and to both the areas grazed and those under the exclusion chambers. Note that no N fertiliser was used

Year	Month	Element (g m^{-2})			
		P	K	S	Mg
1997	November	1.5	9.5	2.7	0.0
1998	August	1.5	9.5	2.8	0.0
1999	June	1.5	9.5	2.7	0.0
1999	September	1.5	9.5	2.7	0.0
2000	May	1.5	9.5	2.7	0.0
2001	December	0.0	8.0	3.8	0.5
2002	September	0.0	0.0	0.8	1.0
2002	November	0.0	4.0	1.7	0.0
2003	August	1.8	1.2	8.2	1.0
2003	October	0.0	1.2	1.3	1.0
2004	July	1.6	3.6	3.1	0.0

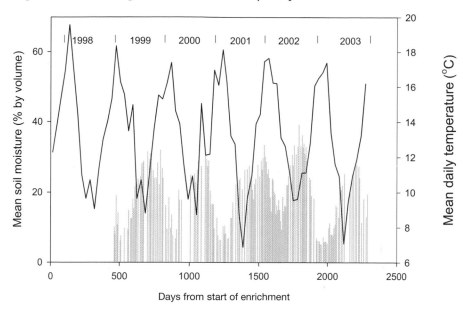

Fig. 9.1 Mean daily temperature at 2 m height (*solid line*) and volumetric soil moisture content to 150 mm depth measured by time domain reflectometry (method given by Newton et al. 2003). Soil moisture measurements started in late 1998

9.3 Experimental Treatments

The e[CO$_2$] rings are enriched with CO$_2$ during the photoperiod to a target concentration of 475 ppm. Enrichment continues all year, even during periods of extended drought. When the above ground biomass reaches 180–200 g m^{-2} dry weight all rings are grazed by sheep to a target residual of 50–70 g m^{-2}, equating to a residual height of approximately 2.5 cm. On occasions, the rings are grazed to remove rank growth, even though herbage mass is below the protocol level. Until April 2001, the rings were mob grazed, with animals having free access to all rings. Thereafter, animals were confined within individual rings for the duration of the grazing period, to ensure that nutrients were recycled by animals back to the site of their removal by grazing. Animal numbers are adjusted according to the herbage on offer, but an average grazing would involve 3–5 sheep ring^{-1} for a period of 4 days. Over 1997–2003, the rings were grazed 3, 5, 4, 7, 4, 8, 5 times year^{-1}, respectively. Two permanent areas (1.0 × 0.5 m) within each ring have exclusion cages to prevent grazing. These areas are cut and sampled with powered hand shears to 2 cm above ground level at the start of each grazing period. Cuts are also taken in the grazed areas pre- and post-grazing to calculate standing biomass, herbage growth rates and to provide samples for botanical dissection and chemical analyses.

9.4 Resource Acquisition

9.4.1 Photosynthesis

There was a small stimulation of net photosynthesis in some species under $e[CO_2]$ but evidence of strong downregulation in others. After 2 years and 3 years of enrichment, the C3 grass *Lolium perenne* had a higher CO_2 assimilation rate at $e[CO_2]$, while there was negligible stimulation of the legumes *Trifolium repens* and *T. subterraneum* (von Caemmerer et al. 2001). The relationship between N content and CO_2 assimilation rate was similar in $e[CO_2]$ and $c[CO_2]$ plants, indicating that the downregulated legumes in $e[CO_2]$ had lower N content (von Caemmerer et al. 2001). This has been a common finding in this experiment (Allard et al. 2003; Newton et al. 2001) and has occurred in other systems (e.g. Zanetti et al. 1996); but it is not an intuitive response, given the ability of legumes to control their N concentration. As the N content of these *Trifolium* species changes with age (particularly in *T. repens*), it may be that differences in N content are generated by a CO_2 effect on ontogeny (von Caemmerer et al. 2001).

Herbage growth in grazed pastures is the accumulation of a series of regrowth periods that follow defoliation. Figure 9.2 tracks the canopy gas exchange during a summer regrowth. There was a small increase in CO_2 fixed per unit ground area early in the regrowth period, but the major difference between the treatments was significantly higher assimilation by the $e[CO_2]$ canopy late in the regrowth. After 35 days, there was no difference in leaf area index (LAI) between treatments (3.42 $c[CO_2]$, 3.30 $e[CO_2]$), nor were LAIs at a value that would be likely to inhibit net canopy photosynthesis (Parsons et al. 1988). Consequently we conclude that the advantage in assimilation for the $e[CO_2]$ was not because $e[CO_2]$ enhanced growth in low light conditions, as shading increased during canopy development (Curtis and Wang 1998). More likely the difference was due to an enhanced capacity at $e[CO_2]$ to fix carbon when soil moisture was restricted. Soil moisture in this regrowth period declined rapidly (Fig. 9.2) and, using large turves of this same soil in a controlled growth room experiment, we observed that the $e[CO_2]$ canopies continued to fix C into a simulated drought, while assimilation stopped under $c[CO_2]$ (Newton et al. 1996). This result is consistent with a general view that interaction with soil moisture content is one of the most important drivers of grassland responses to $e[CO_2]$ (Morgan et al. 2004). Interestingly, despite a

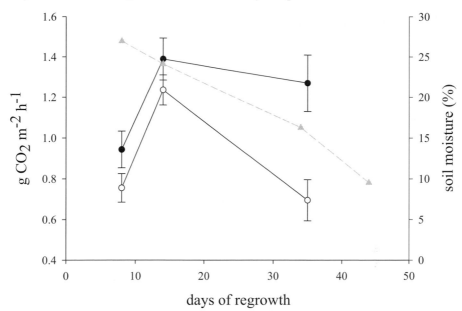

Fig. 9.2 CO$_2$ assimilation rate of canopies in e[CO$_2$] rings (*filled circles*) and c[CO$_2$] rings (*open circles*) over a regrowth period (summer) starting on 7 December 1999 (mean ± sem). The *dotted line* is the average volumetric soil moisture content of c[CO$_2$] and e[CO$_2$] rings. Assimilation was measured with an EGM-2 gas analyser with a CPY-2 canopy assimilation chamber 150 mm in diameter with a volume of 2500 cm^2 (both from PP Systems, Hitchin, UK)

significantly lower rate of CO$_2$ assimilation, there was more herbage harvested in the c[CO$_2$] than in the e[CO$_2$] treatment at the end of the regrowth (mean ± sem, c[CO$_2$] 389 ± 29.8 g m^{-2} cf. e[CO$_2$] 333 ± 25.8 g m^{-2}). This result is indicative of relatively greater allocation of C belowground and we consider this below.

9.4.2 Nutrients

One important pathway of nutrient recycling in grasslands is litter decomposition. The effects of e[CO$_2$] on litter decomposition were assessed at this site and, as expected from the literature (Norby and Cotrufo 1998), no direct effect of e[CO$_2$] on litter decomposition was observed at the plant species scale (Allard et al. 2004a). Nevertheless when CO$_2$ driven changes in botanical composition and in biomass allocation (i.e. increases in root growth) were taken into account, the decomposition rate of plant litter was increased by 15% under e[CO$_2$], while N release rate was increased by 18%. In grazed grassland, plant litter decomposition is only one route by which nutrients are recycled.

Depending on the herbage utilisation rate by the grazers, up to 50 % of the produced herbage is ingested by the animals (Parsons and Chapman 2000) and about 30 % of the ingested C is recycled as faeces, while most of the ingested N returns to the pasture in urine. When this route was taken into account, we found that organic matter decomposition rates were only about 10 % higher in the e[CO_2] treatment, largely because of slower decomposition of dung, particularly in dry conditions (Allard et al. 2004a). In addition, depending on the sink for N in the animals, sheep fed on e[CO_2] herbage either increased (Allard et al. 2003) or decreased (D. Hélary, unpublished data) the proportion of ingested N returned in their urine. Modifying the proportion of the dietary N that is recycled through urine or in the faeces can have a profound influence on the fate of this N as well as the subsequent feedback on the growing vegetation, in particular if a heterogeneous pattern of these fluxes is taken into account.

Nitrogen input, as opposed to recycling, in this system is from biologically fixed N. Analysis of isotopic signatures of the major legumes, *T. repens* and *T. subterraneum* and soil mineral N extracted by in situ ion exchange membranes (S. Bowatte and R.A. Carran, unpublished data) suggest that legume content alone could be used as a guide to relative N inputs. On this basis, the e[CO_2] rings received 2.4 times as much N relative to the c[CO_2] rings over the 1997–2003 period. However, the plant N recovered in the harvested aboveground herbage was not different between treatments; and, in total, the N harvested in non-legumes was 83 ± 25.8 g m^{-2} (mean $\pm$ sem) in c[CO_2] rings and 88 ± 18.6 g m^{-2} in e[CO_2] rings. Over time, direct measurements of plant available N using ion exchange membranes suggests a declining availability of N, particularly NO_3-N in the e[CO_2] treatment (Fig. 9.3). This result is indicative of 'progressive N depletion' (PNL; Luo et al. 2004), a syndrome that may constrain the ability of ecosystems to increase productivity or act as a sink for C. It must be also be recognised that in these grazed pasture a significant portion of yield is derived from N-rich, urine-affected areas and this confounds any average calculation of N inputs. Potassium (K), magnesium (Mg), sulfur (S) and phosphate (PO_4) are all nutrients that require management at this site. These, and N, are recycled through the excreta of the grazing sheep and become available to plants, but are aggregated into patches of excess supply within a larger area of older patches in which nutrient availability is declining toward a background level (Carran and Theobald 1999). This patchiness, which is characteristic of all grazed pastures, is made more complex because those nutrients recycled in faeces (PO_4, Mg) are generally separated from those in urine (Haynes and Williams 1993). Any response to e[CO_2] in plant growth or C sequestration will vary in this experiment as a consequence of patchy nutrient return.

Recent measurements, made after 7 years enrichment, suggest that N depletion is continuing and that PO_4 limitation is also occurring; and a repeated measures analysis of weekly ion exchange membrane sampling

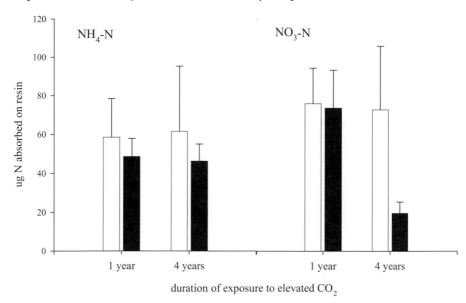

Fig. 9.3 Average weekly accumulation of NH$_4$-N and NO$_3$-N on ion exchange membrane strips over a period of 9 weeks following application of urine at a rate of 300 kg N ha^{-1} to pasture cores (300 mm diameter, 300 mm depth) previously exposed to e[CO$_2$] for 1 year or 4 years. The measurements were made at the same time, the difference in exposure time being generated by introduction of cores at different intervals. There was no difference between treatments or times for NH$_4$-N but the CO$_2$ × duration interaction for NO$_3$-N was $P=0.046$. Values are means and sem

taken from August 2003 until April 2004 showed significantly lower ($P=0.04$) PO$_4$-P absorbed on resins in the e[CO$_2$] rings (mean absorbed values were: c[CO$_2$] 21.9 µg PO$_4$-P week^{-1}, e[CO$_2$] 16.9 µg PO$_4$-P week^{-1}). This response would have profound effects on natural ecosystem responses to e[CO$_2$] (Stöcklin and Körner 1998). It is a less serious issue where management intervention is possible, but does imply that increased rates of fertiliser may be required to maintain the PO$_4$-dependent legume component at a desirable level.

9.4.3 Soil Moisture

Across the 7 years enrichment, there was no significant difference in volumetric soil moisture content to 15 cm between the CO$_2$ treatments (but see later for an example of short-term differences). While increased soil moisture content in CO$_2$ enriched plots has been measured in a number of field experiments with chambers (e.g. Nelson et al. 2004; Niklaus et al. 2003; Zavaleta et al.

2003), this is less frequently observed in FACE experiments [e.g. small grassland plots (Reich et al. 2001), sweetgum (*Liquidambar styraciflua* L.) plantation (Belote et al. 2003), desert (Nowak et al. 2004)]. However, we have observed subtle differences in soil moisture dynamics in our experiment as $e[CO_2]$ soils have developed reduced hydrophobicity compared to $c[CO_2]$ soils (Newton et al. 2003). The consequences of this change are not yet clear, but repellency is known to affect many important processes such as aggregation of soil particles and C sequestration, which have been shown to be sensitive to CO_2 enrichment (e.g. Gill et al. 2002; Niklaus et al. 2003; Rillig et al. 1999).

9.5 Resource Transformation

9.5.1 Aboveground Yield and Species Composition

Total herbage yield from 1997–2003 inclusive was 8.4 % greater in the $e[CO_2]$ treatment but was not statistically higher than $c[CO_2]$ (Fig. 9.4a). This was principally because the most abundant functional group (C3 grasses) did not respond to CO_2 (Fig. 9.4b). However, over the 7 years enrichment, the broadleaf species – legumes and forbs – had significantly greater growth. The principal forb species were *Hypochaeris radicata*, *Leontodon autumnalis* and *Rumex acetosella*; and the CO_2 stimulation for these species was variable between years but remained positive after the first few months of enrichment (Fig. 9.4 c). In contrast, the legumes – principally *T. repens*, *T. subterraneum* and *T. micranthum* – responded strongly initially but this response declined over time (Fig. 9.4 c). Legume content is of critical importance in pastures of this type as they contribute the major input of N and provide animal feed of high quality. There were no $CO_2 \times$ management effects for total yields or for the change in abundance of functional groups. However, differences were apparent for species within functional groupings. Figure 9.5 shows the yield of two legumes at peak biomass under both grazing and cutting. The perennial *T. repens* responded to $[CO_2]$ in a similar way whether grazed or cut, but the annual *T. subterraneum* had a positive response to $[CO_2]$ under grazing but a negative response under cutting. It seems likely that the cutting treatment prevented expression of a $[CO_2]$-induced increase in recruitment from seed (Edwards et al. 2001) which plays a more significant role in the abundance of *T. subterraneum* than that of *T. repens*.

Fig. 9.4 (a) Total annual harvested yield from e[CO$_2$] rings (*filled bars*) and c[CO$_2$] rings (*open bars*; mean, sem). **(b)** Mean annual production of functional groups for the period 1997–2003 from e[CO$_2$] rings (*filled bars*) and c[CO$_2$] rings (*open bars*). Grass values did not differ between treatments but both legumes (*P*=0.008) and forbs (*P*=0.012) were significantly more abundant in e[CO$_2$] rings. Analysis was by repeated measures anova (Genstat 2002). **(c)** Relative yield ({e[CO$_2$] −c[CO$_2$]}/c[CO$_2$]) of functional groups

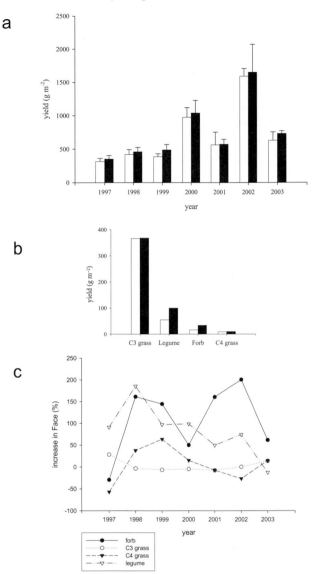

9.5.2 Belowground Yield

Data from in-ground root cores (Allard et al. 2004b), from which the growth of new root biomass was determined at intervals, provided direct evidence for greater C allocation belowground under e[CO$_2$] as well as showing important differences between grazing and cutting management systems. Over a 4-month period during late summer, aboveground biomass was nearly 50 % greater under cutting but was not affected by [CO$_2$] (Fig. 9.6). In contrast, not only was root biomass under grazing double that under cutting, it was greater

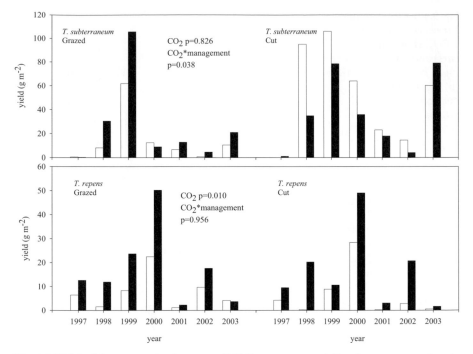

Fig. 9.5 Yield of the two major legumes, *Trifolium repens* and *T. subterraneum* taken in early spring of the years 1997–2003 from c[CO_2] and e[CO_2] rings and from grazed areas and from grazing exclusion cages (cut). Antedependence analysis (Kenward 1987) was used to evaluate treatment effects in the Genstat statistical package (Genstat 2002)

with e[CO_2] but only under grazing. Root biomass responses under the various treatments are likely to have led to observed differences in soil moisture: during this sampling period soils were drier under grazing with mean values for volumetric soil moisture to 15 cm depth being 15.3 % under grazing and 17.9 % under cutting ($P=<0.001$ from repeated measures analysis of variance), possibly because the greater root biomass led to increased water extraction. Interactions with the water sparing effects of elevated CO_2 were also evident – soils tended to be wetter in e[CO_2] plots and more so under grazing – with mean values as follows: c[CO_2] cut 17.8 %, c[CO_2] grazed 14.4 %, e[CO_2] cut 18.0 %, e[CO_2] grazed 16.1 %, the $CO_2 \times$ management interaction being significant ($P=0.05$) from a repeated measures analysis of variance.

In addition to the assimilated C appearing in root biomass, it is also possible that C is exuded directly from roots into the soil. This rhizodeposition could not be assessed directly in the field. Consequently, soil cores were taken from the experiment and a controlled environment experiment was conducted to trace the fate of ^{14}C that was pulse-labelled, using young trans-

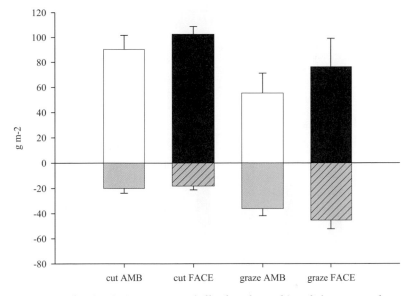

Fig. 9.6 Root production in ingrow cores (Allard et al. 2004b) and aboveground regrowth over late-summer (January–April 2004) of pastures under grazing or cutting and c[CO$_2$] or e[CO$_2$] (mean, sem)

planted ryegrass plants. This experiment showed that e[CO$_2$] significantly increased the amount of C exuded to the soil (Allard et al. 2005). The increased exudation was not due to an increased root biomass under e[CO$_2$] but resulted from a stimulation of the exudation process per se (Allard et al. 2005). While this experiment did not show that enhanced exudation occurred in situ, it does highlight that exudation provides a potentially important mechanism for the introduction of readily decomposable C into the soil. Together with increased rates of root turnover (Allard et al. 2004b), there is the potential for substantial changes in the C dynamics of pastures under e[CO$_2$] and for interactions to occur with the management system (see differences between cutting and grazing in Fig. 9.6). Perhaps the first indication of such changes has been observed in the increased accumulation of coarse fraction soil organic matter in the e[CO$_2$] rings (Allard et al. 2004b), but longer periods of exposure will be necessary to determine the outcome of these changes.

9.5.3 Chemical Composition and Feed Quality

A reduction in the N content of plants is an almost ubiquitous response to e[CO$_2$] (Poorter et al. 1997). In a single species sward this outcome could result in reduced availability of nutrients for animals. However, in a diverse

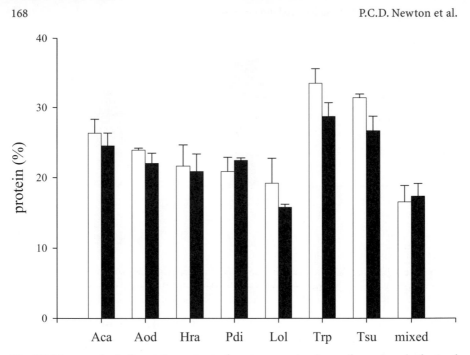

Fig. 9.7 Mean and sd of protein content of pasture species (green leaves) and of mixed herbage (green and dead material) grown at c[CO$_2$] (*open bars*) or in e[CO$_2$] (*closed bars*) after 4 years enrichment (for methods, see Allard et al. 2003). Species are the C3 grasses *Agrostis capillaris* (*Aca*), *Anthoxanthum odoratum* (*Aod*) and *Lolium perenne* (*Lpe*), the forb *Hypochaeris radicata* (*Hra*), the C4 grass *Paspalum dilatatum* (*Pdi*) and the legumes *Trifolium repens* (*Tr*) and *T. subterraneum* (*Tsu*)

sward the potential for species composition change can play a part in modifying this outcome. This effect is shown in Fig. 9.7, where we measured the expected decrease in protein content of individual plants, but the protein content of the mixed herbage was unchanged due to the increasing abundance of the legumes (Fig. 9.4b) – species that contain a higher N content. Herbage from e[CO$_2$] rings frequently has a higher in vitro digestibility than c[CO$_2$] herbage (Allard et al. 2003), which should result in greater nutrient availability to ruminants. In addition, even when e[CO$_2$] and c[CO$_2$] herbage was compared at the same level of digestibility in an indoor feeding trial, we found greater nutrient absorption and retention by sheep feeding on e[CO$_2$] herbage (D. Hélary, unpublished data), perhaps indicating that important differences might have occurred in some minor chemical constituents that have yet to be identified.

9.6 Conclusions

Exposure of a grazed pasture to a CO_2 concentration of 475 ppm resulted in marked changes in plants and soil with evident grazing × CO_2 interactions.

- In this botanically diverse pasture at a $[CO_2]$ concentration expected around 2030, there were no significant changes in the quantity of feed produced nor in its gross nutrient composition.
- However, the way in which this diet was produced was markedly different between $c[CO_2]$ and $e[CO_2]$ treatments. Changes in the chemical composition of plant tissue were offset by changes in plant community composition such that a reduced plant-level N concentration at $e[CO_2]$, of about 11 % in the C3 species, was compensated for at the pasture level by a greater abundance of the high N-containing *Trifolium* species.
- The change in plant community composition resulted in an increase in the rate of litter decomposition at elevated $[CO_2]$ of about 10 % (Allard et al. 2004a).
- CO_2 fixation was greater at $e[CO_2]$ – particularly during periods of low soil moisture – and this additional C was primarily directed belowground resulting in increased root exudation and increased rates of root turnover in the $e[CO_2]$ rings (Allard et al. 2004b). A fourfold increase in the root herbivorous nematode *Longidorus elongatus* in $e[CO_2]$ rings (Yeates et al. 2003) probably reflects this increased resource supply.
- There is evidence for progressively less N and P being available for plant growth at $e[CO_2]$.
- Animals fed $e[CO_2]$ herbage had higher rates of nutrient absorption even when gross nutrient characteristics of the feed were not different.
- There were significant interactions between $[CO_2]$ and management (cutting or grazing) for root growth, soil moisture and plant species composition; in addition there were significant differences in nutrient cycling through animals fed forage grown at $c[CO_2]$ or $e[CO_2]$, emphasising that it is hazardous to draw conclusions about $[CO_2]$ effects on grazed grassland from experiments in which grazing is simulated by cutting.

References

Allard V, Newton PCD, Lieffering M, Clark H, Matthew C, Gray Y (2003) Nutrient cycling in grazed pastures at elevated CO_2: N returns by animals. Global Change Biol 9:1731–1742

Allard V, Newton PCD, Lieffering M, Soussana J-F, Grieu P, Matthew C (2004a) Elevated CO_2 effects on decomposition processes in a grazed grassland. Global Change Biol 10:1553–1564

Allard V, Newton PCD, Soussana J-F, Carran RA, Matthew C (2004b) Increased quantity and quality of coarse soil organic matter fractions at elevated CO_2 in a grazed grassland are a consequence of enhanced root growth and turnover, Plant Soil 276:49–60

Allard V, Robin C, Newton PCD, Lieffering M, Soussana JF (2005) Short- and long-term effects of elevated CO_2 on *Lolium perenne* rhizodeposition and its consequences on soil organic matter turnover and plant N yield. Soil Biol Biochem available at: http://www.sciencedirect.com/science/5163.883069 c330859374d77a/sdarticle.pdf

Belote RT, Weltzin JF, Norby RJ (2003) Response of an understory plant community to elevated $[CO_2]$ depends on differential responses of dominant invasive species and is mediated by soil water availability. New Phytol 161:827–835

Caemmerer S von, Ghannoum O, Conroy JP, Clark H, Newton PCD (2001) Photosynthetic responses of temperate species to free air CO_2 enrichment (FACE) in a grazed New Zealand pasture. Aust J Plant Physiol 28:439–450

Carran RA, Theobald PW (1999) Long-term effects of excreta return on properties of a pasture soil. Nutr Cycl Agroecosyst 56:79–85

Cowie JD, Hall AD (1965) Soils and agriculture of Flock House, Bulls, Manawatu, NZ. New Zealand Soil Bureau report no. 1/1965. Government Printer, Wellington

Curtis P, Wang × (1998) A meta-analysis of elevated CO_2 effects on woody plant mass, form, and physiology. Oecologia 113:299–313

Edwards GR, Clark H, Newton PCD (2001) Carbon dioxide enrichment affects seedling recruitment in an infertile, permanent grassland grazed by sheep Oecologia 127:383–394

Genstat (2002) Genstat 6, release 1.0.20. Lawes Agriculture Trust, IACR, Rothamsted

Gill RA, Polley WA, Johnson HB, Anderson LJ, Maherali H, Jackson RB (2002) Nonlinear grassland responses to past and future atmospheric CO_2. Nature 417:279–282

Haynes RJ, Williams PH (1993) Nutrient cycling and soil fertility in the grazed pasture ecosystem. Adv Agronom 49:119–190

Hovenden MJ, Miglietta F, Zaldei A, Vander Schoor JK, Wills KE, Newton PCD (2005) The TasFACE climate change impacts experiment: design and performance of combined elevated CO_2 and temperature enhancement in a native Tasmanian grassland. Aust J Bot (in press)

Kenward MG (1987) A method for comparing profiles of repeated measurements, Appl Stat 36:296–308

Lewin KF, Hendrey GR, Nagy J, LaMorte RL (1994) Design and application of a free-air carbon dioxide enrichment facility. Agric For Meteorol 70:15–29

Luo Y, Currie WS, Dukes JS, Finzi A, Hartwig U, Hungate B, McMurtrie RE, Oren R, Parton WJ, Pataki DE, Shaw MR, Zak DR, Field CB (2004) Progressive nitrogen limitation of ecosystem responses to rising atmospheric carbon dioxide. Bioscience 54:731–739

Morgan JA, Pataki DE, Grünzweig JM, Körner C, Newton PCD, Niklaus PA, Nippert J, Nowak RS, Parton W, Clark H, Del Grosso SJ, Knapp AK, Mosier AR, Polley W, Shaw R (2004) The role of water relations in grassland and desert ecosystem responses to rising atmospheric CO_2. Oecologia 140:11–25

Nelson JA, Morgan JA, LeCain DR, Mosier AR, Milchunas DG, Parton WG (2004) Elevated CO_2 increases soil moisture and enhances plant water relations in a long-term field study in the semi-arid shortgrass steppe of Northern Colorado. Plant Soil 259:169–179

Newton PCD, Clark H, Bell CC, Glasgow EM (1996) Interaction of soil moisture and elevated CO_2 on the above-ground growth rate, root length density and gas exchange of turves from temperate pasture. J Exp Bot 47:771–779

Newton PCD, Clark H, Edwards GR (2001) CO_2 enrichment of a permanent grassland grazed by sheep using FACE technology. In: Shimizu H (ed) Carbon dioxide and vegetation: advanced international approaches for absorption of CO$_2$ and responses to CO$_2$. Center for Global Environmental Research, Tsukuba, pp 97–105

Newton PCD, Carran RA, Lawrence EJ (2003) Reduced water repellency of a grassland soil under elevated atmospheric CO$_2$. Global Change Biol 10:1–4

Niklaus PA, Alphei J, Ebersberger D, Kandeler E, Tscherko D (2003) Six years of in situ CO$_2$ enrichment evoke changes in soil structure and biota of nutrient-poor grassland. Global Change Biol 9:585–600

Norby RJ, Cotrufo MF (1998) A question of litter quality. Nature 396:17–18

Nowak RS, Zitzer SF, Babcock D, Smith-Longozo V, Charley TN, Coleman JS, Seeman JR, Smith SD (2004) Elevated atmospheric CO$_2$ does not conserve soil water in the Mojave desert. Ecology 85:93–99

Owensby CE, Cochran RC, Auen LM (1996) Effects of elevated carbon dioxide on forage quality for ruminants. In: Körner C, Bazzaz FA (eds) Carbon dioxide, populations and communities. Academic Press, San Diego, pp 363–371

Parsons AJ, Chapman DF (2000) The principles of pasture growth and utilisation. In: Hopkins A (ed) Grass. Blackwell, London, pp 31–89

Parsons AJ, Johnson IR, Harvey A (1988) Use of a model to optimise the interaction between the frequency and severity of intermittent defoliation and to provide a fundamental comparison of the continuous and intermittent defoliation of grass. Grass For Sci 43:49–59

Poorter H, VanBerkel Y, Baxter R , DenHertog J, Dijkstra P, Gifford RM, Griffin KL, Roumet C, Roy J, Wong SC (1997) The effects of elevated CO$_2$ on the chemical composition and construction costs of leaves of 27 C3 species. Plant Cell Environ 20:472–482

Reich PB, Tilman D, Craine J, Ellsworth D, Tjoelker MG, Knops J, Wedin D, Naeem S, Bahauddin D, Goth J, Bengston W, Lee TD (2001) Do species and functional groups differ in acquisition and use of C, N and water under varying atmospheric CO$_2$ and N availability regimes? A field test with 16 grassland species. New Phytol 150:435–448

Rillig MC, Wright SF, Allen MF, Field CB (1999) Rise in carbon dioxide changes soil structure. Nature 400:628

Stöcklin J, Körner C (1998) Interactive effects of CO$_2$, P availability and legume presence on calcareous grassland: results of a glasshouse experiment. Funct Ecol 13:200–209

Yeates GW, Newton PCD, Ross DJ (2003) Significant changes in soil microfauna in grazed pasture under elevated carbon dioxide, Biol Fertil Soils 38:319–326

Zanetti S, Hartwig UA, Lüscher A, Hebeissen T, Frehner M, Fischer BU, Hendrey GR, Blum H, Nösberger J (1996) Stimulation of symbiotic N$_2$ fixation in *Trifolium repens* L. under elevated atmospheric CO$_2$ in a grassland ecosystem. Plant Physiol 112:575–583

Zavaleta ES, Shaw MR, Chiariello NR, Thomas BD, Cleland EE, Field CB, Mooney HA (2003) Responses of a California grassland community to three years of experimental climate change, elevated CO$_2$, and N deposition. Ecol Monogr 73:585–604

10 Responses to Elevated [CO$_2$] of a Short Rotation, Multispecies Poplar Plantation: the POPFACE/EUROFACE Experiment

G. Scarascia-Mugnozza, C. Calfapietra, R. Ceulemans, B. Gielen, M.F. Cotrufo, P. DeAngelis, D. Godbold, M.R. Hoosbeek, O. Kull, M. Lukac, M. Marek, F. Miglietta, A. Polle, C. Raines, M. Sabatti, N. Anselmi, and G. Taylor

10.1 Introduction

10.1.1 Research Leading to This Experiment

Forest and agricultural soils present interesting opportunities to conserve and sequester carbon. Soil C pools may be restored and enlarged by managing agricultural and forest soil and by the use of surplus agricultural land for reforestation or high-yield woody crops for biomass and bioenergy. Recently, bioenergy woody crops were reported to have the greatest potential for C mitigation, for their capability to sequester C and, contemporarily, to substitute fossil fuel carbon (Smith et al. 2000).

10.1.2 Focus on Agroforestry Plantations

The ability of managed forest and agroforestry systems to sequester carbon at the regional and global scale will be, however, strongly influenced by the responses of trees and tree communities to global change, particularly to the predicted increase in atmospheric [CO$_2$] (Norby et al. 1999). Short rotation forestry (SRF) is an appropriate scale of investigation to understand the responses of trees and ecosystems to a changing environment, for its peculiar characteristics of completing a full rotation cycle in a reasonable time interval (3–5 years), while providing the possibility of considering a wide array of

Ecological Studies, Vol. 187
J. Nösberger, S.P. Long, R.J. Norby, M. Stitt,
G.R. Hendrey, H. Blum (Eds.)
Managed Ecosystems and CO$_2$
Case Studies, Processes, and Perspectives
© Springer-Verlag Berlin Heidelberg 2006

management strategies. Quantifying the CO_2 sequestration potential of forest and agro-forestry ecosystems under changing climatic conditions is anyhow very difficult with current scientific tools, unless manipulative studies are conducted at the truly ecosystem scale, as is the case for FACE experiments (Hendrey et al. 1993; see also Chapters 11–13).

10.1.3 Objectives and Hypotheses

The main objectives of the POPFACE experiment are to determine the functional responses of a cultivated, agro-forestry system, namely a poplar plantation, to current and future atmospheric $[CO_2]$ and to assess the interactive effects of this anthropogenic perturbation with the other natural environmental constraints on key biological processes and structures. Additionally, this experiment is intended to yield data relevant to assess the potential for increasing the C-sequestering capacity and woody biomass production within the European Union, using such forest tree plantations.

The general hypotheses underlying the experiment were that: (i) little acclimation of carbon assimilation processes occurs at the leaf and stand level in a fast-growing tree plantation; and, hence, net primary production and carbon accumulation in the soil increase in $e[CO_2]$ and (ii) exposure to $e[CO_2]$ would interact differently with contrasting genotypes and nitrogen availability.

10.2 Site Description

10.2.1 Location and Layout of Experiment

The experimental plantation and FACE facility (Fig. 10.1a) are located in an agricultural region of central Italy, near Viterbo (Tuscania; 42°22' N, 11°48' E, alt. 150 m). The site was covered by woody vegetation until 1950 and has been managed for herbaceous crop production ever since. As the experimental site is only 15 km from the coast, the prevailing winds during the growing season are westerly, favourable for the FACE design. In spring 1999, before establishing a 9-ha poplar plantation on a former agricultural field, six experimental areas, hereafter called "plots" (30 × 30 m), were selected over this land, with a minimum distance between plots of 120 m to avoid cross-contamination. Three of these areas, representing the "control" treatment, were left under untreated conditions, whereas in each of the other three, representing the $e[CO_2]$ treatment, an octagonal ring (22 m diameter) of polyethylene tubes was established. Each plot was divided into two parts by a physical resin/glass

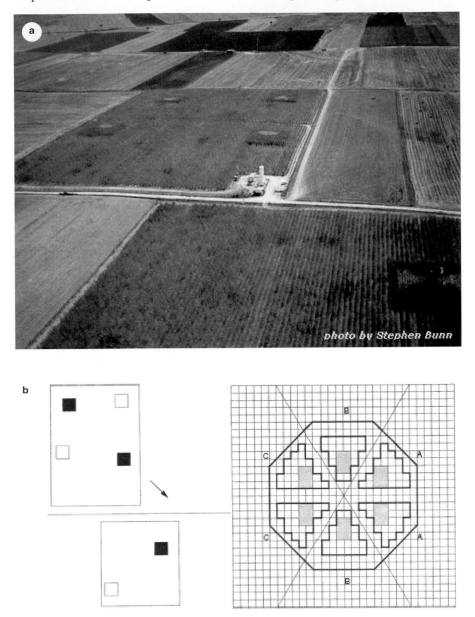

Fig. 10.1 (**a**) Aerial view of the POPFACE site. The plantation is 9 ha large and includes six experimental plots (*octagonal rings*). The distance between adjacent plots is at least 100 m. (**b**) Schematic layout of the experimental plantation (*arrow* points to the North) and an experimental plot (*A, B* and *C* represent the three tested species; from Gielen et al. 2003a)

barrier (1 m deep in the soil) for future nitrogen differential treatments in the two halves of each plot, although no fertilisation treatment was applied during the first 3-year rotation of the experiment. Each half-plot was further divided into three radial sectors, each occupied by a different genotype (Fig. 10.1b). Each plot includes, therefore, six sectors (3 clones × 2 N-treatments) yielding 58 plants per sector within the FACE ring. However, only the 24 inner plants within each sector were utilized for physiological, structural and growth measurements.

Statistical analyses were applied as a randomized complete block design with three replicates, current and elevated CO_2 concentration, two nitrogen levels (only in the second rotation) and three species (genotypes) as split-plot factors.

10.2.2 Soil Types, Fertilisation, Irrigation

A detailed soil survey was carried out in November 1998 to characterize the soil properties of the experimental site, before planting trees and starting the $[CO_2]$ atmospheric enrichment, and also to assess soil spatial variability in order to appropriately locate experimental blocks and plots. Soil was then classified as Pachic Xerumbrept – silt loam agricultural soil, more than 1 m deep, originating from a geological substrate derived from sedimentary material of volcanic origin and marine deposits (Hoosbeek et al. 2004). The soil reaction is acidic (pH 5.00), while the carbon content is poor and nitrogen content moderately rich (respectively, 0.98 % and 0.12 %). Physico-chemical soil characteristics were relatively variable among and within the sampled areas and hence experimental blocks and plots were then selected to minimize this variation.

In the first 3-year rotation cycle, no fertiliser was applied to the plantation or plots because chemical analyses showed a fair availability of N for the poplar trees. In the second rotation cycle, a differential fertilisation treatment was applied to the experimental plots. A drip irrigation system, 50 km long in total, was installed both in the field and in the experimental plots to avoid summer drought stress.

10.2.3 Meteorological Description

The climate of the POPFACE site is typically Mediterranean, with warm and dry summer, mild and humid autumn and winter, mean annual air temperature of 14 °C and mean annual precipitation of about 800 mm. To continuously monitor and record weather parameters at the site, for documenting experimental conditions and for applying ecological models to the poplar stand, a meteorological station was installed in the field.

10.2.4 Stand History and Description

After the land was ploughed, the poplar plantation was established in late spring 1999, just before the starting of CO$_2$ fumigation, using uniform hardwood cuttings (length 25 cm). The entire 9-ha field was planted with *Populus × euramericana* genotype I-214, at a planting density of 5000 trees per hectare (spacing 2 × 1 m). The six experimental plots therein were planted with three different poplar genotypes, at a planting density of 10 000 trees per hectare (spacing 1 × 1 m) in order to have a sufficient number of experimental trees and a closed canopy after a short time. This planting density is common in SRIC poplar plantations (Ceulemans et al. 1992). The three species were *P. × euramericana* Dode (Guinier) (*P. deltoides* Bart. ex Marsh. × *P. nigra* L.), *P. nigra* L. and *P. alba* L.

Growth of trees was rapid over the 3-year rotation cycle: the tree height was 1.4–1.8 m at the end of the first growing season, 6.0–7.0 m in the second year, producing a completely closed tree canopy, and at the end of the third growing season the height of the canopy was almost 10 m. At the end of the third growing season the plantation was harvested, both inside the experimental plots and in the surrounding field. All trees were cut at the base of the stem and a large sample was utilized for aboveground biomass analyses while a smaller group was also excavated to measure belowground biomass. Then, trees regenerated vegetatively from the stumps in the coming spring (2002), producing a multistem, coppice plantation.

10.3 Experimental Treatment

10.3.1 Atmospheric [CO$_2$] Enrichment

In the POPFACE system, pure CO$_2$ is released to the atmosphere (Miglietta et al. 2001) through many small gas jets discharged from tiny holes, laser drilled into PVC pipes, 20 mm wide, that were mounted on telescopic poles, during the first 3-year experimental period, and on meteorological towers in the second 3-year experimental period. These horizontal pipes formed octagons, with diagonals of 22.2 m, around the elevated [CO$_2$] experimental plots. As the trees grew considerably in height during each rotation cycle, more pipes were vertically added to form up to three or four pipe layers, horizontally arranged one above the other, in each FACE experimental plot (see Chapter 2).

Liquid CO$_2$ was regularly supplied from natural CO$_2$ vents located in central Italy. The stable isotope signature of the supplied CO$_2$, measured as δ^{13}C, was –6‰ versus the PDB standard which is close to the current value of about –8‰. Daytime CO$_2$ enrichment was provided from bud burst to leaf fall. The target [CO_2] in the FACE plots was 550 ppm; and the measured CO_2 concentration was within 20% deviation from the pre-set target concentration for 91% of the time to 72.2% of the time, respectively, at the beginning and at the end of each rotation cycle of the plantation. Wind conditions in the POPFACE site were quite regular over the operating years, with episodes of high wind conditions that were repeated a few times during the growing seasons.

10.3.2 Nitrogen Fertilisation

At the beginning of the second rotation cycle, a fertilisation treatment was added to one-half of each experimental plot. The total amount of N supplied was 212 kg ha^{-1} year^{-1} in 2002 and 290 kg ha^{-1} yr^{-1} during 2003 and 2004. The N was supplied in constant weekly amounts with a 4:1 NH_4^+:NO_3^- ratio in 2002, whereas it was supplied in weekly amounts proportional to the growth rate in 2003 and 2004, with a 1:1 NH_4^+:NO_3^- ratio. Each experimental plot was equipped with a plastic 200-l tank (in which the fertiliser was dissolved) and a hydraulic pump (Ferti-injector Amiad; Imago, Italy), connected to the irrigation system.

10.3.3 Species Comparison

Poplars were selected as planting material because of their rapid growth and elevated biomass production that allowed a simulation of the life cycle of a forest plantation, in a relatively short time period. At the same time, their ability to propagate vegetatively made it possible to utilize genetically homogeneous plant material, thus facilitating the interpretation of experimental results. However, three different poplar species were included in the experiment, to represent poplar biodiversity (Table 10.1).

10.3.4 Interactions

Interactions among experimental factors that were specifically investigated were:
1. [CO_2] treatment × poplar species, to quantify the range of species-specific responses to changing environmental factors;
2. [CO_2] treatment × N fertilisation was added to the POPFACE experiment in the second rotation, to modulate plant and ecosystem responses by changing the availability of this decisive and often limiting factor.

Table 10.1 Main characteristics of the poplar genotypes used in the POPFACE experiment (from Calfapietra et al. 2001)

Species	P. alba L.	P. nigra L.	P. × euramericana Dode (Guinier)
Genotype	2AS11	Jean Pourtet	I-214
Sex	Male	Male	Female
Origin	Italy[a]	France[a]	Italy[b]
Rooting	Medium	Very good	Very good
Branching habit	Medium	Very high	Low
Apical control	Good	Good	Very good
Bud-burst[c]	End of March	End of March	End of March
Bud set[c]	End of October	Beginning of October	Middle of September

[a] Seed origin
[b] Origin of the selected hybrid
[c] Indicative dates for central Italy

10.4 Resource Acquisition

10.4.1 Photosynthesis and Respiration

None of the three study clones exhibited any consistent photosynthetic acclimation during the first, 3-year rotation cycle (Bernacchi et al. 2003). This was shown by large increases in leaf photosynthesis at e[CO$_2$] and the lack of any consistent decrease in either $V_{c,max}$ or J_{max}. Although some very small decreases in Rubisco protein level may have occurred, this did not result in any apparent decline in photosynthesis. The lack of significant acclimation may be explained by increased sink capacity for carbohydrate; and the high growth rate of experimental clones maintained the demand for photoassimilate, preventing down-regulation of photosynthesis. This is supported by the soluble sugar content in the leaf tissue. Neither sucrose, fructose or glucose showed higher accumulation in the trees grown at e[CO$_2$]. In contrast, growth at e[CO$_2$] caused increases in leaf starch content, up to 200 % at midday, in summer.

After 3 years of growth at e[CO$_2$], the mean enhancement in midday photosynthesis for all three clones was 55 % (Fig. 10.2). The integrated daily assimilation rate showed an increase under e[CO$_2$], ranging from 28 % for *P. nigra* in May and 86 % for *P. × euramericana* in September. The light-saturated rate of photosynthesis measured by *A/c$_i$* analysis was higher with growth at e[CO$_2$] throughout the three seasons for all clones. There was no apparent progressive decrease in stimulation over time. Even though the absolute values of $V_{c,max}$ were similar for the three clones, the seasonal pattern

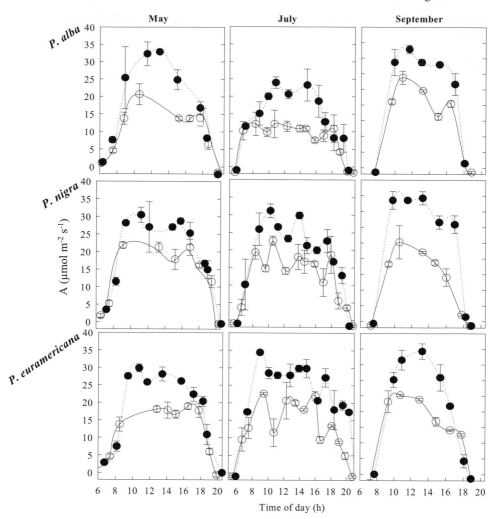

Fig. 10.2 Diurnal time-course of the rate of leaf CO_2 uptake (*A*) in poplar clones grown at c[CO_2] (*open circles*) and e[CO_2] (*closed circles*). Plants were measured in May, July and September, during the second season of growth. Each point is the mean of replicate measurements made within 2 h, in three replicate rings (from Bernacchi et al. 2003)

of $V_{c,max}$ between the clones was different. The response of J_{max} was very similar to that of $V_{c,max}$. Again, there were no consistent decreases in J_{max} with growth at e[CO_2].

It is possible that, after several years of growth at e[CO_2], photosynthetic acclimation becomes greater in magnitude as nutrients, particularly N, become limiting. Although this may decrease photosynthetic rates per se, it is unlikely to fully diminish the stimulation in C uptake. This is because e[CO_2] increases the efficiency of photosynthesis in respect to N, light and water, thereby reducing the probability of a particular environmental factor totally negating the stimulation of C fixation.

In conclusion, the high photosynthetic and growth rates of poplar make it ideal both as a C sink and as a biofuel. Stimulation of photosynthesis was maintained after the first three seasons where there was no indication of photosynthetic down-regulation. However, it has to be verified whether this stimulation will remain over long-term periods, especially considering that frequent coppicing creates considerable variations in the sinks and a critical retranslocation of sugars.

10.4.2 Stomatal Conductance

Measurements of stomatal conductance in this experiment showed that g_s was generally decreased by e[CO$_2$], varying between 16% and 35%, but this reduction was often not statistically significant (Bernacchi et al. 2003; Tricker et al. 2005). Nevertheless, the reduction in stomatal conductance was maintained, even after 5 years of exposure to e[CO$_2$] (Tricker et al. 2005). The ensuing improvements in instantaneous water-use efficiency resulted from altered stomatal function rather than development and they call into question the role of the HIC gene – a negative regulator of stomatal initiation shown to be sensitive to e[CO$_2$] (Gray et al. 2000) – for field-grown trees in long-term experiments. Our on-going research suggests that, despite leaf-level reductions in stomatal conductance in e[CO$_2$], canopy-scale increases in water use are likely in these trees, particularly in *P. × euramericana*, following on from their increased size and leaf area in e[CO$_2$] (G. Taylor, personal communication).

10.4.3 Nitrogen and Other Nutrient Concentrations and Dynamics

Plant N is usually one of the limiting factors when the long-term responses of plants to e[CO$_2$] are analysed. Therefore, fertilisation could be crucial to sustain a faster growth which usually occurs under e[CO$_2$]. The N concentration in leaves was investigated in 2003 (second year of the second rotation cycle), assuming that this parameter is strongly related to the nutrient status of the whole tree (Kozlowski and Pallardy 1997). Leaf N concentration on a mass basis increased under fertilisation by 16–22%, depending on species and [CO$_2$] treatment. However, the foliar N concentration was found to slightly decrease under e[CO$_2$], although not significantly. This apparent decrease was mostly a dilution effect, because it disappeared when N content was expressed on a leaf area basis. Differences between treatments disappeared for all species by the end of the growing season and values of N concentration showed similar values in all treatments around 2%. Moreover, the limiting

levels of PAR by the end of the growing season could have inhibited N accumulation in the leaves, as it was also observed along the vertical profile of the
canopy (Gielen et al. 2003).

10.4.4 LAI and Light Interception

Tree cover developed over the first 3-year rotation, from bare land to complete closure. In fact, the highest annual LAI values for the POPFACE poplar
stands (Fig. 10.3) reached 1 m^2 m^{-2} in the first growing season, 5 m^2 m^{-2} in
the second year and 7 m^2 m^{-2} in the third year, during the first rotation cycle
(Gielen et al. 2001). The effect of e[CO_2] was mostly evident on *P. nigra*, in
the first growing season, as its LAI was stimulated by 252 % and was significantly larger than the LAI of *P. alba* and *P. × euramericana* (significant
treatment × genotype interaction). Nevertheless, LAI was unaffected by
e[CO_2] after canopy closure.

   In the second rotation cycle, during the 2 years after coppicing, LAI was
stimulated up to 45 % by e[CO_2] (Liberloo et al. 2005). An increased number
of shoots and sylleptic branches (especially for *P. nigra*) and increased leaf
sizes could have caused the higher LAI values. Another reason for the
observed increase in leaf production in e[CO_2] could be a longer growing season caused by shifts in phenology, as indicated by delayed leaf fall during
canopy decline in e[CO_2], at least for *P. × euramericana* (Tricker et al. 2004).
Continuous records of photosynthetic photon irradiance (Qp) under the
canopy showed a higher absorbed Qp in the e[CO_2] treatment only during the
period preceding canopy closure. Therefore, results for Qp canopy absorp-
tance fitted those for the evolution of LAI. Additionally, e[CO_2] did not affect
ratios of red to far-red light, which agreed with results for leaf chlorophyll and
specific leaf area.

   The effect of e[CO_2] on single leaf expansion was also investigated in our
experiment, but only in the *P. × euramericana* genotype (Ferris et al. 2001;
Taylor et al. 2001, 2003) and was related to leaf plastochron index (LPI) as a
measure of leaf development. Leaf expansion was stimulated at very early
(LPI 0–3) and late (LPI 6–8) stages in development. Early and late effects of
e[CO_2] were largely the result of increased cell expansion and increased cell
production, respectively. Spatial effects of e[CO_2] were also marked and
increased final leaf size, i.e. leaf area, but not leaf length, demonstrating a
changed leaf shape in response to e[CO_2].

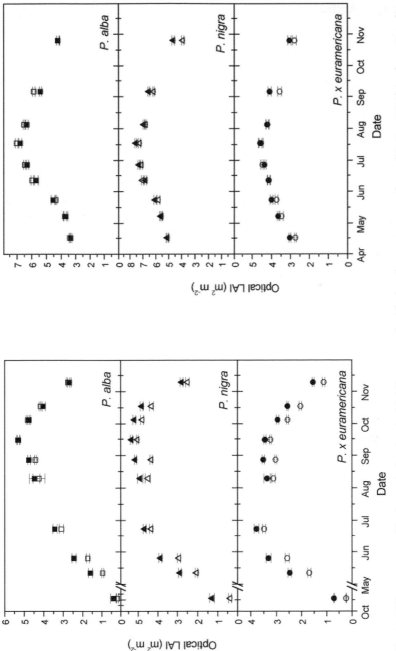
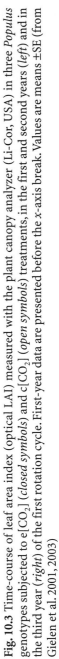

Fig. 10.3 Time-course of leaf area index (optical LAI) measured with the plant canopy analyzer (Li-Cor, USA) in three *Populus* genotypes subjected to e[CO₂] (*closed symbols*) and c[CO₂] (*open symbols*) treatments, in the first and second years (*left*) and in the third year (*right*) of the first rotation cycle. First-year data are presented before the x-axis break. Values are means ±SE (from Gielen et al. 2001, 2003)

10.4.5 Canopy Architecture

Trees in $e[CO_2]$ produced more sylleptic branches on the height growth of the first year (HGI 1) compared to those in $c[CO_2]$. During the first year, the number of branches per unit of tree height increased by 23 % (not significant) for *P. nigra* and more than doubled for *P. × euramericana*. Differences were considerably smaller, or even opposite in sign, for sylleptic branches produced in the second year (sylleptic branches on HGI 2). However, for all three genotypes the canopy was deeper (by 5–15 %) in $e[CO_2]$ at the end of the second year as a result of the first-year stimulation of sylleptic branching. Also, $e[CO_2]$ increased average branch dimensions (diameter and length) in the first year, for all three genotypes. In contrast, no clear effects of CO_2 enrichment were found on branch angles, angle of origin of the branch and angle of termination (Gielen et al. 2002).

In the second rotation cycle, after coppicing, $e[CO_2]$ stimulated the number of shoots per stool, more particularly for *P. nigra* than for the other species, but did not affect dominant shoot height (Liberloo et al. 2005). Fertilisation decreased the number of shoots per stool, possibly because fertilisation strongly increased competition in $e[CO_2]$ and consequently enhanced mortality.

10.4.6 Root Development and Mycorrhizal Colonization

$e[CO_2]$ caused roots to develop deeper into the soil (Lukac et al. 2003). Analysis of the vertical distribution of roots showed increased allocation of biomass into deeper soil horizons (20–40 cm) under $e[CO_2]$ conditions for *P. alba* and *P. nigra*, from 23 % to 36 % and from 20 % to 39 % respectively, but not for *P. × euramericana*, suggesting a genotype-specific response.

Mycorrhizal symbiosis can produce profound effects on the availability of N and other nutrients to plants; and, conversely, an increased supply of C under $e[CO_2]$ conditions can be beneficial for fungal symbionts. The response of arbuscular mycorrhizal colonization to $e[CO_2]$ was found to vary among *Populus* species (Lukac et al. 2003). In both *P. alba* and *P. nigra*, hyphal presence inside fine roots increased (by +29 % and +36 %, respectively), but little effect was observed in *P. × euramericana* (+2 %). A similar species-specific effect was observed for ectomycorrhizae as EM colonization significantly increased only in *P. alba* (+78 %). The fact that poplars were colonized with dissimilar intensities appears to be due to the various strategies of nutrient and water acquisition adopted by the different *Populus* species.

10.5 Resource Transformation

10.5.1 Aboveground Productivity

Responses of poplar agro-forestry systems to atmospheric CO$_2$ enrichment can be grouped into the effect of e[CO$_2$] before and after canopy closure. Initially, e[CO$_2$] stimulated growth of *Populus*, although not to the same extent for all three studied species (Calfapietra et al. 2003a). Stem volume index (D^{2}H), a parameter directly related to biomass, was considerably enhanced by e[CO$_2$] at the end of the first year, especially in *P. nigra* (+121%) and *P. euramericana* (+73%) and to a lesser extent in *P. alba* (+30%). This large stimulation of growth by e[CO$_2$] progressively decreased during the second year (Fig. 10.4), after canopy closure, converging for all species to a common value of approximately +20%. Although these findings point towards a loss of positive response to CO$_2$ enrichment, results must be considered in the perspective of ontogeny. Because trees were larger in e[CO$_2$] after the first year, relative growth was almost inevitably lower in e[CO$_2$]. Since the absolute growth rate of similar-sized trees did not differ between c[CO$_2$] and e[CO$_2$], the growth-stimulating effect of CO$_2$ enrichment was largely a first-year stimulation, sustained over a three-year growth period (Fig. 10.4). Finally, the production of biomass (stem, branches, coarse roots) after 3 years of growth was stimulated under e[CO$_2$] by 24%, averaged across the species (Table 10.2; Calfapietra et al. 2003b).

In the second rotation cycle, after coppicing, an increase in biomass production was observed under e[CO$_2$] (Liberloo et al. 2005); and this increase in biomass as a response to e[CO$_2$] was caused by an initial stimulation of absolute and relative growth rates, which disappeared after the first growing season following coppicing. An ontogenetic decline in growth in e[CO$_2$], together with strong competition inside the dense plantation, may have caused this decrease. Fertilisation did not influence aboveground growth, although some responses to e[CO$_2$] were more pronounced in fertilised trees.

10.5.2 Belowground Productivity

Exposure of all three *Populus* genotypes to e[CO$_2$] resulted in larger trees with greater root systems (Lukac et al. 2003). However, *Populus* genotypes utilized in this research did not increase their root production by the same magnitude under e[CO$_2$]. The smallest increase in standing root biomass induced by e[CO$_2$] occurred in *P. alba* (+47%). *P. nigra* and *P. × euramericana* responded to e[CO$_2$] with increases of +76% and +71%, respectively. All data recorded live root biomass; and only negligible amounts of root necromass were

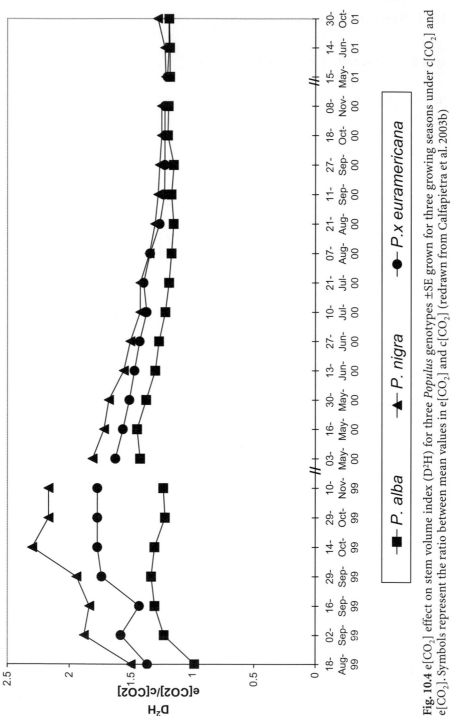

Fig. 10.4 e[CO$_2$] effect on stem volume index (D^{2}H) for three *Populus* genotypes ±SE grown for three growing seasons under c[CO$_2$] and e[CO$_2$]. Symbols represent the ratio between mean values in e[CO$_2$] and c[CO$_2$] (redrawn from Calfapietra et al. 2003b)

Table 10.2 Total biomass (oven dry weight values per hectare) and its distribution within the tree components for three *Populus* genotypes after the final harvest at the end of the first 3-year rotation cycle (from Calfapietra et al. 2003b)

	P. alba		P. nigra		P. x euramericana	
	c[CO$_2$] (Mg ha^{-1})	e[CO2] (Mg ha^{-1})	c[CO$_2$] (Mg ha^{-1})	e[CO2] (Mg ha^{-1})	c[CO$_2$] (Mg ha^{-1})	e[CO$_2$] (Mg ha^{-1})
Syll. branches on HGI 1	1.52 (0.39)	1.70 (0.31)	1.89 (0.32)	3.21 (0.28)	0.60 (0.13)	0.88 (0.30)
Prol. brances on HGI 1	1.58 (0.24)	1.94 (0.35)	1.48 (0.15)	1.62 (0.27)	1.79 (0.29)	2.45 (0.29)
Syll. branches on HGI 2	3.57 (0.52)	3.53 (0.40)	2.52 (0.35)	3.37 (0.64)	0.12 (0.04)	0.44 (0.30)
Prol. brances on HGI 2	1.43 (0.24)	2.04 (0.37)	2.94 (0.37)	2.45 (0.38)	2.66 (0.55)	3.49 (0.73)
Syll. branches on HGI 3	1.33 (0.24)	1.71 (0.42)	1.83 (0.21)	2.08 (0.31)	1.04 (0.32)	1.02 (0.24)
Second order branches	2.03 (0.42)	2.24 (0.48)	0.91 (0.19)	1.22 (0.23)	0.25 (0.08)	0.28 (0.13)
Total branches	11.45 (2.30)	13.15 (1.93)	11.57 (1.35)	13.95 (1.39)	6.45 (1.21)	8.56 (1.36)
Stem	28.87 (3.65)	38.23 (3.28)	42.01 (2.58)	47.78 (2.57)	31.71 (2.68)	39.82 (2.93)
Stump	2.65 (1.16)	4.78 (1.10)	2.73 (0.66)	3.87 (0.73)	2.95 (0.82)	4.46 (0.98)
Coarse roots	4.93 (1.71)	5.68 (1.84)	5.71 (2.46)	6.43 (1.46)	4.92 (1.27)	5.65 (1.37)
Total above-ground biomass	40.32 (4.31)	51.38 (3.81)	53.58 (2.91)	61.73 (2.93)	38.16 (2.94)	48.39 (3.23)
Total below-ground biomass	7.58 (2.80)	10.46 (2.96)	8.44 (2.79)	10.30 (2.07)	7.87 (2.07)	10.11 (2.40)
Total biomass	47.90 (6.41)	61.84 (5.78)	62.02 (4.57)	72.03 (4.39)	46.03 (4.41)	58.50 (4.88)

detected in the soil. This is probably due to a very fast rate of decomposition of dead root biomass in this ecosystem.

Ingrowth core measurements showed a corresponding increase in fine root production. e[CO_2] enhanced fine root growth of *P.alba* by 56 %, *P.nigra* by 97 % and *P. × euramericana* by 73 %. Root turnover was determined according to the model originally proposed by Dahlman and Kucera (1965), who identified root turnover as annual belowground production divided by maximum belowground standing crop. e[CO_2] increased not only the amount of fine roots produced, but also the rate of root turnover in all three genotypes, on average from 1.5 to 2.0. This increase was largest for *P. alba*, which speeded up root turnover under e[CO_2] by 45 %, while *P. nigra* and *P. × euramericana* showed a 27 % increment.

The above-mentioned effects of e[CO_2] on root biomass, fine root production and, also, colonization by mycorrhizal fungi can have profound effects on the amount and quality of C entering the soil. e[CO_2] clearly resulted in an increase in the C transferred belowground in all three *Populus* species. Such boosted investment of assimilated C belowground could be the means by which trees obtain the additional resources necessary to sustain increased growth.

10.5.3 Soil Carbon: Litter Production, Soil Respiration and C-Pools

In all species, e[CO_2] induced a small, and not significant, increase in the annual litter production, from 3 % to 6 %. The atmospheric CO_2 enrichment also affected litter decay rates, with two different mechanisms: (a) by altering litter quality, the decomposition rate of litter was slowed down on average by 7 % after 8 months of incubation, (b) when incubated in the field under e[CO_2], litter decomposition accelerated, especially in the initial stage of the process, possibly as a consequence of increased soil biological activity and soil C input into the rhizosphere environment under e[CO_2] compared to c[CO_2]. Also, these responses were genotype-specific and *P. nigra* was the most affected by the treatment (Cotrufo et al. 2005).

The rate of soil CO_2 efflux was measured in the field twice per month, i.e. nearly 20 times per year, starting from the second year onward. A highly significant treatment effect on soil respiration was observed: annual soil respiration was stimulated by 34 % to 50 % under e[CO_2], depending on genotype and year (P. DeAngelis, unpublished data).

Based on observed increments in above- and belowground biomass under e[CO_2], we expected greater increases in soil C_{new} and C_{total} under e[CO_2] than under c[CO_2] (Hoosbeek et al. 2004). C_{new} is the amount of C taken up by the soil during the experiment and was estimated with the C_3/C_4 stable isotope method (Van Kessel et al. 2000; see also Chapter 21). The old soil C pool (C_{old}) was defined as C_{total} minus C_{new}, while the respired C was calculated as the difference between C$_{old}$ at the beginning and the end of the experiment. C$_{total}$ (C$_{old}$ + C$_{new}$) increased by 12 % and 3 %, i.e. 484 g m^{-2} and 107 g m^{-2}, under the control and e[CO$_2$] respectively. During the same time-span, 704 g m^{-2} and 926 g m^{-2} C$_{new}$ were taken up by the soil under the control and e[CO$_2$]. The old C pool lost relatively more C under e[CO$_2$], resulting in a loss of C by respiration of respectively 220 g m^{-2} and 819 g m^{-2} under c[CO$_2$] and e[CO$_2$]. We hypothesize that the opposite effects of e[CO$_2$] on soil C$_{total}$ and C$_{new}$ were caused by a priming effect of the newly incorporated litter. The priming effect was defined as the stimulation of SOM decomposition caused by the addition of labile substrates (Dalenberg and Jager 1989). This priming effect induced by e[CO$_2$] may have caused increased respiration rates. However, preliminary data (not shown) on soil C collected in the years 2003 and 2004 (second rotation) indicate that the priming effect was a temporary effect. Due to the change of land use at the beginning of the POPFACE project, total soil C content kept increasing but, as opposed to the first rotation, the increase in soil C$_{total}$ under e[CO$_2$] during the second rotation was larger than under c[CO$_2$].

10.5.4 Wood Quality and Biochemical Composition of Wood and Roots

Growth and anatomical wood properties were analysed in secondary sprouts (Luo et al. 2005). In the three poplar clones, most of the growth and anatomical traits showed no uniform response pattern to e[CO$_2$] or N-fertilisation. In cross-sections of young poplar stems, tension wood accounted for 2–10 % of the total area and was not affected by e[CO$_2$]. In *P. nigra*, N-fertilisation caused an about 2-fold increase in tension wood, but not in the other clones. The formation of tension wood was not related to the diameter or height growth of the shoots. In *P. × euramericana*, N-fertilisation resulted in significant reductions in fibre length. In *P. × euramericana* and *P. alba*, e[CO$_2$] caused decreases in wall thickness, but the effect was less pronounced than that caused by N fertilisation. In *P. nigra* and *P. × euramericana*, e[CO$_2$] induced increases in vessel diameters.

10.5.5 Pest and Disease Susceptibility

The results concerning both *Marssonina* spp and *Melampsora* spp attacks on the three poplar species did not show any significant variation between [CO$_2$] treatments. However, it should be noted that rust incidence was always greater in c[CO$_2$] plots than in e[CO$_2$], even though the difference was never statistically significant.

The only clear and direct effect of e[CO$_2$] was an increase in the intensity of attacks of sooty moulds. These could be related to alterations in the tree metabolism affecting the honeydew composition or to different behaviour of phyllomyzous insects that predisposes to sooty mould infections. All these hypotheses should be verified, but it is evident that leaf photosynthetic capacity could be seriously affected if sooty mould infections increase considerably under e[CO_2].

10.6 Consequences and Implications

10.6.1 Forest Management

Effects of e[CO_2] on intensively managed SRF plantations may range from increased biomass production and water use efficiency to a different tree architecture with enhanced branchiness and deeper root systems. These responses will vary with species and genotypes. Therefore it will be relevant to select the appropriate genetic material most responsive to atmospheric CO_2 enrichment. Another important aspect of management is how the stand will regenerate after harvesting: in SRF poplar culture, stand regeneration can be obtained by coppicing. Coppice regeneration has the advantage that it does not require any tree planting or soil disturbance, therefore reducing the risk of abrupt burst of soil C-effluxes at the beginning of a new rotation cycle while maintaining high levels of soil C-sequestration. Coppicing can be repeated without any major negative effect on stump resprouting ability for, at least, three to five times repeatedly, which corresponds to a total time interval of 12–20 years. However, we still do not know whether e[CO_2] will have an impact on the resprouting ability of woody stumps and whether this will interact with different poplar genotypes. Longer-term studies will be helpful to answer these types of question.

Cultural practices such as fertilisation and irrigation are obviously very important for expanding the C-sequestration potential, especially in less-humid environments or with nutrient-demanding woody crops. However, peculiar cultural practices can be implemented (i.e. crop association with N-fixing trees) to make silvicultural management of SRF plantations more sustainable and environment-friendly. In any case, nutrients, particularly N, will be depleted under a CO_2-enriched atmosphere. Therefore, the selection of nutrient-efficient plant material will also be a valid strategy.

10.6.2 Global Carbon Cycle

Planting trees and forests is part of an expansion strategy of C-sinks that can greatly contribute to GHG mitigation by sequestering C into woody biomass and into the soil (IPCC 2000). While natural forests suffer from deforestation, the importance of forest plantations is increasing, because they cover 190 Mha worldwide, with planting rates of 8.5–10.5 Mha year^{-1} and an annual net gain of 1.96 Mha year^{-1} between 1965 and 1990 (FAO 2001).

In the POPFACE experiment, annual net primary productivity during the first rotation cycle was calculated as the sum of the increment of woody biomass, foliage production and investment into roots. The NPP for all three genotypes, at the end of the 3-year rotation, was increased in e[CO$_2$] by 20–36 %, as compared to the control. On a relative basis, fine root production was much more stimulated by e[CO$_2$] than the other components of NPP, particularly for *P. nigra* and *P. × euramericana*. However, stimulations of the aboveground woody component accounted for 67 %, 53 % and 61 % for *P. alba*, *P. nigra* and *P. × euramericana*, respectively, of the total increase in NPP caused by e[CO$_2$]. Therefore, rising atmospheric [CO$_2$] might further expand the potential of forest plantations for C-sequestration (see Chapters 11–13), in the above- and belowground biological components (DeLucia et al. 1999). Particular attention should, however, be devoted to soil processes and their dynamics (Schlesinger and Lichter 2001) because, as our experiment also suggests, C-incorporation into the different soil fractions evolves through short-term effects (i.e. priming effect) that can be reversed on a longer time-scale.

10.6.3 Other Ecosystem Goods and Services

Short rotation poplar plantations are man-made, simplified ecosystems mainly devoted to biomass production. However, they can also provide other goods and services, difficult to quantify in monetary terms but of increasing importance worldwide. Beside C-sequetration by trees and soil, an important environmental service given by SRF is bioenergy, consisting of biomass production substituting for fossil fuel and therefore reducing GHG emissions. Bioenergy production can be increased by e[CO$_2$] in relation to the positive response of aboveground biomass. Another interesting contribution of SRF plantations to environmental amelioration is the ability of trees and plantations to remediate soil and water pollution by extracting, immobilizing or metabolizing various pollutants. Very few experimental data exist on this subject, calling for more research in this field. However, it is reasonable to hypothesize that the large increase in C-assimilation under e[CO$_2$] will improve the phytoremediation properties of fast-growing tree plantations because of their high biomass productivity, high water and nutrient uptake, increased root productivity and greater root exudates, that will also augment the mass and metabolic activities of microbial and fungal soil communities.

Finally, soil erosion control and slope stabilization could be beneficially influenced by e[CO$_2$] because of the large increase in root mass and their penetration into deeper soil layers.

10.7 Conclusions

This FACE infrastructure has made it possible to conduct one of the first European experiments on climate change, at the scale of a planted forest ecosystem. The results that were produced can be summarized as follows.

The three poplar genotypes studied did not exhibit any consistent photosynthetic acclimation during this long-term experiment. This was shown by the large increases in leaf photosynthesis at elevated $[CO_2]$ and the lack of any consistent decrease in either $V_{c,max}$ or J_{max}. The lack of significant acclimation may be explained by increased sink capacity for carbohydrate. Even after 6 years of growth at elevated $[CO_2]$, the mean enhancement in midday photosynthesis for the three studied genotypes was more than 50 %.

Elevated $[CO_2]$ substantially increased, by 20 % on average, the GPP of all three *Populus* species. The stimulation declined sharply over the 3 years, but this was attributed to the transition from open to closed canopy and was not the result of photosynthetic acclimation.

Total annual biomass productivity, averaged over the three genotypes, was 17.3 Mg ha^{-1} year^{-1} and 21.4 Mg ha^{-1} year^{-1}, respectively, under the control and e$[CO_2]$, with a large species-specific effect. Therefore, the average e$[CO_2]$ stimulation of NPP for the three poplar genotypes was greater than 20 %. Also, the elevated $[CO_2]$ treatment had a positive effect (14 %) on NPP/GPP at the end of the first 3-year rotation cycle. Hence, elevated atmospheric $[CO_2]$ enhanced the productivity and light-use efficiency of a poplar SRC ecosystem, but without changing the biomass allocation pattern.

Elevated $[CO_2]$ largely enhanced fine root production of the studied poplar species, by 60–90 %, increased the rate of root turnover in all three genotypes, from 1.5 to 2.0 and also stimulated root colonization by mycorrhizal fungi.

Large increases in soil CO_2 effluxes were indeed observed in the e$[CO_2]$ treatment during the entire experiment. Many factors appeared to contribute to the e$[CO_2]$ stimulation of soil respiration. In fact, e$[CO_2]$ increased both root biomass and specific root respiration, as well as C input to the soil.

Despite the enhancing effect of e$[CO_2]$ on the accumulation of newly produced carbon (C_{new}) in the soil, at the end of the first rotation cycle, the increase in the total soil C pool (C_{total}) was smaller in e$[CO_2]$ as compared to control plots. It is suggested that these contrasting results were caused by a priming effect of the newly incorporated root litter, the priming effect being defined as the stimulation of SOM decomposition caused by the addition of labile substrates.

In conclusion, the e$[CO_2]$ stimulating effect on NPP was mainly attributed to an increase in relatively slow-turnover C pools (wood). Also, more new soil C was added to the total soil C pool under e$[CO_2]$. However, an increased loss of old soil C was observed in the FACE treatment, possibly caused by a priming effect. Therefore, the net ecosystem productivity was not significantly increased by e[CO$_2$] during the first 3 years of the experiment. However, preliminary results from the second 3-year rotation seem to indicate that the priming effect could be reversed under longer time intervals, causing the total soil C under elevated [CO$_2$] to be larger than under ambient [CO$_2$].

Acknowledgements.   Funding was provided by the EC Fifth Framework Program, Environment and Climate RTD Programme, research contract EVR1-CT-2002-40027 (EURO-FACE), by the Centre of Excellence "Forest and Climate" of the Italian Ministry of University and Research (MIUR) and by the Italy–USA Bilateral Programme on Climate Change of the Italian Ministry of Environment. B.Gielen acknowledges the Fund for Scientific Research–Flanders for her post-doctoral fellowship. We wish to thank also all the young researchers, students and technicians who helped us in conducting the experiment and in data collection.

References

Bernacchi C, Calfapietra C, Davey P, Wittig V, Scarascia Mugnozza G, Raines C, Long S (2003) Photosynthesis and stomatal conductance responses of poplars to free-air [CO$_2$] enrichment (PopFACE) during the first growth cycle and immediately following coppice. New Phytol 159:609–621

Calfapietra C, Gielen B, Sabatti M, De Angelis P, Scarascia Mugnozza G, Ceulemans R (2001) Growth performance of *Populus* exposed to "free air carbon dioxide enrichment" during the first growing season in the POPFACE experiment. Ann For Sci 58:819–828

Calfapietra C, Gielen B, Sabatti M, De Angelis P, Miglietta F, Scarascia-Mugnozza G, Ceulemans R (2003a) Do above-ground growth dynamics of poplar change with time under [CO$_2$] enrichment? New Phytol 160:305–318

Calfapietra C, Gielen B, Galema ANJ, Lukac M, De Angelis P, Moscatelli MC, Ceulemans R, Scarascia-Mugnozza G (2003b) Free-air [CO$_2$] enrichment (FACE) enhances biomass production in a short-rotation poplar plantation (POPFACE). Tree Physiol 23:805–814

Ceulemans R, Scarascia-Mugnozza G, Wiard BM, Braatne JH, Hinckley TM, Stettler RF, Isebrands JG, Heilman PE (1992) Production physiology and morphology of *Populus* species and their hybrids grown under short rotation. I. Clonal comparisons of 4-year growth and phenology. Can J For Res 22:1937–1948

Cotrufo MF, De Angelis P, Polle A (2005) Leaf litter production and decomposition in a poplar short rotation coppice exposed to free air CO$_2$ enrichment (POPFACE). Global Change Biol (in press)

Dahlman RC, Kucera CL (1965) Root productivity and turnover in native prairie. Ecology 46:84–89

Dalenberg JW, Jager G (1989) Priming effect of some organic additions to ^{14}C-labelled soil. Soil Biol Biochem 21:443–448

DeLucia EH, Hamilton JG, Naidu SL, Thomas RB, Andrews JA, Finzi A, Lavine M, Matamala R, Mohan JE, Hendrey GR, Schlesinger WR (1999) Net primary production of a forest ecosystem with experimental [CO$_2$] enrichment. Science 284:1177–1179

FAO (2001) Global forest resources assessment 2000. Main report, FAO forestry paper. FAO, Rome

Ferris R, Sabatti M, Miglietta F, Millis R.F, Taylor G (2001) Leaf area is stimulated in *Populus* by free air CO_2 enrichment (POPFACE), through increased cell expansion and production. Plant Cell Environ 24:305–315

Gielen B, Calfapietra C, Sabatti M, Ceulemans R (2001) Leaf area dynamics in a poplar plantation under free-air carbon dioxide enrichment. Tree Physiol 21:1245–1255

Gielen B, Calfapietra C, Claus A, Sabatti M, Ceulemans R (2002) Crown architecture of *Populus* spp is differentially modified by free-air CO_2 enrichment (POPFACE). New Phytol 153:91–99

Gielen B, Liberloo M, Bogaert J, Calfapietra C, De Angelis P, Miglietta F, Scarascia-Mugnozza G, Ceulemans R (2003) Three years of free-air CO_2 enrichment (POPFACE) only slightly affect profiles of light and leaf characteristics in closed canopies of *Populus*. Global Change Biol 9:1022–1037

Gray JE, Holroyd GH, Lee FM van der, Bahrami AR, Sijmons PC, Woodward FI, Schuch W, Hetherington AM (2000) The HIC signalling pathway links [CO_2] perception to stomatal development. Nature 408:713–716

Hendrey GR, Lewin KF, Nagy J (1993) Free air carbon dioxide enrichment: development, progress, results. Vegetatio 104/105:17–31

Hoosbeek MR, Lukac M, van Dam D, Godbold DL, Velthorst E.J, Biondi FA, Peressotti A, Cotrufo MF, De Angelis P, Scarascia-Mugnozza GE (2004) More new carbon in the mineral soil of a poplar plantation under free air carbon enrichment (POPFACE): cause of increased priming effect? Global Biogeochem Cycles 18:1–7

IPCC 2000 (2000) Land use, land use change and forestry. Report for the intergovernmental panel on climate change. Cambridge University Press, Cambridge

Kozlowski TT, Pallardy SG (1997) Physiology of woody plants, 2nd edn. Academic Press, San Diego

Liberloo M, Gillen SY, Calfapietra C, Marinari S, Luo ZB, DeAngelis P, Ceulemans R (2005) Elevated CO_2 concentration, fertilization and their interaction: growth stimulation in a short-rotation poplar coppice (EUROFACE). Tree Physiol 25:179–189

Lukac M, Calfapietra C, Godbold D (2003) Production, turnover and mycorrhizal colonisation of root systems of three *Populus* species grown under elevated [CO_2] (POPFACE). Global Change Biol 9:838–848

Luo ZB, Langenfeld-Heyser R, Calfapietra C, Polle A (2005). Influence of free air CO_2 enrichment (EUROFACE) and nitrogen fertilisation on the anatomy of juvenile wood of three poplar species after coppicing. Trees 19:109–118

Miglietta F, Peressotti A, Vaccari F, Zaldei A, DeAngelis P, Scarascia Mugnozza G (2001) Free air CO_2 enrichment (FACE) of a poplar plantation: the POPFACE fumigation system. New Phytol 150:465–476

Norby RJ, Wullschleger SD, Gunderson CA, Johnson DW, Ceulemans R (1999) Tree responses to rising [CO_2] in field experiments: implications for the future forest. Plant Cell Environ 22:683–714

Schlesinger WH, Lichter J (2001) Limited carbon storage in soil and litter of experimental forest plots under increased atmospheric [CO_2]. Nature 411:466–469

Smith P, Powlson DS, Smith JU, Falloon P, Coleman K (2000) Meeting Europe's climate change commitments: quantitative estimates of the potential for carbon mitigation by agriculture. Global Change Biol 6:525–539

Taylor G, Ceulemans R, Ferris R, Gardner SDL, Shao BY (2001) Increased leaf area expansion of hybrid poplar in elevated [CO_2]. From controlled environment to open-top chambers and to FACE. Environ Pollut 115:463–472

Taylor G, Tricker PJ, Zhang FZ, Alston VJ, Miglietta F, Kuzminsky E (2003) Spatial and temporal effects of free-air CO_2 enrichment (POPFACE) on leaf growth, cell expansion, and cell production in a closed canopy of poplar. Plant Physiol 131:177–185

Tricker, PJ, Calfapietra C, Kuzminsky E, Puleggi R, Ferris R, Nathoo M, Pleasants LJ, Alston V, De Angelis P, Taylor G(2004) Long-term acclimation of leaf production, development, longevity and quality following 3 yr exposure to free-air CO$_2$ enrichment during canopy closure in *Populus*. New Phytol 162:413–426

Tricker PJ, Trewin H, Kull O, Clarkson GJ, Eensalu E, Tallis MJ, Colella A, Doncaster CP, Sabatti M, Taylor G (2005) Stomatal conductance and not stomatal density determines the long-term reduction in leaf transpiration of poplar in elevated [CO$_2$]. Oecologia (in press)

Van Kessel C, Nitschelm J, Horwath WR, Harris D, Walley F, Luscher A, Hartwig U (2000) Carbon-13 input and turn-over in a pasture soil exposed to long-term elevated atmospheric [CO$_2$]. Global Change Biol 6:123–135

11 The Duke Forest FACE Experiment: CO_2 Enrichment of a Loblolly Pine Forest

W.H. Schlesinger, E.S. Bernhardt, E. H. DeLucia, D.S. Ellsworth, A.C. Finzi, G. R. Hendrey, K.S. Hofmockel, J Lichter, R. Matamala, D. Moore, R. Oren, J.S. Pippen, and R.B. Thomas

11.1 Introduction

The free-air CO_2 enrichment (FACE) experiment in the Duke Forest tests how a forest will respond to future, higher levels of CO_2 in Earth's atmosphere. The experiment is focused on changes in tree growth, water use, and the sequestration of carbon in wood and soils. CO_2 fumigation began in a prototype plot in 1994 and in three additional experimental plots in 1996 in a 16-year-old stand of loblolly pine (*Pinus taeda*). Loblolly pine is a widespread early successional tree in the southeastern United States (Oosting 1942), where it is also a major commercial species dominating >10 × 10[6] ha (Burns and Honkala 1990; Harlow et al. 1991). Growth of pine plantations on abandoned agricultural land is thought to yield a substantial carbon sink in the southeast (Delcourt and Harris 1980; Caspersen et al. 2000). This reforestation may account for part of the large putative carbon sink in North America (Tans et al. 1990; Fan et al. 1998). Some investigators believe that an enhanced C sink due to CO_2 fertilization of forests might ultimately slow the rise of CO_2 in Earth's atmosphere (Idso et al. 1991), but that hypothesis remains untested in nature (but see Schimel et al. 2000).

Prior work in open-top chambers and glass-house experiments had shown large increases in the biomass of loblolly pine in response to growth at elevated CO_2 concentrations ($e[CO_2]$) with ample soil nutrients (Thomas et al. 1994; Tissue et al. 1996, 1997). An initial motivation for the FACE experiment in Duke Forest was to examine this growth response in natural conditions, where trees experience competition, drought, nutrient limitations, pests, and pathogens. The Duke FACE experiment seeks to answer a critical question for

Ecological Studies, Vol. 187
J. Nösberger, S.P. Long, R.J. Norby, M. Stitt,
G.R. Hendrey, H. Blum (Eds.)
Managed Ecosystems and CO_2
Case Studies, Processes, and Perspectives
© Springer-Verlag Berlin Heidelberg 2006

foresters and policy makers: Can we expect more growth and carbon sequestration in these forests in the future? FACE technology allows us to answer
that question today, by applying e[CO_2] to experimental plots with unaltered,
natural levels of other growth parameters (see Chapter 1).

11.2 Site Description

Located in Orange County, North Carolina, USA (35°97' N, 79°09' W), the
experimental site was clear-cut in 1982, to remove a 40- to 60-year-old mixed
pine forest. The site was drum-chopped and burned prior to tree planting in
1983. The 32-ha experimental forest of loblolly pine is derived from 3-year-
old, half-sibling seedlings planted in 2.4 × 2.4 m spacing. In 1994, when CO_2
enrichment commenced in the prototype plot, the pine trees had grown to
12 m height, reaching 14 m height in 1996 at the start of the formal experiment. In 1998, the leaf area index of pine was 3.7 m^2 m^{-2} (Schäfer et al. 2002) and
the stand was just beginning to enter a stage of competitive self-thinning
(Peet and Christensen 1980; Christensen and Peet 1981). Pine composed 98 %
of the canopy. Deciduous tree species, which dominate the understory, include
sweetgum (*Liquidambar styraciflua*), red maple (*Acer rubrum*), winged elm
(*Ulmus alata*), red bud (*Cercis canadensis*), and dogwood (*Cornus florida*)
that have sprouted from stumps or dispersed from the surrounding vegetation. A few of these individuals, most often sweetgum and tulip polar (*Liriodendron tulipifera*), reach the canopy.

The soils are clay loams, classified as low-fertility Ustic Hapludalfs of the
Enon series, which are typical of many upland areas in the southeastern
United States. Throughout this region, these soils supported many decades of
cotton and tobacco agriculture before abandonment to silviculture in the
early 1900s (J. Edeburn, personal communication). The soils are relatively
homogeneous, derive from mafic bedrock, and exhibit acidic (pH 5.75), well
developed profiles of mixed clay mineralogy. Boreholes show up to 1 m of topsoil underlain by 5 m of saprolite, above a highly fractured granodiorite or
diorite bedrock. Variation in elevation ranges up to 15 m across the site, but
topographic relief is generally less than 1°. The static water table lies at 6 m
depth, but the site drains poorly and surface soils often become saturated in
the spring. The mean annual temperature is 15.5 °C and the mean annual precipitation is 1140 mm.

The FACE experiment consists of seven circular plots, 30 m in diameter
(Fig. 11.1). Four of the seven plots are fumigated with CO_2 to maintain an
atmospheric concentration 200 ppm above current CO_2 levels (c[CO_2]), to
simulate the Earth's atmosphere in the year 2050 (Hendrey et al. 1999; Chapter 2). The three remaining, control plots are identical to the e[CO_2] plots,
except that they are fumigated with ambient air (i.e., c[CO_2]). The formal

Fig. 11.1 The Duke Forest free-air CO$_2$ enrichment (FACE) project, shown looking south, with the six experimental plots in the foreground and the prototype plot in the distance. Photo by Will Owens

experiment was begun on 27 August 1996; and the e[CO$_2$] was maintained continuously when the air temperature was >5 °C during the first 6 years, except for brief periods during Hurricane Fran in 1996 and Hurricane Floyd in 1999. Beginning in 2003, the fumigation has been maintained only during daylight hours and when the ambient air temperature is >5 °C, similar to the fumigation protocol at the FACE prototype plot.

The CO$_2$ used for fumigation is derived from natural gas and consequently is strongly depleted in ^{13}C relative to ^{12}C (the δ^{13}C value is –43.0±0.6). Raising the atmospheric CO$_2$ concentration by 200 ppm with this source of CO$_2$ reduces the δ^{13}C ratio of the atmosphere in the fumigated plots from –8‰ to –20‰. One would expect that new photosynthate produced from this atmosphere would carry a $\delta^{13}C$ ratio of approximately $-40\permil$ (Farquhar et al. 1982). In fact, we have measured values of $-39\permil$ to $-42\permil$ in new pine needles and fine roots grown in FACE conditions (Ellsworth 1999; Finzi et al. 2001; Matamala et al. 2003). We track the incorporation of this isotopic signature into soil organic matter to estimate the turnover of soil organic fractions (Schlesinger and Lichter 2001; Lichter et al. 2005).

Prior to the experiment, samples of the upper mineral soil at 0–12 cm depth were collected from 119 locations spanning the 32-ha research site. These samples were analyzed for C, N, P, Ca, Mg, and K; and the results were tested for spatial autocorrelation. This analysis indicated that the spatial dependence of each soil parameter was less than the shortest distance between any two experimental plots (i.e., 85 m), implying that the plots were independent replicate samples of this pine forest. At the start of the experiment, the plots were paired, based on subjective criteria of similarity, and one member of each pair was assigned to control or fumigated status. As the experiment has unfolded, we have realized the importance of subtle variations in forest and soil conditions across the site, requiring the use of pretreatment conditions as a covariate in analysis of variance, with $n=3$ in each category, analyzing the prototype data separately. Some values cited here differ from those published earlier as a result of this reanalysis of data.

11.3 Results

11.3.1 Resource Acquisition

Photosynthetic rates of canopy leaves (i.e., pine needles) are directly related to leaf nitrogen content (Fig. 11.2). Growth in $e[CO_2]$ slightly reduces the concentration of N in canopy leaves (Oren et al. 2001; Finzi et al. 2004), but throughout the experiment, the canopy loblolly pine have shown a 40–50 % increase in photosynthesis (Ellsworth 1999; Myers et al. 1999; Schäfer et al, 2003; Springer et al. 2005; Ellsworth and Klimas 2005). The increase in photosynthesis is not accompanied by a reduction in the amount of water used by canopy pine trees (Ellsworth et al. 1995; Ellsworth 1999; Schäfer et al. 2002).

Understory species show a variable response to growth at $e[CO_2]$ (DeLucia and Thomas 2000), but no understory species contributes a significant amount to forest production or carbon storage. There is some evidence for a reduction in water use by sweetgum (*Liquidambar styraciflua*; Schäfer et al. 2002; Herrick et al. 2004).

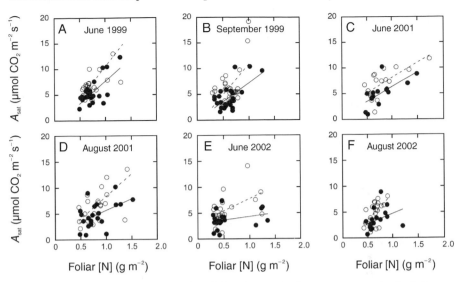

Fig. 11.2 Photosynthetic rate (*A*) of loblolly pine and other canopy species in full sunlight, shown as a function of foliage N content (g N m^{-2} leaf area) in the Duke Forest CO₂ enrichment (FACE) experiment. *Open circles* are from c[CO₂] plots, *closed circles* are from e[CO₂] (Springer et al. 2005)

11.3.2 Resource Transformation

We made repeated measures of tree diameter growth and litterfall and used these to estimate the effect of growth at e[CO₂] on net primary production (NPP) in the loblolly pine plantation (DeLucia et al. 1999; Hamilton et al. 2002; Moore et al. 2005). Elevated CO₂ increased the annual basal area increment (BAI) of individual canopy pine trees by 13–27 % during the first 8 years of the experiment (Moore et al. 2005). BAI in ambient and fumigated plots was positively correlated with growing season temperature and the amount of rainfall. Exposure to e[CO₂] increased the rate but not the duration of growth in most years. The stimulation was largely confined to emergent and dominant individuals and was not evident in sub-canopy trees. However, an understory vine, poison ivy (*Toxicodendron radicans*), showed a 77 % increase in growth over the same interval (Mohan et al. 2005). Exposure to e[CO₂] caused an increase in the biomass increment of 108 g C m^{-2} year^{-1} (27 %) in the pine trees, contributing nearly half of the 185 g C m^{-2} year^{-1} (32 %) increase in net carbon storage in this forest.

Each year, a substantial portion of the forest NPP returned to the soil as litterfall, which was 21 % greater in fumigated plots during 1998–2000 (Finzi et al. 2002). Early studies indicated that the leaf area index is similar in control and fumigated plots (DeLucia et al. 2002; Schäfer et al. 2002), but more recent work suggests an increase in LAI in e[CO$_2$] plots (McCarthy et al., personal communication). Although pine cones and seeds account for only a small fraction of all litterfall (<1 %; Finzi et al. 2002), these tissues showed a surprising, disproportionate increase (200 %) in response to tree growth at e[CO$_2$] (LaDeau 2005). Production of fine roots also accounted for only a small fraction (5–7 %) of NPP, but fine root growth was consistently greater, as much as 86 %, during the summer in plots maintained at e[CO$_2$] (Matamala and Schlesinger 2000; Pritchard et al. 2001). There were no long-term changes in the winter-time biomass of fine roots, so the higher rate of fine root growth was accompanied by greater root death, resulting in a greater absolute turnover, but no change in percentage turnover, of root tissues as a result of growth at e[CO$_2$] (Matamala et al. 2003). The mean residence time for carbon in fine roots was 4.2 years – much longer than traditionally thought but consistent with recent measurements of the radiocarbon content of roots in various eastern forests (Gaudinski et al. 2001).

Across the six experimental plots, NPP was directly related to soil nitrogen mineralization, but the rate of production at a given level of nitrogen availability was substantially higher in e[CO$_2$] plots (Fig. 11.3). Greater growth of perennial tissue (i.e., wood) and greater turnover of foliage, roots, and reproductive tissues is necessarily associated with a greater nutrient demand by the forest (Table 11.1). For instance, as a consequence of the significant increase in foliage biomass, there were significant increases in canopy N and P content at e[CO$_2$] (Finzi et al. 2002, 2004). Some of the nutrient demand can be met through increases in nutrient-use efficiency, such as by the resorption of nutrients from senescent foliage before it drops. The remainder of the nutrient demand must be met by greater uptake from the soil. As each cohort of pine needles age over their 19-month lifetime (Zhang and Allen 1996), the content of N and P declines, suggesting that the demand set by rapid forest growth under e[CO$_2$] is not matched by nutrient uptake

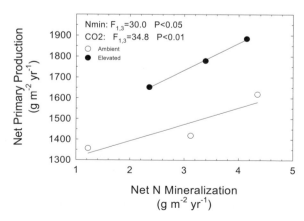

Fig. 11.3 The relationship between NPP and annual rate of net nitrogen mineralization during the second year of the experiment. *Open circles* are from c[CO$_2$] plots, *closed circles* are from e[CO$_2$] (from Finzi et al. 2002)

Table 11.1 Nutrient budget of control (ambient) and fumigated (elevated CO_2) plots of the Duke Forest FACE experiment, average of 1998–2000 (from Finzi et al. 2002)

Component	Control plot (kg ha^{-1} year^{-1})	Fumigated plot (kg ha^{-1} year^{-1})	Change (%)
Plant requirement	51.9	61.5	18.5
Retranslocation	28.6	31.5	10.2
Uptake from soil	23.3	30.0	28.9
Nitrogen-use efficiency (NPP/requirement)	292	311	6.6

from the soil (Finzi et al. 2004). We also measured slightly greater retranslocation of nitrogen and phosphorus from foliage before abscission, conferring greater nutrient-use efficiency in production (i.e., NPP per unit nitrogen uptake from the soil).

We have independent evidence from several experiments that microbial activity in the soil is limited by the supply of organic carbon, such that one might expect an increase in microbial activity with the greater inputs of dead plant materials in plots maintained at e[CO_2] (Gallardo and Schlesinger 1994; Allen and Schlesinger 2004). Nevertheless, in comparisons of control and fumigated plots at the FACE experiment, there is no evidence of substantial changes in soil microbial activity that allow an easy identification of the source of additional nutrients for plant uptake. There are no significant changes in soil microbial biomass, gross nitrogen mineralization, and absolute or specific nitrogen immobilization by soil microbes, at this and other FACE sites (Finzi and Schlesinger 2003; Zak et al. 2003). Decomposition of foliage is similar between control and fumigated plots (Finzi et al. 2001). Rates of asymbiotic nitrogen fixation in the soil are low and not significantly different among plots (Hofmockel and Schlesinger 2005); and we have not been able to establish nitrogenase activity in *Cercis canadensis*, despite observations of nitrogen fixation by other members of this genus in Europe (Bryan et al. 1996). Thus, increased nitrogen demand must be satisfied by greater root growth and exploration of the soil.

Evidence for a greater activity of roots is seen in higher root respiration (George et al. 2003), higher accumulations of CO_2 in the soil pore space, and greater CO_2 efflux from the soil surface (Andrews and Schlesinger 2001; Bernhardt et al. 2005). Root activity is estimated to account for 55 % of the respiration at the soil surface (Andrews et al. 1999). Greater CO_2 in the soil pore space is associated with greater faster rates of chemical weathering of soil minerals via the formation of carbonic acid. We have measured higher concentrations of weathering products in the soil solution collected at 200 cm depth, with proportional increases of cations and bicarbonate much greater than for Cl, which increases with depth as a result of the plant uptake of water (Table 11.2). Some of the greater flux of cations may derive from the pool held on the cation exchange capacity in these soils, but the greater flux of silicon indicates direct chemical weathering of soil minerals. The greater flux of bicarbonate in seepage waters results in a small, incremental net sink for carbon (5 g m^{-2} year^{-1}) in the forest maintained at e[CO_2]. This flux is consistent with the increasing content of alkalinity in North American riverwaters and a small global sink for carbon via this pathway during the past century (Raymond and Cole 2003).

In control plots, carbon in undecomposed plant debris on the forest floor accumulated at a rate of 78 g m^{-2} year^{-1}, similar to forest floor accumulations in other aggrading loblolly pine plantations in this region (Richter et al. 1999; Johnson et al. 2003). Observation of a significant increase in the mass of the forest floor in e[CO_2] plots is consistent with our observations that there have been few significant changes in soil microbial activity as a result of plant growth at e[CO_2]. During the first 6 years of the Duke Forest FACE experiment, organic C accumulated in the forest floor of the e[CO_2] plots at a rate which was 52 ± 16 g C m^{-2} year^{-1} faster than in c[CO_2] plots (Lichter et al. 2005).

In contrast to the forest floor, we detected no statistically significant incremental storage of carbon in the soil organic matter of the e[CO_2] plots relative to c[CO_2] plots (Lichter et al. 2005). Carbon sequestration in the soil (0–30 cm depth) was 89 ± 44 g C m^{-2} year^{-1}, averaged across e[CO_2] and c[CO_2] plots during the first 6 years of the experiment. The accumulation of soil organic carbon in the Duke forest plots is greater than average rates reported for other loblolly pine forests in the Southeast, including values of 28 g C m^{-2} year^{-1} in Virginia (Schiffman and Johnson 1989), 4 g C m^{-2} year^{-1} in South Carolina (Richter et al. 1999), and virtually no change in the content of soil organic matter in a forest in Tennessee (Johnson et al. 2003). We found substantial changes in the δ^{13}C of labile fractions of soil organic matter in the fumigated

Table 11.2 Chemistry of the soil solution at 15 cm and 200 cm depth in control and fumigated plots of the Duke forest FACE experiment (from Andrews and Schlesinger 2001)

Chemical	Depth (cm)	c[CO_2]	e[CO_2]
Cl	15	0.04	0.05
	200	0.10	0.17
Cations	15	0.36	0.41
	200	0.76	2.82
HCO3 – alkalinity	15	0.13	0.13
	200	0.64	1.68
Dissolved silicon	15	0.07	0.08
	200	0.28	0.35

plots, associated with inputs of new photosynthate under FACE conditions. Similar to Hoosbeck et al. (2004; see Chapter 10), we postulate a priming effect of labile carbon, increasing the decomposition of existing organic matter which is replaced by the new inputs, so there is little increment to storage.

11.3.3 Nitrogen Limitation

With the direct relation between forest growth and nitrogen availability (Fig. 11.3) and only indirect evidence that soil nitrogen turnover may have increased, we ask the question: is the CO$_2$ response that we observe in Duke forest sustainable over many years? Alternatively, one might expect that the higher growth rates at e[CO$_2$] would lead to an increasing nutrient deficiency in this forest. Increased rates of fine root growth, such as at the Duke FACE site, are often associated with tree growth in nutrient-deficient soils (Waring and Schlesinger 1985). Norby et al. (2002; see Chapter 13) report a sustained increase in NPP, largely seen in roots, in the deciduous forest FACE experiment at Oak Ridge; and Oren et al. (2001) show that low soil fertility reduced the rate of carbon sequestration in woody biomass in the FACE prototype plot at Duke after 3 years. To date, our evidence on this point is equivocal. Mea-

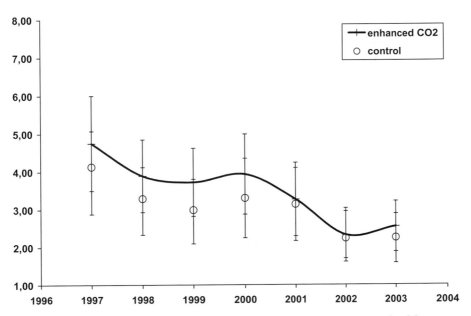

Fig. 11.4 Mean annual wood ring increment in three cores taken from each of five trees in each of the three plots at c[CO$_2$] and e[CO$_2$]. *Open circles* are from c[CO$_2$] plots, *closed circles* are from e[CO$_2$]. Unpublished data of J.S. Pippen and A. Ballentyne (Duke University)

surements of tree rings in 32 trees from the Duke forest experiment show the
expected age-related downtrend in wood increment among trees in the control plots (Fig. 11.4). The wood increment of trees in the fumigated plots was
24 % greater than controls in 1999, declining to an insignificant difference in
2004 (Ballentyne, personal communication). However, among a larger sample
of trees instrumented with dendrometer bands, estimates of basal area increment among the canopy pines, which comprise the majority of plant biomass,
show no loss of stimulation with time (Moore et al. 2005). Calculated estimates of total stand production and increment (Fig. 11.5) show a decline with
time, although a strong percent CO_2 stimulation of growth still persists today.
A large stimulation in 2002 is associated with low rates of production, and

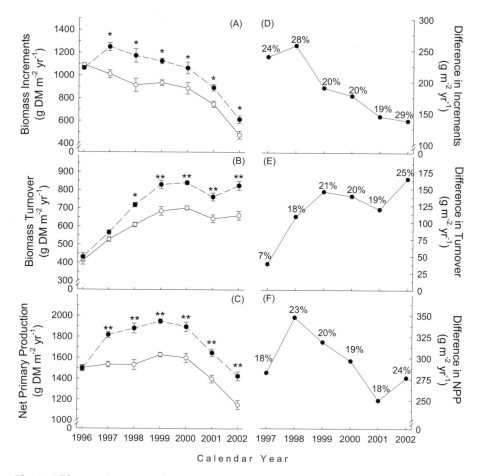

Fig. 11.5 Biomass increment (largely wood), biomass turnover (largely litterfall) and net
primary production in the Duke Forest free air CO_2 enrichment (FACE) experiment; and
trends in the difference between high CO_2 plots and ambient in each parameter. *Open
circles* are from c[CO_2] plots, *closed circles* are from e[CO_2] (from Finzi et al. 2005)

presumably low nitrogen demand, during an exceptionally dry year. Various theoretical models suggest that progressive nitrogen limitation should reduce the CO_2 stimulation of plant production (Comins and McMurtrie 1993; Luo et al. 2003). The experimental addition of nitrogen at the prototype plot and in a study of *P. taeda* of similar age growing on sand, indicates that the time-lag for the onset of limitation can vary from zero (i.e., an immediate response to $e[CO_2]$) in the nutrient-poor sand to 3 years in the moderately nitrogen-poor FACE prototype (Oren et al. 2001). A direct test of this hypothesis awaits field fertilization trials which are scheduled to begin in the formal, replicated FACE experiment in January 2005.

11.4 Estimated Global Carbon Sink in Forests

We can use the data from the Duke Forest FACE experiment to estimate the maximum increment to the carbon sink in forests that might be expected from growth at future, higher levels of atmospheric CO_2. The incremental sink for carbon in forests grown at $e[CO_2]$ is185 g C m^{-2} year^{-1} (Table 11.3), somewhat less than the estimate of 272 g C m^{-2} year^{-1} using a variety of coupled models (Schäfer et al. 2003). Houghton (2003) estimates a terrestrial carbon sink of 2.9×10^{15} g C year^{-1} in undisturbed forests worldwide during the 1990s. If the world's forests were all to respond similarly to Duke Forest, showing a 32 % increment to the carbon sink (Table 11.3), these forests would accumulate an additional 0.93×10^{15} g C in the year 2050, when atmospheric CO_2 would reach the levels of our experiment. This additional sink for carbon would amount to 6.2 % of the estimated emissions from fossil fuels in 2050 (15 $\times 10^{15}$ g C year^{-1}; IPCC 2001) or 11.6 % of the incremental emissions compared to today. These represent an upper limit for the incremental sink in forests, as we should expect that many forests will be less responsive than young loblolly pine, as will be forests growing on nutrient-poor soils (Oren et al. 2001). Our

Table 11.3 Estimated carbon sink in $c[CO_2]$ and $e[CO_2]$ plots of the Duke Forest FACE experiment. All data are g C m^{-2} year^{-1}; from Finzi et al. (2002), Lichter et al. (2005), Andrews and Schlesinger (2001)

Component	$c[CO_2]$	$e[CO_2]$	Difference	%
Stem biomass	397	505	108	27
Forest floor	78	130	52	67
Soil organic matter	89	89	0	
Groundwater flux	16	21	5	31
Total	580	765	185	32

work supports the conclusions of Schimel et al. (2000), who suggest that a much greater effect is seen from reforestation than from the direct effects of CO_2 on growing forests. As for the expected sink in agricultural soils, an enhanced sink for carbon in forests seems unlikely to solve the global warming problem; and we must also look hard to cut emissions from fossil fuel combustion (Jackson and Schlesinger 2004).

11.5 Conclusions

Free-air CO_2 enrichment (FACE) in the Duke Forest provides a whole-ecosystem arena in which to examine the response of a temperate coniferous forest to high, future levels of atmospheric CO_2. At the end of 8 years of the experiment, we conclude:

- Photosynthetic rates by canopy foliage have increased up to 50 % over controls.
- Basal area increment has been stimulated 13–27 % versus that in control plots, with interannual variation due to variations in temperature and moisture during the growing season.
- Biomass increment has increased by 108 g C m^{-2} $year^{-1}$ (27 %) over that in control plots.
- Growth and respiration of roots are higher in CO_2 fumigated plots.
- Litterfall is greater in high CO_2 plots and forest floor accumulation has increased.
- There has been little or no change in the total amount of soil organic matter as a result of CO_2 fumigations.
- While the stimulation of growth by high CO_2 persists after 8 years of fumigation, there is evidence of nitrogen limitation in the fumigated plots.

References

Allen AS, Schlesinger WH (2004) Nutrient limitations to soil microbial biomass and activity in loblolly pine forests. Soil Biol Biochem 36:581–589

Andrews JA, Schlesinger WH (2001) Soil CO_2 dynamics, acidification, and chemical weathering in a temperate forest with experimental CO_2 enrichment. Global Biogeochem Cycles 15:149–162

Andrews JA, Harrison KG, Matamala R, Schlesinger WH (1999) Separating of root respiration from total soil respiration using carbon-13 labeling during free-air carbon dioxide enrichment (FACE). Soil Sci Soc Am J 63:1429–1435

Bernhardt ES, Barber JJ, Pippen JS, Taneva, L, Andrews JA, Schlesinger WH (2005). Long-term effects of free air CO_2 enrichment (FACE on soil respiration. Biogeochemistry (in press)

Bryan JA, Beryln GP, Gordon JC (1996) Toward a new concept of the evolution of nitrogen fixation in the Leguminosae. Plant Soil 186:151–159

Burns RM, Honkala BH (1990) Silvics of North American trees, vol 2. Hardwoods. US Department of Agriculture, Washington, D.C.

Caspersen JP, Pacala, SW, Jenkins JC, Hurtt GC, Moorcroft PR, Birdsey RA (2000) Contributions of land-use history to carbon accumulation in U.S. forests. Science 290:1148–1151

Christensen NL, Peet RK (1981) Secondary forest succession on the North Carolina piedmont. In: West DC, Shugart HH, Botkin DB (eds) Forest succession: concepts and applications. Springer, Berlin Heidelberg New York, pp 230–245

Comins HN, McMurtrie RE (1993) Long-term response of nutrient-limited forests to CO_2 enrichment: equilibrium behavior of a plant–soil model. Ecol Appl 3:666–681

Delcourt HF, Harris WF. (1980) Carbon budget of the southeastern U.S. biota: analysis of historical change in trend from source to sink. Science 210:321–323

DeLucia EH, Thomas RB (2000) Photosynthetic responses to CO_2 enrichment of four hardwood species in a forest understory. Oecologia 122:11–19

DeLucia EH, Hamilton JG, Naidu SL, Thomas RB, Andrews JA, Finzi A, Lavine M, Matamala R, Mohan JE, Hendrey GR, Schlesinger WH (1999) Net primary production of a forest ecosystem under experimental CO_2 enrichment. Science 284:1177–1179

DeLucia EH, George K, Hamilton JG (2002) Radiation-use efficiency of a forest exposed to elevated concentrations of atmospheric carbon dioxide. Tree Physiology 22:1003–1010

Ellsworth DS (1999) CO_2 enrichment in a maturing pine forest: are CO_2 exchange and water status in the canopy affected? Plant Cell Environ 22:461–472

Ellsworth DS, Klimas C (2005) Photosynthetic-temperature responses and seasonal photosynthetic enhancement in an evergreen forest during three years of CO_2 enrichment. Tree Physiol (in press)

Ellsworth DS, Oren R, Huang C, Phillips N, Hendrey GR (1995) Leaf and canopy response to elevated CO_2 in a pine forests under free-air CO_2 enrichment. Oecologia 104:139–146

Fan S, Gloor M, Mahlman J, Pacala S, Sarmiento J, Takahashi T, Tans P (1998) A large terrestrial carbon sink in North America implied by atmospheric and oceanic carbon dioxide data and models. Science 282:442–446

Farquhar GD, O'Leary MH, Berry JA (1982) On the relationship between carbon isotope discrimination and the intercellular carbon dioxide concentration in leaves. Aust J Plant Physiol 9:121–137

Finzi AC, Schlesinger WH (2003) Soil-nitrogen cycling in a pine forest exposed to 5 years of elevated carbon dioxide. Ecosystems 6:444–456

Finzi AC, Allen AS, DeLucia EH, Ellsworth DS, Schlesinger WH (2001) Forest litter production, chemistry, and decomposition following two years of free-air CO_2 enrichment. Ecology 82:470–484

Finzi AC, DeLucia EH, Hamilton JG, Richter DD, Schlesinger WH (2002) The nitrogen budget of a pine forest under free air CO_2 enrichment. Oecologia 132:567–578

Finzi AC, DeLucia EH, Schlesinger WH (2004) Canopy N and P dynamics of a southeastern US pine forest under elevated CO_2. Biogeochemistry 69:363–378

Finzi AC, DeLucia EH, Lichter J, McCarthy H, Moore D, Oren R, Pippen, JS, Schlesinger WH (2005) Progressive nitrogen limitation of ecosystem processes under elevated CO_2 in a warm-temperate forest. Ecology (in review)

GallardoA, Schlesinger WH (1994) Factors limiting microbial biomass in the mineral soil and forest floor of a warm-temperate forest. Soil Biol Biochem 26:1409–1415

Gaudinski JB, Trumbore SE, Davidson EA, Cook AC, Markewitz, D, Richter DD (2001) The age of fine-root carbon in three forests of the eastern United States measured by radiocarbon. Oecologia 129:420–429

George K, Norby RJ, Hamilton JG, DeLucia EH (2003) Fine-root respirationin a loblolly pine and sweetgum forest growing in elevated CO_2. New Phytol 160:511–522

Hamilton JG, DeLucia EH, George K, Naidu SL, Finzi AC, Schlesinger WH (2002) Forest carbon balance under elevated CO_2. Oecologia 131:250–260

Harlow WM, Harrar ES, Hardin JW, White FM (1991) Textbook of dendrology, 7th edn. McGraw–Hill, New York.

Hendrey GR, Ellsworth DS, Lewin KF, Nagy J (1999) A free-air enrichment system for exposing tall forest vegetation to elevated atmospheric CO_2. Global Change Biol 5:293–309

Herrick JD, Maherali H, Thomas RB (2004) Reduced stomatal conductance in sweetgum (*Liquidambar styraciflua*) sustained over long-term CO_2 enrichment. New Phytol 162:387–396

Hofmockel KS, Schlesinger WH (2005) Water limitations to asymbiotic N fixation in a temperate pine plantation. (in preparation)

Hoosbeck MR, Lukac M, van Dam D, Godbold DL, Velthorst EJ, Biondi FA, Peressotti A, Cotrufo MF, De Angelis P, Scarascia-Mugnozza G (2004) More new carbon in the mineral soil of a poplar plantation under free air carbon enrichment (POPFACE): cause of increased priming effect? Global Biogeochem Cycles 18, doi:10.1029/2003 GB002127.

Houghton RA (2003) Revised estimates of the annual net flux of carbon to the atmosphere from changes in land use and land management 1850–2000. Tellus 55B:378–390

Idso SB, Kimball BA, Allen SG (1991) CO_2 enrichment of sour orange trees: two and a half years into a long-term experiment. Plant Cell Environ 14:351–353

IPCC (2001) Working group one, third assessment report. Cambridge University Press, Cambridge.

Jackson RB, Schlesinger WH (2004) Curbing the U.S. carbon deficit. Proc Natl Acad Sci USA 101:15827–15829

Johnson DW, Todd DE, Tolbert VR (2003) Changes in ecosystem carbon and nitrogen in a loblolly pine plantation over the first 18 years. Soil Sci Soc Am J 67:1594–1601

LaDeau SL (2005) The reproductive ecology of *Pinus taeda* growing in elevated CO_2. PhD thesis, Duke University, Durham, N.C.

Lichter J, Barron SH, Bevacqua CE, Finzi AC, Irving KF, Roberts MT, Stemmler EA, Schlesinger WH (2005) Soil carbon sequestration and turnover in a pine forest after six years of atmospheric CO_2 enrichment. Ecology 86: 1835–1847

Luo Y, White L, Canadell JG, DeLucia EH, Ellsworth DS, Finzi AC, Lichter J, Schlesinger WH (2003) Sustainability of terrestrial carbon sequestration: a case study in Duke Forest with inversion approach. Global Biogeochem Cycles 17, doi:10.1029/2002 GB001923

Matamala R, Schlesinger WH (2000) Effects of elevated atmospheric CO_2 on fine root production and activity in an intact temperate forest ecosystem. Global Change Biol 6:967–979

Matamala R, Gonzalez-Meler MA, Jastrow JD, Norby RJ, Schlesinger WH (2003) Impacts of fine root turnover on forest NPP and soil c sequestration potential. Science 302:1385–1387

Mohan JE, Ziska LH, Thomas RB, Sicher RC, George K, Clark JS, Schlesinger WH (2005) Poison ivy grows larger and more poisonous at elevated atmospheric CO_2. (in preparation)

Moore DJ, Aref S, HoRM, Pippen JS, Hamilton J, DeLucia EH (2005) Inter-annual varia-
    tion in the response of *Pinus taeda* tree growth to long term Free Air Carbon Dioxide
    Enrichment (FACE). Global Change Biol (in review)

Myers DA, Thomas RB,.DeLucia EH (1999) Photosynthetic capacity of loblolly pine
    (*Pinus taeda* L.) trees during the first year of carbon dioxide enrichment in a forest
    ecosystem. Plant Cell Environ 22:473–481

Norby RJ, Hanson PJ, O'Neill EG, Tschaplinski TT, Weltzin JF, Hansen RA,. Cheng W,
    Wullschleger SD, Gunderson CA, Edwards NT, Johnson DW (2002) Net primary pro-
    ductivity of a CO_2-enriched deciduous forest and the implications for carbon storage.
    Ecol Appl 12:1261–1266

Oosting HJ (1942) An ecological analysis of the plant communities of Piedmont, North
    Carolina. Am Midl Nat 28:1–126

Oren R, Ellsworth DS, Johnsen KH, Phillips N, Ewers BE, Maier C, Schafer KVR,
    McCarthy H, Hendrey GR, McNulty SG, Katul G (2001) Soil fertility limits carbon
    sequestration by forest ecosystems in a CO_2-enriched atmosphere. Nature 411:469–
    472

Peet RK, Christensen NL (1980) Succession: a population process. Vegetatio 43:131–140

Pritchard SG, Rogers HH, Davis MA, Van Santan E, Prior SA, Schlesinger WH (2001) The
    influence of elevated atmospheric CO_2 on fine root dynamics in an intact temperate
    forest. Global Change Biol 7:829–837

Raymond PA, Cole JJ (2003) Increase in the export of alkalinity from North America's
    largest river. Science 301:88–91

Richter DD, Markewitz D, Trumbore SE, Wells CG (1999) Rapid accumulation and
    turnover of soil carbon in a re-establishing forest. Nature 400:56–58

Schäfer KVR, Oren R, Lai, C-T, Katul GG (2002) Hydrologic balance in an intact temper-
    ate forest ecosystem under ambient and elevated atmospheric CO_2 concentration.
    Global Change Biol 8:895–911

Schäfer KVR, Oren R, Ellsworth DS, Lai C-T, Herrick JD, Finzi AC, Richter DD, Katul GG
    (2003) Exposure to an enriched CO_2 atmosphere alters carbon assimilation and allo-
    cation in a pine forest ecosystem. Global Change Biol 9:1378–1400

Schiffman PM, Johnson WC (1989) Phytomass and detrital carbon storage during forest
    regrowth in the southeastern United States piedmont. Can J For Res 19:69–78

Schimel DS, Melillo JM, Tian H, McGuire AD, Kicklighter D, Kittel T, Rosenbloom N, Run-
    ning SW, Thornton P, Ojima D, Parton W, Kelly R, Sykes M, Neilson R, Rizzo B (2000)
    Contribution of increasing CO_2 and climate to carbon storage by ecosystems in the
    United States. Science 287:2004–2006

Schlesinger WH, Lichter J (2001) Limited carbon storage in soil and litter of experimen-
    tal forest plots under increased atmospheric CO_2. Nature 411:466–469

Springer CJ, DeLucia EH, Thomas RB (2005) Relationship between net photosynthesis
    and foliar nitrogen concentration in a loblolly pine forest ecosystem grown with ele-
    vated atmospheric carbon dioxide. Tree Physiol (in press)

Tans PP Fung IY. Takahashi T (1990) Observational constraints on the global atmos-
    pheric CO_2 budget. Science 247:1431–1438

Thomas RB, Lewis JD, Strain BR (1994) Effects of leaf nutrient status on photosynthetic
    capacity in loblolly pine (*Pinus taeda*) seedlings grown in elevated atmospheric CO_2.
    Tree Physiol 14:947–960

Tissue DT, Thomas RB, Strain BR (1996) Growth and photosynthesis of loblolly pine
    (*Pinus taeda*) after exposure to elevated CO_2 for 19 months in the field. Tree Physiol
    16:49–59

Tissue DT, Thomas RB, Strain BR (1997) Atmospheric CO_2 enrichment increases growth
    and photosynthesis of *Pinus taeda* L.: a four-year experiment in the field. Plant Cell
    Environ 20:1123–1134

Waring RH, Schlesinger WH (1985) Forest ecosystems: concepts and management. Academic Press, San Diego

Zak DR, Holmes WE, Finzi AC, Norby RJ, Schlesinger WH (2003) Soil nitrogen cycling under elevated CO_2: a synthesis of forest FACE experiments. Ecol Appl 13:1508–1514

Zhang S, Allen HL (1996) Foliar nutrient dynamics of 11-year-old loblolly pine (*Pinus taeda*) following nitrogen fertilization. Can J For Res 26:1426–1439

12 Impacts of Elevated Atmospheric $[CO_2]$ and $[O_3]$ on Northern Temperate Forest Ecosystems: Results from the Aspen FACE Experiment

D.F. KARNOSKY and K.S. PREGITZER

12.1 Introduction

Globally, mean atmospheric carbon dioxide $[CO_2]$ and tropospheric ozone $[O_3]$ have risen 30–36% since pre-industrial times (IPCC 2001). These increases in $[CO_2]$ are largely due to increased emissions from fossil fuel burning (Beedlow et al. 2004), while the increases in $[O_3]$ are primarily related to increasing emissions of oxidized nitrogen (NO_x) and volatile organic emissions from fossil fuel combustion (Felzer et al. 2004). Nearly 25% of the Earth's forests are currently at risk from $[O_3]$ where peak concentrations exceed 60 ppb (Fowler et al. 1999). Those authors predict that half of the Earth's forests will be subjected to peak concentrations exceeding 60 ppb by the year 2100. Little is known about how forest ecosystems will respond to these co-occurring pollutants (Gower 2003).

The Aspen FACE (free air CO_2 enrichment) experiment was established in 1997 in northern Wisconsin to examine the impacts of elevated tropospheric O_3 ($e[O_3]$), alone and in combination with elevated atmospheric CO_2 ($e[CO_2]$), on the structure and functioning of a northern forest ecosystem dominated by the rapid-growing, pioneer species trembling aspen (*Populus tremuloides* Michx.) but including also another rapid-growing, pioneer species paper birch (*Betula papyrifera* Marsh) and the slower-growing, later-successional species sugar maple (*Acer saccharum* Marsh). Trembling aspen is the most widely distributed tree species in North America. Aspen forest types make up over 8.8×10^6 ha in the United States and 17.8×10^6 ha in Canada. In Wisconsin alone, where this experiment is located, aspen, birch and maple stands comprise over 50% of the State's vast forest resource. Aspen and birch make up some 70% of the pulpwood harvested in the Lake States (Piva 1996).

Ecological Studies, Vol. 187
J. Nösberger, S.P. Long, R.J. Norby, M. Stitt,
G.R. Hendrey, H. Blum (Eds.)
Managed Ecosystems and CO_2
Case Studies, Processes, and Perspectives
© Springer-Verlag Berlin Heidelberg 2006

The main objective of the Aspen FACE has been to examine how e[CO$_2$] and e[O$_3$] affect carbon and nitrogen cycles and the ecological interactions of forests (Dickson et al. 2000). Specifically, we are studying the impacts of these co-occurring greenhouse gases on aggrading northern forests in terms of physiological processes, growth and productivity, carbon sequestration, competitive interactions and stand dynamics, interactions with pests and ecosystem processes, such as foliar decomposition, mineral weathering and nutrient cycling (Karnosky et al. 2003, 2005). Furthermore, we have been interested in how temporal changes in ecosystem structure and functioning occur from establishment phase through to crown closure and beyond. We hypothesized that the ecosystem-level responses to e[CO$_2$] and e[O$_3$] would be driven by the responsiveness of the keystone tree species and that ecophysiological responses of the keystone species would cascade through the ecosystems in a predictable manner.

12.2 Site Description

The Aspen FACE project was established in 1997 with 4- to 6-month-old aspen rooted cuttings of five genotypes and 4- to 6-month-old sugar maple and paper birch seedlings planted on a sandy loam soil on a relatively flat-terrain, old-field site at 490 m elevation on a U.S. Forest Service site which had previously been used for poplar genetic trials (Dickson et al. 2000). This experiment consists of 12 30-m diameter rings (Fig. 12.1), assigned to factorial treatments of [CO$_2$] (current and elevated) and [O$_3$] (current and elevated) during daylight hours throughout the growing season. Treatments are arranged in a randomized complete block design ($n=3$). In one half of each ring, we planted five trembling aspen genotypes of differing CO$_2$ and O$_3$ responsiveness. The other half of each ring is further divided into two quarters; one is planted with aspen and sugar maple and the other is planted with aspen and paper birch; and each FACE ring was planted at 1 × 1 m spacing. Approximately 20 000 hybrid poplar unrooted cuttings were planted around the 12 rings to improve uniformity of fetch into each ring. Trees were irrigated for the first growing season and weeds were controlled through a combination of hand hoeing and herbicides for the first two growing seasons. No supplemental fertilization was conducted and pest control practices were only conducted in the establishment phase of this experiment when pests were deemed to be threatening survival of the experiment.

The site is located in a continental climate with a frost-free growing season of approximately 120 days. Summer temperatures average 16.1 °C, reaching highs of about 32 °C; and winter temperatures average –6.7 °C, but reach –20 °C. Details of the Aspen FACE micrometeorology can be found at http://www.ncrs.fs.fed.us/4401/focus/face/ meteorology/.

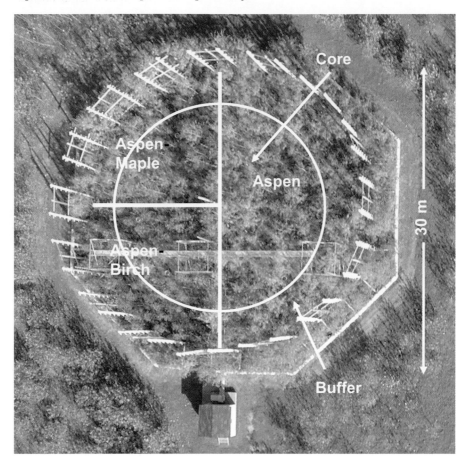

Fig. 12.1 Each 30-m diameter Aspen FACE ring is divided into three communities. Growth measurements are taken from the trees in the core of each ring (Modified from Karnosky 2005)

12.3 Experimental Treatments

Carbon dioxide and O_3 are delivered via a computer-controlled system modified from Hendrey et al. (1999) during the daylight hours, with our target $[CO_2]$ being 560 ppm, which is about 200 ppm above the daylight current CO_2 ($c[CO_2]$) concentration. Ozone was applied at a target of 1.5 × current and was not delivered during days when the maximum temperatures were projected to be less than 15 °C or when plants were wet from fog, dew, or rain events. During the past 7 years, $c[CO_2]$ has averaged 347–362 ppm and our $e[CO_2]$ treatments have averaged 530–548 ppm, current O_3 ($c[O_3]$) has averaged 39–41 ppb and our $e[O_3]$ treatments averaged 49–54 ppb. Additional details of the experimental design and pollutant generation and monitoring can be found in

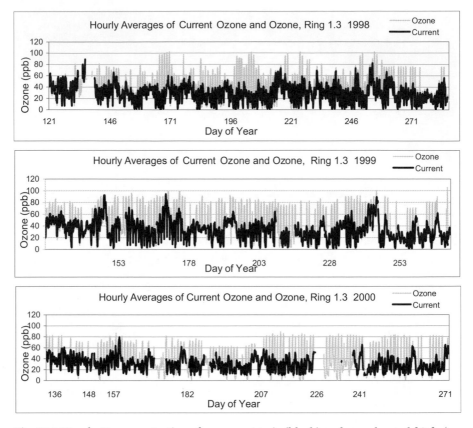

Fig. 12.2 Hourly O_3 concentrations from current air (black) and one elevated $[O_3]$ ring (gray) in the Aspen FACE project during 1998–2003 (Modified from Karnosky et al. 2005)

Karnosky et al. (2003). Treatment summaries for $[CO_2]$ and $[O_3]$ were published in Karnosky et al. (2003, 2005) and Sharma et al. (2003). Hourly $[O_3]$ values for one current and one $e[O_3]$ ring are shown in Fig. 12.2.

12.4 Resource Acquisition

12.4.1 Photosynthesis and Conductance

The effects of $e[CO_2]$ and $e[O_3]$ on photosynthesis at Aspen FACE varied by species, crown position, shoot type and leaf age. However, $e[CO_2]$ generally enhanced photosynthesis of the early successional species [20–33 % in aspen (Noormets et al. 2001a, b; Karnosky et al. 2003; Sharma, 2003; Ellsworth et al.

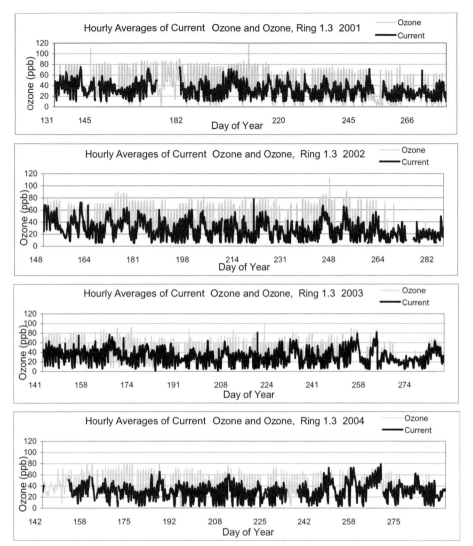

2004) and 50–70 % in birch (Takeuchi et al. 2001; Karnosky et al. 2003; Ellsworth et al. 2004)] but did not alter photosynthesis in sugar maple (Karnosky et al. 2003; Sharma et al. 2003).

Elevated [O$_3$] had little or no negative effect on either birch or maple photosynthesis but decreased photosynthesis some 29–40 % in aspen in four of five aspen clones. When e[O$_3$] was combined with e[CO$_2$], photosynthesis was similar to control rates for all three species (Noormets et al. 2001a, b; Karnosky et al. 2003; Sharma et al. 2003). Studies of A_{max} of two aspen clones suggest that photosynthetic responses over time have remained largely unchanged.

12.4.2 Respiration

In our cross-site comparison using O_2 uptake (rather than CO_2 loss), we found evidence that $e[CO_2]$ has little effect on instantaneous nighttime foliar respiration (Davey et al. 2004).

12.4.3 Nitrogen Dynamics

Elevated $[CO_2]$ and $[O_3]$ are known to alter the chemical composition of foliage which, in turn, can have many implications for carbon acquisition and allocation. Decreased foliar nitrogen per unit leaf mass has been documented in all three of our species under $e[CO_2]$ (Ellsworth et al. 2004). Furthermore, foliar nitrogen has been correlated with a decline in photosynthesis under $e[CO_2]$ at our site as well as at several other sites (Ellsworth et al. 2004). Long et al. (2004) argue the fact that there is no loss of stimulation of A_{sat} or maximum quantum yield of CO_2 uptake and suggest that foliar N change regulates photosynthesis at $e[CO_2]$ and this physiological response should be regarded as acclimation to higher levels of $[CO_2]$.

The fact that $e[CO_2]$ alters both foliar N (on both a mass and an area basis) and defense compounds has also been implicated in changes in performance of herbivorous insects for both aspen (Holton et al. 2003; Kopper and Lindroth 2003a, b) and birch (Kopper et al. 2001). These effects were sometimes positive and sometimes negative.

Litter quality was also shown to be affected by treatment at our site (Parsons et al. 2004). Elevated $[CO_2]$ resulted in poorer quality (higher C/N ratios, lignin/N and tannins), regardless of $[O_3]$ treatment. These changes in litter quality affected litter decomposition, with litter from $e[CO_2]$ treatments breaking down more slowly than those from $c[CO_2]$ (Parsons et al. 2004).

We have also examined soil nitrogen transformations to determine whether either $e[CO_2]$ or $e[O_3]$ could affect forest N cycling in this way (Holmes et al. 2003). While $e[O_3]$ significantly decreased gross N mineralization and microbial biomass N, $e[CO_2]$ did not (Holmes et al. 2003).

12.4.4 Leaf Area

Leaf area and leaf area duration can also contribute substantially to C gain in forest trees. In our study of aggrading aspen, aspen–birch and aspen–maple communities, $e[CO_2]$ consistently resulted in larger leaf area index (LAI) values throughout much of each growing season, as shown in Fig. 12.3. This increase in LAI is at least partially attributable to the slightly larger leaves at our site under $e[CO_2]$ (Oksanen et al. 2001).

Plot–level mean of three different methods.

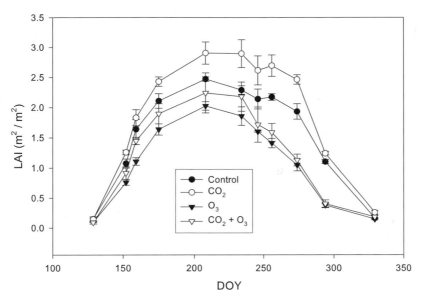

Fig. 12.3 Estimated seasonal trends for LAI for aggrading aspen stands in the Aspen FACE experiment in northern Wisconsin during the 2002 growing season, based on plot-level means in the aspen one-half of each ring for litter-fall, optical (hemispherical fish-eye approach) and ceptometer data. Values are means ±SE. (From Giardina et al., unpublished data)

Ozone also has large and consistent impacts on LAI at the Aspen FACE site. We have documented consistent delays in spring bud break, decreased mid-season leaf area index and accelerated senescence and leaf abscission in the autumn under e[O$_3$] (Karnosky et al. 2003, 2005). Again, leaf size appears to be impacted by [O$_3$], as we have generally found smaller leaves under e[O$_3$] (Oksanen et al. 2001). Elevated [CO$_2$] generally offsets some, but not all, of the adverse effects of O$_3$] on LAI (Karnosky et al. 2003, 2005) and leaf size (Oksanen et al. 2001).

12.4.5 Root Development

After three growing seasons, live fine-root biomass (<1.0 mm) averaged over community types in Aspen FACE was 263 g m^{-2}, of which 81 % was composed of roots <0.5 mm in diameter (King et al. 2001). Elevated [CO$_2$] stimulated fine-root production, almost doubling maximal seasonal standing crop (96 % increase) compared to the control treatment. This result is consistent with our previous open-top chamber research (Pregitzer et al. 2000). The increase in fine root biomass at e[CO_2] corresponded with an increase in soil respiration as would be expected (King et al. 2001). Elevated [O_3] resulted in a decrease in fine root biomass, which was marginally significant after three growing seasons (King et al. 2001). On-going studies document a continued significant increase in production of all size classes of roots at e[CO_2], with no apparent shift in carbon allocation to the different sizes of roots (King et al. 2004). Studies of fine root dynamics in Aspen FACE have documented that e[CO_2] results in rates of fine root turnover at 2.0 year^{-1} (Pregitzer, unpublished data), essentially the same as those reported previously from open-top chamber research (Pregitzer et al. 2000) and not different from the control treatment. Elevated [O_3] has increased fine root turnover to 2.5 year^{-1} (Pregitzer, unpublished data), a result consistent with other studies of the effects of e[O_3] on *Populus* fine root dynamics (Coleman et al. 1996). Taken together, root studies at Aspen FACE suggest that e[CO_2] increases the flux of carbon from root systems to the soil, while e[O_3] alters whole plant source–sink relationships, resulting in more rapid root turnover and a smaller crop of standing fine root biomass. Whether or not these changes in carbon allocation to roots are accompanied with a change in root tissue quality is an on-going line of investigation.

12.5 Resource Transformation

12.5.1 Growth and Productivity

Aboveground growth estimates (diameter, height) and biomass production (sample harvests in 2000 and 2002) both show similar trends in which the dominant plant responses are driving ecosystem composition and function (Isebrands et al. 2001; McDonald et al. 2002; Percy et al. 2002; Karnosky et al. 2003, 2005). Species and genotypes within species (aspen) are highly variable in these responses. The general trends for aspen consist of significantly increased growth and productivity under e[CO_2] and significantly decreased responses under e[O_3]. Elevated [O_3] generally offset the growth and productivity enhancement by e[CO_2]. While long-term growth enhancement has been reported to be unsustainable in some systems, our growth enhancement has continued through the 6 years of our study, as particularly evidenced by the large stimulation still shown by paper birch. Interestingly, sugar maple has not been enhanced by e[CO_2] in our study. Responsiveness of our species to [CO_2] are (from most to least enhanced by [CO_2]): birch>aspen>maple. For e[O_3], aspen is sensitive to [O_3] while birch and maple are more tolerant. In the long term, the combined treatment has resulted in the strongest growth decrease in sugar maple (Karnosky 2005).

Whether or not above- and belowground carbon allocation patterns will change under prolonged exposure to e[CO$_2$] and e[O$_3$] remain active research questions. We have detected no changes in allometry in our study (King et al. 2005). While we see stimulation of aboveground growth in both aspen and birch, we see similar enhancement of root growth under e[CO$_2$]. Similarly, we see nearly identical shoot and root growth reductions for aspen exposed to e[O$_3$].

12.5.2 Soil Carbon

Increased carbon inputs to the soil under e[CO$_2$] correspond with a significant increase in total soil respiration (King et al. 2001, 2004). Stable soil carbon formation under e[CO$_2$] was 50 % greater than under e[CO$_2$] + e[O$_3$] after only 4 years of exposure, indicating that changes in NPP induced by changes in the atmosphere may have a significant impact on the formation of stable soil carbon (Loya et al. 2003). These findings, which link changes in the Earth's atmosphere with the formation of soil carbon, have begun to influence air pollution control strategies in Europe (see Grennfelt 2004). It may be that increased flux of carbon through the soil will eventually lead to the formation of more stable soil carbon under e[CO$_2$].

12.5.3 Wood Quality

Stem wood properties of all three of our species were examined after 3 years and 5 years (Kaakinen et al. 2004; Kaakinen, personal communication). As in many of the growth responses, aspen was most sensitive to changes in wood properties and maple was least responsive. Lignin concentration of wood generally decreased under e[O$_3$] for aspen and birch (Kaakinen et al. 2004), as would be expected from previous research at our site suggesting increased PAL transcripts under e[O$_3$] (Wustman et al. 2001). This may indicate a change of carbon allocation under e[O$_3$]. The decreased lignin under e[O$_3$] was ameliorated by e[CO$_2$] (Kaakinen et al. 2004). Elevated [CO$_2$] had little effect on stem wood properties, with slight alternations of a-cellulose and hemicellulose concentrations (Kaakinen et al. 2004).

12.5.4 Pest, Disease and Herbivore Susceptibility

The epicuticular waxes are located on the outermost surfaces of plant leaves and are in direct contact with the atmosphere. The role of these waxes in plant defense are well established. Mankovska et al. (1998, 2003, 2005) documented wax structure changes and increased stomatal occlusion under all treatments, but the largest occlusion has occurred repeatedly in the combination treatment. Percy et al. (2002) and Karnosky et al. (2002) reported significant increases in wax deposits due to $e[O_3]$ as well as changes in wax chemistry. These structural changes also alter wettability (Karnosky et al. 2002; Percy et al. 2003) and we believe this causes a predisposition to certain foliar surface fungi.

We found evidence of increased *Melampsora* rust occurrence on aspen under $e[O_3]$ (Karnosky et al. 2002; Percy et al. 2002). We also documented an increased abundance of aphids and a decrease in their natural enemies in aspen under $e[O_3]$ (Percy et al. 2002). Forest tent caterpillar, a cyclic pest which annually defoliates millions of hectares of aspen and birch forests was found to be impacted by both $e[O_3]$ and $e[CO_2]$. Forest tent caterpillar pupal masses increased under $e[O_3]$ (Percy et al. 2002; Kopper et al. 2003b) and egg mass parasitism decreased and egg mass foam protection increased under $e[CO_2]$ (Mattson, unpublished data). Elevated $[CO_2]$ stimulated production of phenolic compounds in birch, suggesting resistance to lagomorph herbivores may be increased under $e[CO_2]$ (Mattson et al. 2004).

12.6 Consequences and Implications

The Aspen FACE project has demonstrated that $e[O_3]$ at moderate levels can dramatically impact the response of forest ecosystems to $e[CO_2]$ during early stand development (Table 12.1). As $e[O_3]$ will continue to rise downwind of major metropolitan regions around the world (Fowler et al. 1999; IPCC 2001), it is important for global carbon cycle models to take into account $e[O_3]$ in projections of carbon uptake and sequestration under $e[CO_2]$.

It is critically important to determine whether the trends Aspen FACE has shown with the early growth phase for aspen, birch and maple will continue as these young stands equilibrate with respect to $e[CO_2]$ and $e[O_3]$. This study is unique among forest FACE experiments, as we have the opportunity to examine responses from establishment through stand maturity, to examine how these interacting greenhouse gases affect ecosystem structure and functioning over the entire life history of these stands.

Table 12.1 Summary of responses of trembling aspen to e[CO$_2$] (530–548 ppm), e[O$_3$] (1.5x current), or e[CO$_2$]+e[O$_3$] compared with control during 7 years of treatments at the Aspen FACE project. This table is modified from Karnosky et al. (2003) and Karnosky (2005)

	CO$_2$	O$_3$	CO$_2$ + O$_3$	Source
Foliar gene expression and biochemistry				
Rubisco	↓[a]	→	→→	Wustman et al. (2001), Noormets et al. (2001a)
RbcS[b] transcripts	→	→	→→	Wustman et al. (2001)
PAL transcripts	→	↑	→	Wustman et al. (2001)
Ascorbate peroxidase	→	n.s.	→	Wustman et al. (2001)
Catalase, Acc oxidase	→	↑	→	Wustman et al. (2001), Oksanen et al. (2003)
Glutathione reductase	n.s.	↑	→	Wustman et al. (2001)
H$_2$O$_2$ accumulation	n.s.	↑↑	n.s.	Oksanen et al. (2003)
Phenolic glycosides	n.s.	↑→	n.s.	Lindroth et al. (2002), Kopper and Lindroth (2003a, b)
Tannins	n.s.	↓	↑	Lindroth et al. (2002), Kopper and Lindroth (2003a, b)
Foliar nitrogen	↓→	↓	n.s.↓	Lindroth et al. (2001), Kopper and Lindroth (2003a, b), Holton et al. (2003)
C:N ratio of foliage	↑	→	↑↑	Lindroth et al. (2001)
Starch	→	n.s.→	n.s.	Wustman et al. (2001)
Gas exchange				
A_{max} lower canopy	n.s.	↓↓	↑ (young) ↓ (older)	Noormets et al. (2001a), Sharma et al. (2003)
A_{max} upper canopy	↑↑	↓↓	n.s.	Noormets et al. (2001b), Ellsworth et al. (2004)
Stomatal limitation	→	n.s.↓	→	Noormets et al. (2001a)
Stomatal conductance	→	↓↑	→	Noormets et al. (2001a), Sharma et al. (2003)
Foliar respiration	n.s.	↑	n.s.	Takeuchi et al. (2001), Noormets (2001a), Davey et al. (2004)
Soil respiration	↑↑	→	n.s.	King et al. (2001, 2004)
Microbial respiration	↑↑	n.s.	n.s.	Phillips et al. (2002)
Stomatal density	n.s.	n.s.	n.s.	Percy et al. (2002), Karnosky et al. (2003)
Chlorophyll content	↓→	↓→	→	Wustman et al. (2001)
Chloroplast structure	↓	→	→	Oksanen et al. (2001), Wustman et al. (2001)

Table 12.1 (*Continued*)

	CO_2	O_3	$CO_2 + O_3$	Source
Peroxisome number	n.s.	↑↑	n.s.	Oksanen et al. (2003)
O_3 flux	→	↑↑	↑	Noormets et al. (2001a)
Growth and productivity				
Leaf thickness	↑	n.s.	n.s.	Oksanen et al. (2001)
Leaf size	↑	→	→	Wustman et al. (2001)
Leaf area	↑	→	n.s.	Noormets et al. (2001b)
LAI	↑	→	n.s.	Karnosky et al. (2003, 2005)
Height growth	↑	→	n.s.	Isebrands et al. (2001), Percy et al. (2002), Karnosky et al. (2003, 2005)
Diameter growth	↑	→	n.s.	Isebrands et al. (2001), Percy et al. (2002), Karnosky et al. (2003, 2005)
Volume growth	↑↑	→→	n.s.	Isebrands et al. (2001)
Fine root biomass	↑	→→	n.s.	King et al. (2001)
Fine root turnover	↑	n.s.	n.s.	King et al. (2001, 2004)
Spring budbreak	n.s.	Delayed	n.s.	Karnosky et al. (2005)
Autumn budset	Delayed	Early	n.s.	Karnosky et al. (2005)
Foliar retention – autumn	↑↑	→→	n.s.	Karnosky et al. (2005)
Fine root biomass	↑	n.s.	n.s.	King et al. (2001)
Fine root turnover	n.s.	n.s.	n.s.	King et al. (2001, 2004)
Spring budbreak	Delayed	Delayed	n.s.	Karnosky et al. (2005)
Autumn budset	Delayed	Early	n.s.	Karnosky et al. (2005)
Foliar retention – autumn	↑↑	→→	n.s.	Karnosky et al. (2005)
Wood				
Pith to bark distance	↑	→		Kaakinen et al. (2004)
Vessel lumen diameter	n.s.	→	n.s.	Kaakinen et al. (2004)
Lignin	n.s.	←	n.s.	Kaakinen et al. (2004)
Cellulose	n.s.	n.s.	n.s.	Kaakinen et al. (2004)
Hemicellulose	→	n.s.	←	Kaakinen et al. (2004)

	CO$_2$	O$_3$	CO$_2$ + O$_3$	Reference
Leaf surfaces				
Crystalline wax structure	→	→↓	↓↓	Karnosky et al. (1999, 2002)
Stomatal occlusion	←	←	↑↑	Karnosky et al. (1999), Mankovska et al. (2003)
Wax chemical composition	n.s.	Change	n.s.	Karnosky et al. (2002)
Wettability	n.s.	↑↑	↑	Karnosky et al. (2002)
Trophic interactions				
Melampsora leaf rust	n.s.	↑↑	↑↑	Karnosky et al. (2002), Percy et al. (2002)
Aphids	n.s.	←	n.s.	Percy et al. (2002)
Aphid dispersal	↓↓	↑↑	↑	Mondor et al. (2004a)
Proportion of winged aphid offspring	→↓	→↓	n.s.	Mondor et al. (2004b)
Blotch leaf miner	↓↓	↓↓	↓↓	Kopper and Lindroth (2003a)
Forest tent caterpillar	n.s.	←	n.s.	Kopper and Lindroth (2003b)
Ecosystem level				
NPP	↑↑	↓↓	n.s.	King et al. (2005)
Soil carbon formation	↑↑	↓↓	↓↓	Loya et al. (2003)
Nitrogen mineralization	n.s.	↓↓	n.s.	Holmes et al. (2003)
Litter decomposition (k-value)	→↓	n.s.	→↓	Parsons et al. (2004)
Competitive indices	←	→	↓↓	McDonald et al. (2002)
Soil invertebrate diversity	→↑	→	n.s.	Loranger et al. (2004)
Microbial enzymes	←	n.s.	n.s.	Phillips et al. (2002), Larson et al. (2002)
Microbial biomass	←	n.s.	n.s.	Phillips et al. (2002), Larson et al. (2002)

a Responses are shown as: small but statistically significant increases (↑), large and statistically significant increases (↑↑), small but statistically significant decreases (↓), large and statistically significant decreases (↓↓), nonsignificant effects (n.s.) compared to trees grown in control rings with current CO$_2$ and O$_3$. Foliar analyses and leaf surface properties were largely determined from recently mature leaves of all three species during mid-season. Gas exchange data were taken from all leaf ages and throughout the growing season.

b Abbreviations: RbcS = small subunit of Rubisco; PAL = phenylalanine ammonialyase; SOD = super oxide dismutase; ACC = 1-aminocyclopropane-1-carboxylic acid; C = carbon; N = nitrogen; A_{max} = maximum photosynthesis rate; LAI = leaf area index; NPP = net primary productivity

12.7 Conclusions

Aspen FACE has documented genetic variation in responses at both the inter-specific and intraspecific levels for long-term exposure of young northern trees to e[CO_2] and e[O_3].

- Elevated [CO_2] increased total biomass by 25, 45 and 60 % in the aspen, aspen–birch and aspen–maple communities (King et al. 2005).
- Elevated [O_3] decreased these communities by some 23, 13 and 14 %, respectively, while the combined treatments resulted in total biomass responses of –7.8, +8.4 and +24.3 %, respectively (King et al. 2005).
- For e[CO_2], the three species ranked (from most to least responsive; Karnosky et al. 2005) as birch>aspen>maple.
- For e[O_3], aspen was more responsive than either birch or maple; and amongst the aspen clones, clone 8L was most tolerant, followed by clones 271, 216, 42E and 259 (Karnosky et al. 2005).
- Stable soil carbon formation was 50 % greater under e[CO_2] than under e[CO_2] + e[O_3] (Loya et al. 2003).
- Pest dynamics in Aspen FACE were altered by treatment. Under e[O_3], for example, *Melampsora* leaf rust was 3–5 × higher, regardless of [CO_2] level (Karnosky et al. 2002; Percy et al. 2002).

Acknowledgements. This research was partially supported by the Office of Science (BER), U.S. Department of Energy, Grant No. DE-FG02-95ER62125, USDA Forest Service Northern Global Change Program, the USDA North Central Research Station, Michigan Technological University, the Praxair Foundation and the McIntire–Stennis Program.

References

Beedlow PA, Tingey DT, Phillips DL, Hogsett WE, Olszyk DM (2004) Rising atmospheric CO_2 and carbon sequestration in forests. Front Ecol Environ 2:315–322

Coleman MD, Dickson RE, Isebrands JG, Karnosky DF (1996) Root growth and physiology of potted and field-grown trembling aspen exposed to tropospheric ozone. Tree Physiol 16:145–152

Davey PA, Hunt S, Hymus GJ, DeLucia EH, Drake BG, Karnosky DF, Long SP (2004) Respiratory oxygen uptake is not decreased by an instantaneous elevation of [CO_2], but is increased with long-term growth in the field at elevated [CO_2]. Plant Physiol 134:520–527

Dickson RE, Lewin KF, Isebrands JG, Coleman MD, Heilman WE, Riemenschneider DE, Sober J, Host GE, Hendrey GR, Pregitzer KS, Karnosky DF (2000) Forest atmosphere carbon transfer storage-II (FACTS II) – the aspen free-air CO_2 and O_3 enrichment (FACE) project in an overview. General Technical Report NC-214. USDA Forest Service North Central Experiment Station, St Paul, Minn., 68 pp

Ellsworth DS, Reich PB, Naumburg ES, Koch GW, Kubiske ME, Smith SD (2004) Photosynthesis, carboxylation and leaf nitrogen responses of 16 species to elevated pCO$_2$ across four free-air CO$_2$ enrichment experiments in forest, grassland and desert. Global Change Biol 10:2121–2138

Felzer B, Kicklighter D, Melillo J, Wang C, Zhuang Q, Prinn R (2004) Effects of ozone on net primary production and carbon sequestration in the conterminous United States using a biogeochemistry model. Tellus 56B:230–248

Fowler D, Cape JN, Coyle M, Flechard C, Kuylenstierna J, Hicks K, Derwent D, Johnson C, Stevenson D (1999) The global exposure of forests to air pollutants. J Water Air Soil Pollut 116:5–32

Gower ST (2003) Patterns and mechanisms of the forest carbon cycle. Annu Rev Environ Resour 28:169–204

Grennfelt P (2004) New directions: recent research findings may change ozone control policies. Atmos Environ 38:2215–2216

Hendrey GR, Ellsworth DS, Lewis FK, Nagy J (1999) A free-air enrichment system for exposing tall forest vegetation to elevated atmospheric CO$_2$. Global Change Biol 5:293–309

Holmes WE, Zak DR, Pregitzer KS, King JS (2003) Soil nitrogen trans-formations under *Populus tremuloides*, *Betula papyrifera* and *Acer saccharum* following 3 years exposure to elevated CO$_2$ and O$_3$. Global Change Biol 9:1743–1750

Holton MK, Lindroth RL, Nordheim EV (2003) Foliar quality influences tree–herbivore–parasitoid interactions: effects of elevated CO$_2$, O$_3$, and genotype. Oecologia 137:233–244

IPCC (2001) A report of working group I of the intergovernmental panel on climate change. http://www.ipcc.ch/

Isebrands JG, McDonald EP, Kruger E, Hendrey G, Pregitzer K, Percy K, Sober J, Karnosky DF (2001) Growth responses of *Populus tremuloides* clones to interacting carbon dioxide and tropospheric ozone. Environ Pollut 115:359–371

Kaakinen S, Kostiainen K, Ek F, Saranpää P, Kubiske ME, Sober J, Karnosky DF, Vapaavuori E (2004) Stem wood properties of *Populus tremuloides*, *Betula papyrifera* and *Acer saccharum* saplings after three years of treatments to elevated carbon dioxide and ozone. Global Change Biol 10:1513–1525

Karnosky DF (2005) Ozone effects on forest ecosystems under a changing global environment. J Agric Meteorol 60:353–358

Karnosky DF, Mankovska B, Percy K, Dickson RE, Podila GK, Sober J, Noormets A, Hendrey G, Coleman MD, Kubiske M, Pregitzer KS, Isebrands JG (1999) Effects of tropospheric O$_3$ and interaction with CO$_2$: results from an O$_3$-gradient and a FACE experiment. J Water Air Soil Pollut 116:311–322

Karnosky DF, Percy KE, Xiang B, Callan B, Noormets A, Mankovska B, Hopkin A, Sober J, Jones W, Dickson RE, Isebrands JG (2002) Interacting elevated CO$_2$ and tropospheric O$_3$ predisposes aspen (*Populus tremuloides* Michx.) to infection by rust (*Melampsora medusae* f.sp. *tremuloidae*). Global Change Biol 8:329–338

Karnosky DF, Zak DR, Pregitzer KS, Awmack CS, Bockheim JG, Dickson RE, Hendrey GR, Host GE, King JS, Kopper BJ, Kruger EL, Kubiske ME, Lindroth RL, Mattson WJ, McDonald EP, Noormets A, Oksanen E, Parsons WFJ, Percy KE, Podila GK, Riemenschneider DE, Sharma P, Thakur RC, Sober A, Sober J, Jones WS, Anttonen S, Vapaavuori E, Mankovska B, Heilman WE, Isebrands JG (2003) Tropospheric O$_3$ moderates responses of temperate hardwood forests to elevated CO$_2$: a synthesis of molecular to ecosystem results from the Aspen FACE project. Funct Ecol 17:289–304

Karnosky DF, Pregitzer KS, Zak DR, Kubiske ME, Hendrey GR, Weinstein D, Nosal M, Percy KE (2005) Scaling ozone responses of forest trees to the ecosystem level in a changing climate. Plant Cell Environ 28:965–981

King JS, Pregitzer KS, Zak DR, Karnosky DF, Isebrands JG, Dickson RE, Hendrey, GR, Sober J (2001) Fine root biomass and fluxes of soil carbon in young stands of paper birch and trembling aspen as affected by elevated atmospheric CO_2 and tropospheric O_3. Oecologia 128:237–250

King JS, Hanson PJ, Bernhardt E, DeAngelis P, Norby RJ, Pregitzer KS (2004) A multi-year synthesis of soil respiration responses to elevated atmospheric CO_2 from four forest FACE experiments. Global Change Biol 10:1027–1042

King JS, Kubiske ME, Pregitzer KS, Hendrey GR, McDonald EP, Giardina CP, Quinn VS, Karnosky DF (2005) Tropospheric O_3 compromises net primary production in young stands of trembling aspen, paper birch and sugar maple in response to elevated atmospheric CO_2. New Phytol 168:623–636

Kopper BJ, Lindroth RL (2003a) Responses of trembling aspen (*Populus tremuloides*) phytochemistry and aspen blotch leafminer (*Phyllonorycter tremuloidiella*) performance to elevated levels of atmospheric CO_2 and O_3. Agric For Entomol 5:17–26

Kopper BJ, Lindroth RL (2003b) Effects of elevated carbon dioxide and ozone on the phytochemistry of aspen and performance of an herbivore. Oecologia 134:95–103

Kopper BJ, Lindroth RL, Nordheim EV (2001) CO_2 and O_3 effects on paper birch (Betulaceae: *Betula papyrifera*) phytochemistry and whitemarked tussock moth (Lymantriidae: *Orgyia leucostigma*) performance. Environ Entomol 30:1119–1126

Larson JL, Zak DR, Sinsabaugh RL (2002) Microbial activity beneath temperate trees growing under elevated CO_2 and O_3. Soil Sci Soc Amer J 66:1848–1856

Lindroth RL, Kopper BJ, Parsons WFJ, Bockheim JG, Sober J, Hendrey GR, Pregitzer KS, Isebrands JG, Karnosky DF (2001) Effects of elevated carbon dioxide and ozone on foliar chemical composition and dynamics in trembling aspen (*Populus tremuloides*) and paper birch (*Betula papyrifera*). Environ Pollut 115:395–404

Lindroth RL, Wood SA, Kopper BJ (2002) Response of quaking aspen genotypes to enriched CO_2: foliar chemistry and insect performance. Agric For Entomol 4:315–323

Long SP, Ainsworth EA, Rogers A, Ort DR (2004) Rising atmospheric carbon dioxide: plants FACE the future. Annu Rev Plant Biol 55:591–628

Loranger GI, Pregitzer KS, King JS (2004) Elevated CO_2 and O_3 concentrations differentially affect selected groups of the fauna in temperate forest soils. Soil Biol Biochem 36:521–1524

Loya WM, Pregitzer KS, Karberg NJ, King JS, Giardina CP (2003) Reduction of soil carbon formation by tropospheric ozone under elevated carbon dioxide. Nature 425:705–707

Mankovska B, Percy K, Karnosky DF (1998) Impact of ambient tropospheric O_3, CO_2, and particulates on the epicuticular waxes of aspen clones differing in O_3 tolerance. Ekologia (Bratislava) 18:200–210

Mankovska B, Percy K, Karnosky DF (2003) Impact of greenhouse gases on epicuticular waxes of *Populus tremuloides* Michx.: results from an open-air exposure and a natural O_3 gradient. Ekologia (Bratislava) 22:182–194

Mankovska B., Percy E, Karnosky DF (2005) Impacts of greenhouse gases on epicuticular waxes of *Populus tremuloides* Michx.: results from an open-air exposure and a natural O_3 gradient. Environ Pollut 137:580–586

Mattson WJ, Kuokkanen K, Niemelä P, Julkunen-Tiitto R, Kellomäki S, Tahvanainen J (2004) Elevated CO_2 alters birch resistance to *Lagomorpha* herbivores. Global Change Biol 10:1402–1413

McDonald EP, Kruger EL, Riemenschneider DE, Isebrands JG (2002) Competitive status influences tree-growth responses to elevated CO_2 and O_3 in aggrading aspen stands. Funct Ecol 16:792–801

Mondor EB, Tremblay MN, Awmack CS, Lindroth RL (2004a) Divergent pheromone-mediated insect behaviour under global atmospheric change. Global Change Biol 10:1820–1824

Mondor EB, Tremblay MN, Lindroth RL (2004b) Transgenerational phenotypic plasticity under future atmospheric conditions. Ecol Letters 7:941–946

Noormets A, Sober A, Pell EJ, Dickson RE, Podila GK, Sober J, Isebrands JG, Karnosky DF (2001a) Stomatal and non-stomatal limitation to photosynthesis in two trembling aspen (*Populus tremuloides* Michx.) clones exposed to elevated CO$_2$ and/or O$_3$. Plant Cell Environ 24:327–336

Noormets A, McDonald EP, Kruger EL, Sober A, Isebrands JG, Dickson RE, Karnosky DF (2001b) The effect of elevated carbon dioxide and ozone on leaf- and branch-level photosynthesis and potential plant-level carbon gain in aspen. Trees 15:262–270

Oksanen E, Sober J, Karnosky DF (2001) Interactions of elevated CO$_2$ and ozone in leaf morphology of aspen (*Populus tremuloides*) and birch (*Betula papyrifera*) in aspen FACE experiment. Environ Pollut 115:437–446

Oksanen E, Häikiö E, Sober J, Karnosky DF (2003) Ozone-induced H$_2$O$_2$ accumulation in field-grown aspen and birch is linked to foliar ultrastructure and peroxisomal activity. New Phytol 161:791–799

Parsons WFJ, Lindroth RL, Bockheim JG (2004) Decomposition of *Betula papyrifera* leaf litter under the independent and interactive effects of elevated CO$_2$ and O$_3$. Global Change Biol 10:1666–1677

Percy KE, Awmack CS, Lindroth RL, Kubiske ME, Kopper BJ, Isebrands JG, Pregitzer KS, Hendrey GR, Dickson RE, Zak DR, Oksanen E, Sober J, Harrington R, Karnosky DF (2002) Altered performance of forest pests under CO$_2$- and O$_3$-enriched atmospheres. Nature 420:403–407

Percy KE, Mankovska B, Hopkin A, Callan B, Karnosky DF (2003) Ozone affects leaf surface pest interactions. In: Karnosky DF, Percy KE, Chappelka AH, Simpson C, Pikkarainen JM (eds) Air pollution, global change and forests in the new millennium. Elsevier Press, Amsterdam, pp 247–258

Phillips RL, Zak DR, Holmes WE, White DC (2002) Microbial community composition and function beneath temperate trees exposed to elevated atmospheric carbon dioxide and ozone. Oecologia 131:236–244

Piva RJ (1996) Pulpwood production in the Lake States, 1995. Research Note NC-370. USDA Forest Service North Central Forest Experiment Station, St Paul, Minn., 5 pp

Pregitzer KS, Zak DR, Maziasz J, DeForest J, Curtis PS, Lussenhop J (2000) Interactive effects of atmospheric CO$_2$ and soil-N availability on fine roots of *Populus tremuloides*. Ecol Appl 10:1833

Sharma P, Sober A, Sober J, Podila GK, Kubiske ME, Mattson WJ, Isebrands JG, Karnosky DF (2003) Moderation of [CO$_2$]-induced gas exchange responses by elevated tropospheric O$_3$ in trembling aspen and sugar maple. Ekologia (Bratislava) 22:304–317

Takeuchi Y, Kubiske ME, Isebrands JG, Pregitzer KS, Hendrey G, Karnosky DF (2001) Photosynthesis, light and nitrogen relationships in a young deciduous forest canopy under open-air CO$_2$ enrichment. Plant Cell Environ 24:1257–1268

Wustman BA, Oksanen E, Karnosky DF, Sober J, Isebrands JG, Hendrey GR, Pregitzer KS, Podila GK (2001) Effects of elevated CO$_2$ and O$_3$ on aspen clones varying in O$_3$ sensitivity: can CO$_2$ ameliorate the harmful effects of O$_3$? Environ Pollut 115:473–481

13 CO$_2$ Enrichment of a Deciduous Forest: The Oak Ridge FACE Experiment

R.J. Norby, S.D. Wullschleger, P.J. Hanson, C.A. Gunderson, T.J. Tschaplinski, and J.D. Jastrow

13.1 Introduction

The free-air CO$_2$ enrichment (FACE) experiment on the Oak Ridge National Environmental Research Park (Tenn., USA) is part of a long-standing effort to understand how the eastern North American deciduous forest will be affected by CO$_2$ enrichment of the atmosphere and what the feedbacks are from the forest to the atmosphere and the global carbon cycle budget. This is a goal we have been working toward for many years by studying *components* of the deciduous forest system (e.g., individual small trees, isolated processes), but the size and complexity of the forest have heretofore precluded measurement of the *integration* of those components. FACE technology permits us to take the critical leap to measuring the integrated response of an intact forest ecosystem with a focus on stand-level mechanisms.

The Oak Ridge FACE experiment was established in 1997 in a closed-canopy, monoculture plantation of the deciduous hardwood tree, sweetgum (*Liquidambar styraciflua* L.). This sweetgum plantation offers the opportunity for rigorous tests of hypotheses that address the essential features of a forest stand and how they could influence the responses to CO$_2$ (Norby et al. 1999). These features include: (a) the closed canopy, which constrains growth responses, (b) full occupancy of the soil by the root system, which constrains the nutrient cycle, (c) the larger scale of the trees compared to saplings in open-top chambers, which changes the functional relationships of carbon cycling and water use, and (d) the longer time-scale that can be addressed, permitting studies of soil carbon changes.

Ecological Studies, Vol. 187
J. Nösberger, S.P. Long, R.J. Norby, M. Stitt,
G.R. Hendrey, H. Blum (Eds.)
Managed Ecosystems and CO$_2$
Case Studies, Processes, and Perspectives
© Springer-Verlag Berlin Heidelberg 2006

13.2 Site Description

13.2.1 Physical

The experiment is located on the Oak Ridge National Environmental Research Park in Roane County, Tenn. (35° 54' N, 84° 20' W) in southeastern United States (Fig. 13.1). The 1.7-ha plantation was established in 1988 on an old terrace of the Clinch River (elevation 230 m). Six 25-m diameter plots were laid out in 1996 and construction of the FACE facility began thereafter, following the design employed at the loblolly pine FACE experiment in North Carolina (Hendrey et al. 1999). Subsequently, one plot was removed from the experiment because of substantial differences in soil characteristics from the other five plots (Norby et al. 2001). Two plots receive air with an elevated concentration of CO_2 ($e[CO_2]$) and three receive air with close to the current CO_2 concentration ($c[CO_2]$). Two of the $c[CO_2]$ plots have towers, vent pipes, and blowers identical to the $e[CO_2]$ plots; the other $c[CO_2]$ plot does not have any FACE apparatus. After accounting for a buffer zone adjacent to the vent pipes, the effective plot size is 20 m diameter (314 m²). The experimental unit is considered to be the whole plot; there is no blocking.

Fig. 13.1 The Oak Ridge FACE experiment is located in a sweetgum (*Liquidambar styraciflua*) plantation on the Oak Ridge National Environmental Research Park in Tennessee, USA

13.2.2 Soil Types

The soil at the site, which is classified as an Aquic Hapludult, developed in alluvium washed from upland soils derived from a variety of rocks including dolomite, sandstone, and shale. It has a silty clay loam texture and is moderately well drained (Van Miegroet et al. 1994). The soil is slightly acid (water pH approximately 5.5–6.0) with high base saturation largely dominated by exchangeable Ca. Bulk density is 1.5 g cm^{-3}, C content is 74 Mg ha^{-1}, and N content is 11 Mg ha^{-1}. When the plantation was established in 1988, the soil was disked; and herbicide was used in 1989 and 1990 to control competition from weeds. No fertilizer has been added and there has been no additional soil disturbance, except that associated with sample collection.

13.2.3 Meteorological Description

The climate is typical of the humid southern Appalachian region. Mean annual temperature (1962–1993) is 13.9 °C and mean annual precipitation is 1371 mm, with a generally even distribution of precipitation throughout the year, although droughts do occur during the growing season (Gunderson et al. 2002). Weather records during the experiment are reported by Riggs et al. (2003a).

13.2.4 Stand Description

When the plantation was established in 1988, one-year-old, bare-rooted sweetgum seedlings from a commercial nursery were planted at a spacing of 2.3 × 1.2 m. Based on analysis of tree rings measured on trees removed during the construction of FACE apparatus, tree growth was in an exponential growth phase until approximately 1993, when it became linear (i.e., basal area increment approximately equal each year.) When pretreatment baseline measurements were initiated in 1997, stand basal area was 29.0 m^2 ha^{-1} with an average tree height of 12.4 m and stem diameter of 11.3 cm. The canopy was closed and LAI was 5.5. By 2003, stand basal area had increased to 42.1 m^2 ha^{-1}, average tree height was 16.7 m, and stem diameter was 14.4 cm. Leaf area index did not change, but as the top of the canopy moved upward, lower branches were cast off and canopy depth remained constant. Initially there were about 90 trees per plot, but an average of one suppressed tree per plot died each year. As of 2004, the trees had not produced fruit.

The understory was very sparse when the experiment was started in 1997 but gradually increased. Important species include an invasive C$_4$ annual grass (*Microstegium vimineum*), non-native, invasive woody plants (*Lonicera japonica, Ligustrum sinense*), and other taxa. Tree seedlings, including *Acer negundo, A. rubra, Liriodendron tulipifera*, and *Quercus alba*, are sparse (Belote et al. 2004).

13.3 Experimental Treatments

Exposure to $e[CO_2]$ commenced in two plots in April 1998 and has continued during the growing season (April–November) since then. The CO_2 is a byproduct of the natural gas used in ammonia production and has a $\delta^{13}C$ signature of approximately –50‰. The $[CO_2]$ set-point in 1998 was a constant 565 ppm. In 1999 and 2000, a dual set-point (565 ppm day and 645 ppm night, with the beginning of day and night defined by a solar angle of 0°) was used to better represent the diurnal variation in $c[CO_2]$. Nighttime fumigation was discontinued in 2001 because it interfered with soil respiration measurements. The average daytime $[CO_2]$ during the 1998–2003 growing seasons was 544 ppm in the two CO_2-enriched plots, including periods when the exposure system was not functioning, and 391 ppm in $c[CO_2]$ plots. The "current" concentration is higher than the global average because of high values in early morning hours when the wind is low. Contamination of $c[CO_2]$ plots by adjacent $e[CO_2]$ plots was approximately 10 ppm, based on comparison with $[CO_2]$ measured distant from the FACE array (Norby et al. 2001). Hourly records of $[CO_2]$ are given by Riggs et al. (2003b). There are no other treatments applied to the plots.

13.4 Resource Acquisition

13.4.1 CO_2 Effects on Physiological Functions and Metabolites

13.4.1.1 Carbon

Photosynthetic and stomatal responses to $e[CO_2]$ were measured over six growing seasons in upper and mid-crown foliage to evaluate the impacts of environmental variation, exposure duration, and changes in foliar biochemistry. Measurements were taken at saturating irradiance, encompassing the full range of variability in temperature, vapor pressure deficit (VPD), and soil water potential occurring within and across seasons. Photosynthetic CO_2 assimilation (A) averaged 46 % higher in $e[CO_2]$ (Fig. 13.2), in both mid- and upper canopy foliage (Gunderson et al. 2002); this response was sustained over the 6-year period. The stimulation of A by CO_2 enrichment was greatest

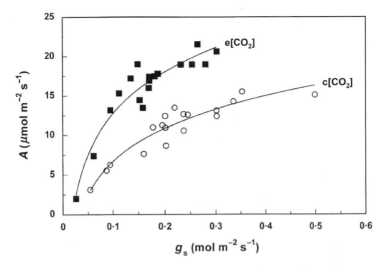

Fig. 13.2 Upper canopy photosynthesis as a function of stomatal conductance to water vapor (g_s) on 23 dates over six summers. Modified from Gunderson et al. (2002), with permission of New Phytologist Trust

at sample times when g_s was least responsive to [CO_2]. Late-season droughts in 1998 and 1999 led to dry soils and high VPD, reducing A in both treatments, through both stomatal and non-stomatal limitations (Gunderson et al. 2002). Absolute treatment differences were noticeably diminished whenever g_s was below 0.15 mol m^{-2} s^{-1}, although relative effects (elevated:current) varied greatly under those conditions. In all cases, CO_2 responses returned when atmospheric and soil moisture conditions improved. The ratio of intercellular [CO_2] to ambient [CO_2] (C_i/C_a) in the upper canopy was 0.67 in c[CO_2] and 0.64 in e[CO_2]. Measurement of A as a function of C_i revealed no significant CO_2 treatment effects on photosynthetic or biochemical capacity (i.e., no change in A_{max}, V_{cmax} or J_{max}), although starch concentration was increased (+16%) and N concentration was reduced (–10%; Sholtis et al. 2004). In contrast, soluble sugar concentrations (mass basis) were not affected by CO_2 (Sholtis et al. 2004; Tschaplinski, unpublished data), despite the consistently higher A. These results suggest that established sweetgum trees in closed-canopy forests would sustain a long-term positive response to e[CO_2] without reductions in photosynthetic capacity, subject only to seasonal variability and constraints associated with environmental conditions.

Nighttime respiration (R_N) was measured on leaves in the upper and lower canopy in 1999 and 2000; and leaf respiration in the light (R_L) was estimated in the upper canopy in 2000 (Tissue et al. 2002). There were no significant short-term effects of e[CO_2] on R_N or long-term effects on R_N or R_L, when expressed on an area, mass or nitrogen basis. Upper-canopy leaves had 54%

higher R_N (area basis) than lower-canopy leaves. CO_2 enrichment significantly increased the number of leaf mitochondria by 62 %. Growth in e[CO_2] did not affect the relationships between R_N and any measured leaf structural or chemical characteristic.

Stem respiration rates (per unit stem volume) of individual trees in August 2002 ranged over 33–66 µmol CO_2 m^{-3} s^{-1} in control plots and 40–94 µmol CO_2 m^{-3} s^{-1} in CO_2-enriched plots (Edwards et al. 2002). Respiration rates were consistently higher under CO_2 enrichment than at c[CO_2] throughout both the growing season and the dormant season. Averaged over an entire year, stem respiration was increased 33 % by CO_2 enrichment because of a 23 % increase in growth respiration and a 48 % increase in maintenance respiration. Respiration rates of the upper stem and small branches were 4 × to 6 × higher than that of the lower bole, which is an important consideration in scaling these volume-specific rates to whole-tree and whole-plot respiration (Edwards et al. 2002). The CO_2-induced increase in stem respiration rates occurred concomitantly with elevated (28 %) sucrose concentrations in stems, whereas the concentrations of other soluble carbohydrates were not different between treatments (Edwards et al. 2002). Furthermore, the treatment difference in stem sucrose concentration was eliminated within 4 days of turning off the CO_2 fumigation temporarily in June 2001. There were, however, no treatment differences in stem soluble carbohydrate concentrations in the dormant season when differences in respiration were still evident, suggesting that substrates other than sucrose may drive the higher respiration rate.

Fine-root respiration (R_T) was partitioned between maintenance (R_M), growth (R_G), and N uptake respiration (R_N; George et al. 2003). Maintenance respiration was the majority (86 %) of R_T. There was no significant effect of CO_2 enrichment on instantaneous R_M, whether expressed on a mass or nitrogen basis, and no effect on tissue construction cost. Specific rates were scaled to annual, whole-plot rates using data on fine-root production, fine-root standing crop, and the temperature-sensitivity of the response (George et al. 2003).

13.4.1.2 Water

Mid-day measurements of stomatal conductance for leaves sampled in plots with e[CO_2] were as much as 44 % lower at elevated than at c[CO_2] (Wullschleger et al. 2002). Leaves in the upper canopy showed the strongest response to CO_2 enrichment, with no significant differences observed for leaves located in the middle and lower portions of the canopy. Estimates of canopy conductance averaged over the growing season were 14 % lower in stands exposed to CO_2 enrichment, although the magnitude of this response was dependent on vapor pressure deficit, radiation, and soil water potential (Wullschleger et al. 2002). The compensated heat-pulse technique was used to measure rates of sap velocity in 1999 (Wullschleger and Norby 2001). Sap velocity averaged 13 % less for trees in e[CO$_2$] compared with c[CO$_2$].

13.4.1.3 Nitrogen

Whole-canopy nitrogen concentration (N$_M$) was 16.8 mg g^{-1} in c[CO$_2$] and 11 % lower (14.9 mg g^{-1}) in e[CO$_2$] (Norby and Iversen 2006). Except in the first year of treatment (1998), the difference in canopy N$_M$ was also observed in leaf litter. N$_M$ was lower in e[CO$_2$] than in c[CO$_2$] plots at every depth of the canopy, but the CO$_2$ effect was greater toward the top because of a dilution effect. Leaf mass per unit area (LM$_A$) was higher in e[CO$_2$] only at the top of the canopy because of increased leaf density, which was related in part to a higher content of nonstructural carbohydrates (Sholtis et al. 2004). However, N expressed on a leaf area basis (N$_A$) was lower in e[CO$_2$] in the middle layers of the canopy, indicating a real effect on N content (Norby and Iversen 2006). Fine root N$_M$ was more variable than leaf N$_M$ and there was no significant effect of [CO$_2$] (Norby and Iversen 2006).

13.4.2 CO$_2$ Effects on Tree and Stand Structure

13.4.2.1 Leaf Area Index

Leaf area index (LAI) and its seasonal dynamics are key determinants of terrestrial productivity and, therefore, of the response of ecosystems to a rising atmospheric [CO$_2$]. Despite the central importance of LAI, there is very little evidence from which to assess how forest LAI will respond to increasing [CO$_2$]. LAI throughout the 1999–2002 growing seasons was assessed using a combination of data on transmittance of photosynthetically active radiation (PAR), mass of litter collected in traps, and LM$_A$ (Norby et al. 2003). There was no effect of [CO$_2$] on any expression of leaf area, including peak LAI, average LAI, or leaf area duration. Peak canopy mass was increased 8 % by CO$_2$ enrichment, reflecting a similar increase in LM$_A$.

13.4.2.2 Root System Structure

The root system of the sweetgum trees comprises a woody heart root, woody lateral roots that extend several meters from the trunk, and smaller and fine roots. Excavation of some trees revealed that most (90 %) of the root biomass was in the heart root and woody lateral roots, and based on allometric analysis, woody root biomass was not affected by CO$_2$ enrichment. Minirhizotron analysis indicated that approximately 80 % of fine-root length was in roots <0.5 mm diameter and less than 5 % was in roots >1 mm diameter (Norby et al. 2004). The fine-root standing crop was usually at a maximum in mid-July in $c[CO_2]$ and in August–September in $e[CO_2]$. Beginning in 2000, the peak standing fine-root mass in CO_2-enriched plots was more than doubled that in $c[CO_2]$ plots. In 1998, most of the root length was in the upper soil: 40 % in the top 15 cm and 79 % in the top 30 cm. Five years later, 63 % of the root length was still at 0–30 cm depth in $c[CO_2]$, but the root distribution was distinctly different in $e[CO_2]$ plots. CO_2 enrichment increased root length in the upper profile, but the largest increases occurred in deeper soil: 3-fold more length at 30–45 cm and 4-fold more at 45–60 cm.

13.4.3 Structure–Function Integration

13.4.3.1 Carbon Uptake

Gross primary productivity (GPP) is a function of leaf-level photosynthesis and LAI. Since leaf-level photosynthesis increased in $e[CO_2]$ and there was no change in LAI, an increase in GPP is implied. Based on calculations from canopy conductance (Wullschleger et al. 2002), GPP in 1999 was 27 % higher in $e[CO_2]$ (Norby et al. 2002). A similar estimate (22 %) of the response of GPP to CO_2 enrichment was generated by adding net primary productivity (NPP, see Section 13.5.1.3) to annual whole-tree (autotrophic) respiration (DeLucia et al. 2005). Autotrophic respiration in 2000 was calculated by summing leaf, wood, and fine-root respiration across the whole plot and for the whole year (DeLucia et al. 2005). Respiration was 28 % higher in $e[CO_2]$ and consumed 48 % of GPP in $c[CO_2]$ and 51 % of GPP in $e[CO_2]$. Consistent with the lack of effect of $[CO_2]$ on LAI, there also was no effect on absorbed photosynthetically active radiation (APAR). Hence, the observed increase in C uptake (GPP and NPP) was attributed to an increase in light-use efficiency (LUE; Norby et al. 2003) and reflected the sustained response of leaf-level photosynthesis to CO_2 enrichment. The current evidence from this and other experiments seems convincing that LAI of non-expanding forest stands will not be different in a future CO_2-enriched atmosphere (Körner et al. 2005, 2006) and that increases in LUE and productivity in $e[CO_2]$ are driven primarily by functional responses (e.g., increased photosynthesis) rather than by structural changes.

13.4.3.2 Stand Water Use

There is widespread belief that CO_2-induced reductions in stomatal conductance will have important consequences for forest water use, and in turn, ecosystem-scale processes that depend on water availability. Simple measurements of water use by seedlings or saplings growing in isolation, however, are insufficient to capture the complex temporal and spatial control of transpiration that inevitably takes place in closed-canopy stands. Whole-stand transpiration in the FACE experiment was estimated as a function of measured sap velocity, total stand sapwood area, and the fraction of sapwood functional in water transport (Wullschleger and Norby 2001). Maximum daily rates of stand transpiration during the 1999 growing season were 5.6 mm day^{-1} in c[CO$_2$] and 4.4 mm day^{-1} in e[CO$_2$], a 21 % reduction. Averaged across the entire growing season, the relative effect of [CO$_2$] was only a 10 % reduction: 3.1 mm day^{-1} and 2.8 mm day^{-1} in c[CO$_2$] and e[CO$_2$], respectively. Similar patterns were observed again in 2004 (Wullschleger, unpublished data; Fig. 13.3). The largest differences in 1999 occurred during May when stand water use was 104 mm in c[CO$_2$], but only 84 mm in e[CO$_2$], a 19 % reduction (Wullschleger and Norby 2001). In 2004 e[CO$_2$] reduced transpiration 20 % in both June and July (Fig. 3).

When transpiration of the trees was combined with calculations of evaporation from soil, estimates of annual evapotranspiration showed relatively small reductions due to atmospheric CO$_2$ enrichment (Wullschleger et al. 2002). The attenuation of CO$_2$ response from the scale of the leaf to tree to stand to ecosystem illustrates that the hydrological response of a closed-canopy plantation to e[CO$_2$] depends on the temporal and spatial scale of observation. These observations emphasize the importance of interacting variables (e.g., soil moisture, VPD) and confirm that integration of measurements over space and time reduce what, at the leaf level, might otherwise appear to be a large and significant response.

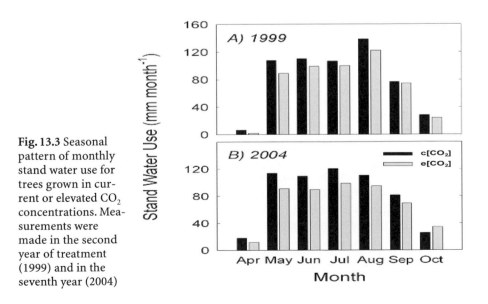

Fig. 13.3 Seasonal pattern of monthly stand water use for trees grown in current or elevated CO$_2$ concentrations. Measurements were made in the second year of treatment (1999) and in the seventh year (2004)

13.4.3.3 Nitrogen Cycling

Nitrogen cycling in mature forests is characterized by tight internal cycles that are not replicated in CO_2 enrichment experiments with tree seedlings or saplings in chambers (Johnson et al. 2004). In the FACE experiment, increased leaf mass production (Section 13.4.2.1) in CO_2-enriched canopies was offset by the reduced concentration of N, such that there was no effect on peak canopy N mass (Norby and Iversen 2006). The total amount of N used in fine-root production varied with year and increased significantly with CO_2 enrichment, reflecting an increase in root production (Section 13.5.1.2) with no effect on N concentration in roots (Section 13.4.1.3). Annual N uptake, which comprises the N content of leaf litter, wood increment, fine-root production, and throughfall minus deposition (Johnson et al. 2004), was significantly higher in e[CO_2], with an average increase over the period 1998–2003 of 29 %. Most of the difference was in the fine-root component. Uptake accounted for 74 % of annual N requirement in c[CO_2] and a significantly higher fraction (79 %) in e[CO_2], with retranslocation accounting for the remaining fraction of requirement (Norby and Iversen 2006). Annual N uptake increased linearly with fine-root length duration (the integration of root length over the growing season), suggesting that the greater investment in fine-root production in e[CO_2] increased the access to available soil N and thereby precluded the development of N limitations (Norby et al. 2004).

13.5 Resource Transformation

13.5.1 Productivity

13.5.1.1 Aboveground Production

Stem growth was assessed using allometric relationships developed on harvested trees from outside the experimental plots (Norby et al. 2001). Aboveground wood production was 35 % greater in CO_2-enriched plots during the first year of exposure (1998). In the second year, however, the difference in growth was reduced to 15 % and was no longer statistically significant, with further reductions in the third and subsequent years. The total dry matter increment of aboveground woody tissue from 1998 to 2003 was 7.2 kg m^{-2} in c[CO_2] and 7.7 kg m^{-2} in e[CO_2]; this difference of 6.9 % is not statistically significant. Leaf mass production was 7.4 % higher in e[CO_2] (Norby et al. 2003).

13.5.1.2 Belowground Production

Fine-root production and mortality were measured through analysis of minirhizotron images, which were collected biweekly from 1998 through 2003 (Norby et al. 2004). Annual production of fine roots more than doubled in response to CO$_2$ enrichment; and the nearly continuous 6-year record revealed the highly dynamic responses of the root system and its changing distribution in the soil profile. Although the seasonal pattern of fine-root production varied year to year, root productivity generally was higher in March through June than in July through October. The effect of [CO$_2$] on annual production, which was first observed during the third year, was highly significant, with production 2.2-fold higher in CO$_2$-enriched plots from 2000 to 2003. Mortality was highest in late summer and fall. Annually, mortality matched production in both c[CO$_2$] and e[CO$_2$]; hence, net production was close to zero. Root turnover, which is the fraction of the root population that is replaced during a year, was calculated as annual production divided by the maximum standing crop. Averaged over the 6 years of observation, root turnover was 1.7 year^{-1}, corresponding to mean residence time (MRT) of 0.62 year. Root turnover rate was not affected by [CO$_2$]. Cohort analysis also was used to estimate fine-root longevity (Norby et al. 2004), yielding an MRT ranging from 0.81 year to 1.4 year, with no effect of [CO$_2$] and in agreement with an analysis based on the replacement of pre-treatment roots with new roots with the distinct ^{13}C signature of the added CO$_2$ (Matamala et al. 2003). The absence of an effect of [CO$_2$] on turnover rate indicates that the increase in root mortality and concomitant input of root C into the soil was a direct result of increased root production and not an alteration of root physiology.

13.5.1.3 Ecosystem Productivity

Aboveground productivity of the forest understory was estimated in 2001 and 2002. The productivity of *Lonicera japonica* was consistently greater under e[CO$_2$], whereas the response of *Microstegium vimineum* to CO$_2$ enrichment differed between wet and dry years and mediated total community response (Belote et al. 2004). The understory accounted for less than 5 % of ecosystem NPP (Norby et al. 2002).

NPP of the forest plots was strongly dominated by the sweetgum trees. In c[CO$_2$], sweetgum NPP ranged over 852–1062 g C m^{-2} year^{-1} (1.85–2.33 kg dry matter m^{-2} year^{-1}), with wood production (bole, branches, and woody root) accounting for 65 %, leaf litter 21 %, and fine-root production 14 % (Fig. 13.4). Averaged over 6 years (1998–2003), NPP was 22 % higher in e[CO$_2$] (24 % if expressed in dry matter units). The response of NPP to CO$_2$ enrichment varied from 16 % to 38 % in different years, but showed no trend through time. The

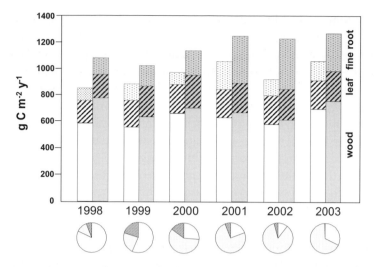

Fig. 13.4 Net primary productivity of sweetgum trees and the relative allocation of the additional C taken up in response to CO_2 enrichment. For each set of bars, unshaded bars on the left are for plots in $c[CO_2]$ and shaded bars on the right are $e[CO_2]$; lowest section is wood, middle section (striped) is leaf, and upper section (stippled) is fine roots. Pie charts show the relative distribution between wood (clear), leaves (striped), and fine roots (stippled) of the additional C taken up in CO_2-enriched plots

additional C taken up and converted to organic matter by trees in CO_2-enriched plots was allocated primarily to fine-root production. Excluding the first year of treatment (to exclude pre-treatment influences), 65 % of the pool of additional C was allocated to fine roots, 28 % to wood, and 7 % to leaves (Fig. 13.4). The additional allocation to wood did not produce a significant increase in wood increment.

Annual NEP was estimated from NPP and soil respiration (Norby et al. 2002). Elevated atmospheric $[CO_2]$ stimulated soil respiration (King et al. 2004). From early spring to mid-summer, soil respiration increased approximately four-fold, from 1 µmol m^{-2} s^{-1} to 4 µmol m^{-2} s^{-1}, respectively, over a soil temperature range of 7–23 °C, and stimulation of CO_2 efflux at 20 °C averaged 12.0 %. Most of the stimulation occurred mid growing season when fine-root activity was greatest. The seasonal Q_{10} relationship ranged from 1.89 to 2.60. Total soil C efflux in $c[CO_2]$ plots increased from 600 g C m^{-2} $year^{-1}$ in 1997 to 996 g C m^{-2} $year^{-1}$ in 2000, but then dropped to 698 g C m^{-2} $year^{-1}$ in 2001. Stimulation of cumulative soil C efflux by eCO_2 ranged from 8.3 % to 17.1 % and averaged only 12.0 % for all years. The temperature sensitivity of bulk soil respiration appeared to be unaffected by $[CO_2]$.

The heterotrophic component of soil respiration was estimated based on analysis of the ^{13}C content of soil CO_2 efflux in relation to seasonal variation in fine-root standing crop (Norby et al. 2002). R_H was higher in $e[CO_2]$, offsetting some of the gain in NPP, but total NEP for the years 1998–2001 was 679 g C m^{-2} higher in CO$_2$-enriched plots (170 g C m^{-2} year^{-1}). Half of this gain in ecosystem C was in wood, but this fraction was declining steeply (Fig. 13.4), and the overall contribution of wood to C sequestration was becoming progressively smaller as the experiment continued. Over the longer term, if present trends continue, an increase in NEP in e[CO$_2$] in this forest will be associated with changes in soil C.

13.5.2 Soil C

13.5.2.1 Carbon Input and Decomposition

The preferential allocation of additional C to fine roots rather than to woody biomass has significant implications for the potential of this forest to sequester C. Fine roots in this sweetgum stand have a mean residence time of approximately 1 year (Matamala et al. 2003, Norby et al. 2004), so the C that is allocated to them is not sequestered in plant biomass as is the C allocated to wood. Annually, 131 g C m^{-2} are added to soil through fine-root mortality in c[CO$_2$], and almost twice that amount in e[CO$_2$] (average over 1998–2003). After fine-root C is transferred to soil pools, it might be returned rapidly to the atmosphere through microbial respiration or stored as longer-lived organic compounds. Hence, it is important to understand the fate of the C in dead fine roots. In both laboratory and field incubations, fine roots lost 35–42 % of their dry mass over 360 days; and there was no effect of the CO$_2$ concentration in which they had been produced (Johnson et al. 2004).

Leaf litter accounts for an annual addition to soil of 214 g C m^{-2} in c[CO$_2$] and 230 g C m^{-2} in e[CO$_2$]. Leaf litter does not accumulate in this ecosystem due to the activity of earthworms: usually one year's cohort of leaf litter has disappeared by the time the next year's cohort falls (Johnson et al. 2004). Although the lower N concentration of litter in e[CO$_2$] would suggest a slower decomposition rate, the rapid disappearance renders moot any difference with respect to C storage or N availability.

13.5.2.2 Carbon Pools

Mineral soil was sampled in depth increments of 0–5, 5–15, and 15–30 cm after removing surface litter. Soils were fractionated into stable microaggregates (53–250 µm diam.) and non-microaggregated soil by using a microaggregate isolator (Six et al. 2000) and analyzed for organic C and C isotopic composition (Jastrow et al. 2005). Soil C in the surface 5 cm increased linearly during the first 5 years of exposure to e[CO$_2$], but remained constant in the c[CO_2] plots. No significant changes in soil C were found for either e[CO_2] or c[CO_2] at deeper depths. Isotopic data confirmed that net C inputs in the surface 5 cm were more than double those at 5–15 cm and over four times the inputs at 15–30 cm. Soil C stocks increased by 220±45 g C m^{-2} during the initial 5 years of exposure to e[CO_2]. Hence, the average accrual rate of 44±9 g C m^{-2} year^{-1} corresponded to more than half of the estimated annual NEP not accounted for by woody increment. (The remaining fraction might be in deeper soil, which was not sampled or lost in the error terms.) A portion of accrued soil C was stabilized by association with soil minerals. The proportion of C in microaggregated soil averaged 58 % under both e[CO_2] and c[CO_2] and was unchanged over time, suggesting that additional inputs derived from CO_2 enrichment were processed and protected in much the same manner as in soil in c[CO_2] plots with little apparent saturation of this protection mechanism (Jastrow et al. 2005). The formation and cycling of microaggregates is a key factor in physically protecting detrital inputs from rapid decomposition and helps to create conditions wherein microbial residues and breakdown products can become stabilized in organomineral complexes (Christensen 2001).

13.5.2.3 Microbial Activity and Nutrient Cycling

The indirect responses of soil microbiota to e[CO_2] have the potential to alter nutrient availability and soil carbon storage. However, despite the substantial increase in C deposition to soil in CO_2-enriched plots, there have been no detectable changes in microbial activity. Substrate utilization by soil bacteria and extracellular enzyme activities associated with bulk soil and fine-root rhizoplanes were measured over 1999–2001 (Sinsabaugh et al. 2003). Rhizoplane enzyme activity was similar to that of bulk soil; and there was no response to e[CO_2] in any of the measured variables. Johnson et al. (2004) noted a weak trend of increased soil microbial activity in e[CO_2] (5–23 %) and possibly increased microbial immobilization of some mineral nutrients, including N. The additional amount immobilized into microbial biomass in e[CO_2] represented approximately 4–22 kg N ha^{-1} in the 0–20 cm top soil layer, which is similar in magnitude to the amount of N needed for net woody tissue increment. However, no effect of [CO_2] on microbial N, gross N mineralization, microbial N immobilization, or specific microbial N immobilization was observed by Zak et al. (2003). Overall, there is little evidence that changes in plant litter production under e[CO_2] will initially slow soil N availability and produce a negative feedback on NPP; but our ability to predict long-term changes in soil N availability and hence whether greater NPP will be sustained in a CO_2-enriched atmosphere depends on an understanding of the time-scale over which greater plant production modifies microbial N demand (Zak et al. 2003).

13.5.3 Products

Although this sweetgum plantation was initially established to evaluate the species' potential as a biofuel (Van Miegroet et al. 1994), it has not been managed in that way; and questions about the effect of e[CO$_2$] on the production of biofuel feedstock or other forest products are not relevant to the objectives of the FACE experiment. Wood density was measured on a subset of trees as part of the allometric estimation of dry matter increment (Norby et al. 2001); and there has been no indication of a CO$_2$ effect on wood density. The trees in the FACE plots have not flowered or produced fruit as of 2004.

13.5.4 Biotic Interactions

Based on sweepnet and pitfall techniques, Sanders et al. (2004) determined that total arthropod abundance and richness across all trophic groups did not differ between c[CO$_2$] and e[CO$_2$] plots. However, particular trophic groups, especially the predators, were more abundant and had higher levels of species richness in c[CO$_2$] than in e[CO$_2$] plots. There were no distinct treatment effects on arthropod community composition, but there were strong temporal effects on community composition.

Neher et al (2004) extracted and enumerated nematodes in soil cores collected in 1999 and 2000. CO$_2$ enrichment decreased the total abundance of nematodes and decreased the respiration and biomass of the nematode community. Composition of the community changed, with a higher relative abundance of fungivores in e[CO$_2$] and a lower abundance of herbivores, bacterivores, and predators. Estimated annual productivity of soil nematode communities ranged from 0.6 g C m^{-2} year^{-1} to 4.7 g C m^{-2} year^{-1}, representing ca. 1 % of ecosystem NPP.

13.6 Consequences and Implications

13.6.1 Forest Management

Carbon sequestration in forests could be an important strategy for managing the global carbon cycle and slowing the inexorable increase in atmospheric [CO$_2$]. An increased rate of sequestration in response to e[CO$_2$], as observed in this experiment, would appear to be an important factor to incorporate into the design or evaluation of forest C sequestration projects; and forest FACE experiments are sometimes justified on this basis. However, extrapolations from perturbation experiments such as these are difficult because ecosystem

C sequestration rates are projected to respond differently to gradual versus step increases in atmospheric CO_2; and transient responses are expected (Luo et al. 2003). Furthermore, the relatively small increment in C storage in the sweetgum forest in response to the atmospheric CO_2 concentration that will be attained in about 2050 suggests a much smaller increment within the period that a C sequestration project might be implemented.

Converting unproductive land to forests or improving the silviculture of existing forests is likely to yield far greater rates of C sequestration that render the CO_2 fertilization effect trivial. Nevertheless, the intensive and integrated evaluation of C metabolism and cycling in the FACE experiment should help to inform forest C management. For example, the large response of fine roots to CO_2 enrichment in this experiment and the resulting deposition of increased C to deeper soil profiles suggests that forests may have more potential for C sequestration than may be apparent from aboveground analysis (e.g., Casperson et al. 2000), but this observation must be tempered with the formidable difficulty in measuring belowground processes (Norby et al. 2004).

13.6.2 Global C Cycle

The implications of the Oak Ridge FACE experiment reside primarily in the arena of the global carbon cycle and the potential for a negative feedback between increasing atmospheric CO_2 and forests to alter the rate at which greenhouse warming occurs. The sustained response of NPP to 542 ppm CO_2 in this experiment (a 24 % increase) is consistent with projections from earlier experiments with younger trees (Norby et al. 1999) and ecosystem and global models (e.g., Cramer et al. 2001). The basis for the response of NPP is the sustained stimulation of leaf-level photosynthesis and stand-level LUE. There has been no evidence for acclimation of photosynthesis to $e[CO_2]$, which has sometimes been given as a reason to discount CO_2 fertilization. There also has been no indication of the development of a progressive N limitation, which has been proposed as a negative feedback on the response of NPP to rising $[CO_2]$ (Luo et al. 2004), although we cannot discount the possibility of an N limitation developing in the future. Hence, the Oak Ridge FACE experiment provides support for the inclusion of a CO_2 fertilization effect in models of the future trajectory of the global C cycle.

The question arises, however, as to whether the responses to CO_2 enrichment observed in these young, fast-growing plantation trees are predictive of the response of older and larger, mature forest trees. Körner et al. (2005) used a new web-FACE design to expose ten mature, 30-m tall trees in a deciduous forest to 530 ppm for 4 years. The physiological responses of these trees (e.g., photosynthesis, foliar N and non-structural carbohydrate concentrations) were very similar to the responses of the Oak Ridge sweetgum trees, thereby increasing our confidence that our FACE results are relevant to questions about temperate forests in general. As in the sweetgum experiment, there was no significant increase in basal area increment, although with only ten trees in the web-FACE experiment, the power to detect CO$_2$ effects on growth was very low. The web-FACE design did not permit analysis of NPP, so there is no basis for comparative analysis of this critical ecosystem-scale response.

The FACE experiment can provide additional guidance for ecosystem models being used to evaluate CO$_2$ responses. LAI of the sweetgum stand has not increased and there is now strong evidence that LAI of non-expanding forest stands will not be different in a future CO$_2$-enriched atmosphere. Model routines that assume increased LAI should be avoided because they alter estimates of N demand and water use. The assumption that CO$_2$ effects on stomatal conductance result in reductions in forest water use is incorporated into some models, but this has not been supported statistically because of attenuation of the response across scales coupled with the low replication of FACE experiments. Models will be especially challenged to represent C allocation such that the dramatic response of fine-root production observed in this experiment is reproduced. The fine-root response of the sweetgum trees in the FACE experiment has important implications for carbon, water, and nutrient cycles; and understanding how different ecosystems allocate C to fast- or slow-turnover pools may be key to predicting their integrated response to atmospheric CO$_2$ enrichment. The allocation response is the primary difference between this experiment and the similar FACE experiment in a *Pinus taeda* stand (Chapter 11). NPP and NEP responded similarly to CO$_2$ enrichment in the two experiments (DeLucia et al. 2005), but in the pines the additional productivity was recovered primarily in wood.

The response of plants to e[CO$_2$] should also be considered in the context of other changing environmental variables. Future enhancements of GPP driven by increasing [CO$_2$] will be attenuated to varying degrees by increased respiration from warming and by loss of photosynthesis from exposure to drought or air pollutants. Hanson et al. (2005) used the responses to e[CO$_2$] in this FACE experiment, combined with other observations of forest responses to increased temperature, ozone, and altered precipitation patterns, in a model-based analysis of an upland-oak forest in 2100. The dominant effects of e[CO$_2$] on the simulated forest were reduced when combined with effects of warming, precipitation change, ozone exposure, tissue acclimation, and changes in biomass and element stocks. The analysis of the impact of e[CO$_2$] on the global carbon cycle using results from FACE experiments must expand to consider the influence of multiple interacting environmental variables.

13.7 Conclusions

Physiological and stand-level responses observed in the closed-canopy *Liquidambar styraciflua* forest in the Oak Ridge FACE experiment are contributing to our understanding of how the eastern North American deciduous forest will be affected by CO_2 enrichment of the atmosphere.

- Photosynthetic CO_2 assimilation averaged 46% higher in $e[CO_2]$. The response was sustained throughout the canopy; and there was no loss of photosynthetic capacity over time.
- Stomatal conductance in upper canopy leaves was reduced as much as 44% in $e[CO_2]$, but the responses of canopy conductance and stand transpiration averaged over the growing season were much less (14% and 10%, respectively).
- NPP during the first 6 years of treatment averaged 967 g C m^{-2} year^{-1} in $c[CO_2]$ and 1164 g C m^{-2} year^{-1} in $e[CO_2]$, an increase of 22%. There was no effect of $[CO_2]$ on LAI or APAR; hence, the increase in NPP was attributed to greater LUE.
- The increased C uptake in $e[CO_2]$ was partitioned primarily to fine-root production, which was more than doubled in years 3–6. There was no significant increase in wood production after the first year of treatment.
- Annual N uptake increased linearly with fine-root length duration and was significantly greater in $e[CO_2]$. Leaf and litter N concentrations were reduced in $e[CO_2]$, but there was no indication of a negative feedback of N availability on NPP.
- Increased fine-root mortality added more C to the soil in $e[CO_2]$. C efflux from soil also increased; nevertheless, $e[CO_2]$ caused a significant increase in C accrual in the surface 5 cm of soil. The average accrual rate was 44 g C m^{-2} year^{-1}.

Acknowledgments. The Oak Ridge FACE experiment has been sponsored by the US Department of Energy, Office of Science, Biological and Environmental Research. Oak Ridge National Laboratory is managed by UT-Battelle, LLC, for the US Department of Energy under contract DE-AC05-00OR22725; and the University of Chicago operates Argonne National Laboratory under contract W-31-109-ENG-38 with the US Department of Energy.

References

Belote RT, Weltzin JF, Norby RJ (2004) Response of an understory plant community to elevated [CO$_2$] depends on differential responses of dominant invasive species and is mediated by soil water availability. New Phytol 161:827–835

Caspersen JP, Pacala SW, Jenkins JC, Hurtt GC, Moorcroft PR, Birdsey RA (2000) Contributions of land-use history to carbon accumulation in US forests. Science 290:1148–1151

Christensen B (2001) Physical fractionation of soil and structural and functional complexity in organic matter turnover. Eur J Soil Sci 52:345–353

Cramer W, Bondeau A, Woodward FI, Prentice IC, Betts RA, Brovkin V, Cox PM, Fisher V, Foley JA, Friend AD, Kucharik C, Lomas MR, Ramankutty N, Sitch S, Smith B, White A, Young-Molling C (2001) Global response of terrestrial ecosystem structure and function to CO$_2$ and climate change: results from six dynamic global vegetation models. Global Change Biol 7:357–373

DeLucia EH, Moore DJ, Norby RJ (2005) Contrasting responses of forest ecosystems to rising atmospheric CO$_2$: implications for the global C cycle. Global Biogeochem Cycles 19:GB3006.

Edwards NT, Tschaplinski TJ, Norby RJ (2002) Stem respiration increases in CO$_2$-enriched trees. New Phytol 155:239–248

George K, Norby RJ, Hamilton JG, DeLucia EH (2003) Fine-root respiration in a loblolly pine and sweetgum forest growing in elevated CO$_2$. New Phytol 160:511–522

Gunderson CA, Sholtis JD, Wullschleger SD, Tissue DT, Hanson PJ, Norby RJ (2002) Environmental and stomatal control of photosynthetic enhancement in the canopy of a sweetgum (*Liquidambar styraciflua* L) plantation during three years of CO$_2$ enrichment. Plant Cell Environ 25:379–393

Hanson PJ, Wullschleger SD, Norby RJ, Tschaplinski TJ, Gunderson CA (2005) Importance of changing CO$_2$, temperature, precipitation, and ozone on carbon and water cycles of an upland-oak forest: incorporating experimental results into model simulations. Global Change Biol 11:1402–1423

Hendrey GR, Ellsworth DS, Lewin KF, Nagy J (1999) A free-air enrichment system for exposing tall forest vegetation to elevated atmospheric CO$_2$. Global Change Biol 5:293–309

Jastrow JD, Miller RM, Matamala R, Norby RJ, Boutton TW, Rice CW, Owensby CE (2005) Elevated atmospheric CO$_2$ increases soil carbon. Global Change Biol 11:2057–2064

Johnson DW, Cheng W, Joslin JD, Norby RJ, Edwards NT, Todd DE Jr (2004) Effects of elevated CO$_2$ on nutrient cycling in a sweetgum plantation. Biogeochemistry 69:379–403

King JS, Hanson PJ, Bernhardt E, DeAngelis P, Norby RJ, Pregitzer KS (2004) A multi-year synthesis of soil respiration responses to elevated atmospheric CO$_2$ from four forest FACE experiments. Global Change Biol 10:1027–1042

Körner C, Asshoff R, Bignucolo O, Hättenschwiler S, Keel SG, Pelaez-Riedl S, Pepin S, Siegwolf RTW, Zotz G (2005) Carbon flux and growth in mature deciduous forest trees exposed to elevated CO$_2$. Science 309:1360–1362

Körner C, Morgan J, Norby RJ (2006) CO$_2$ enrichment: from plot responses to landscape consequences. In: Canadell J, Pataki D, Pitelka L (eds) Terrestrial ecosystems in a changing world. Springer, Berlin Heidelberg New York (in press)

Luo Y, White LW, Canadell JG, DeLucia EH, Ellsworth DS, Finzi A, Lichter J, Schlesinger WH (2003) Sustainability of terrestrial carbon sequestration. A case study in Duke Forest with inversion approach. Global Biogeochem Cycles 17:1021

Luo Y, Su B, Currie WS, Dukes JS, Finzi A, Hartwig U, Hungate B, McMurtrie RE, Oren R, Parton WJ, Pataki DE, Shaw MR, Zak DR, Field CB (2004) Progressive nitrogen limitation of ecosystem responses to rising atmospheric carbon dioxide. BioScience 54:731–739

Matamala R, Gonzàlez-Meler MA, Jastrow JD, Norby RJ, Schlesinger WH (2003) Impacts of fine root turnover on forest NPP and soil C sequestration potential. Science 302:1385–1387

Neher DA, Weicht TR, Moorhead DL, Sinsabaugh RL (2004) Elevated CO_2 alters functional attributes of nematode communities in forest soils. Funct Ecol 18:584–591

Norby RJ, Iversen CM (2006) Nitrogen uptake, distribution and turnover in a CO_2-enriched sweetgum forest. Ecology 57:5–14

Norby RJ, Wullschleger SD, Gunderson CA, Johnson DW, Ceulemans R (1999) Tree responses to rising CO_2: implications for the future forest. Plant Cell Environ 22:683–714

Norby RJ, Todd DE, Fults J, Johnson DW (2001) Allometric determination of tree growth in a CO_2-enriched sweetgum stand. New Phytol 150:477–487

Norby RJ, Hanson PJ, O'Neill EG, Tschaplinski TJ, Weltzin JF, Hansen RT, Cheng W, Wullschleger SD, Gunderson CA, Edwards NT, Johnson DW (2002) Net primary productivity of a CO_2-enriched deciduous forest and the implications for carbon storage. Ecol Appl 12:1261–1266

Norby RJ, Sholtis JD, Gunderson CA, Jawdy SS (2003) Leaf dynamics of a deciduous forest canopy: no response to elevated CO_2. Oecologia 136:574–584

Norby RJ, Ledford J, Reilly CD, Miller NE, O'Neill EG (2004) Fine-root production dominates response of a deciduous forest to atmospheric CO_2 enrichment. Proc Natl Acad Sci USA 101:9689–9693

Riggs JS, Tharp ML, Norby RJ (2003a) ORNL FACE weather data (http://cdiac.ornl.gov/programs/FACE/ornldata/weatherfiles.html). Carbon Dioxide Information Analysis Center, Oak Ridge, Tenn.

Riggs JS, Tharp ML, Norby RJ (2003b) ORNL FACE CO_2 data (http://cdiac.ornl.gov/programs/FACE/ornldata/CO_2files.html). Carbon Dioxide Information Analysis Center, Oak Ridge, Tenn.

Sanders NJ, Belote RT, Weltzin JF (2004) Multi-trophic effects of elevated CO_2 on understory plant and arthropod communities. Environ Entomol 33:1609–1616

Sholtis JD, Gunderson CA, Norby RJ, Tissue DT (2004) Persistent stimulation of photosynthesis by elevated CO_2 in a sweetgum (*Liquidambar styraciflua* L.) forest stand. New Phytol 162:343–354

Sinsabaugh RL, Saiya-Cork K, Long T, Osgood MP, Neher DA, Zak DR, Norby RJ (2003) Soil microbial activity in a *Liquidambar* plantation unresponsive to CO_2-driven increases in primary productivity. Appl Soil Ecol 24:263–270

Six J, Elliott ET, Paustian K (2000) Soil macroaggregate turnover and microaggregate formational mechanism for C sequestration under no-tillage agriculture. Soil Biol Biochem 32:2099–2103

Tissue DT, Lewis JD, Wullschleger SD, Amthor JS, Griffin KL, Anderson OR (2002) Leaf respiration at different canopy positions in sweetgum (*Liquidambar styraciflua*) grown in ambient and elevated concentrations of carbon dioxide in the field. Tree Physiol 22:1157–1166

Van Miegroet H, Norby RJ, Tschaplinski TJ (1994) Optimum nitrogen fertilization in a short-rotation sycamore plantation. For Ecol Manage 64:25–40

Wullschleger SD, Norby RJ (2001) Sap velocity and canopy transpiration for a 12-year-old sweetgum stand exposed to free-air CO_2 enrichment. New Phytol 150:489–498

Wullschleger SD, Gunderson CA, Hanson PJ, Wilson KB, Norby RJ (2002) Sensitivity of stomatal and canopy conductance to elevated CO$_2$ concentration – interacting variables and perspectives of scale. New Phytol 153:485–496

Zak DR, Holmes WE, Finzi AC, Norby RJ, Schlesinger WH (2003) Soil nitrogen cycling under elevated CO$_2$: a synthesis of forest FACE experiments. Ecol Appl 13:1508–1514

14 Long-Term Responses of Photosynthesis and Stomata to Elevated [CO$_2$] in Managed Systems

S.P. Long, E.A. Ainsworth, C.J. Bernacchi, P.A. Davey, G.J. Hymus, A.D.B. Leakey, P.B. Morgan, and C.P. Osborne

14.1 Introduction

14.1.1 The Theory of Responses of Photosynthesis and Stomatal Conductance to Elevated [CO$_2$]

Plants can only perceive a change in atmospheric [CO$_2$] through tissues that are exposed to the open air. The protective cuticle of higher-plant shoots means that, with the exception of the stigma and germinating pollen tube, only the inner surfaces of the guard cells of stomata and the photosynthetic mesophyll cell wall surfaces exposed to the atmosphere of the internal air spaces can directly sense a change in atmospheric [CO$_2$]. While many steps in metabolism utilize or respond to CO$_2$, the only sites where there is convincing evidence for a response in the concentration range of relevance (240–1000 ppm) are ribulose-1:5-bisphosphate carboxylase/oxygenase (Rubisco) and a yet undefined metabolic step affecting stomatal aperture, that may also involve Rubisco (Buckley et al. 2003; Long et al. 2004). Although leaf respiration was once thought to respond directly to variation in [CO$_2$] over this range, this is now shown to be an artifact of earlier measurement techniques (Amthor et al. 2001; Davey et al. 2004; Jahnke 2001), and unsupported by examination of the [CO$_2$] sensitivity of metabolic steps controlling dark respiration (Gonzalez-Meler and Siedow 1999; see Chapter 15). Photosynthesis and stomatal movement, the subject of this chapter, are therefore the primary points of response to elevated (e)[CO$_2$]; and all other changes in the system follow on from the response at this level.

The direct causes of the instantaneous increase in C$_3$ photosynthesis with elevation of [CO$_2$] are two properties of the primary carboxylase of C$_3$ photosynthesis: ribulose-1:5-bisphosphate carboxylase/oxygenase (Rubisco);

Ecological Studies, Vol. 187
J. Nösberger, S.P. Long, R.J. Norby, M. Stitt,
G.R. Hendrey, H. Blum (Eds.)
Managed Ecosystems and CO$_2$
Case Studies, Processes, and Perspectives
© Springer-Verlag Berlin Heidelberg 2006

1. The enzyme is not saturated by present atmospheric $[CO_2]$, and so $e[CO_2]$ will increase the velocity of carboxylation and net photosynthesis.
2. CO_2 is a competitive inhibitor of the oxygenation reaction, which leads to photorespiratory release of CO_2.

Photorespiration typically releases 20–40 % of recent photosynthate as CO_2 (Long et al. 2004). This significantly reduces net photosynthesis of C_3 crops and will be suppressed in favor of greater carbon gain by rising $[CO_2]$. Because the kinetic properties of Rubisco are highly conserved across C_3 crops, the improvement in photosynthetic gain with rising $[CO_2]$ can be calculated for all C_3 crops with some confidence (Long et al. 2004). An increase in atmospheric $[CO_2]$ from today's 372 ppm to 550 ppm would increase net leaf photosynthesis by 12–36 %, while elevation to 700 ppm would generate a stimulation of 18–63 %, for a leaf temperature of 25 °C. The lower end of these ranges represents light-limited photosynthesis, while the greatest stimulation occurs in theory under light-saturated conditions, when the amount of Rubisco is assumed to be limiting. Since crop canopies in the field gain their carbon in roughly equal quantities from light-limited and light-saturated photosynthesis, actual increases in canopy carbon uptake are likely to be in the middle of these ranges (Long 1991; Long et al. 2004). This theoretical stimulation increases with temperature because both the specificity of Rubsico for CO_2 and the solubility of CO_2 relative to O_2 declines with temperature (Long 1991; Long et al. 2004).

A new mechanistic model of the response of stomatal conductance (g_s) to the environment assumes that the photosynthetic rate in the guard cell may be predicted via the biochemical photosynthetic model of Farquhar et al. (1980) and that aperture is determined by ATP concentration (Buckley et al. 2003). The model has proved exceptional in predicting a wide range of stomatal responses to different environmental perturbations. It follows from this model, and many preceding observations, that g_s increases with light and humidity, and decreases with declining leaf water potential and with rising $[CO_2]$. Further, at a given humidity and leaf water status, conductance varies to maintain the ratio of external (c_a) to intercellular (c_i) $[CO_2]$ constant. This model, as with its predecessors, predicts therefore that g_s decreases on transfer of leaves from $c[CO_2]$ to $e[CO_2]$. Will this decline, and therefore the increased restriction on CO_2 diffusion into the mesophyll, lessen the response of net leaf photosynthesis (A) to $e[CO_2]$?

For a constant A, any decrease in g_s causes a linearly proportional increase in the diffusion gradient ($c_a–c_i$). From the response of A to c_i, it is possible to deduce the decrease in A that results from the decline in $[CO_2]$ across the stomata (i.e., $c_a–c_i$). Figure 14.1 shows this response and the effect of g_s on c_i. For leaves in the current atmospheric $[CO_2]$ of 372 ppm, if $A = 0$ and the stomata are open, then $c_i = c_a$. This is illustrated by the intercept of the left-hand dotted line and the x-axis. As A increases, c_i declines linearly and in inverse proportion to g_s; and the point at which this line (the supply function) intercepts the response curve of A to c_i (solid black line; the demand function) gives the operating c_i. If there were no diffusive barrier (i.e., $g_s = 8$), c_i would equal the external [CO$_2$], as indicated by the vertical dashed line originating from 372 ppm on the x-axis. If A (marked on Fig. 14.1) is the actual rate at the actual c_i, the limitation imposed by the stomata (l) is given by $(A^\circ - A)/A^\circ$ (in this example 0.136). At an e[CO$_2$] of 572 ppm, g_s is assumed to be decreased to 0.65, the value at the c[CO$_2$], following the expectation that g_s is inversely proportional to [CO$_2$] and that c_i/c_a remains constant. This is represented by the more negative slope of the supply function by the dotted line originating from the x-axis at 572 ppm (i.e., current plus 200 ppm). However, because the slope of the demand function (dA/dc_i) diminishes with increase in c_i, stomatal limitation (l) in this example is just 0.045. Therefore, and despite partial closure, the limitation that the stomata place on photosynthesis is substantially less at e[CO$_2$] than at c[CO$_2$]. If there was no decrease in stomatal aperture and therefore conductance, how much greater would the increase in A be on increasing [CO$_2$] from 372 ppm to 572 ppm? For the same example, if g_s was held constant on increasing [CO$_2$], then l would be 0.025, compared to 0.045. Extrapolating from Fig. 14.1, a decrease in g_s lowers A by only 1.5%. But, because transpiration is linearly proportional to g_s, it lowers transpiration by

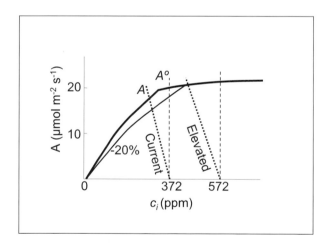

Fig. 14.1 The response of light-saturated leaf net CO$_2$ uptake (A) to intercellular CO$_2$ concentration (c_i), or demand function, as predicted from the leaf biochemical model of photosynthesis of Farquhar et al. (1980). The *dotted black lines* labeled "Current" and "Elevated" show the supply functions, i.e., the decline in c_i that occurs with increasing A, at both the current atmospheric concentration (372 ppm) and a future elevated concentration (572 ppm). The slope of each line is inversely proportional to stomatal conductance. The *vertical dashed lines* show the supply functions assuming an infinite stomatal conductance. The lower initial slope of the demand function (*thinner line*) shows how A would be decreased by a 20% decrease in Rubisco activity in the absence of any change in capacity for regeneration of RubP. Figure redrawn from Long et al. (2004)

35 %. Hence, this change would in theory greatly improve the efficiency of water use with a negligible reduction in potential A.

The preceding analysis of photosynthetic and stomatal responses considers the instantaneous response, i.e. how do A and g_s respond when a plant grown at $c[CO_2]$ is transferred to $e[CO_2]$? However, growth at $e[CO_2]$ does alter the photosynthetic apparatus, so to what extent does this alter the expectations developed above for the instantaneous effects of transfer to $e[CO_2]$?

14.1.2 Chamber Acclimation and Down-Regulation of Photosynthesis

As free air CO_2 enrichment (FACE) technology was emerging, many studies had already been completed which had grown plants at $e[CO_2]$ and examined acclimation. Thousands of experimental studies have evaluated the response of crops to the increases in atmospheric $[CO_2]$ expected to occur this century (for reviews, see Ainsworth et al. 2002; Amthor 2001; Drake et al. 1997; Jablonski et al. 2002; Kimball 1983). Drake et al. (1997), across 60 separate studies in which plants were grown at $e[CO_2]$ in large pots or in the field, showed that the long-term increase in A at an average $e[CO_2]$ of 680 ppm was 58 %, but was only 28 % for the 105 studies in which plants had been grown in small pots (<10 l). Other studies, particularly in the ecological arena with parallel systems and via modeling, developed the paradigm that response to CO_2 would be short-lived and often limited by nitrogen. Oechel et al. (1994) reported that, with 3 years of growth at an $e[CO_2]$ of 680 ppm, there was a complete loss of the initial stimulation of ecosystem CO_2 uptake seen when the system was first placed in $e[CO_2]$. The authors noted: "*These are the first field evidence for the complete homeostatic adjustment of CO_2 flux (in a native ecosystem) to elevated atmospheric CO_2, and support conclusions from a variety of sources. This other work includes phytotron experiments on Arctic plants, Arctic ecosystem microcosms, and non-Arctic plants, greenhouse experiments on a simulated tropical ecosystem, and open-top field chamber experiments on grasses and pine trees.*" This essentially stated that genetic and resource limitations would prevent a prolonged stimulation by $e[CO_2]$. For managed ecosystems, these limitations may be expected to be absent. However, an alternative viewpoint may be that selection of annual crops with a narrow period of grain-filling to allow uniform maturation for harvest could impose a greater genetic constraint. Extensively managed grasslands or areas where legislative or financial restrictions prevent high rates of fertilizer addition could be constrained by nitrogen, as in natural systems. Further, Diaz et al. (1993) reported from chamber experiments, using species exhibiting a wide range of growth strategies, that down-regulation of growth response could occur on fertile soils. This was attributed to increased sequestration of nutrients by an increased loss of organic carbon from roots of plants grown in $e[CO_2]$. The two highly influential studies of Oechel et al. (1994) and Diaz et al. (1993) cast doubt even on sustained stimulation of photosynthesis in managed ecosystems. A considerable divergence of opinion and results, from large sustained stimulations of photosynthesis to a complete loss of stimulation, existed as FACE experiments became established.

14.1.3 A Purpose to Down-Regulation of Photosynthesis and Stomatal Conductance?

Restriction on growth response to increased photosynthesis, either genetic or nutritional, would lead to an increase in non-structural carbohydrate concentrations. Expression of a number of genes is known to be sensitive to soluble carbohydrates, including glucose and sucrose (see Chapter 16). Those genes that show up-regulated transcription are termed feast genes, versus those down-regulated, famine genes (Koch 1996). Feast genes include several involved in secondary metabolism, sucrose transport and synthesis of storage carbohydrates. Famine genes include those coding for photosynthetic proteins, in particular the small sub-unit of Rubisco. This is consistent with the most prominent change in the leaf photosynthetic apparatus that has been observed, a decline in the amount and activity of Rubisco. This has been suggested as the basis of the decrease in response of production to e[CO$_2$] and one that would inherently preclude or at least decrease a response in the long term. Bunce and Sicher and Bunce (1997) observed a loss of Rubisco and Rubisco activity from leaves of wheat and barley grown under season-long e[CO$_2$] in open-top chambers (OTCs) in the field and showed a linear relation between this loss and a loss of photosynthetic capacity. An alternative perspective is that the decline in Rubisco reflects a decreased requirement for Rubisco at e[CO$_2$]. The response of A to c_i, as described by the well tested mechanistic model of Farquhar et al. (1980), is biphasic (Fig. 14.1). As c_i increases from zero, A increases steeply where Rubisco activity is limiting to a transition point beyond which RubP is limiting and dA/dc_i is small and approaches zero (Fig. 14.1). At light saturation, this transition is commonly at the c_i that occurs in the present atmospheric [CO$_2$]; typically c_i is about 0.7 × atmospheric [CO$_2$] (Drake et al. 1997; Long et al. 2004). This implies that there is a balance between the amount of active Rubisco, represented by the maximum velocity of carboxylation ($V_{c,max}$), and the capacity for RubP regeneration, represented by the maximum whole-chain electron transport rate (J_{max}). If atmospheric [CO$_2$] increases, c_i is expected to rise proportionately, since the ratio c_i/c_a in C$_3$ plants appears unaffected (for a review, see Drake et al. 1997). If c_i increases by 140 ppm, resulting from elevation of the external [CO$_2$] by 200 ppm, photosynthesis would be limited solely by RubP regeneration; and a substantial (20 %) loss of Rubisco activity could occur without affecting photosynthesis. This is illustrated in Fig. 14.1 by the slope marked "–20 %". While A would be insensitive to a large decrease in Rubisco and therefore $V_{c,max}$, any change in capacity for regeneration of RubP (J_{max}) would cause a decrease in A. Since total quantities of protein invested in Rubisco and in the apparatus for regeneration of RuBP are similar and collectively account for over 50 % of leaf protein, a selective decrease in Rubisco relative to other leaf proteins would be necessary to achieve the theoretical scenario illustrated by Fig. 14.1. Such a selective decrease would result in the nitrogen that would otherwise be sequestered into excess Rubisco being available for growth of additional tissue, so partially or wholly alleviating any sink limitation. Thus down-regulation of Rubisco would serve to increase efficiency of resource use, in particular nitrogen. However Drake et al. (1997), reviewing prior chamber studies, showed that the loss of Rubisco protein with growth in e[CO_2] relative to c[CO_2] was ca. 15 % and directly proportional to the 15 % decrease in total leaf protein. There was then no evidence of the selective loss of Rubisco that would theoretically confer an advantage.

If there was a significant independent acclimation of stomatal response to e[CO_2], this would be evident in a change in c_i/c_a during light-saturated photosynthesis. Drake et al. (1997) showed that, across 33 different studies of plants grown at a mean e[CO_2] of 680 ppm, g_s was consistently ca. 20 % less but c_i/c_a was just 1 %, and not significantly, lower than in plants grown at c[CO_2]. This appeared compelling evidence of a lack of any independent acclimation of stomatal response. By maintaining constant c_i/c_a, a substantial gain in efficiency of water use would result.

14.1.4 Expectations of FACE

In summary, expectations of photosynthesis and stomatal conductance for plants of managed ecosystems grown in e[CO_2] derived from chamber studies were that: (1) the initial stimulation of photosynthesis would not persist particularly under extensive management, (2) while theoretically advantageous, a selective loss of Rubisco activity does not occur, and (3) stomatal conductance does not acclimate independently of photosynthesis to e[CO_2]. A further expectation from theory, rarely investigated in field chamber studies, is that stimulation would be greatest during the middle of the day when light levels are saturating and least in the hours immediately following sunrise and immediately preceding sunset, when photosynthesis is light-limited.

14.2 Why FACE for Photosynthesis and Conductance?

Most information about photosynthetic responses of crops to e[CO_2] has been derived from experimental studies that have used greenhouses, artificially illuminated controlled environmental chambers, transparent field enclosures or OTCs (Ainsworth et al. 2002; Drake et al. 1997; Hendrey 1992; Jablonski et al. 2002; Kimball 1983). These systems have both theoretical and practical limitation in assessing the response of photosynthesis to growth at e[CO$_2$] (Hendrey 1992; McLeod and Long 1999; see Chapter 3). While enclosure methods provide an atmosphere with enriched [CO$_2$], they also significantly alter other aspects of the environment surrounding the plant. Many of these studies, including some field studies, used plants grown in pots. Arp (1991) showed that rooting volume altered the response to plants to e[CO$_2$]; and further experiments reported a strong feedback when roots encounter a physical barrier (Arp 1991; Masle et al. 1990; Thomas and Strain 1991). Ainsworth et al. (2002) showed that the response of soybean to e[CO$_2$] observed in studies where plants were rooted in the ground was more than double that observed in studies which used pots of >9 liters rooting volume.

Most field studies of crops have used OTCs, transparent walled chambers, of up to 2 m diameter. Despite the fact that the top of the chamber is open to the atmosphere, there are environmental differences between even the best engineered OTCs and the adjacent unenclosed crop. The effect of the OTC itself may exceed that of elevation of [CO$_2$] (Day et al. 1996; Drake et al. 1989). Whitehead et al. (1995) compared microclimatic conditions within and outside OTCs. When the outside photosynthetic photon irradiance was 1600 μmol m^{-2} s^{-1} (about 75 % of full sunlight in summer at mid-latitudes), air temperature within the chamber was 4.3 °C higher and water vapor pressure deficit (VPD) 0.8 kPa higher. The transmission of total solar irradiance into the chambers was lower and the ratio of diffuse to total solar irradiance in the chambers was altered. From a theoretical perspective, each of these differences could have profound effects on photosynthesis and stomatal conductance, and alter the response to e[CO$_2$]. Higher temperature exaggerates the stimulation of photosynthesis by e[CO$_2$] (Long 1991). Similarly, higher humidity may result in higher conductances and also exaggerate the response of photosynthesis to e[CO$_2$]. In contrast, decreased direct sunlight may diminish the response of photosynthesis to e[CO$_2$] by decreasing the amount of time during the day in which the canopy or parts of the canopy are light-saturated (McLeod and Long 1999). Therefore, even if chamber effects do not change the direction of a response, they alter its magnitude.

Additionally, small isolated plots in agronomic trials often overestimate treatment effects on biomass, production and yields (Roberts et al. 1993). Gaps caused by sampling within a small area exacerbate this problem. Increased radiation interception at the edges of small plots can exaggerate the effect of a treatment. Development of the photosynthetic apparatus in young leaves is dependent on the amount and quality of light. Both are artificially altered in small plots. A typical 2-m diameter OTC would have a ground surface area of <3.1 m^2. Therefore, in a 2-m diameter chamber, more than 50 % of the vegetation is less than 30 cm from the chamber wall and 75 % is within 50 cm of the wall. The recommended border or buffer area for agricultural trials is typically twice the vegetation height (Roberts et al. 1993). Therefore, even a 50-cm high short-season soybean crop would require a 1-m buffer zone; and thus no area within an OTC would be free from edge effects (Long et al. 2004). Consequently, knowledge of photosynthetic responses of crops to $e[CO_2]$ is currently derived from experiments that are considered unacceptable in standard agronomic trials (McLeod and Long 1999).

Size and enclosure have practical limitations for the investigation of photosynthesis and conductance. A key question in understanding crop responses to $e[CO_2]$ is how are the responses of photosynthesis and transpiration affected over the diurnal course? For example, does stimulation diminish in the afternoon as carbohydrates accumulate and the possibility of feedback limitation develops? Is stimulation less at the beginning and end of the day, as predicted from theory? Investigation of these questions requires diurnal sampling of photosynthetic rates within the canopy. In OTCs, this would require frequent removal of side panels to allow access and a concurrent disruption of the $[CO_2]$ treatment. FACE presents no such barriers to the investigation of diurnal changes in photosynthesis and stomatal conductance. To investigate mechanisms underlying variation in diurnal response, such as accumulation of carbohydrates, activation of Rubisco and gene expression, requires harvesting of significant numbers of leaves for protein, metabolite and nucleic acid analysis. If we assume that no more than 5 % of a canopy should be removed by destructive analyses within a year to avoid significant alteration of canopy structure and lighting, this would provide just 400 cm^2 in a 1-m diameter OTC, compared to 160 000 cm^2 in 20-m diameter FACE plots. An area of 400 cm^2 would limit several biochemical and molecular biological analyses and may not allow repeated sampling. This would be essential, for example, to understand the seasonal progression of photosynthetic acclimation.

14.3 Which FACE?

This review of the effects of $e[CO_2]$ is limited to the annual crops and herbaceous managed ecosystems that have been studied in FACE (Table 14.1), although some comparisons are made to studies of trees in FACE. Mini-FACE systems as small as 1-m diameter have been developed (Miglietta et al. 1996), but they do not escape all of the problems of enclosures outlined above, in particular scale. For example, substantial differences in the photosynthetic response of wheat to $e[CO_2]$ were observed in a mini-FACE (Miglietta et al. 1996) versus a full-size FACE system (Nie et al. 1995a, b; Wall et al. 2000). This review is therefore limited to full-size FACE systems of >8-m diameter plots. It is also limited to fully replicated experiments, since unreplicated FACE studies have the risk of confounding $e[CO_2]$ in the single replicate with unidentified differences with the $c[CO_2]$ of the control plot(s).

Table 14.1 A meta-analysis of FACE effects on light-saturated CO$_2$ uptake (A_{sat}), diurnal carbon assimilation (A'), stomatal conductance (g_s), maximum carboxylation rate or in vivo Rubisco activity ($V_{c,max}$), maximum rate of whole-chain electron transport, reflecting capacity for RubP regeneration (J_{max}), and ratio of $V_{cmax}{:}J_{max}$. *df* Degrees of freedom, *CI* confidence interval. Methods follow those outlined by Ainsworth and Long (2005) and are based on the results reported from different FACE studies (Ainsworth et al. 2003a; Ainsworth et al. 2003b; Ainsworth et al. 2004; Anten et al. 2004; Bernacchi et al. 2005; Garcia et al. 1998; Hileman et al. 1992; Idso et al. 1994; Leakey et al. 2004; Miglietta et al. 1998; Osborne et al. 1998; Rogers et al. 1998; Seneweera et al. 2002; Von Caemmerer et al. 2001; Wall et al. 2000; see also Chapters 3, 4, 5, 8, 9)

Variable	Group	df	Effect size (E)	Lower 95 % CI	Upper 95 % CI
A_{sat}		143	1.25	1.22	1.29
	C$_3$ cereal	34	1.15	1.08	1.22
	Herbage C$_3$ grass	49	1.40	1.33	1.48
	Herbage legume	7	1.37	1.17	1.59
	Soybean	25	1.19	1.12	1.27
	C$_4$ cereal	17	1.17	1.08	1.27
	Cotton	3	1.28	0.86	1.70
A'		92	1.24	1.21	1.28
	C$_3$ cereal	38	1.20	1.14	1.26
	Herbage C$_3$ grass	37	1.37	1.29	1.44
	Herbage legume	6	1.22	1.07	1.38
	C$_4$ cereal	4	1.09	0.93	1.28
	Cotton	3	1.11	0.87	1.42
g_s		78	0.75	0.72	0.80
	C$_3$ cereal	20	0.60	0.52	0.68
	Herbage C$_3$ grass	3	0.82	0.58	1.17
	Herbage legume	3	0.77	0.60	0.99
	Soybean	20	0.84	0.75	0.93
	C$_4$ cereal	7	0.71	0.57	0.88
	Cotton	12	0.77	0.66	0.90
$V_{c,max}$		148	0.85	0.82	0.87
	C$_3$ cereal	16	0.85	0.78	0.93
	Herbage C$_3$ grass	94	0.82	0.79	0.86
	Herbage legume	11	0.85	0.75	0.98
	Soybean	20	0.93	0.86	1.00
J_{max}		116	0.93	0.90	0.95
	C$_3$ cereal	10	0.90	0.81	0.98
	Herbage C$_3$ grass	69	0.92	0.88	0.95
	Herbage legume	11	0.88	0.79	0.98
	Soybean	19	0.98	0.92	1.05
$V_{c,max}{:}J_{max}$		44	0.95	0.92	0.97
	C$_3$ cereal	10	0.93	0.86	1.00
	Herbage legume	11	0.97	0.93	1.00
	Soybean	20	0.94	0.91	0.97

14.4 Have Findings From FACE Altered Perspectives?

14.4.1 Photosynthesis is Increased Less and Stomatal Conductance Decreased More in FACE

FACE experiments with crops have $e[CO_2]$ to between 550 ppm and 600 ppm, compared to 680–700 ppm in many chamber experiments (for reviews, see Ainsworth and Long 2005; Drake et al. 1997; Kimball 1983; Long et al. 2004). The mean value used in these FACE experiments, ca. 575 ppm, simulates the increase anticipated for the middle of this century as a mean of the different emissions scenarios considered by the Intergovernmental Panel on Climate Change (IPCC; Prentice et al. 2001). Table 14.1 presents a meta-analysis of the responses of photosynthesis and stomatal conductance of crops to this increase as observed in FACE. The mean increase in light-saturated photosynthesis (A_{sat}) is 25 %, although substantially less if the annual C_3 crops are considered in isolation (17 %). Drake et al. (1997) reported an average increase of 58 % across 60 chamber studies in which $[CO_2]$ was increased on average to 680 ppm. Considering the non-linearity of the response of A_{sat} to $[CO_2]$, this would equate to an increase of ca. 41 % at 575 ppm, substantially more than the 25 % observed in FACE. This supports the contention (Section 14.1.2) that elevated humidity and temperature in chambers may exaggerate the stimulatory effect of $e[CO_2]$ in chambers. This is further supported by observations of stomatal conductance. Averaged across 41 studies, Drake et al. (1997) found an average 20 % decrease in g_s with growth at an $e[CO_2]$ of 680 ppm equating to 13 % at 575 ppm. This is about half of the actual decrease in g_s of 25 % observed in FACE. This suggests the impact of $e[CO_2]$ on stomatal conductance is substantially more under fully open-air conditions than has previously been observed in chamber experiments; and this in part accounts for the smaller increase in A_{sat} (Table 14.1).

14.4.2 Photosynthesis is Stimulated Less at the Beginning and End of the Day

A very consistent effect observed across all FACE studies which have conducted diurnal analyses of A in upper canopy leaves is that the proportionate increase is greatest during the period when leaves are light-saturated. Fig. 14.2 shows the diurnal courses of leaf photosynthesis at different stages in the life cycle of spring wheat in FACE at Maricopa, Ariz. (Garcia et al. 1998). At this site, clear skies on each measurement date resulted in light-saturating conditions throughout the day, except the 2 h after sunrise and the 2 h prior to sunset. Little difference in A at $c[CO_2]$ and $e[CO_2]$ was apparent at these times of day, while a highly significant increase was observed during the periods of

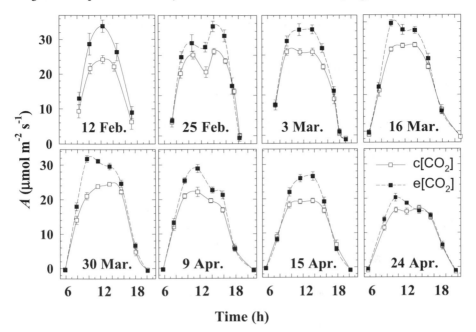

Fig. 14.2 Diurnal courses of leaf net photosynthetic CO$_2$ uptake (A) in spring wheat grown in e[CO$_2$] (550 ppm, *closed symbols*) and c[CO$_2$] (368 ppm, *open symbols*). Each illustrated point is the mean (±1 SD) of replicate plots measured at that point in time (n = 4). Each of the four individual plot values was the pooled average of five leaves sampled within the plot at that point in time. The eight diurnal courses illustrated are therefore from measurements of 2480 leaves over the course of the season. The eight days in sequence represent the following stages of development in sequence: tillering, early stem elongation, late stem elongation, inflorescence emergence, anthesis, early grain fill, mid-grain fill and completion of grain fill. Redrawn from Garcia et al. (1998)

light saturation (Fig. 14.2). Very similar patterns were reported for diurnal courses of A in *Glycine max* (Rogers et al. 2004), *Lolium perenne* (Ainsworth et al. 2003a) and *Trifolium repens* (Ainsworth et al. 2003b).

14.4.3 Stimulation of Photosynthesis is Sustained and Little Affected by Nitrogen Supply

While the average stimulation of photosynthesis in FACE is less than that observed in chamber studies, there is no evidence of a diminution of response with time. In annual crops, the stimulation observed in the early vegetative stage was maintained throughout the reproductive stage (Garcia et al. 1998; Rogers et al. 2004). The only exception in wheat was on the final date of measurement (Fig. 14.2). Here, grain-filling was complete in e[CO$_2$], but not in c[CO_2]; and leaf senescence was more advanced (Garcia et al. 1998). The grassland swards at Eschikon, Switzerland, were grown for 10 years at e[CO_2] in FACE. Perennial ryegrass (*Lolium perenne* L. cv. Bastion) was grown under an e[CO_2] of 600 ppm, with two contrasting nitrogen levels and abrupt changes in the source:sink ratio following periodic harvests. More than 3000 measurements characterized the response of leaf photosynthesis and stomatal conductance to e[CO_2] across each growing season for the duration of the experiment. Over the 10 years as a whole, growth at e[CO_2] resulted in a 43 % higher rate of light-saturated leaf photosynthesis and a 37 % increase in daily integral of *A* (Fig. 14.3). Photosynthetic stimulation was maintained despite a 30 % decrease in stomatal conductance and significant decreases in both $V_{c,max}$ amd J_{max}. In contrast with theoretical expectations and the results of shorter duration experiments, the experiment provided no evidence of significant change in leaf photosynthetic stimulation across a 10-year period in the extensive, low-nitrogen, treatment (Table 14.2; Ainsworth et al. 2003a). Indeed, stimulation of the daily integral of *A* in the low-nitrogen treatment on the first day of study was remarkably similar to that on the last day, 10 years later (Table 14.2). The study of Ainsworth et al. (2003a) appears the most com-

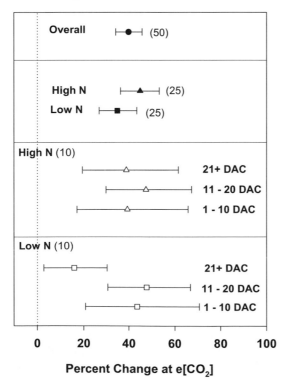

Daily Net Carbon Assimilation

Fig. 14.3 The percent change of daily integrated carbon assimilation of upper canopy leaves of *Lolium perenne* at Eschikon with growth at e[CO_2]. Sample sizes () for each categorical treatment are given in parenthesis. The mean effect sizes (*symbols*) are surrounded by 95 % confidence intervals. *DAC* indicates the number of days since the sward was cut, such that source:sink ratio increases with DAC. Two levels of nitrogen (N) representing extensive (low N) and intensive (high N) were maintained. Adapted from Ainsworth et al. (2003a)

Table 14.2 Increase in the daily integral of leaf CO$_2$ uptake for upper canopy leaves of *Lolium perenne* on different days throughout the 10 years of the FACE experiment at Eschikon (adapted from Ainsworth et al. 2003a)

Date of measurement			Daily mean temperature (°C)	Global radiation (MJ day^{-1})	Percent increase in daily integral of A (high N – low N)	
2	August	1993	20.5	24.5	30.7	38.4
23	June	1994	19.9	27.0	36.5	37.5
25	June	1994	21.6	24.9	−0.3	−12.3
22	July	1994	20.8	25.7	50.0	34.2
28	July	1994	21.8	21.3	66.9	100.4
24	May	1995	19.4	21.4	36.0	42.9
20	June	1995	20.8	28.0	27.5	41.6
9	July	1995	22.6	26.5	47.4	49.9
18	October	1995	13.0	8.5	54.0	50.7
2	July	1996	12.5	9.7	49.2	46.1
14	July	1996	20.6	26.1	60.5	52.0
25	June	1997	12.7	18.8	30.5	34.4
2	July	1997	17.2	22.8	49.2	46.1
9	July	1997	17.7	28.7	42.3	28.2
12	July	1997	18.6	23.0	39.9	33.4
16	July	1997	18.5	28.2	52.8	39.1
30	July	1997	19.9	26.3	40.0	41.4
10	August	1997	20.1	26.3	60.5	52.0
19	August	1997	18.6	20.4	50.2	37.2
23	September	2000	11.8	16.3	43.2	27.6
28	April	2001	11.6	18.0	16.8	4.2
24	May	2001	19.4	21.4	27.9	33.0
9	May	2002	15.9	19.0	25.0	30.5
20	May	2002	14.3	29.5	29.2	39.2

prehensive study of the effect of long-term growth at e[CO$_2$] on leaf photosynthesis in any system. It provides no support for the contention that stimulation of leaf photosynthesis declines with time or that it is diminished by a low nitrogen supply in managed systems.

14.4.4 In Vivo Rubisco Activity is Decreased More than Capacity for RubP Regeneration

Figure 14.1 reasoned from theory that resource use efficiency in e[CO$_2$] would be maximized by decreasing Rubisco content, effecting a decrease in $V_{c,max}$, but with no decrease in J_{max}. This would be acclimation in the sense that physiological adjustment in response to the e[CO$_2$] better fits the plant to the new environment by decreasing $V_{c,max}/J_{max}$. Table 14.1 shows that, in each study reporting the ratio, $V_{c,max}/J_{max}$ is significantly decreased, indicating a preferential loss of Rubisco activity. This contrasts to the survey of chamber studies by Drake et al. (1997), which indicated no preferential loss of Rubisco. In theory, at an $e[CO_2]$ of 572 ppm and temperature of 25 °C, $V_{c,max}$ could be decreased by 20 % without affecting A_{sat}, providing that the $V_{c,max}$ at $c[CO_2]$ was just sufficient to support the observed rate of A_{sat} (Fig. 14.1). This is also consistent with the survey of FACE experiments as a whole, which showed a significant decline in $V_{c,max}/J_{max}$ and showed that the decrease in leaf nitrogen content was almost entirely explained by the observed decrease in Rubisco content (Ainsworth and Long 2005). The mean decrease in $V_{c,max}$ across the managed ecosystems in FACE analyzed here was 15 % (Table 14.1), suggesting that decreased Rubisco was simply an elimination of capacity in excess at $e[CO_2]$, but not $c[CO_2]$. In wheat at Maricopa, Ariz., a selective loss of Rubisco relative to other proteins was observed as flag leaves matured in $e[CO_2]$ and was related to changes in transcript levels (Nie et al. 1995a, b). In *L. perenne* at Eschikon, $V_{c,max}$ and J_{max} were significantly lower in $e[CO_2]$ than in $c[CO_2]$ in the low-nitrogen treatment immediately prior to the periodic (every 4–8 weeks) decrease in canopy due to harvesting of this herbage crop. This difference was smaller after the cut, suggesting a dependence upon the balance between the sources and sinks for carbon (see Chapter 16) and consistent with the expectation that decreased investment in Rubisco at $e[CO_2]$ would be most advantageous when the nitrogen supply is low. However, while the decrease was smaller, J_{max} was also significantly decreased by 7 % and unlike the decrease in $V_{c,max}$ would cause a proportionate decrease in A_{sat}. This decrease in J_{max}, coupled with a greater than expected decrease in g_s, explains most of the 11 % shortfall in observed (25 %) versus theoretical stimulation (36 %) that would occur in the absence of any acclimation.

14.5 Conclusion

FACE have provided the only fully open-air treatment of crops with the elevations of $[CO_2]$ anticipated to occur by the middle of this century. Observed effects differ substantially from prior observations in chambers and from theoretical expectations.

The increase in light-saturated photosynthesis is substantially less than that observed in chambers or expected from theory. This is particularly marked in annual C_3 grain crops, where the mean increase of 18 % was just half the theoretical expectation. This may explain the much smaller increases in yield of these crops observed in FACE versus chambers (Long et al. 2004).

For annual crops over the growing season and for perennial herbage crops grown for 10 years, there is no evidence of any long-term loss of stimulation of photosynthesis by $e[CO_2]$ at either high or low nitrogen supplies.

Consistent with theoretical expectation, increase in photosynthesis is pronounced during periods of light saturation around mid-day, but minimal under the light-limiting conditions following dawn and preceding dusk.

Decrease in stomatal conductance is almost double that observed in chamber experiments.

In line with theoretical expectation on changes that would increase resource use efficiency in e[CO$_2$] but in contrast to chamber studies, a significant selective loss of in vivo Rubisco activity and content has been observed. While smaller than the loss in Rubisco activity, a decrease in capacity for regeneration of RubP is also observed and this explains most of the short-fall in observed versus theoretical increase in leaf photosynthesis.

References

Ainsworth EA, Long SP (2005) What have we learned from 15 years of free-air CO$_2$ enrichment (FACE)? A meta-analytic review of the responses of photosynthesis, canopy. New Phytol 165:351–371

Ainsworth EA, Davey PA, Bernacchi CJ, Dermody OJ, Heaton EA, Moore DJ, Morgan PB, Naidu SL, Yoo Ra HS, Zhu XG, Curtis PS, Long SP (2002) A meta-analysis of elevated [CO$_2$] effects on soybean (*Glycine max*) physiology, growth and yield. Global Change Biol 8:695–709

Ainsworth EA, Davey PA, Hymus GJ, Osborne CE, Rogers A, Blum H, Nösberger J, Long SP (2003a) Is stimulation of leaf photosynthesis by elevated carbon dioxide concentration maintained in the long term? A test with *Lolium perenne* grown for 10 years at two nitrogen fertilization levels under free air CO$_2$ enrichment (FACE). Plant Cell Environ 26:705–714

Ainsworth EA, Rogers A, Blum H, Nösberger J, Long SP (2003b) Variation in acclimation of photosynthesis in *Trifolium repens* after eight years of exposure to free air CO$_2$ enrichment (FACE). J Exp Bot 54:2769–2774

Ainsworth EA, Rogers A, Nelson R, Long SP (2004) Testing the „source-sink" hypothesis of down-regulation of photosynthesis in elevated CO$_2$ in the field with single gene substitutions in *Glycine max*. Agric For Meteorol 122:85–94

Amthor JS (2001) Effects of atmospheric CO$_2$ concentration on wheat yield: review of results from experiments using various approaches to control CO$_2$ concentration. Field Crops Res 73:1–34

Amthor JS, Koch GW, Willms JR, Layzell DB (2001) Leaf O$_2$ uptake in the dark is independent of coincident CO$_2$ partial pressure. J Exp Bot 52:2235–2238

Anten NPR, Hirose T, Onoda Y, Kinugasa T, Kim HY, Okada M, Kobayashi K (2004) Elevated CO$_2$ and nitrogen availability have interactive effects on canopy carbon gain in rice. New Phytol 161:459–471

Arp WJ (1991) Effects of source–sink relations on photosynthetic acclimation to elevated CO$_2$. Plant Cell Environ 14:869–875

Bernacchi CJ, Morgan PB, Ort DR, Long SP (2005) The growth of soybean under free air CO$_2$ enrichment (FACE) stimulates photosynthesis while decreasing in vivo Rubisco capacity. Planta 220:434–446

Buckley T, Mott K, Farquhar G (2003) A hydromechanical and biochemical model of stomatal conductance. Plant Cell Environ 26:1767–1785

Davey P, Hunt S, Hymus G, Drake B, DeLucia E, Karnosky D, Long S (2004) Respiratory oxygen uptake is not decreased by an instantaneous elevation of $[CO_2]$, but is increased by long-term growth in the field at elevated $[CO_2]$. Plant Physiol 134:520–527

Day FP, Weber EP, Hinkle CR, Drake BG (1996) Effects of elevated atmospheric CO_2 on fine root length and distribution in an oak–palmetto scrub ecosystem in central Florida. Global Change Biol 2:143–148

Diaz S, Grime JP, Harris J, McPherson E (1993) Evidence of a feedback mechanism limiting plant-response to elevated carbon-dioxide. Nature 364:616–617

Drake BG, Leadley PW, Arp WJ, Nassiry D, Curtis PS (1989) An open top chamber for field studies of elevated atmospheric CO_2 concentration on saltmarsh vegetation. Funct Ecol 3:363–371

Drake BG, GonzalezMeler MA, Long SP (1997) More efficient plants: a consequence of rising atmospheric CO_2? Annu Rev Plant Physiol Plant Mol Biol 48:609–639

Farquhar GD, Von Caemmerer S, Berry JA (1980) A biochemical model of photosynthetic CO_2 assimilation in leaves of C_3 species. Planta 149:78–90

Garcia RL, Long SP, Wall GW, Osborne CP, Kimball BA, Nie GY, Pinter PJ, Lamorte RL, Wechsung F (1998) Photosynthesis and conductance of spring-wheat leaves: field response to continuous free-air atmospheric CO_2 enrichment. Plant Cell Environ 21:659–669

Gonzalez-Meler MA, Siedow JN (1999) Direct inhibition of mitochondrial respiratory enzymes by elevated CO_2: does it matter at the tissue or whole-plant level? Tree Physiol 19:253–259

Hendrey GR (1992) Global greenhouse studies – need for a new approach to ecosystem manipulation. Crit Rev Plant Sci 11:61–74

Hileman DR, Bhattacharya NC, Ghosh PP, Biswas PK, Lewin KF, Hendrey GR (1992) Responses of photosynthesis and stomatal conductance to elevated carbon-dioxide in field-grown cotton. Crit Rev Plant Sci 11:227–231

Idso SB, Kimball BA, Wall GW, Garcia RL, Lamorte R, Pinter PJ, Mauney JR, Hendrey GR, Lewin K, Nagy J (1994) Effects of free-air CO_2 enrichment on the light response curve of net photosynthesis in cotton leaves. Agric For Meteorol 70:183–188

Jablonski LM, Wang XZ, Curtis PS (2002) Plant reproduction under elevated CO_2 conditions: a meta-analysis of reports on 79 crop and wild species. New Phytol 156:9–26

Jahnke S (2001) Atmospheric CO_2 concentration does not directly affect leaf respiration in bean or poplar. Plant Cell Environ 24:1139–1151

Kimball BA (1983) Carbon-dioxide and agricultural yield – an assemblage and analysis of 430 prior observations. Agron J 75:779–788

Koch KE (1996) Carbohydrate-modulated gene expression in plants. Annu Rev Plant Physiol Plant Mol Biol 47:509–540

Leakey ADB, Bernacchi CJ, Dohleman FG, Ort DR, Long SP (2004) Will photosynthesis of maize (*Zea mays*) in the US Corn Belt increase in future CO_2 rich atmospheres? An analysis of diurnal courses of CO_2 uptake under free-air concentration enrichment (FACE). Global Change Biol 10:951–962

Long SP (1991) Modification of the response of photosynthetic productivity to rising temperature by atmospheric CO_2 concentrations – has its importance been underestimated. Plant Cell Environ 14:729–739

Long SP, Ainsworth EA, Rogers A, Ort DR (2004) Rising atmospheric carbon dioxide: plants face the future. Annu Rev Plant Biol 55:591–628

Masle J, Farquhar GD, Gifford RM (1990) Growth and carbon economy of wheat seedlings as affected by soil resistance to penetration and ambient partial-pressure of CO_2. Aust J Plant Physiol 17:465–487

McLeod A, Long S (1999) Free-air carbon dioxide enrichment (FACE) in global change research: a review. Adv Ecol Res 28:1–55

Miglietta F, Giuntoli A, Bindi M (1996) The effect of free air carbon dioxide enrichment (FACE) and soil nitrogen availability on the photosynthetic capacity of wheat. Photosynth Res 47:281–290

Miglietta F, Magliulo V, Bindi M, Cerio L, Vaccari FP, Loduca V, Peressotti A (1998) Free air CO$_2$ enrichment of potato (*Solanum tuberosum* L.): development, growth and yield. Global Change Biol 4:163–172

Nie GY, Hendrix DL, Webber AN, Kimball BA, Long SP (1995a) Increased accumulation of carbohydrates and decreased photosynthetic gene transcript levels in wheat grown at an elevated CO$_2$ concentration in the field. Plant Physiol 108:975–983

Nie GY, Long SP, Garcia RL, Kimball BA, Lamorte RL, Pinter PJ, Wall GW, Webber AN (1995b) Effects of free-air CO$_2$ enrichment on the development of the photosynthetic apparatus in wheat, as indicated by changes in leaf proteins. Plant Cell Environ 18:855–864

Oechel WC, Cowles S, Grulke N, Hastings SJ, Lawrence B, Prudhomme T, Riechers G, Strain B, Tissue D, Vourlitis G (1994) Transient nature of CO$_2$ fertilization in arctic tundra. Nature 371:500–503

Osborne CP, LaRoche J, Garcia RL, Kimball BA, Wall GW, Pinter PJ, LaMorte RL, Hendrey GR, Long SP (1998) Does leaf position within a canopy affect acclimation of photosynthesis to elevated CO$_2$? Analysis of a wheat crop under free-air CO$_2$ enrichment. Plant Physiol 117:1037–1045

Prentice I, Farquhar G, Fasham M, Goulden M, Heinmann M, Jaramillo V, Kheshgi H, Le Quere C, Scholes R, Wallace D (2001) The carbon cycle and atmospheric carbon dioxide. In: Houghton J, Ding Y, Griggs D, Noguer M, Linden P van der, Dai X, Maskell K (eds) Climate change 2001: the scientific basis. contributions of working group I to the third assessment report of the intergovernmental panel on climate change. Cambridge University Press, Cambridge, pp 183–238

Roberts MJ, Long SP, Tieszen LL, Beadle CL (1993) Measurement of plant biomass and net primary production of herbaceous vegetation. In: Hall DO, Scurlock JMO, Bolhàr-Nordenkampf HR, Leegood RC, Long SP (eds) Photosynthesis and production in a changing environment: a field and laboratory manual. Chapman & Hall, London, pp 1–21

Rogers A, Fischer BU, Bryant J, Frehner M, Blum H, Raines CA, Long SP (1998) Acclimation of photosynthesis to elevated CO$_2$ under low-nitrogen nutrition is affected by the capacity for assimilate utilization. Perennial ryegrass under free-air CO$_2$ enrichment. Plant Physiol 118:683–689

Rogers A, Allen DJ, Davey PA, Morgan PB, Ainsworth EA, Bernacchi CJ, Cornic G, Dermody O, Dohleman FG, Heaton EA, Mahoney J, Zhu XG, Delucia EH, Ort DR, Long SP (2004) Leaf photosynthesis and carbohydrate dynamics of soybeans grown throughout their life-cycle under free-air carbon dioxide enrichment. Plant Cell Environ 27:449–458

Seneweera SP, Conroy JP, Ishimaru K, Ghannoum O, Okada M, Lieffering M, Kim HY, Kobayashi K (2002) Changes in source–sink relations during development influence photosynthetic acclimation of rice to free air CO$_2$ enrichment (FACE). Funct Plant Biol 29:945–953

Sicher RC, Bunce JA (1997) Relationship of photosynthetic acclimation to changes of Rubisco activity in field-grown winter wheat and barley during growth in elevated carbon dioxide. Photosynth Res 52:27–38

Thomas RB, Strain BR (1991) Root restriction as a factor in photosynthetic acclimation of cotton seedlings grown in elevated carbon-dioxide. Plant Physiol 96:627–634

Von Caemmerer S, Ghannoum O, Conroy JP, Clark H, Newton PCD (2001) Photosynthetic responses of temperate species to free air CO_2 enrichment (FACE) in a grazed New Zealand pasture. Aust J Plant Physiol 28:439–450

Wall GW, Adam NR, Brooks TJ, Kimball BA, Pinter PJ, LaMorte RL, Adamsen FJ, Hunsaker DJ, Wechsung G, Wechsung F, Grossman-Clarke S, Leavitt SW, Matthias AD, Webber AN (2000) Acclimation response of spring wheat in a free-air CO_2 enrichment (FACE) atmosphere with variable soil nitrogen regimes. 2. Net assimilation and stomatal conductance of leaves. Photosynth Res 66:79–95

Whitehead D, Hogan KP, Rogers GND, Byers JN, Hunt JE, McSeveny TM, Hollinger DY, Dungan RJ, Earl WB, Bourke MP (1995) Performance of large open-top chambers for long-term field investigations of tree response to elevated carbon-dioxide concentration. J Biogeogr 22:307–313

15 Carbon Partitioning and Respiration – Their Control and Role in Plants at High CO_2

P.W. Hill, J.F. Farrar, E.L. Boddy, A.M. Gray, and D.L. Jones

15.1 Introduction

How does $e[CO_2]$ affect processes downstream of photosynthesis? This question cuts to the heart of our knowledge of plant growth. No fair answer to it can avoid emphasising our large areas of ignorance. Sometimes our ignorance is so profound that even the questions are hard to phrase with precision. Yet the question demands a good answer, since it is also seminal to questions about responses to climate change, including carbon sequestration. Although forest wood production is often considered to be the most significant terrestrial store of C, globally four times as much C resides in soils as in plant biomass, with only tropical forests storing approximately equal quantities of C in plant biomass and in soil (IPCC 2001). The partitioning of carbon and dry matter within plants affects – indeed controls – the way in which C enters the soil (and wood biomass) and thus has profound consequences for C sequestration. We focus on this consequence, arguing that we cannot understand C sequestration until we know about plant allocation and respiration. We consider what has been done in FACE, but also set the context by reference to laboratory studies, since it is only when we can reconcile data from the full range of experimental and observational systems that we are able to say we understand how plants respond to $e[CO_2]$. We suggest that laboratory studies are sufficiently advanced to provide a preliminary extended hypothesis of how C partitioning works, but that this hypothesis needs detailed and rigorous testing in the field and in FACE.

Ecological Studies, Vol. 187
J. Nösberger, S.P. Long, R.J. Norby, M. Stitt,
G.R. Hendrey, H. Blum (Eds.)
Managed Ecosystems and CO_2
Case Studies, Processes, and Perspectives
© Springer-Verlag Berlin Heidelberg 2006

15.2 A Brief Background to Partitioning of Dry Matter and Carbon

The progression of partitioning of dry matter can be described relatively easily in young plants in a constant environment: it is allometric. That is, double log plots which relate the increase of weight during growth of two plant parts are linear (Farrar and Gunn 1998). Factors that alter partitioning – the transition to flowering, the rate of supply of nitrogen or phosphate, the temperature (Farrar and Gunn 1998) – simply alter the allometric coefficient relating the relative weights of two plant parts. Pleasingly, the allometric relationship has a simple underlying cause: it is the result of an unchanged ratio of the instantaneous rate of partitioning between two plant parts (Farrar and Gunn 1998). Although there is an historical focus on variability of shoot:root ratio in response to environmental variables, carbon (and thus dry matter) partitioning and shoot:root ratio are relatively conserved (Farrar and Jones 2003). That net partitioning between shoot and root or, better, between leaf and fine root (Körner 1991) is conserved is perhaps not surprising since to produce a plant within a stable C:N ratio means that the relative amounts of organs which acquire each of these resources need to be conserved. Thus whilst we know that variables such as nitrogen supply and light dose alter shoot:root partitioning, they do so within rather narrow limits.

However it is compounds of carbon – usually sucrose – which constitute the bulk of what is actually being partitioned at any moment. Since plant composition is conserved, the gross partitioning of carbon and that of dry weight are quite tightly related. Gross and net partitioning of carbon are distinguished by the latter, excluding the respiratory loss of part of the carbon which has entered a particular organ. Respiration provides the energy to drive biosynthesis and transport; and there is stoichiometry between tissue and whole plant respiration and the rates of growth, maintenance (turnover of non-growing tissue) and net ion uptake (Lambers 1987).

If we wish to test our understanding of the regulation of partitioning in plants, we can ask a simple question: can we explain the descriptive allometric relationship and the change in the allometric coefficient with external conditions, mechanistically? No, we cannot. We argue below, and others have argued too, that we can give partial answers to questions about regulating partitioning in specific instances.

A simple approach to partitioning immediately demonstrates that questions of regulation can be resolved into two types. First, what regulates the flux of C through the whole plant? Since the rate of entry of C into the plant is equal to the sum of fluxes through it, this question can be made more specific – what regulates the rate of whole-plant photosynthesis? The second type of question concerns what regulates partitioning, the relative flux at branch points in the plant system.

An important principle underlies any answer to the first question. The regulation of photosynthetic rate is shared between all parts of plant carbon metabolism, according to the principles of metabolic control analysis. We have convincing evidence that these principles are right, from photosynthetic metabolism in leaves (Stitt 1996) to carbon partitioning in whole plants (Sweetlove et al. 1998; Farrar 1999a). Significant regulation of photosynthetic rate is exerted by processes within source leaves, but sinks also exercise significant control – perhaps about half. In theory, we should be able to attribute control to individual gene products, but we are some way from doing this in practice.

We have far less idea about how carbon and dry matter are partitioned. Whilst the identification of some genes which are involved in regulation (Heineke et al. 1999; Hellmann et al. 2000) is a welcome step forward, we do not yet have an adequate body of knowledge, probably because the partitioning of C is a complex process divisible into a suite of metabolic, transport and biophysical components (Farrar 1999b).

15.3 Export From Source Leaves

The impact of e[CO$_2$] on photosynthesis and total carbohydrate pools in the leaf is considered elsewhere in this volume (see Chapters 14 and 16). Under many circumstances, the rate of net photosynthesis and thus the rate of production of non-structural carbohydrate is sustainably enhanced. This extra carbon is partitioned and used in plant growth. However when growth is primarily limited by other variables such as nitrogen supply, e[CO$_2$] has much less effect. A subtler response to a step-change in e[CO$_2$] is that the stimulatory effect on photosynthesis and relative growth rate can be transient, possibly because relative source and sink control of growth have re-established themselves. But, because plant size has been increased by the transient stimulation in e[CO$_2$], the absolute growth rate remains higher than in the control. Like any treatment that changes size, the effect of size per se must be removed in comparing relative growth rate or any of its components such as whole-plant photosynthesis or respiration; either allometry, or comparison at the same developmental stage or size, should be used to establish a treatment effect, rather than comparison at the same age (Farrar and Gunn 1998).

Carbon fixed into (usually) sucrose in the mesophyll is partitioned between storage, maintenance, and export. Each of these has a respiratory cost. What determines export? The key conclusion is that the rate of export is not determined in the leaf alone (Farrar 1999a; Gunn et al. 1999b). Direct evidence from flux analysis (Gunn and Farrar, unpublished data), considerations of how phloem works by establishing turgor pressure differences between source leaves and sinks, and knowledge of feedback control at the gene level together demonstrate that events remote from the source leaf have a role in determining the rate of export (Farrar 1999b; Farrar and Jones 2003). Sugar concentration may be important in the context of CO_2 – it may be that high export fluxes need a high concentration in the leaf. Since material not used in maintenance or export is stored, storage too must be partly dependent on remote events. No work on plants grown at $e[CO_2]$ has examined carbon flux and metabolism at the single-cell level (e.g., Pollock et al. 2003), which is essential for a complete understanding of flux control.

Since plants in FACE typically photosynthesise and grow faster than plants at $c[CO_2]$, it is implicit that their leaves are exporting more sucrose. There are few direct demonstrations. When *Lolium perenne* swards grown in FACE were labelled with $^{14}CO_2$, plants growing in $e[CO_2]$ fixed 127 % more C than those growing in $c[CO_2]$, irrespective of N treatment. N status affected C allocation: plants growing with 140 kg N ha^{-1} $year^{-1}$ exported 23 % more of the C fixed from source leaves than those growing at 560 kg N ha^{-1} $year^{-1}$ during the first 48 h after pulse-labelling. The difference in export due to N supply was slightly greater in $e[CO_2]$ (Hill et al. 2006). Conversely, *Dactylis glomerata* grown in FACE for 16 months photosynthesised less, exported a smaller proportion of recently fixed carbon at the same speed of phloem transport, and stored less carbohydrate over a diel cycle (Farrar and Gunn unpublished; Table 15.1). This species behaves quite differently when grown for short periods in controlled environments: then it accumulates more sugars because its photosynthetic rate is higher and the export from its source leaves is consequently faster (Gunn and Farrar unpublished; Fig. 15.1).

Table 15.1 Carbon flux in leaves of *Dactylis glomerata* grown in FACE. Regions near the tips of mature leaves of *D. glomerata* in the Swiss FACE rings were supplied with $^{14}CO_2$ at constant specific activity for 20 min. Export was measured by comparing the ^{14}C in the leaves immediately and 3 h after the pulse. At the same times, leaves plus sheaths were sampled and the spatial distribution of ^{14}C along them measured; speed of export was measured by the displacement of the semi-logarithmic spatial profiles. Soluble carbohydrates were sampled at 4-h intervals through a single 24-h cycle (Gunn and Farrar, unpublished data)

	350 ppm	600 ppm
Net photosynthesis (μmol m^{-2} s^{-1})	9.8	8.8
Soluble carbohydrates (g m^{-2})	3.7	3.1
Carbohydrate increase during day (g m^{-2})	5.1	4.2
Export of ^{14}C (% in 3 h)	69.0	57.0
Speed of ^{14}C export (cm h^{-1})	75.0	86.0

Fig. 15.1 Compartmental analysis of source leaf carbon flux in *Dactylis glomerata* after growth in $c[CO_2]$ or $e[CO_2]$. *D. glomerata* was grown in controlled environment cabinets and leaf five was supplied with $^{14}CO_2$ at constant specific activity for 20 min whilst held over the end window of a Geiger–Muller tube. Export of ^{14}C over the next 24 h was monitored and the data used for compartmental analysis, ascribing non-structural carbohydrate to either vacuoles or a labile pool which is the immediate source for transport (Farrar 1999a; Gunn and Farrar, unpublished data)

Top: 350 ppm CO_2
Bottom: 700 ppm CO_2

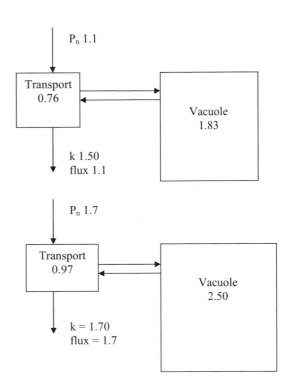

15.4 Whole-Plant Partitioning

Sugars exported from leaves in the phloem are partitioned between the sinks in which they are metabolised or stored. Theory and evidence combine to the view that each step in the partitioning process contributes a share of the control of flux. The driving force for movement is a turgor gradient along the sieve tubes. Branch points are the key to partitioning. The proximate determinants of partitioning are thus the turgor gradients in the branched sieve-tube system, but how the turgor is determined in the sieve tubes within sinks is far from understood. The turgor gradient from a branch point to each of its ends in a sink will depend on distance as well as turgor in each sink. Since plants are typically larger in $e[CO_2]$, identifying a CO_2-dependent change in partitioning means allowing for ontogenetic drift, using a method such as allometry or comparing plants of equal size or developmental stage rather than age (Farrar and Gunn 1996).

FACE experiments have found e[CO$_2$]-induced increases in tree wood, although increases appear to be due mainly to increased overall plant biomass; and C partitioning to wood may even be relatively lower than in c[CO$_2$] (DeLucia et al. 1999; Norby et al. 2002; Nowak et al. 2004). Many studies claim greater partitioning to roots at e[CO$_2$] (Cotrufo and Gorissen 1997; Norby et al. 2002; Suter et al. 2002). Careful studies of young plants grown hydroponically in controlled environments and analysed allometrically however demonstrate that plants are simply bigger – they have not altered their S:R partitioning (Gunn et al. 1999a). FACE in the field can be different (Suter et al. 2002) – roots are often relatively substantially larger perhaps because, following canopy closure, plants change their net allocation of dry matter to favor them more at e[CO$_2$] than at c[CO$_2$]. We lack a theoretical basis for understanding shoot:root partitioning in closed canopies.

In the Swiss FACE experiment, *Lolium perenne* increased root weight relatively more than shoot weight (Hebeisen et al. 1997; Suter et al. 2002). Relatively more root than shoot at the end of a growth period could be because more C has been partitioned to the roots, or because the roots are turning over more slowly at e[CO$_2$]. Evidence from labelling shoots in FACE with ^{14}CO$_2$ demonstrates greater partitioning of recent assimilate to roots (it does not eliminate the possibility of altered root turnover). e[CO$_2$] increased the ratio of ^{14}C in the root to that in the shoot after 6 days, indicating that plants grown in e[CO$_2$] allocated proportionately more photosynthate below ground. This ratio was reduced in plants growing at high N, so the plants with the higher N supply allocated less C below ground than those receiving the lower N supply. Plants grown in e[CO$_2$] allocated less ^{14}C below ground per unit root weight, but owing to larger root systems still allocated more ^{14}C below ground. The overall effect of differences in fixation, partitioning, and respiration was that plants grown in e[CO$_2$] retained more ^{14}C below ground after 6 days than plants grown in c[CO$_2$], with maximal amounts in the lower N treatment (Hill et al. 2006).

It might be gross partitioning, or root or leaf turnover, that is different in FACE. An alternative explanation for the difference between FACE and controlled environments is that FACE does not offer a simple, single-factor treatment. Since plants are bigger, they transpire more and take up more N, in spite of being more efficient at both N and water use (Nowak 2004). Their roots may therefore be in a low-N, dry environment compared with plants at c[CO$_2$] and may be responding as plants do in such circumstances – they detect the lower availability of these resources and respond by allocating more C to root growth. A direct experimental approach to this idea would be of value in interpreting many field and FACE experiments.

However, other studies suggest little or no change in S:R but a decrease in harvest index. A meta-analysis of soybean at high CO$_2$ shows clearly that the larger plants are unchanged in S:R whether nodulated or not, but seed yield accounts for a smaller part of final dry weight, so harvest index is reduced by

CO_2 (Ainsworth et al. 2002). Specific leaf area (SLA) is reduced in soybean (Ainsworth et al. 2002), as indeed it is across a range of species subjected to a meta-analysis (Poorter and Navas 2003) where the fraction of plant weight in leaves was unaltered by CO_2. Leaf area was reduced per unit of weight, but it is not clear whether another part of the plants was reduced in weight to preserve leaf area per se.

Meta-analysis has also been applied to a range of species, both C3 and C4, grown in FACE, where SLA is again reduced but only in C3 species. Leaf area index is not significantly affected (Long et al. 2004). Seed yield is increased (by about the same degree as total dry weight), suggesting no change in the proportional partitioning to reproduction, a conclusion echoed in another meta-analysis of FACE experiments (Ainsworth and Long 2005). Soybean may therefore be the exception.

Is this relative invariance in partitioning a function of regulation by the plant? An experiment with *Dactylis glomerata* suggests that it is. When half the leaves are removed from young plants in hydroponics, they regrow by partitioning carbon preferentially to the shoots so that the partially defoliated plants rapidly return to control values of S:R (Farrar 1996). The final value of S:R is not altered by e[CO_2], but the regrowth occurs much faster.

15.4.1 Growth and Development

Growth and development are inextricably linked: since carbon can be partitioned to existing organs or used in the initiation of new ones, continued increase in weight must be accompanied by the initiation of new organs and development needs substrates. Even if developmental events such as the transition to flowering have specific triggers, they still consume resources and are in part an exercise in partitioning.

There are numerous studies showing that e[CO_2] alters development. What is needed is evidence which unequivocally separates faster development from faster growth, either by using allometry or by examining the development indexed against a particular growth stage or against dry weight. An example is given for *Dactylis glomerata* in Table 15.2, where plants completely expanded leaf 5.2 days earlier when grown in e[CO_2], but were also heavier (development and growth were partly uncoupled) and more tiller sites were filled. Wheat grown in FACE reaches its maximal rate of grain filling about 100 accumulated thermal units earlier than controls (Li et al. 2000).

A meta-analysis of many experiments on soybean (Ainsworth et al. 2002) with an average increase in plant weight of over 40 % due to e[CO_2] shows that development and growth are partly uncoupled just as for *Dactylis glomerata* – increase in weight is greater than increase in branch or leaf number, so that the extra weight is only partly in more developmental units – it is partly in

Table 15.2 Biomass allocation and partial uncoupling of growth and development in *Dactylis glomerata* plants grown in c[CO_2] and e[CO_2]. *D. glomerata* was grown from seed in controlled environment cabinets and sampled 2 days after leaf five had fully expanded, irrespective of the calendar date

	350 ppm	700 ppm
Number of tillers	4.6	6.1
Weight of main stem leaf blades (mg)	54.0	93.0
Weight of leaf five (mg)	16.0	27.0
Area of leaf five (cm²)	4.4	4.9
Root weight (mg)	61.0	89.0
Shoot:root ratio	2.5	2.6
Total weight (mg)	193.0	348.0

some units being heavier. Since SLA is lower but there are no recorded changes in leaf size, presumably each leaf is heavier at e[CO_2].

A meta-analysis of FACE experiments shows a clear increase in leaf number in C3 plants, but this increase is less than that in plant weight and is accompanied by reduced SLA (Long et al. 2004). Again, this is compatible with each leaf being heavier and representing a partial uncoupling of growth and development. A second meta-analysis (Ainsworth and Long 2005) shows a more modest increase in leaf number than in dry matter, just as for plants in controlled conditions. Stem diameter and height are increased less than dry matter, but since the geometry of a growing woody stem is roughly that of a cone, proportionality would not be expected.

It might be idle to talk of 'uncoupling growth and development' if the increase in weight of individual leaves were entirely due to the accumulation of storage carbohydrates, for then it might be argued that the growth of *structure* remained tightly coupled to development (although we would still not know why or how). This appears not to be the case: in *D. glomerata*, only part of the increased weight of leaves is attributable to stored carbohydrate (Farrar et al., unpublished data).

15.5 Within Root Partitioning

Partitioning within plants affects – indeed controls – the way in which C and other elements enter the soil and thus has profound consequences for C sequestration. Indeed we can define *potential sequestration* as that C which enters the soil from plants and is not rapidly (say within 24 h) lost by the respiratory activity of soil microorganisms. It is a concept which places an upper limit on the *potential* rate of C sequestration and begins to link short-term process to long-term C sequestration.

15.5.1 Roots are a Sink for Photosynthetically Fixed C

Roots are C investments which gain a return in mineral nutrients and water absorbed from the soil. C from above-ground photosynthesis is used for the construction and maintenance of roots and to energise the active uptake of nutrients; and some C is lost passively from roots to the soil solution. C exports from above to below ground eventually pass to the soil in varying forms (Fig. 15.2), the proportion of which can have significant effects on soil microfauna, decomposition of SOM, N mineralisation and C sequestration, which can feed back to alter whole-plant and within-root partitioning.

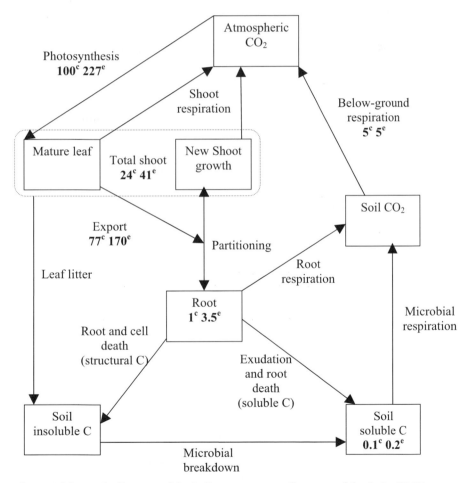

Fig. 15.2 Schematic diagram of the *Lolium perenne*–soil system of the Swiss FACE experiment. Estimates of pools and fluxes show the fate of ^{14}C fixed in a 20-min pulse. Fluxes (*arrows*) are the quantity of ^{14}C allocated over a 6-day period. Pools (*boxes*) are the quantity of ^{14}C remaining in pools after 6 days. In both cases, *units* refer to 100 fixed in c[CO$_2$] controls. Superscript *c* and *e* denote c[CO$_2$] and e[CO$_2$] treatments, respectively

15.5.2 Root Growth

The rate of root growth is roughly described by the flux of C from shoots and is determined by each of the partial processes which contribute to that flux – it is thus partly determined by events and processes in the shoot and in the root itself (Farrar and Jones 2003). One mechanism by which the rate of root growth may be controlled is sugar-controlled gene expression – high sugar concentrations and fluxes turn on key genes associated with the processes and fluxes that determine growth, as long as permissive features such as sufficient N supply and adequate temperature are satisfied (Koch 1997). It is easy to interpret the faster root growth at e[CO_2] in terms of this model, generated by the photosynthetic response to e[CO_2] but potentially constrained by the status of the root itself.

The C cost of constructing roots depends on both species and environment and is linked to a number of root characteristics that help determine root lifespan, such as tissue density and root diameter (Schläpfer and Ryser 1996; Van der Krift and Berendse 2002). Architectural plasticity may represent an alternative to plasticity of biomass allocation as a response to the spatial heterogeneity of nutrients (Fitter 1994). In general, coarse and woody roots grow faster and turn over more slowly, but have higher C construction costs than fine roots. However, the partitioning of new assimilate between fine and woody roots is not understood.

15.5.3   Exudation, Mucilage, and Cell Death

Exudation is thought to account for approximately 0.5–5.0 % of net fixed C (Farrar and Jones 2003; Fig. 2) and provides the most readily utilisable substrate for microbial growth. Some soluble C, including C fixed only a few hours previously (Dilkes et al. 2004), is lost by diffusion down the concentration gradient between root cell cytoplasm and the soil solution. The net flux may be modified by microbial utilisation in the rhizosphere and reuptake by plants (Jones and Darrah 1996).

Roots also secrete mucilage-containing detached root border cells, which are metabolically active and may be involved in signalling (Brigham et al. 1998). Cells are sloughed from the cortices of living roots; and lysates consisting of polymeric C compounds and enzymes enter the rhizosphere, providing further substrate for microorganisms (Farrar et al. 2003). Exudation, secretion, and cell death are largely functions of root growth and represent relatively constant proportions of root C import unless the environment changes, although this assertion from laboratory studies needs rigorous testing in the field.

15.5.4 Root Death and Turnover

Root turnover is a dynamic process, giving a significant flux of C from roots to the soil, although the magnitude of this flux remains unpredictable due to the difficulty of simultaneously measuring root growth and death. The bulk of this input consists of structural C sources, such as cell walls. Roots differ considerably in their decomposability as a function of their structure and chemical composition; and the quantity and quality of decomposing litter determine the rate of nitrogen mineralisation in soils.

Thus C partitioning within and from roots has profound effects on C sequestration, soil C and N fluxes, and future plant growth. Increased growth of plants at e[CO_2] may create an enhanced demand for mineral nutrients (Campbell and Sage 2002; Calfapietra et al. 2003; Johnson et al. 2004), which could lead to depletion of soil nutrient pools. Such effects, we suggest, both affect and are affected by within-root partitioning.

Should conditions be altered, both whole-plant (see above) and within-root partitioning could be modified. In nutrient-poor conditions, uptake and mobilisation of nutrients may be enhanced by stimulation of organic acid exudation (Farrar and Jones 2003; Sicher 2005). Both phosphorus deficiency and e[CO_2] increased non-structural carbohydrate pools in barley roots (Sicher 2005). This shift in partitioning towards labile C pools in conjunction with a decrease in absolute C transfer below ground (van Ginkel et al. 1997) could decrease potential C sequestration. Further responses to nutrient depletion could include changes in root architecture and partitioning between fine and coarse roots (Derner et al. 2001), which would alter turnover and decomposition of roots. Root demographic and architectural responses to e[CO_2] in conjunction with nutrient deficiency are poorly understood. The net effect of interacting factors on C partitioning and sequestration is far from clear, even in monocultures.

15.5.5 Elevated CO_2 and FACE Experiments

In FACE experiments, e[CO_2] can stimulate below-ground productivity in both grassland and forest ecosystems (Allen et al. 2000; Matamala and Schlesinger 2000; Calfapietra et al. 2003; Johnson et al. 2004; Hill et al. 2006). Regardless of whether whole-plant partitioning is altered, or plants are simply larger, this leads to a greater flux of C from roots to the soil. Our current ability to predict the fate of this additional C input is limited by technical difficulties in making accurate measurements on below-ground processes, and lack of understanding of ecosystem-level feedbacks between plants, soil microbes, and soil organic matter. Vital questions remain: will additional soil C storage occur; and will soil nutrients support long-term increases in plant productivity?

The relative partitioning of additional C within roots under e[CO_2] is crucial in determining its residence time in the soil. If C is predominantly partitioned to pools with fast turnover times (exudates, non-structural carbohydrates), little C storage occurs (Allen et al. 2000). FACE studies show that e[CO_2] results in increases in root biomass in both forest (Calfapietra et al. 2003; Lukac et al. 2003) and grassland ecosystems (Suter et al. 2002; Hill et al. 2006), accompanied by stimulated below-ground respiration (King et al. 2004), suggesting that flux of labile C to soil microorganisms also increases. We do not know if this increase is proportional to biomass increase, or whether relative partitioning within the root has changed.

In growth cabinet experiments (Cotrufo and Gorissen 1997; Hill et al. 2004) the root biomass of grass species is the main determinant of below-ground C flux. In a FACE experiment using 16 grassland species, soil CO_2 flux was correlated with above-ground biomass, again suggesting that it is proportional to the flux of C from above ground and that partitioning of this flux between labile and structural C is not significantly altered (Craine et al. 2001). However, growth in e[CO_2] may result in subtle differences in the relationship between below-ground CO_2 fluxes and plant biomass (discussed in Section 15.6).

15.6 Respiration

Respiration is the major loss of C from the plant–soil system. Gaseous losses of volatile organic compounds have been investigated in relation to e[CO_2], but have not been found to alter systematically; and they constitute a very small proportion of fixed C in most plant species (Hansen et al. 1997; Vuorinen et al. 2005). Respiration is the unavoidable cost of synthesising and maintaining tissues and the acquisition of nutrients (Wullschleger et al. 1994; Farrar 1999b) and results in large fluxes of CO_2 into the air and soil (King et al. 2004).

15.6.1 Direct and Indirect Effects of CO_2

CO_2 has been claimed to be a direct inhibitor of plant respiration, the inhibition being reversible and starting at concentrations as low as 500 ppm (Amthor 1991). Very careful work has shown that direct inhibition of CO_2 evolution is an artefact of the experimental methods used by the original investigators and there is no direct inhibition by CO_2 (Jahnke 2001; Jahnke and Krewitt 2002), with a parallel lack of effect on respiratory oxygen uptake (Davey et al. 2004). This is a relief, since it otherwise would be hard to see how roots respired in soil where concentrations of CO_2 in the soil air can reach 20 000 ppm.

Respiration rate of plants grown at $e[CO_2]$ has been reported to be both faster, and slower, than controls (Drake et al. 1998; Davey et al. 2004; Gonzalez-Meler et al. 2004). Perhaps not surprisingly, the variance of this estimate is large although one review concludes that, over a range of species, the proportion of fixed C lost in respiration is not significantly changed by growth at $e[CO_2]$ (Poorter and Navas 2003).

It is useful to distinguish carefully between the respiration of source leaves and sinks. Since mature source leaves of plants at high CO_2 are photosynthesising faster and exporting faster than plants at $c[CO_2]$, the expectation would be of faster respiration rates to support these activities, directly energising phloem transport and maintaining the photosynthetic system. Conversely, lower respiration would result from lower N and protein concentrations and thus reduced maintenance costs (Bunce 1994; Wullschleger et al. 1994; Bunce and Ziska 1996; Polley et al. 1999). Thus a range of responses to CO_2 is understandable, depending on just which processes are most affected in the leaf being measured. It might also follow the base of data expression, since leaves are commonly heavier per unit area at $e[CO_2]$ (see above). Expression per unit leaf area makes comparison with photosynthetic fluxes easier and removes artefacts due to transient storage. $e[CO_2]$ has little effect on leaf architecture or chlorophyll concentration in *Populus* stands, so changes in respiration are not attributable to structural changes (Gielen et al. 2003a).

Sinks generally grow faster at high CO_2 and this should be reflected in higher specific respiration rates of their growing regions. Sometimes this occurs: root apices of barley, for example, respire faster when the plants are at $e[CO_2]$ (Collis et al. 1996). A whole root system may be different – since about one-third of root respiration energises ion uptake (Lambers 1987) – and the uptake and partial local reduction of nitrate can be a major part of that respiratory cost. Plants growing with higher nitrogen use efficiency at $e[CO_2]$ (Nowak 2004) may have slower root respiration associated with N acquisition. Again, a range of responses can be understood, but for neither sources nor sinks do we have good examples of respiration rate at $e[CO_2]$ being linked to rates of underlying processes.

These expectations do not take mechanism into account. Some of the genes which underlie both respiration itself, and some of the key processes it energises, are regulated by the sugar and nitrogen status of the tissue (Koch 1996; Smeekens 2000). We still await a good linking of whole-plant response to CO_2 with the models of control of gene expression emerging from the laboratory.

15.6.2 Above-Ground Respiration and FACE

Although total respiratory losses from shoots may be enhanced by $e[CO_2]$ due to larger plants, FACE studies, in the relatively few reports, have generally not found leaf or stem respiration per unit tissue to be changed (Hamilton et al. 2001; Tissue et al. 2002; Gielen et al. 2003b). However, measurements of O_2 uptake on a range of species subject to long-term CO_2 enrichment by Davey et al. (2004) have recently shown that leaf dark respiration rates were increased on both an area (11 %) and a mass (7 %) basis.

15.6.3 Roots in Soil

Soil respiration is a compilation of three sources of CO_2: plant roots, microorganisms using substrates derived from plant exudation or recent death, and microorganisms using old soil organic C as substrate (Fig. 15.2). The partitioning of photosynthate by the plant determines the relative magnitudes of the major C inputs (above-ground litter, root exudation, and root death) and thus root respiration (Högberg et al. 2001; Kuzyakov and Cheng 2001). Different plant parts decompose at different rates and thus alter the rate of below-ground respiration. For instance, leaves generally decompose faster than roots (Palviainen et al. 2004). Furthermore, growth in $e[CO_2]$ generally reduces the decomposition rate of all plant tissues, although this response may be modified by both plant N status and soil conditions (Gorissen and Cotrufo 2000; Van Ginkel et al. 2000; Van Groenigen et al. 2005).

For a grass sward at steady state, about half of gross plant C is allocated to roots, perhaps two-thirds of that is respired by the roots, and half of what remains in the shoot enters structure and then litter (Lambers 1987; Farrar 1999a, b; Farrar et al. 2003). Several techniques converge to suggest that roots and root + rhizosphere account for about half of soil respiration (Hanson et al. 2000; Högberg et al. 2001; Kuzyakov 2002; Søe et al. 2004). Microbial metabolism of new C sources in the soil (from exudation and secretion) is rapid, with half-lives of less than 1 h for low molecular weight C (Jones et al. 2004).

Root respiration at steady state is an invariate proportion of root C flow (Farrar 1999a) – at least in systems so far examined carefully, which means in hydroponics. This is because root respiration energises growth, maintenance, and nutrient acquisition, which is quantitatively linked in constant conditions. Therefore this statement needs to be subject to test in the field, where more variation is to be expected. Root respiration is rapidly responsive to shoot status; and new photosynthate is rapidly respired from roots within a few hours, emphasising the close connection between them (Dilkes et al. 2004). Indeed, in a FACE study with *Beta vulgaris*, assimilation and soil respiration were linked. Soil respiration was 34 % higher under e[CO_2] during a period of high, but was not increased by CO_2 at low, irradiance (Søe et al. 2004). Root respiration is the most responsive component to e[CO_2], due to a greater degree of flexibility in architecture and growth rates (Butnor et al. 2003), but due to the high CO_2 concentrations in soil, effects of CO_2 on below-ground (root and rhizosphere) respiration will necessarily be indirect.

15.6.4 Below-Ground Respiration and FACE

Below-ground respiration increases in many FACE experiments across a variety of plant–soil systems (Craine et al. 2001; Pendall et al. 2001; Butnor et al. 2003; King et al. 2004; Ross et al. 2004). A reduction in in situ below-ground respiration was reported for the *Lolium perenne* swards of the Swiss FACE experiment (Ineson et al. 1998), although during the incubation of cores from the same swards, Xie et al. (2005) found respiration to be greater in soil subjected to e[CO_2]. The magnitude of reported effects of e[CO_2] in FACE range from a 10 % reduction to a 70 % increase in below-ground respiration (Ineson et al. 1998; Pendall et al. 2001). It is likely that increased plant size (Daepp et al. 2000) or below-ground allocation (Suter et al. 2002) under e[CO_2] are largely responsible for increases in below-ground respiration, which frequently correlates with plant biomass (see above). However, in the laboratory, specific root-dependent respiration (respiration from root and rhizosphere per unit root biomass) can be increased under e[CO_2] in *L. perenne* (Van Ginkel et al. 1997; Hill et al. 2004).

Our measurements from pulse labelling in the *L. perenne* swards of the Swiss FACE experiment found the reverse to be true, such that below-ground losses of $^{14}CO_2$ per unit root were 35 % lower in e[CO_2] than in c[CO_2], which represented a 77 % smaller proportion of fixed ^{14}C. Thus, although plants grown in e[CO_2] had 83 % more root biomass, total below-ground losses of $^{14}CO_2$ were unaffected by CO_2 and represented a 57 % smaller proportion of photosynthetically fixed ^{14}C. Such a decrease in specific root-dependent respiration might explain how below-ground respiration could be lower in the *L. perenne* swards of the Swiss FACE experiment under e[CO_2] than under c[CO_2], despite larger root biomass in e[CO_2] (Ineson et al. 1998; Suter et al. 2002). There are two possible alternative explanations for the difference between laboratory and field determinations of specific root-dependent respiration:

1. Due to the presence of dead root, measurements made after long-term plant growth in the field may overestimate root biomass more in e[CO_2] than in c[CO_2] if root decomposition is retarded after growth at e[CO_2]. This discrepancy would be likely to be more pronounced in the field than in the laboratory due to the relatively short-term nature of most laboratory experiments.

2. Our determination of specific root-dependent respiration in the Swiss FACE was only for the respiration of recent C (^{14}C), so an increase in the average age of C respired in e[CO_2] relative to that respired in c[CO_2] (e.g. due to label dilution) could potentially explain the discrepancy. However, neither of the alternatives can explain the lower below ground respiration found by Ineson et al. (1998).

If correct, the combination of lower total below-ground respiration and lower specific root-dependent respiration suggests that extra mineralisation of existing soil C due to the priming effect, thought to be occurring in POP-FACE (Hoosbeek et al. 2004), did not occur in the *L. perenne* swards of the Swiss FACE. This difference in the response of below-ground C to e[CO_2] may reflect either plant and soil differences or differences in the duration of exposure of plants to e[CO_2] (>5 years for *L. perenne*, 2 years for poplar) or a combination of the two.

Clearly we can neither understand not predict the impacts of e[CO_2] on below-ground respiration. We need to move from description, via an understanding of simple systems, to testing hypotheses based on those systems in the field.

15.7 Conclusion

- Plants grown in e[CO_2] generally fix up to 300 % more C than those grown in c[CO_2] (Ainsworth et al. 2003; Long et al. 2004; Hill et al. 2006). Combined with increased allocation of C below ground of around 50 % due to shifts in whole plant partitioning (Suter et al. 2002; Hoosbeek et al. 2004; Hill et al. 2006), this should result in greater potential sequestration of C.
- We do not know whether e[CO_2] is the direct cause of shifts in whole plant partitioning. Increased productivity at e[CO_2] leads to increased demand for nutrients and water. In low-nutrient conditions, some plants allocate a greater proportion of fixed C below ground, which would also lead to greater sequestration of C although within root partitioning may shift towards labile C fluxes (Sicher 2005). Microbial utilisation of plant-derived C also determines sequestration.
- Our field measurements found a >50 % decrease in the proportion of fixed C to be respired below ground in e[CO_2] than in c[CO_2] (Hill et al. 2006). However, some laboratory measurements have found a >60 % increase, whilst still reporting increases in total C remaining below ground (Van Ginkel et al. 1997).
- Many FACE experiments have reported increases in below-ground respiration, but it is often not clear whether this is as a consequence of larger plants having more root, greater allocation of photosynthate to roots with unchanged function, or more subtle changes in root function.

- Although e[CO_2] has been found to alter below-ground respiration per unit root biomass (Van Ginkel et al. 1997; Hill et al. 2004, 2006), it remains unknown whether this is due to changes in root respiration or the rhizodeposition of C which is subsequently respired by soil microbes. No technique exists to reliably separate these two contributions to the below-ground respiration flux. Thus it is not obvious how prolonged exposure to e[CO_2] will affect C flow from plant to soil.
- Many investigations of C flow in the plant–soil system have predicted that soil C sequestration should increase under e[CO_2] (Lutze and Gifford 1995; Van Ginkel et al. 1997; Hill et al. 2006). However, direct increases in soil C were not found in the Swiss FACE experiment, even after eight years' CO_2 enrichment (Van Groenigen et al. 2002). It may be that, contrary to predictions, C sequestration is not increased under e[CO_2]; but it may also be that statistically significant differences are difficult to find when increases in soil C are relatively small in comparison with the large and variable pool of existing soil C.
- Our current ability to predict the fate of the additional C input from plant to soil is limited by technical difficulties in making accurate measurements on below-ground processes. It is also limited by our poor understanding of underlying processes, so that we cannot create good models which would act as hypotheses for experiments conducted in the field or FACE; we are closest to this aspiration for single plants and furthest for the soil.
- Only close collaboration between plant and soil scientists, working in both laboratory and field, is likely to produce a real understanding of plant responses to a high CO_2 world.

References

Ainsworth EA, Long SP (2005) What have we learned from 15 years of free-air CO_2 enrichment (FACE)? A meta-analytic review of the responses of photosynthesis, canopy properties and plant production to rising CO_2. New Phytol 165:351–372

Ainsworth EA, Davey PA, Bernacchi CJ, Dermody OC, Heaton EA, Moore DJ, Morgan PB, Naidu SL, Yoo Ra H-S, Zhu X-G, Curtis PS, Long SP (2002) A meta-analysis of elevated [CO_2] effects on soybean (*Glycine max*) physiology, growth and yield. Global Change Biol 8:695–709

Ainsworth EA, Davey PA, Hymus GJ, Osborne CP, Rogers A, Blum H, Nosberger J, Long SP (2003) Is stimulation of leaf photosynthesis by elevated carbon dioxide concentration maintained in the long term? A test with *Lolium perenne* grown for 10 years at two nitrogen fertilization levels under free air CO_2 enrichment (FACE). Plant Cell Environ 26:705–714

Allen AS, Andrews JA, Finzi AC, Matamala R, Richter DD, Schlesinger WH (2000) Effects of free-air CO_2 enrichment (FACE) on belowground processes in a *Pinus taeda* forest. Ecol Appl 10:437–448

Amthor JS (1991) Respiration in a future higher carbon dioxide world. Plant Cell Environ 14:13–20

Brigham LA, Woo HH, Wen F, Hawes M (1998) Meristem-specific suppression of mitosis and a global switch in gene expression in the root cap of pea by endogenous signals. Plant Physiol 118:1223–1231

Bunce JA (1994) Responses of respiration to increasing atmospheric carbon dioxide concentrations. Physiol Plant 90:427–430

Bunce JA, Ziska LH (1996) Responses of respiration to increases in carbon dioxide concentration and temperature in three soybean cultivars. Ann Bot 77:507–514

Butnor JR, Johnsen KH, Oren R, Katul GG (2003) Reduction of forest floor respiration by fertilisation on both carbon dioxide-enriched and reference 17-year-old Loblolly pine stands. Global Change Biol 9:849–861

Calfapietra C, Gielen B, Galema ANJ, Lukac M, De Angelis P, Moscatelli MC, Ceulemans R, Scarascia-Mugnozza G (2003) Free-air CO_2 enrichment (FACE) enhances biomass production in a short-rotation poplar plantation. Tree Physiol 23:805–814

Campbell CD, Sage RF (2002) Interactions between atmospheric CO_2 concentration and phosphorous nutrition on the formation of proteoid roots in white lupin. Plant Cell Environ 25:1051–1059

Collis BE, Plum SA, Farrar JF, Pollock CJ (1996) Root growth of barley at elevated CO_2. Aspects Appl Biol 45:181–185

Cotrufo MF, Gorissen A (1997) Elevated CO_2 enhances below-ground allocation in three perennial grass species at different levels of N availability. New Phytol 137:421–431

Craine JM, Wedin DA, Reich PB (2001) Grassland species effects on soil CO_2 flux track the effects of elevated CO_2 and nitrogen. New Phytol 150:425–434

Daepp M, Suter D, Almeida JPF, Isopp H, Hartwig UA, Frehner M, Blum H, Nösberger J, Lüscher A (2000) Yield response of *Lolium perenne* swards to free air CO_2 enrichment increased over six years in a high nitrogen input system on fertile soil. Global Change Biol 6:805–816

Davey PA, Hunt S, Hymus GJ, DeLucia EH, Drake BG, Karnosky DF, Long SP (2004) Respiratory oxygen uptake is not decreased by an instantaneous elevation of [CO_2], but is increased with long-term growth in the field at elevated [CO_2]. Plant Physiol 134:520–527

DeLucia EH, Hamilton JG, Naidu SL, Thomas RB, Andrews JA, Finzi A, Lavine M, Matamala R, Mohan JE, Hendrey GR, Schlesinger WH (1999) Net primary production of a forest ecosystem with experimental CO_2 enrichment. Science 284:1177–1179

Derner JD, Polley HW, Johnson HB, Tischler CR (2001) Root system responses of C_4 grass seedlings to CO_2 and soil water. Plant Soil 231:97–104

Dilkes NB, Jones DL, Farrar J (2004) Temporal dynamics of carbon partitioning and rhizodeposition in wheat. Plant Physiol 134:706–715

Drake BG, Jacob J, Gonzàlez-Meler MA (1998). Photosynthesis, respiration, and global climate change. In: Rhagavendra AS (ed) Photosynthesis: a comprehensive treatise. Cambridge University Press, Cambridge pp 273–282

Farrar JF (1996) Regulation of root weight ratio is mediated by sucrose: opinion. Plant Soil 185:13–19

Farrar JF (1999a) Carbohydrate: where does it comes from, where does it go? In: Bryant JA, Burrell MM, Kruger NJ (eds) Plant carbohydrate biochemistry. Bios, Oxford, pp 29–46.

Farrar JF (1999b) Acquisition, partitioning and loss of carbon. In: Press MC, Scholes JD (eds) Advances in plant physiological ecology. Blackwell, Oxford, pp 25–43.

Farrar JF, Gunn S (1996) Effects of temperature and atmospheric carbon dioxide on source–sink relations in the context of climate change. In: Zamski E, Scheffer AA (eds) Photoassimilate distribution in plants and crops. Dekker, New York, pp 389–406

Farrar JF, Gunn S (1998) Allocation: allometry, acclimation – and alchemy? In: Lambers H, Poorter H, Van Vuuren MMI (eds) Inherent variation in plant growth. Backhuys, Leiden, pp 183–198

Farrar JF, Jones DL (2003) The control of carbon acquisition by and growth of roots. In: Kroon H de, Visser EJW (eds) Root ecology. Springer, Berlin Heidelberg New York, pp 90–124

Farrar J, Hawes M, Jones D, Lindow S (2003) How roots control the flux of carbon to the rhizosphere. Ecology 84:827–837

Fitter AH (1994) Architecture and biomass allocation as components of the plastic response of root systems to soil heterogeneity. In: Caldwell MM, Pearcy RW (eds) Ecophysiological processes above- and belowground. Academic Press, New York, pp 305–323

Gielen B, Liberloo M, Bogaert J, Calfapiertra C, De Angelis P, Miglietta F, Scarascia-Mugnozza G, Ceulemans R (2003a) Three years of free-air CO_2 enrichment (POPFACE) only slightly affect profiles of light and leaf characteristics in closed canopies of *Populus*. Global Change Biol 9:1022–1037

Gielen B, Scarascia-Mugnozza G, Ceulemans R (2003b) Stem respiration of *Populus* species in the third year of free-air CO_2 enrichment. Physiol Plant 117:500–507

Gonzales-Meler MA, Taneva L, Trueman RJ (2004) Plant respiration elevated atmospheric CO_2 concentration: cellular responses and global significance. Ann Bot 94:647–656

Gorissen A, Cotrufo MF (2000) Decomposition of leaf and root tissue of three perennial grass species grown at two levels of atmospheric CO_2 and N supply. Plant Soil 224:75–84

Gunn S, Bailey SJ, Farrar JF (1999a) Partitioning of dry mass and leaf area within plants of three grown at elevated CO_2. Funct Ecol 13:3–11

Gunn S, Farrar JF, Collis BE, Nason M (1999b) Specific leaf area in barley: individual leaves versus whole plants. New Phytol 145:45–51

Hamilton JG, Thomas RB, Delucia EH (2001) Direct and indirect effects of elevated CO_2 on leaf respiration in a forest ecosystem. Plant Cell Environ 24:975–982

Hansen U, Eijk J van, Bertin N, Staudt M, Kotzias D, Seufert G, Fugit J-L, Torres L, Cecinato A, Brancaleoni E, Ciccioli P, Bomboi T (1997) Biogenic emissions and CO_2 exchange investigated on four Mediterranean shrubs. Atmos Environ 31:157–166

Hanson PJ, Edwards NT, Garten CT, Andrews JA (2000) Separating root and soil microbial contributions to soil respiration: a review of methods and observations. Biogeochemistry 48:115–146

Hebeisen T, Lüscher A, Zanetti S, Fischer BU, Hartwig UA, Frehner M, Hendrey GR, Blum H, Nösberger J (1997) Growth response of *Trifolium repens* L. and *Lolium perenne* L. as monocultures and bi-species mixture to free air CO_2 enrichment and management. Global Change Biol 3:149–160

Heineke D, Kauder F, Frommer W (1999) Application of transgenic plants in understanding responses to atmospheric change. Plant Cell Environ 22:623–628

Hellmann H, Barker L, Funch D, Frommer WB (2000) The regulation of assimilate allocation and transport. Aust J Plant Physiol 27:583–594

Hill P, Marshall C, Jones DL, Farrar J (2004) Carbon sequestration: do N inputs and elevated atmospheric CO_2 alter soil chemistry and respiratory C losses? Water Air Soil Pollut Focus 4:177–186

Hill P, Marshall C, Williams G, Blum H, Jones DL and Farrar JF (2006) The fate of photosynthetically-fixed carbon in *Lolium perenne*-Eutric Cambisol grassland as modified by elevated CO_2, nitrogen and sward cutting. Global Change Biol. Submitted

Högberg P, Nordgren A, Buchmann N, Taylor AFS, Ekblad A, Högberg MN, Nyberg G, Ottosson-Löfvenius M, Read DJ (2001) Large-scale forest girdling shows that current photosynthesis drives soil respiration. Nature 411:789–792

Hoosbeek MR, Lukac M, Dam D van, Godbold DL, Velthorst EJ, Biondi FA, Peressotti A, Cotrufo MF, Angelis P de, Scarascia-Mugnozza G (2004) More new carbon in the mineral soil of a poplar plantation under free air carbon enrichment (POPFACE): cause of increased priming effect? Global Biogeochem Cycles 18:1040–1047

Ineson P, Coward PA, Hartwig UA (1998) Soil gas fluxes of N_2O, CH_4 and CO_2 beneath *Lolium perenne* under elevated CO_2: the Swiss free air carbon dioxide enrichment experiment. Plant Soil 198:89–95

IPCC (2001) Intergovernmental panel on climate change. In: Houghton JT, Ding Y, Griggs DJ, Noguer M, Linden PJ van der, Dai X, Maskell K, Johnson CA (eds) Climate change: the scientific basis. Cambridge University Press, Cambridge, p. 192

Jahnke S (2001) Atmospheric CO_2 concentration does not directly affect leaf respiration in bean or poplar. Plant Cell Environ 24:1139–1151

Jahnke S, Krewitt M (2002) Atmospheric CO_2 concentration may directly affect leaf respiration measurement in tobacco, but not respiration itself. Plant Cell Environ 25:641–651

Johnson DW, Cheng W, Joslin JD, Norby RJ, Edwards NT, Todd DE Jr (2004) Effects of elevated CO_2 on nutrient cycling in a sweetgum plantation. Biogeochemistry 69:379–405

Jones DL, Darrah PR (1996) Re-sorption of organic compounds by roots of *Zea mays* L. and its consequences in the rhizosphere. III. Spatial, kinetic and selectivity characteristics of sugar influx and the factors controlling efflux. Plant Soil 178:153–160

Jones DL, Shannon D, Murphy DV, Farrar JF (2004) Role of dissolved organic nitrogen (DON) in soil N cycling in grassland soils. Soil Biol Biochem 36:749–756

King JS., Hanson PJ, Bernhardt E, De Angelis P, Norby RJ, Pregitzer KS (2004) A multiyear synthesis of soil respiration responses to elevated atmospheric CO_2 from four forest FACE experiments. Global Change Biol 10:1027–1042

Koch KE (1996) Carbohydrate-modulated gene expression in plants. Annu Rev Plant Physiol Plant Mol Biol 47:509–540

Koch KE (1997) Molecular crosstalk and the regulation of C- and N-responsive genes. In: Foyer CH, Quick WP (eds) A molecular approach to primary metabolism in higher plants. Taylor and Francis, London, pp 105–124

Körner C (1991) Some often overlooked plant characteristics as determinants of plant-growth – a reconsideration. Funct Ecol 5:162–173

Kuzyakov Y (2002) Separating microbial respiration of exudates from root respiration in non-sterile soils: a comparison of four methods. Soil Biol Biochem 34:1621–1631

Kuzyakov Y, Cheng W (2001) Photosynthesis controls of rhizosphere respiration and organic matter decomposition. Soil Biol Biochem 33:1915–1925

Lambers H (1987) Growth, respiration, exudation and symbiotic association: the fate of carbon translocated to roots. In: Gregory PJ, Lake JV, Rose DA (eds) Root development and function. Cambridge University Press, Cambridge, pp 125–146

Li AG, Hou YS, Wall GW, Trent A, Kimball BA, Pinter PJ (2000) Free-air CO_2 enrichment and drought stress effects on grain filling rate and duration in spring wheat. Crop Sci 40:1263–1270

Long SP, Ainsworth EA, Rogers A, Ort DR (2004) Rising atmospheric carbon dioxide: plants FACE the future. Annu Rev Plant Biol 55:591–628

Lukac M, Calfapietra C, Godbold DL (2003) Production, turnover and mycorrhizal colonization of root systems of three *Populus* species grown under elevated CO_2 (POPFACE). Global Change Biol 9:838–848

Lutze JL, Gifford RM (1995) Carbon storage and productivity of a carbon dioxide enriched nitrogen limited grass sward after one year's growth. J Biogeogr 22:227–233

Matamala M, Schlesinger WH (2000) Effects of elevated atmospheric CO_2 on fine root production and activity in an intact temperate forest ecosystem. Global Change Biol 6:967–979

Norby RJ, Hanson PJ, O'Neill EG, Tschaplinski TJ, Weltzin JF, Hansen RA, Cheng W, Wullschleger SD, Gunderson CA, Edwards NT, Johnson DW (2002) Net primary productivity of a CO_2-enriched deciduous forest and the implications for carbon storage. Ecol Appl 12:1261–1266

Nowak RS, Ellsworth DS, Smith SD (2004) Functional responses of plants to elevated atmospheric CO_2 – do photosynthetic and productivity data from FACE experiments support early predictions? New Phytol 162:253–280

Palviainen M, Finer L, Kurka AM, Mannerkoski H, Piirainen S, Starr M (2004) Decomposition and nutrient release from logging residues after clear-cutting of mixed boreal forest. Plant Soil 263:53–67

Pendall E, Leavitt SW, Brooks T, Kimball BA, Pinter PJ, Wall GW, LaMorte RL, Wechsung G, Wechsung F, Adamsen F, Matthias AD, Thompson TL (2001) Elevated CO_2 stimulates soil respiration in a FACE wheat field. Basic Appl Ecol 2:193–201

Polley HW, Johnson HB, Tischler CR, Torbert HA (1999) Links between transpiration and plant nitrogen: variation with atmospheric CO_2 concentration and nitrogen availability. Int J Plant Sci 160:535–542

Pollock C, Farrar J, Tomes D, Gallagher J, Lu CG, Koroleva O (2003) Balancing supply and demand: the spatial regulation of carbon metabolism in grass and cereal leaves. J Exp Bot 54:489–494

Poorter H, Navas M-L (2003) Plant growth and competition at elevated CO_2: on winners, losers and functional groups. New Phytol 157:175–198

Ross DJ, Newton PCD, Tate KR (2004) Elevated [CO_2] effects on herbage production and soil carbon and nitrogen pools and mineralization in a species-rich, grazed pasture on a seasonally dry sand. Plant Soil 260:183–196

Schläpfer B, Ryser P (1996) Leaf and root turnover of three ecologically contrasting grass species in relation to their performance along a productivity gradient. Oikos 75:398–406

Sicher RC (2005) Interactive effects of inorganic phosphate nutrition and carbon dioxide enrichment on assimilate partitioning in barley roots. Physiol Plant 123:219–226

Smeekens S (2000) Sugar-induced signal transduction in plants. Annu Rev Plant Physiol Plant Mol Biol 51:49–82

Søe ARB, Giesemann A, Anderson T-H, Weigel H-J, Buchmann N (2004) Soil respiration under elevated CO_2 and its partitioning into recently assimilated and older carbon sources. Plant Soil 262:85–94

Stitt M (1996) Metabolic regulation of photosynthesis. In: Baker NR (ed) photosynthesis and the environment. Kluwer, Dordrecht, pp 151–190

Suter D, Frehner M, Fischer BU, Nösberger J, Lüscher A (2002) Elevated CO_2 increases carbon allocation to the roots of *Lolium perenne* under free-air CO_2 enrichment but not in a controlled environment. New Phytol 154:65–75

Sweetlove LJ, Kossmann J, Reismeier JW, Trethewey RN, Hill SA (1998) The control of source to sink carbon flux during tuber development in potato. Plant J 15:697–706

Tissue DT, Lewis JD, Wullschleger SD, Amthor JS, Griffin KL, Anderson R (2002) Leaf respiration at different canopy positions in sweetgum (*Liquidambar styraciflua*) grown in ambient and elevated concentrations of carbon dioxide in the field. Tree Physiol 22:1157–1166

Van der Krift TAJ, Berendse F (2002) Root life spans of four grass species from habitats differing in nutrient availability. Funct Ecol 16:198–203

Van Ginkel JH, Gorissen A, Van Veen JA (1997) Carbon and nitrogen allocation in *Lolium perenne* in response to elevated atmospheric CO_2 with emphasis on soil carbon dynamics. Plant Soil 188:299–308

Van Ginkel JH, Gorissen A, Polci D (2000) Elevated atmospheric carbon dioxide concentration: effects of increased carbon input in a *Lolium perenne* soil on microorganisms and decomposition. Soil Biol Biochem 32:449–456

Van Groenigen K-J, Gorissen, A, Six J, Harris D, Kuikman PJ, Van Groenigen JW, Van Kessel C (2005) Decomposition of ^{14}C-labelled roots in a pasture soil exposed to 10 years of elevated CO_2. Soil Biol Biochem 37:497–506

Vuorinen T, Nerg A-M, Vapaavuori E, Holopainen JK (2005) Emission of volatile organic compounds from two silver birch (*Betula pendula* Roth) clones grown under ambient and elevated CO_2 and different O_3 concentrations. Atmos Environ 39:1185–1197

Wullschleger SD, Ziska LH, Bunce JA (1994) Respiratory responses pf higher plants to atmospheric CO_2 enrichment. Physiol Plant 90:221–229

Xie Z, Cadisch G, Edwards G, Baggs EM, Blum H (2005) Carbon dynamics in a temperate grassland soil after 9 years exposure to elevated CO_2 (Swiss FACE). Soil Biol Biochem 37:1387–1395

16 The Response of Foliar Carbohydrates to Elevated [CO$_2$]

ALISTAIR ROGERS and ELIZABETH A. AINSWORTH

16.1 Introduction

Accumulation of foliar carbohydrates is one of the most pronounced and universal changes observed in the leaves of C$_3$ plants grown at elevated CO$_2$ concentration (e[CO$_2$]). Carbohydrates are both the product of photosynthetic cells and the substrate for sink metabolism. However, carbohydrates are not just substrates; and the role of carbohydrates in regulation of the expression of many plant genes, and the activity of many key enzymes, is well established. As free air CO$_2$ enrichment (FACE) technology was emerging, understanding of the link between carbohydrates and plant responses to growth at e[CO$_2$] was increasing. However, it remained unclear whether the hypotheses that were being refined in model systems would hold up when tested in open-air field experiments. More than a decade of FACE experiments have provided the answer.

16.1.1 Why is it Important to Understand the Response of Foliar Carbohydrates to Growth at e[CO$_2$]?

Sucrose is the main product of photosynthesis and in most plants is the main form of translocated carbon (Farrar et al. 2000). Starch and (in the Gramineae) fructan are transitory foliar storage pools for photosynthate, although in some species vacuolar sucrose is the dominant storage carbohydrate (Chatterton et al. 1989; Pollock and Cairns 1991; Zeeman et al. 2004). Due to the relative instability of glucose and fructose and the osmotic problems associated with storing large quantities of hexose, the levels of sucrose and the storage polysaccharides are generally much higher than the levels of free hexose (Isopp et al. 2000b; Rogers et al. 2004).

Ecological Studies, Vol. 187
J. Nösberger, S.P. Long, R.J. Norby, M. Stitt,
G.R. Hendrey, H. Blum (Eds.)
Managed Ecosystems and CO$_2$
Case Studies, Processes, and Perspectives
© Springer-Verlag Berlin Heidelberg 2006

Sugars are more than merely substrates and products; and they have important signaling functions throughout all stages of the plant's life cycle. The evidence for a regulatory role for hexoses and sucrose is overwhelming (Sheen 1990; Koch 1996; Bush 1999; Moore et al. 1999; Farrar et al. 2000; Smeekens 2000). Of particular significance is the well characterized feedback inhibition of photosynthetic genes by glucose and sucrose (Krapp et al. 1993; Van Oosten and Besford 1994; Jones et al. 1996; Pego et al. 2000).

The carbohydrate composition of foliage can also have effects beyond the plant. High carbohydrate content is often associated with high flavanoid content (Lindroth 1996). This has implications for plant–herbivore interactions since flavanoids are feeding deterrents for many herbivores. In wheat grown at e[CO_2] using FACE technology, Estiarte et al. (1999) reported that growth at e[CO_2] led to an increased foliar carbohydrate content and elevated flavanoid levels. Hendrix et al. (1994) found increased carbohydrate levels in cotton grown at e[CO_2] and also observed evidence of elevated flavanoid levels. Conversely, Hamilton et al. (2005) showed that soybean grown at e[CO_2] had a higher sugar content and was more susceptible to herbivory by Japanese beetles, for which sugars are a known phagostimulant. Also of particular significance to managed ecosystems is evidence that carbohydrate content may effect herbage palatability. Increased water-soluble carbohydrate content in herbage grown at e[CO_2] correlated positively with an increased organic matter digestibility in ruminants (Allard et al. 2003).

It is clear that an understanding of the response of foliar carbohydrates to growth at e[CO_2] is important if we aim to increase our knowledge of how ecosystems will respond to future e[CO_2] environments.

16.1.2 What Were the Known Effects of e[CO_2] on Foliar Carbohydrates Before FACE?

Accumulation of foliar carbohydrates is one of the most marked and widely observed changes in the leaves of C_3 plants grown at e[CO_2] (Farrar and Williams 1991). The most pronounced increases are in the levels of sucrose and the transient storage polysaccharides, starch and fructan. Long and Drake (1992) summarized the response of foliar carbohydrates to growth at e[CO_2] and found large and significant increases in sucrose and starch content in plants grown at e[CO_2].

In the early 1990s, strong evidence was emerging for the role of sugars in the down-regulation of photosynthetic genes (Sheen 1990); and this mechanism offered an attractive explanation for the emerging reports of a loss of photosynthetic capacity observed in plants grown for extended periods at e[CO_2] where there was also a large accumulation of sucrose (Long and Drake 1992). At this time, a special issue of *Plant Cell and Environment* (vol 14(8), 1991) was published that summarized the current, and predominantly pre-

FACE, knowledge of the response of plants to e[CO$_2$]. Contributions to this special issue from Stitt (1991), Farrar and Williams (1991) and Arp (1991) summarized the current knowledge of the response of plant carbohydrates to e[CO$_2$] and emphasized the importance of understanding the role of source–sink relations. Stitt (1991) assessed the evidence for a sink limitation of photosynthesis at e[CO$_2$] and concluded that the long-term ability of a leaf to maintain high photosynthetic rates is dependent on the source–sink status of the whole plant, i.e. a sustained stimulation of photosynthesis at e[CO$_2$] is dependent on an adequate sink capacity for the extra photosynthate produced at e[CO$_2$]. In the same special issue, Arp (1991) also examined the link between source–sink relations and photosynthetic acclimation and concluded that photosynthetic down-regulation was likely an artifact resulting from growing plants with a restricted rooting volume. There was also evidence that physical restriction of root development can cause these feedbacks (Masle et al. 1990; Thomas and Strain 1991). Implicit in Arp's conclusion was the assumption that a marked accumulation of carbohydrates at e[CO$_2$] was also an artifact of a restricted sink capacity.

One of the major problems in confidently extrapolating results from controlled environments to the field was the problem of the "pot effect" (Arp 1991). Whilst carbohydrate accumulation was less marked in plants grown with larger rooting volumes (Long and Drake 1992), it was unclear whether the response of foliar carbohydrates to e[CO$_2$] would prevail in an unlimited rooting volume. Since carbohydrate feedback mechanisms were thought to underlie some important responses of plants to e[CO$_2$], a truly realistic growth environment was needed in order to test hypotheses developed in controlled environments. The central hypothesis around which much of the uncertainty rested was: *The accumulation of foliar carbohydrates at e[CO$_2$] is the result of an insufficient sink capacity to utilize the extra photosynthate produced at e[CO$_2$].*

The advent of FACE technology allowed this hypothesis to be tested in the field in fully open-air conditions where plants lack the constraints that have been implicated as artifacts in many controlled environment studies (Long et al. 2004).

16.2 Do Carbohydrates Accumulate in the Leaves of Plants Grown in the Field Using FACE Technology?

Recent meta-analyses of plant responses to growth at e[CO$_2$] using FACE technology included an analysis of the response of foliar carbohydrates (Fig. 16.1; Ainsworth and Long 2005; Long et al. 2004). Despite an unrestricted rooting volume, plants grown at e[CO$_2$] accumulated significantly more sugars and starch than those plants grown at current (c)[CO$_2$]. In a review that

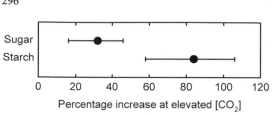

Fig. 16.1 Meta-analysis. The percent increase in the sugar and
starch content per unit leaf area
in plants grown at e[CO_2] using
FACE technology ±95 % confidence interval, $n=30$ independent observations. Figure
redrawn from Long et al. (2004)

preceded most of the work that emanated from FACE studies, Long and Drake
(1992) examined the response of plants to e[CO_2] and summarized the impact
of rooting volume on carbohydrate accumulation. They compared the ratio of
starch in plants grown at e[CO_2] to starch in plants grown at c[CO_2]. They
found that plants grown at e[CO_2] had a markedly higher starch content when
grown in small pots compared with large pots (small pots, e/c = 3.4; large
pots, e/c = 2.2; Long and Drake 1992). The meta-analytical summary of
Ainsworth and Long (2005) found sugar and starch accumulation to be
markedly lower in plants grown using FACE than in the plants grown in large
pots (>10 dm³) in Long and Drake's (1992) study. This observation is consistent with the results of Robbins and Pharr (1988), who showed that plants
grown in a small rooting volume accumulated more starch. However, the
trend for reduced carbohydrate accumulation with increased rooting volume
may reflect the higher [CO_2] in the studies summarized in Long and Drake's
review (range 500–2000 µmol mol⁻¹) compared with that of Ainsworth and
Long (range 475–600 µmol mol⁻¹). Another confounding factor is the data in
Long and Drake (1992) were expressed on a dry mass basis, which tends to
underestimate carbohydrate content. However, despite growing plants in the
field where roots were free to develop and forage for nutrients, there was still
a marked and significant increase in foliar carbohydrate content in plants
grown at e[CO_2].

Soybean offers perhaps the ultimate test of Arp's prediction that field-
grown plants will not become sink-limited (Arp 1991). In addition to the
unrestricted root development possible in the field, soybean may have an
indeterminate vegetative growth pattern and an association with nitrogen-
fixing bacteria that significantly increases the sink for photosynthate (Walsh
et al. 1987; Vessey et al. 1988). Despite these strong sinks for photosynthate,
soybean grown under e[CO_2] in the field still accumulated significantly more
glucose, sucrose and most markedly starch ($P<0.05$, data not shown). Figure 16.2 shows the level of total non-structural carbohydrate (TNC; sum of
glucose, fructose, sucrose and starch) in mature soybean leaves sampled at six
stages of development within the growth season. Despite a near constant
stimulation of diurnal photosynthetic CO_2 uptake (Bernacchi et al., unpublished data), there was a clear trend in TNC content, peaking at the beginning

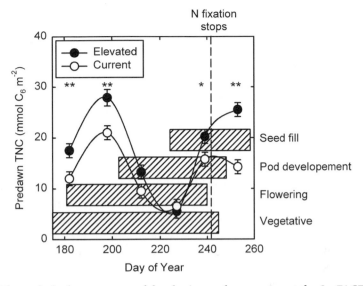

Fig. 16.2 The carbohydrate content of developing soybean grown at the SoyFACE experiment. TNC content calculated as the sum of glucose, fructose, sucrose and starch measured in the lateral leaflets of the most recently fully expanded trifoliate leaves of soybeans grown at e[CO₂] in the field using FACE technology. Samples were taken just before sunrise on three consecutive mornings; and the pre-dawn values for each replicate plot were pooled to give a mean pre-dawn TNC content for each period of measurement (n=4 replicate rings). Horizontal bars indicate the periods of vegetative growth, flowering, pod fill and seed fill. The broken vertical line indicates the stage in development when N-fixation stops. Across the season, there was a significant 39 % increase in TNC (F=62.73, P=0.0042). The date of measurement was significant (F=42.35, $P<0.0001$) and there was also a significant interaction between CO₂ treatment and day of year (F=4.24, P=0.0051). *Asterisks* indicate a significant pairwise comparison between elevated and ambient [CO₂] treatments on a specific day of year (* $P<0.05$, ** $P<0.01$)

and end of the growth season. At these stages of development, the increase in foliar carbohydrate content at e[CO₂] was maximal. The higher amounts of TNC at the beginning and end of the growth season and the significant CO₂ effect at these times correspond to developmental changes in the source–sink balance. In the middle of the season when TNC levels are lowest and there is no effect of growth at e[CO₂], there is a strong sink for photosynthate. Nitrogen fixation is peaking, vegetative development, flowering and pod set are still underway and seed fill is just beginning (Ritchie et al. 1997). The results presented here are supported by data from the same field site taken in a preceding year on a different cultivar (Rogers et al. 2004). It is clear that carbohydrate accumulation at e[CO₂] occurs even in a plant that lacks the constraints thought to exacerbate carbohydrate accumulation in controlled environments. It is also evident from Fig. 16.2 that sink capacity may be determining the extent and timing of carbohydrate accumulation.

Trees have large sinks for photosynthate and may be expected to avoid foliar carbohydrate accumulation at e[CO_2]. Developing loblolly pines experiencing a step change in [CO_2] at the Duke Forest FACE experiment (Hendrey et al. 1999) did not show an accumulation of carbohydrates when measured at multiple stages during the first season of CO_2 exposure (Myers et al. 1999). Rogers and Ellsworth (2002) did report foliar carbohydrate accumulation later in the experiment, but this was confined to old needles at two points in the season. Herrick and Thomas (2001) did not report carbohydrate accumulation in sun or shade leaves of *Liquidambar styraciflua* (sweetgum) growing at e[CO_2] in the understory at the Duke Forest FACE site (Herrick and Thomas 2001). However, Tissue et al. (2002) did report carbohydrate accumulation at e[CO_2] in the same species at the Oak Ridge National Laboratory FACE site (Norby et al. 2001); and Singaas et al. (2000) reported carbohydrate accumulation in *Acer rubrum*, *Ceris canadensis* and *L. styraciflua* at the Duke site. Clearly carbohydrate accumulation in trees is highly variable and our understanding of the mechanisms underlying the response of foliar carbohydrates to e[CO_2] in trees needs to be increased.

16.3 Manipulations of Source–Sink Balance

One way to test the hypothesis that insufficient sink capacity is causing foliar carbohydrate accumulation is to artificially manipulate the source–sink ratio in order to increase the demand for photosynthate. A simple way to do this is to remove source tissue. By decreasing the amount of photosynthetic tissue, the demand for photosynthate will need to be met by fewer source leaves and the remaining leaves will experience an increase in sink strength. The management practice at the Swiss FACE site (Zanetti et al. 1996) afforded an opportunity to study carbohydrate dynamics following partial defoliation.

Perennial ryegrass is a major C_3 pasture grass of humid and temperate regions that has been selected to be grazed and therefore survive periodic partial defoliation. At the Swiss FACE site, ryegrass was managed as a frequently cut herbage crop. The periodic defoliation abruptly decreased the ratio of source (i.e. photosynthetic tissue) to sink (i.e. roots and pseudostems) and in addition led to an increased demand for photosynthate during the regrowth period.

Immediately following partial defoliation, the carbohydrate content in the remaining leaf tissue was markedly reduced; and this reduction continued until about 4 days after defoliation, at which time levels began to rise until they peaked just before the next cutting cycle (Fischer et al. 1997). The effect of source–sink manipulation on carbohydrate accumulation is clear; and the trend is exacerbated at e[CO_2]. Foliar carbohydrate accumulation was significantly greater at e[CO_2] immediately before partial defoliation, but following defoliation the newly developed foliage showed no carbohydrate accumulation at e[CO$_2$], consistent with the greater demand for photosynthate following defoliation (Fischer et al. 1997; Rogers et al. 1998). Isopp et al. (2000b) showed that the diurnal changes in TNC were largely associated with changes in sucrose content, but the long-term increases in TNC immediately before defoliation, that were exacerbated at e[CO$_2$], were associated with a marked accumulation of fructan, indicative of an insufficient demand for photosynthate (Fig. 16.3; Isopp et al. 2000b). Rogers et al. (1998) cut one section of the sward early and measured carbohydrate content in both defoliated and undefoliated swards on the same day. They confirmed that the difference in carbohydrate content following a cut was not due to different meteorological conditions on or preceding the day of measurement (Rogers et al. 1998).

Further support for the major role of sink capacity in determining the response of foliar carbohydrates to e[CO$_2$] comes from the SoyFACE experiment (see Chapter 4), where Ainsworth et al. (2004) grew isogenic lines of soybean that varied by a single gene altering their capacity to utilize photosynthate. Indeterminate soybean cultivars continue vegetative growth after flowering has begun but determinate cultivars do not. Since continued vegetative growth will provide an additional sink for photosynthate, cultivars with a

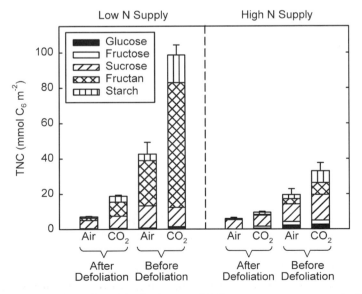

Fig. 16.3 Composition of TNC in the leaves of *Lolium perenne* grown at the Swiss FACE site. Leaves were sampled from plants grown with a high and low nitrogen supply in control (Air) and elevated [CO$_2$] plots (CO2). Samples were taken at 0700 hours, shortly after and shortly before a planned partial defoliation. Plants harvested shortly after a cut had a markedly lower source:sink ratio than those harvested shortly before defoliation. Error bars represent ±SE of mean TNC (n=3 plots). Figure redrawn from Isopp et al. (2000b)

determinate growth form would be predicted to accumulate greater amounts of foliar carbohydrate. Ainsworth et al. (2004) examined the response of foliar carbohydrates to growth at e[CO_2] in a determinate genotype (Williams *dt1*) of a cultivar with an indeterminate growth form (Williams) and an indeterminate genotype (Elf *Dt1*) of a cultivar with determinate growth form (Elf). All plants with determinate growth forms (Williams *dt1* and Elf) accumulated significantly more sugars at e[CO_2], whereas indeterminate plants (Williams and Elf *Dt1*) showed no additional accumulation at e[CO_2]. Starch content was significantly higher in all plants grown at e[CO_2]. Figure 16.4 shows the levels of TNC in these plants. A single gene mutation to change the indeterminate growth form of Williams to a determinate growth form (Williams *dt1*) resulted in a doubling of the amount of extra carbohydrate accumulated at e[CO_2]. However, the opposite single gene substitution to convert the determinate variety Elf to an indeterminate variety (Elf *Dt1*) did not lead to an exacerbated accu-

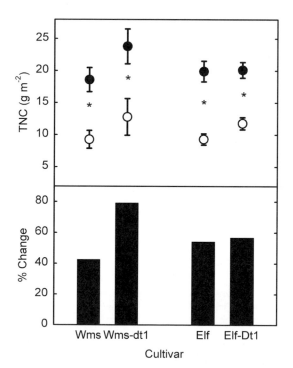

Fig. 16.4 Sink manipulation at the SoyFACE experiment. TNC content of four soybean cultivars ±SE, sampled during pod-fill. Williams (Wms) has an indeterminate growth form and Williams *dt1* (Wms-dt1) is a determinate form of Williams, Elf (Elf) has a determinate growth form and Elf *Dt1* (Elf-Dt1) is an indeterminate form of Elf. The bar charts show percent change in carbohydrate with growth at e[CO_2], i.e. (FACE –control)/control × 100. Pre-planned comparisons of [CO_2] treatments within cultivars were made using linear contrast statements. * $P < 0.05$. Figure redrawn from Ainsworth et al. (2004)

mulation of TNC at e[CO$_2$] (Ainsworth et al. 2004). This difference in the response of the two determinate lines may lie in the fact that Elf was developed as a determinate variety. For a determinate variety to be competitive, it is likely that breeders selected lines with sufficient potential for pod formation to ensure that yield would not be compromised by an insufficient sink capacity. Therefore, an indeterminate line of Elf may not offer a significant advantage at e[CO$_2$].

Together these two FACE experiments have provided strong evidence that sink capacity is a key factor in determining the response of foliar carbohydrates to growth at e[CO$_2$]. The picture is not always so clear. Rogers and Ellsworth (2002) investigated TNC accumulation in the needles of loblolly pine grown at e[CO$_2$] at the Duke Forest FACE site. They anticipated that when there was a strong proximal sink for carbohydrate (developing buds and new needles), they would not observe carbohydrate accumulation at e[CO$_2$]. However, carbohydrate accumulation was maximal at this time and minimal when predicted proximal sink activity was lowest. It is possible that other distal carbohydrate sinks may have played a more significant role than the adjacent developing shoots and needles. Developing shoots and needles are nitrogen sinks as well as carbon sinks; and if nitrogen were limiting needle development, then it is possible that the supply of photosynthate at e[CO$_2$] may have been in excess of sink requirements (Rogers and Ellsworth 2002).

Carbohydrate accumulation in the leaves of trees grown under e[CO$_2$] in FACE experiments has been difficult to predict and is highly variable. Much of this variation may be due to developmental heterogeneity within the tree and the problems associated with selecting comparable leaf tissue for analysis. More accurate methods of measuring leaf development are now available and should help clarify the responses to e[CO$_2$] (Schmundt et al. 1998; Taylor et al. 2003). In addition, carbon sinks such as wood formation, and in some species the emission of volatile organic compounds, complicate the understanding of source–sink relations.

16.4 The Effect of Nitrogen Supply on Sink Capacity

When plant growth is limited by nitrogen supply, carbon is in excess and surplus photosynthate often accumulates in leaves (Rogers et al. 1996). The interaction of e[CO$_2$] and nitrogen supply has been the subject of many studies (Stitt and Krapp 1999). Prior to FACE, many of these studies had been conducted in pots or containers. In addition to the physical constraint imposed by container walls (Arp 1991), enhanced growth under e[CO$_2$] may lead to more rapid exhaustion of the available nitrogen. In this case, plants growing at e[CO$_2$] will experience nitrogen limitation sooner, or to a greater extent than the plants growing at $c[CO_2]$ (Stitt and Krapp 1999; Körner 2003). In the field there is no restriction on root development, and increased exploration of the soil with accelerated growth at $e[CO_2]$ would allow the plant to utilize additional sources of nitrogen as it develops. The FACE experiments provide the opportunity to examine plant responses to $e[CO_2]$ in an open-air environment without the confounding effects of potentially exaggerated nitrogen limitation.

The concept that a low-nitrogen supply could lead to a sink limitation and that this could be exacerbated at $e[CO_2]$ has been investigated at the Swiss FACE site (Fischer et al. 1997; Rogers et al. 1998; Isopp et al. 2000b). In addition to the effect of partial defoliation, Fig. 16.3 also shows the response of foliar carbohydrates to a low- and high-nitrogen supply (Isopp et al. 2000b). High-nitrogen supply reduced foliar carbohydrate content at both $c[CO_2]$ and $e[CO_2]$. In plants harvested shortly before periodic defoliation, the combination of $e[CO_2]$ and a low-nitrogen supply led to a marked increase in foliar carbohydrate content. This trend was also reported by Fischer et al. (1997) and Rogers et al. (1998). The large increase in carbohydrate content at $e[CO_2]$ and low nitrogen immediately prior to planned periodic partial defoliation was consistent with a severe sink limitation. This severe sink limitation was investigated further towards the end of a growth cycle where day-to-day accumulation of foliar carbohydrate in source leaves was examined. Plant grown with a high-nitrogen supply did not show a significant accumulation of carbohydrate between successive days. However, plants with a low-nitrogen supply accumulated significant amounts of carbon over a 24-h time-course. Figure 16.5 shows the extent of this carbon accumulation in two successive years

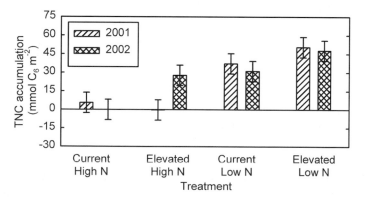

Fig. 16.5 The effect of low nitrogen on sink limitation at the Swiss FACE experiment. Immediately prior to a scheduled defoliation, the accumulation of foliar TNC over a 24-h period was determined in *Lolium perenne* grown at $e[CO_2]$ and $c[CO_2]$ in both low- and high-nitrogen plots. Accumulation was 5-fold greater in plants grown with a low-nitrogen supply ($F=7.96$, $P<0.05$). Error bars represent mean $\pm$SE of TNC accumulation ($n=3$ rings)

of investigation. Together with the results from Fischer et al. (1997), Rogers et al. (1998) and Isopp et al. (2000a, b), it is clear that the low-nitrogen supply treatment at the Swiss FACE site led to a markedly reduced sink capacity and to the accumulation of foliar carbohydrates, a phenomenon that was exacerbated at e[CO$_2$].

16.5 What Are the Signs of a Limited Sink Capacity?

A higher carbohydrate content at e[CO$_2$] is not necessarily an indication of replete sinks. According to the Münch hypothesis, if plants are to match increased photosynthetic rates with increased export rates, they need to increase their capacity to load phloem. The plant has to poise its sucrose levels on the trigger of photosynthetic repression to drive symplastic transport at a maximal level (see Chapter 15). So, if an increased carbohydrate content may not be a clear indicator of a source–sink imbalance, what are the signals of sink limitation and how else might the leaf sense an inadequate capacity in the plant for the utilization of additional photosynthate?

The type of carbohydrate accumulated can communicate the capacity of a plant to utilize current photosynthate. When the capacity of sinks to utilize photosynthate decreases, excess carbohydrate is stored in the leaf as either starch or fructan. Figure 16.1 shows the response of sucrose and starch to growth at e[CO$_2$] using FACE technology. In agreement with prior reports (Farrar and Williams 1991), starch is the main component of the increase in leaf carbohydrates observed in plants grown at e[CO$_2$]. In the Gramineae, the alternative storage polysaccharide fructan also showed marked accumulation at e[CO$_2$]. Figure 16.3 clearly shows fructan storage in sink-limited ryegrass grown at the Swiss FACE site (Isopp et al. 2000b).

Another key indicator of a source–sink imbalance is the accumulation: fixation ratio (Rogers et al. 2004). Additional TNC accumulation during a photoperiod at e[CO$_2$] can be indicative of a limited capacity to utilize photosynthate, particularly if this accumulation is carried over to the next day (Fig. 16.5). However, accumulation during the photoperiod may simply reflect higher photosynthetic rates and a larger transport pool. Accumulation:fixation ratio is a more useful diagnostic parameter. In soybean grown at e[CO$_2$], Rogers et al. (2004) reported that soybean exported ca. 90 % of fixed carbon, but on one occasion during the season, associated with low temperature and developmental reductions in sink capacity, plants grown at c[CO$_2$] retained ca. 20 % of their fixed carbon and plants at e[CO$_2$] retained ca. 50 % (Fig. 16.6).

Moore et al. (1999) offers perhaps the best explanation of how a photosynthetic cell can sense and respond to a source–sink imbalance (Long et al. 2004). Excess sucrose from photosynthesis that accumulates in the vacuole

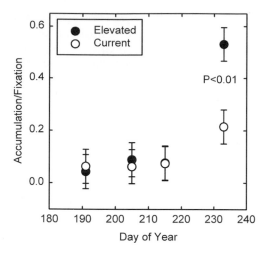

Fig. 16.6 Ratio of foliar carbon accumulation to carbon fixation in the terminal leaflets of the most recently fully expanded trifoliate leaves of soybeans grown in the field at $e[CO_2]$ and $c[CO_2]$ measured on four occasions during development. Carbon accumulation was calculated by subtracting the TNC content at the beginning of the photoperiod from the TNC content at the end of the photoperiod. Carbon fixation is the daily integral of net CO_2 assimilation. There was a significant effect of CO_2 treatment ($P<0.1$), day of year ($P<0.05$) and the interaction of CO_2 × day of year ($P<0.05$). Data are least-square means ±SE of the difference in means; and $P<0.01$ indicates a highly significant difference on day of year = 233, based on a linear pairwise contrast. Figure redrawn from Rogers et al. (2004)

when sinks are unable to utilize all available photosynthate is hydrolyzed by vacuolar invertases to yield glucose and fructose. These hexoses then enter a futile cycle of sucrose synthesis and degradation. Hexokinase catalyzes a key step in this cycle and has a secondary role as a flux sensor, thereby communicating the source–sink imbalance to the cell (Moore et al. 1999). Since there is a good correlation between invertase activity and hexose:sucrose ratio, the end-of-day hexose:sucrose ratio is potentially a useful diagnostic marker for a source–sink imbalance (Moore et al. 1999). Rogers et al. (2004) determined the hexose:sucrose ratio in sink-limited soybean (day of year = 254, Fig. 16.6) and reported a significant and markedly higher hexose:sucrose ratio in plants grown at $e[CO_2]$ compared with those grown at $c[CO_2]$, suggesting that Moore's model may translate to field-grown plants. Further investigation of this model as a possible mechanism that describes how plants sense a source–sink imbalance is required, particularly since a significant proportion of the measured foliar sucrose is in the phloem, where it is not available to a mesophyll-based flux sensor.

16.6 Conclusion

Prior to FACE experiments, there was uncertainty over whether the hypotheses that were being developed from experiments conducted in controlled environments would hold up when tested under fully open-air field conditions.

- Despite predictions that foliar carbohydrates would not accumulate in the leaves of plants grown at e[CO$_2$] in the field where roots are free to develop and explore the soil for nutrients, plants grown in the field using FACE technology still accumulated carbohydrate. Even N-fixing species with large sink capacities exhibited exacerbated carbohydrate accumulation at e[CO$_2$] and perennial ryegrass showed evidence of a severe sink limitation.
- FACE experiments confirmed the importance of sink capacity in determining the timing and extent of foliar carbohydrate accumulation; and they provide a valuable field test for key indicators of insufficient sink capacity.
- Many uncertainties still remain. Cross-talk between carbon and nitrogen metabolism in the leaf is extensive and well documented (Stitt and Krapp 1999) and growth at e[CO$_2$] will have a major impact on carbon and nitrogen metabolism. A full and more mechanistic understanding of the response of foliar carbohydrates to growth at e[CO$_2$] cannot be realized without parallel and comprehensive investigations of nitrogen metabolism.

Acknowledgements. A.R. was supported by the U.S. Department of Energy Office of Science contract No. DE-AC02-98CH10886 to Brookhaven National Laboratory (BNL). This work was supported in part by a Laboratory Directed Research and Development grant from BNL. E.A.A. was supported by the Alexander von Humboldt Foundation and the Juelich Research Center, ICG-III. The authors would also like to acknowledge assistance with fieldwork at the SoyFACE experiment from C.J. Bernacchi, V.E. Wittig and M.R. Harrison.

References

Ainsworth EA, Long SP (2005) What have we learned from 15 years of free-air CO$_2$ enrichment (FACE)? A meta-analytic review of responses to rising CO$_2$ in photosynthesis, canopy properties and plant production. New Phytol 165:351–372

Ainsworth EA, Rogers A, Nelson R, Long SP (2004) Testing the „source–sink" hypothesis of down-regulation of photosynthesis in elevated [CO$_2$] in the field with single gene substitutions in *Glycine max*. Agric For Meteorol 122:85–94

Allard V, Newton PCD, Lieffering M, Clark H, Matthew C, Soussana J-F, Gray YS (2003) Nitrogen cycling in grazed pastures at elevated CO$_2$: N returns by ruminants. Global Change Biol 9:1731–1742

Arp WJ (1991) Effects of source–sink relations on photosynthetic acclimation to elevated CO$_2$. Plant Cell Environ 14:869–875

Bush DR (1999) Sugar transporters in plant biology. Curr Opin Plant Biol 2:187–191

Chatterton NJ, Harrison PA, Thornley WR, Asay KH (1989) Carbohydrate partitioning in 185 accessions of Gramineae grown under warm and cool temperatures. J Plant Physiol 134:169–179

Estiarte M, Penuelas J, Kimball BA, Hendrix DL, Pinter PJ, Wall GW, LaMorte RL, Hunsaker DJ (1999) Free-air CO_2 enrichment of wheat: leaf flavonoid concentration throughout the growth cycle. Physiol Plant 105:423–433

Farrar JF, Williams ML (1991) The effects of increased atmospheric carbon dioxide and temperature on carbon partitioning, source–sink relations and respiration: commissioned review. Plant Cell Environ 14:819–830

Farrar JF, Pollock CJ, Gallagher J (2000) Sucrose and the integration of metabolism in vascular plants. Plant Sci 154:1–11

Fischer BU, Frehner M, Hebeisen T, Zanetti S, Stadelmann F, Lüscher A, Hartwig UA, Hendrey GR, Blum H, Nösberger J (1997) Source–sink relations in *Lolium perrene* L. as reflected by carbohydrate concentrations in leaves and pseudo-stems during regrowth in a free air carbon dioxide enrichment (FACE) experiment. Plant Cell Environ 20:945–952

Hamilton JG, Dermody O, Aldea M, Zangerl AR, Rogers A, Berenbaum M, DeLucia EH (2005) Anthropogenic changes in tropospheric composition increase susceptibility of soybean to insect herbivory. Environ Entomol 34:479–485

Hendrey GR, Ellsworth DS, Lewin KF, Nagy J (1999) A free-air enrichment system for exposing tall forest vegetation to elevated atmospheric CO_2. Global Change Biol 5:293–309

Hendrix DL, Mauney JR, Kimball BA, Lewin K, Nagy J, Hendrey GR (1994) Influence of elevated CO_2 and mild water stress on nonstructural carbohydrates in field-grown cotton tissues. Agric For Meteorol 70:153–162

Herrick JD Thomas RB (2001) No photosynthetic down-regulation in sweetgum trees (*Liquidambar styraciflua* L.) after three years of CO_2 enrichment at the Duke forest FACE experiment. Plant Cell Environ 24:53–64

Isopp H, Frehner M, Almeida JPF, Blum H, Daepp M, Hartwig UA, Lüscher A, Suter D, Nösberger J (2000a) Nitrogen plays a mjor role in leaves when source-sink relations change: C and N metabolism in *Lolium perrene* growing under free air CO_2 enrichment. Aust J Plant Pysiol 27:851–858

Isopp H, Frehner M, Long SP, Nösberger J (2000b) Sucorse-phosphate synthase responds differently to source–sink relations and to photosynthetic rate: *Lolium perenne* L. growing at elevated pCO_2 in the field. Plant Cell Environ 23:597–607

Jones PG, Lloyd JC, Raines CAR (1996) Glucose feeding of intact wheat plants represses the expression of a number of Calvin cycle genes. Plant Cell Environ 19:231–236

Koch KE (1996) Carbohydrate-modulated gene expression in plants. Annu Rev Plant Physiol Plant Mol Biol 47:509–540

Körner C. (2003) Nutrients and sink activity drive plant CO_2 responses – caution with literature-based analysis. New Phytol 159:537–538

Krapp A, Hofmann B, Schafer C, Stitt M (1993) Regulation of the expression of *Rbcs* and other photosynthetic genes by carbohydrates: a mechanism for the sink regulation of photosynthesis. Plant J 3:817–828

Lindroth RL (1996) CO_2-meadiated changes in tree chemistry and tree–lepidoptera interactions. In: Koch GW, Mooney HA (eds) Carbon dioxide and terrestrial ecosystems. Academic, San Diego, pp105–120

Long SP, Drake BG (1992) Photosynthetic CO_2 assimilation and rising atmospheric CO_2 concentrations. In: Baker NR, Thomas H (eds) Crop photosynthesis spatial and temporal determinants. Elsevier, Amsterdam, pp 69–103

Long SP, Ainsworth EA, Rogers A, Ort DR (2004) Rising atmospheric carbon dioxide: plants FACE the future. Annu Rev Plant Biol 55:591–628

Masle J, Farquhar GD, Gifford RM (1990) Growth and carbon economy of wheat seedlings as affected by soil resistance to penetration and ambient partial-pressure of CO$_2$. Aust J Plant Physiol 17:465–487

Moore BD, Cheng SH, Sims D, Seemann JR (1999) The biochemical and molecular basis for photosynthetic acclimation to elevated atmospheric CO$_2$. Plant Cell Environ 22:567–582

Myers DA, Thomas RB, DeLucia EH (1999) Photosynthetic capcity of loblolly pine (*Pinus taeda* L.) trees during the first year of carbon dioxide enrichmant in a forest ecosystem. Plant Cell Environ 22:473–482

Norby RJ, Todd DE, Fults J, Johnson DW (2001) Allometric determination of tree growth in a CO$_2$-enriched sweetgum stand. New Phytol 150:447–487

Pego JV, Kortstee AJ, Huijser C, Smeekens SCM (2000) Photosynthesis, sugars and the regulation of gene expression. J Exp Bot 51:407–416

Pollock CJ, Cairns AJ (1991) Fructan metabolism in grasses and cereals. Annu Rev Plant Physiol Plant Mol Biol 42:77–101

Ritchie S, Hanaway J, Thompson H, Benson G (1997) How a soybean plant develops – special report no. 53. Iowa State University, Ames

Robbins NS, Pharr DM (1988) Effect of restricted root growth on carbohydrate metabolism and whole plant growth of *Cucumis sativus* L. Plant Physiol 87:409–413

Rogers A, Ellsworth DS (2002) Photosynthetic acclimation of *Pinus taeda* (loblolly pine) to long-term growth in elevated *p*CO$_2$ (FACE). Plant Cell Environ 25:851–858

Rogers A, Fischer BU, Bryant J, Frehner M, Blum H, Raines CA, Long SP (1998) Acclimation of photosynthesis to elevated CO$_2$ under low N nutrition is effected by the capacity for assimilate utilization. Perennial ryegrass under free-air-CO$_2$ enrichment (FACE). Plant Physiol 118:683–689

Rogers A, Allen DJ, Davey PA, Morgan PB, Ainsworth EA, Bernacchi CJ, Cornic G, Dermody O, Heaton EA, Mahoney J, Zhu X-G, DeLucia EH, Ort DR, Long SP (2004) Leaf photosynthesis and carbohydrate dynamics of soybeans grown throughout their lifecycle under free-air carbon dioxide enrichment. Plant Cell Environ 27:449–458

Rogers GS Milham PJ Gillings M, Conroy JP (1996) Sink strength may be the key to growth and nitrogen responses in N-deficient wheat at elevated CO$_2$. Aust J Plant Physiol 23:253–264

Schmundt D, Stitt, M, Jähne B, Schurr U (1998) Quantitative analysis of the local rates of growth of dicot leaves at a high temporal and spatial resolution, using image sequence analysis. Plant J 16:505–514

Sheen J (1990) Metabolic repression of transcription in higher-plants. Plant Cell 2:1027–1038

Singsaas EL, Ort DR, DeLucia EH (2000) Diurnal regulation of photosynthesis in understory saplings. New Phytol 145:39–49

Smeekens S (2000) Sugar-induced signal transduction in plants. Annu Rev Plant Physiol Plant Mol Biol 51:49–81

Stitt M (1991) Rising CO$_2$ levels and their potential significance for carbon flow in photosynthetic cells. Plant Cell Environ 14:741–762

Stitt M, Krapp A (1999) The interaction between elevated carbon dioxide and nitrogen nutrition: the physiological and molecular background. Plant Cell Environ 22:583–628

Taylor G, Trickler PJ, Zhang FZ, Alston VJ, Miglietta F, Kuzminsky E (2003) Spatial and temporal effects of free-air CO$_2$ enrichment (POPFACE) on leaf growth, cell expansion, and cell production in a closed canopy of poplar. Plant Physiol 131:177–185

Thomas RB, Strain BR (1991) Root restriction as a factor in photosynthetic acclimation of cotton seedlings grown in elevated carbon-dioxide. Plant Physiol 96:627–634

Tissue DT, Lewis JD Wullschleger SD, Amthor JS, Griffin KL, Anderson OR (2002) Leaf respiration at different canopy positions in sweetgum (*Liquidambar styraciflua*) grown in ambient and elevated concentrations of carbon dioxide in the field. Tree Physiol 22:1157–1166

Van Oosten JJ, Besford RT (1994) Sugar feeding mimics effect of acclimation to high CO_2 – rapid down regulation of RubisCO small subunit transcripts but not of the large subunit transcripts. J Plant Physiol 143:306–312

Vessey JK, Walsh KB, Layzell DB (1988) Oxygen limitation of N_2 fixation in stem-girdled and nitrate-treated soybean. Physiol Plant 73:113–121

Walsh KB, V essey JK, Layzell DB (1987) Carbohydrate supply and N_2 fixation in soybean: the effect of varied daylength and stem girdling. Plant Physiol 85:137–144

Zanetti S, Hartwig UA, Lüscher A, Hebeisen T, Frehner M, Fischer BU, Hendrey GR, Blum H, Nösberger JA (1996) Stimulation of symbiotic N2 fixation in *Trifolium repens* L. under elevated atmospheric pCO$_2$ in a grassland ecosystem. Plant Physiol 112:575–583

Zeeman SC Smith SM, Smith AM (2004) The breakdown of starch in leaves. New Phytol 163:247–261

Part C   Processes

17 Evapotranspiration, Canopy Temperature, and Plant Water Relations

B.A. Kimball and C.J. Bernacchi

17.1 Introduction

Elevated CO_2 concentrations ($e[CO_2]$) cause partial stomatal closure (e.g. see Chapter 14). Such changes in stomatal aperture cause reductions in the rate of transpiration from leaves, which affect the exchange of water vapor and energy from whole plant canopies. In turn, soil water content can also be affected, which ultimately impacts plant water relations.

The free-air CO_2 enrichment (FACE) approach is especially advantageous for assessing the impacts of $e[CO_2]$ on microclimatic processes because there are no walls to alter wind flow or to shade the plant canopies. Therefore, this chapter primarily focuses on the results from such FACE experiments which included a microclimatic component. Observations of the effects of $e[CO_2]$ at concentrations of about 550 ppm (parts per million by volume = $\mu mol\ mol^{-1}$; about 200 ppm above current concentrations) on canopy temperatures, evapotranspiration, soil water content, and plant water relations are presented.

17.2 Canopy Temperature

When plants are grown at $e[CO_2]$, stomatal conductance is reduced (e.g. see Chapter 14), and therefore, transpirational cooling of the plant leaves is also reduced, consequently causing daytime leaf temperatures to rise. Wheat experienced an average canopy temperature increase of 0.6 °C when exposed to FACE at 550 ppm with ample water and nitrogen (Fig. 17.1). Similarly, the canopy temperature of paddy rice, another C_3 grass, rose 0.4 °C . In contrast, the temperature of sorghum, a C_4 grass, rose more than twice as much as wheat, 1.7 °C. Cotton and poplar, woody perennials, appear to be intermediate in canopy temperature response between the C_3 and C_4 grasses. Potato, a C_3

Ecological Studies, Vol. 187
J. Nösberger, S.P. Long, R.J. Norby, M. Stitt,
G.R. Hendrey, H. Blum (Eds.)
Managed Ecosystems and CO_2
Case Studies, Processes, and Perspectives
© Springer-Verlag Berlin Heidelberg 2006

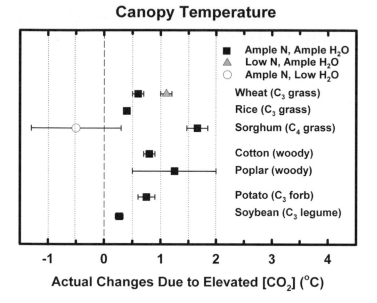

Fig. 17.1 Changes in daytime canopy temperature due to elevated CO_2 (e[CO_2]) from free-air CO_2 enrichment (FACE) experiments at concentrations of about 550 ppm (parts per million by volume = μmol mol^{-1}; about 200 ppm above current concentrations). Wheat and cotton data from Table 2 of Kimball et al. (2002); rice datum from Yoshimoto et al. (2005); sorghum datum from Triggs et al. (2004); poplar datum from Tommasi et al. (2002); potato datum from Magliulo et al. (2003); soybean datum from infrared thermometer measurements in SoyFACE Project, Urbana, Ill

forb, exhibited a temperature rise similar to that of wheat. Soybean also increased in canopy temperature. However, this observed seasonal average increase was less than that of other crops (only about 0.3 °C), probably because it was not irrigated, and during the growing season, it experienced some water stress and consequent stomatal closure in both control and FACE plots, which overwhelmed the CO_2 effect at times.

Compared to the case with ample N, when N was limited, the wheat canopy temperature increase was approximately doubled, to about 1.1 °C (Fig. 17. 1). Such a larger temperature rise is consistent with a greater reduction in stomatal conductance under limited N (e.g. see Chapter 14).

When water is severely limited, stomata close, and [CO_2] no longer has an effect on stomatal conductance or canopy temperature. The low-water point in Fig. 17.1 for sorghum indicates that e[CO_2] caused an average cooling of 0.5 °C but with high variability. Such apparent cooling due to e[CO_2] resulted from the dynamic changes that occurred with time through the growing season (Triggs et al. 2004). Indeed, following irrigation events, canopy temperatures in all FACE plots were higher than the ambient-CO_2 controls. However,

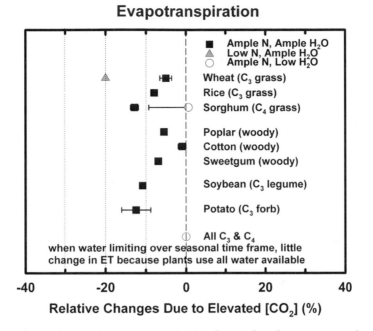

Fig. 17.2 Relative changes in evapotranspiration due to e[CO_2]. Data sources for particular species same as Fig. 17.1, except sweetgum from Wullschleger et al. (2001)

the water conservation resulting from lower evapotranspiration (ET; Fig. 17.2) allowed the FACE plots to continue to function for a few days after the control plots at current CO_2 concentrations (c[CO_2]) exhausted their soil water supply; and during this period of time, the FACE canopies were much cooler than the controls. After both the e[CO_2] and c[CO_2] plants had exhausted their soil water, canopy temperatures rose above air temperature during daytime, but [CO_2] no longer had an effect. Thus, the average e[CO_2] effect was a 0.5 °C cooling, but with much deviation from the mean depending on time after the irrigation events.

The increases in canopy temperature that have been observed in FACE experiments (Fig. 17. 1) are comparable to those predicted for global warming during the middle of this century (IPCC 2001); and they are very likely to occur regardless of whether there is any climate warming or not. These temperature increases imply that the optimal geographic climate ranges over which crops and native species grow likely will shift in the future, even in the absence of any change in global air temperature.

17.3 Evapotranspiration

As already mentioned, because there are no walls to alter wind flow or to shade the plant canopies, the FACE approach produces the most natural conditions possible for assessing the impacts of e[CO_2]on microclimatic processes, such as ET. Elevated [CO_2] causes a decrease in stomatal conductance (e.g. see Chapter 14) which reduces transpiration per unit of leaf area while canopy temperature is increased (Fig. 17.1). The increase in temperature raises the water vapor pressure inside the leaves, which tends to increase leaf transpiration, thereby negating some of the reduction due to the decrease in stomatal conductance (e.g. Kimball et al. 1999). At the same time, e[CO_2] generally stimulates plant growth (e.g., Kimball et al. 2002; see also Case Studies, Chapters 3–13), which results in larger plants with greater leaf area. Thus, the resultant effect of e[CO_2] on ET is a combination of individual effects of the [CO_2] on decreasing stomatal conductance, increasing leaf area, and increasing canopy temperature.

17.3.1 Changes in ET with e[CO_2]

Elevated [CO_2] had little effect (<2% reduction) on the ET of cotton (Fig. 17.2), suggesting that the effects of the e[CO_2] on reducing stomatal conductance and increasing leaf area must have exactly compensated. However, modest (5–8%) reductions in ET were observed in wheat, rice, poplar, and sweetgum, showing that stomatal effects can predominate (Fig. 17.2). Larger reductions, of about 12%, were recorded for soybean, potato, and sorghum at ample water and nitrogen (Fig. 17.2). Using a flow-through chamber technique, rather than the residual energy balance approach of most of the Fig. 17.2 data, Weigel et al. (e.g., see Chapter 7) found ET reductions of 8, 20, and 3% for barley, sugar beet, and wheat, respectively.

When nitrogen was limited, the energy balance approach indicated water savings of about 20% for wheat (Fig. 17.2; Kimball et al. 1999). Such a large reduction in ET is surprising, especially when estimates from soil water balance were much smaller (Table 2 of Kimball et al. 2002; Hunsaker et al. 2000). However, simulations with the *ecosys* model by Grant et al. (2001) predicted a reduction in ET of 16% at low nitrogen, caused by reductions in Rubisco activity and concentration, which forced greater reductions in stomatal conductance (e.g., see Chapter 14) in order to maintain constant C_i:C_a ratio (ratio of internal leaf [CO_2] to that of outside air). Thus, the energy balance result of a 20% reduction in ET of wheat at low N seems reasonable.

When seasonal water supply is severely growth-limiting, one would expect plants to utilize all the available water, so that effects of e[CO_2] on seasonal ET would be minimal. CO_2-enriched plants with more robust root systems, however, might extract and use more water. The latter phenomenon might have happened for wheat in 1993 (Table 2 of Kimball et al. 2002; Hunsaker et al. 1996), but generally the observed effects of FACE on the ET of cotton, wheat, and sorghum under limited water have been inconsistent and small (Table 2 of Kimball et al. 2002). An important variable is the length of time after an irrigation or rainfall event before the soil water supply is exhausted, i.e., those plants which have their ET rate reduced due to e[CO_2] can sustain photosynthesis and growth further into a drought cycle than plants at c[CO_2]. However, the total water used by both e[CO_2]- and c[CO_2]-grown plants will be nearly the same.

17.3.2 Correlations of ET with Canopy Temperature and Shoot Biomass Changes

Samarakoon and Gifford (1995) conducted an interspecific comparison in glasshouses of the effects of e[CO_2] on cotton, wheat, and maize that nicely illustrated the importance of relative changes in leaf area and stomatal conductance in determining the relative effects of [CO_2] on ET. Their e[CO_2]-grown cotton had a large increase in leaf area and a small change in stomatal conductance, so water use per pot actually increased. In the FACE experiments, cotton also had a large growth increase (Fig. 17.3a) and a modest reduction in conductance (as indicated by the modest temperature rise in Fig. 17.1), which resulted in no significant net change in ET (Fig. 17.2). Maize, a C_4 plant, had little photosynthetic or leaf area response in the Samarakoon and Gifford experiment, so the reduction in conductance resulted in significant water conservation. Likewise, in the FACE experiments on sorghum, which also is a C_4 plant, there was no growth response at ample water (Fig. 17.3a), so the decrease in stomatal conductance due to the higher CO_2 resulted in a 13% reduction in ET (Fig. 17.2). Wheat was intermediate between the other two species in both the Samarakoon and Gifford (1995) and the FACE experiments (Fig. 17.2).

While the cotton–wheat–C_4 interspecific comparison in the previous paragraph appears to clearly show the importance of relative changes in stomatal conductance and leaf area with growth in e[CO_2], the inclusion of additional data introduces more variability into the relationship. Overall, a decrease in ET occurred with a rise in canopy temperature, but the correlation is rather low, with an r^2 of only 0.16 (Fig. 17.3b). Of course canopy temperature rise is not a direct surrogate for stomatal conductance reduction, because differences among sites in air vapor pressure deficit and other factors confound it. However, these near-season-long canopy temperature data are more available because they have generally been continuously recorded by those researchers using energy balance techniques to evaluate ET, whereas stomatal conductances were measured only sporadically during the season. In contrast, the

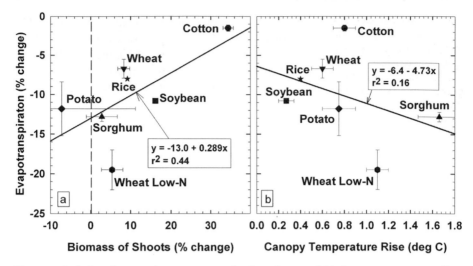

Fig. 17.3 Relative changes in evapotranspiration due to e[CO$_2$] versus corresponding changes in (**a**) shoot biomass [data sources for particular species same as Fig. 17.1, except rice and potato data from Table 2 of Kimball et al. (2002) and soybean biomass datum from Morgan et al. (2005)] and (**b**) canopy temperature (data from Figs. 17.1, 17.2)

correlation between the reduction in ET and the stimulation of shoot biomass due to e[CO$_2$] is stronger ($r^2 = 0.44$, Fig. 17.3a). When the data from Fig. 17.3a and b were combined in a three-dimensional graph (Fig. 17.4) and a multiple regression of change in ET on change in shoot biomass and canopy temperature rise was computed, the correlation was increased to an r^2 of 0.50. Moreover, the r^2 increased to 0.69 when the wheat-low-N point was excluded from the regression. Therefore 50 % of the variation in the observed changes in ET due to e[CO$_2$] can be explained by the effects of the FACE treatment on shoot biomass stimulation and canopy temperature rise (or 69 % if the low-N case is not considered).

The numerous differences among the sites, researchers, and other sources of variability lower the degrees of correlation among seasonal ET reduction due to e[CO$_2$], shoot biomass accumulation, and canopy temperature rise (Figs. 17.3, 17.4). For annual crops, one source of such variability is the dynamically varying effects of e[CO$_2$] through the course of the growing season. Early in the season, before canopy closure, the effects of e[CO$_2$] on growth can predominate, so that the e[CO$_2$]-grown plants may actually require more water than ambient-grown plants. After canopy closure, ET becomes relatively insensitive to leaf area, and then stomatal effects predominate. Thus, crops growing in the future high-[CO$_2$] world may require comparatively more water early in season, but their water requirements likely will decrease after canopy closure.

Fig. 17.4 Relative changes in evapotranspiration due to e[CO_2] versus corresponding changes in shoot biomass and canopy temperature plotted in three dimensions. The r^2 for a multiple linear regression of ET on shoot biomass and canopy temperature changes is 0.50. Same data as Fig. 17.3a, b

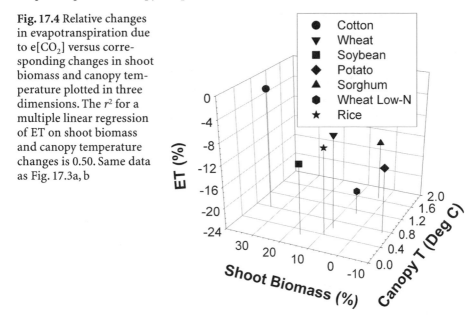

17.3.3 Applicability of Plot-Scale ET Measurements to Regional Scales

A well known theory of Jarvis and McNaughton (1986) postulates that, on a regional scale, there is no control of stomatal resistance on evapotranspiration. They argue that if e[CO_2] or some other factor alters stomatal resistance, then humidity profiles adjust within the planetary boundary layer (PBL), which is a negative feedback that counters stomatal closure. In other words, if the stomata partially close so that there is more resistance to water loss, then the PBL becomes drier and increases the water vapor concentration gradient from inside the leaves to the upper atmosphere, thereby negating any effect of changing the stomata. They also argue that stomatal resistance is only one of a series of resistances to water loss from vegetation, and changing [CO_2] will not affect these other resistances.

However, as reviewed by Kimball et al. (1999), Triggs et al. (2004), and Kimball (2006), several studies using general circulation climate models (GCMs; for predicting Earth's climate) with relatively realistic simulations of land surface processes indicate that the stomatal effects observed at the field scale likely will have significant effects at regional scales. For example, Sellers et al. (1996) examined the effects of doubled [CO_2] on stomatal conductance and net photosynthesis, as well as on climate, using a GCM with a biosphere submodel (SiB2). They considered both short-term effects of e[CO_2] on plant physiology (27 % reduction of stomatal conductance) and longer-term $A_{down-regulated}$ physiology, whereby photosynthesis reverted to values close to those

for the original [CO_2] but stomatal conductance was reduced even more (49 %
overall reduction). Averaged over all land, ET was reduced 2.3 % and 3.5 % due
to the 27 % and 49 % stomatal reductions, respectively, for today's climate.
Based on predictions for a CO_2-induced increase in global temperature, the
respective ET changes were increases of 3.3 % and 0.3 % compared to the
c[CO_2], present-climate case. Thus, the physiological effects of the e[CO_2] on
vegetation approximately compensated those of global warming on ET.

17.3.4 Combined Physiological and Global-Warming Effects of e[CO_2] on ET

The previous sections discuss the effects of e[CO_2] on the ET of open-field-
grown vegetation. However, global warming is predicted to occur (IPCC
2001); and warming will increase the vapor pressure of water inside the
leaves of the plants, thereby increasing the vapor pressure gradient from
inside plant leaves to the outside air, which would increase ET. Global warming is expected to increase global average precipitation and absolute humidity (IPCC 2001). However, relative humidity levels are expected to remain
somewhat constant.

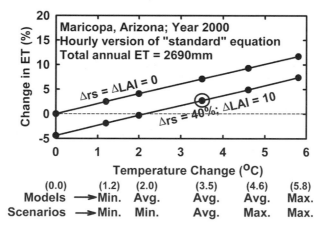

Fig. 17.5 Percent changes in annual ET versus temperature change for an alfalfa crop
using base weather data for the year 2000 at Maricopa, Ariz., an irrigated area in a hot
arid region where the base total ET was 2690 mm. The *upper curve* is for global warming
at constant relative humidity by the amounts shown for the minimums, averages, and
maximums predicted by coupled atmosphere-ocean climate models (IPCC 2001), but
with no accounting for the effects of e[CO_2] on plant leaf area or stomatal resistance. The
lower curve accounts for warming and incorporates the physiological effects of e[CO_2]
by a 40 % increase in stomatal resistance (Δr_s) and a 10 % increase in leaf area (ΔLAI).
Thus, the "bull's-eye" point is the most plausible. The hourly computations were done by
Kimball (2006) using a standard Penman–Monteith equation under consideration for
adoption by the American Society of Civil Engineers (Walter et al. 2000)

Using a "standard" Penman–Monteith equation (Walter et al. 2000), Kimball (2006) computed the sensitivity of ET to both climate and plant physiological variables. The likely effects of global change on ET were calculated using hourly weather data from Maricopa, Ariz., for alfalfa, a crop that grows nearly the whole year at this irrigated site in a hot, arid region. Based on the average of all the climate models and all the emissions scenarios, global warming at constant relative humidity would increase the annual alfalfa ET by about 7% if there were no effects of $e[CO_2]$ on the plant's physiology (Fig. 17.5). However, assuming increases of 40% in stomatal resistance (which is the reciprocal of stomatal conductance) and 10% in leaf area due to $e[CO_2]$, the corresponding net increase in ET would be smaller, about 3%, which is probably the most plausible estimate.

17.4 Soil Water Content

Of course, one result of a reduction in ET due to $e[CO_2]$ (Fig. 17.2) is that, for a closed canopy, the soil will dry more slowly after an irrigation or rainfall event. Therefore, the water content of the soil in FACE plots has often been higher than that in a corresponding plot at today's $c[CO_2]$ [Hunsaker et al. (1996, 2000) for wheat; Ellsworth (1999) for pine forest; Conley et al. (2001) for sorghum; Kammann et al. (2005) for grassland]. Indeed, the slower rate of soil moisture depletion is the means by which changes in ET rate have been determined from soil water content measurements (Hunsaker et al. 1994, 1996, 2000; Conley et al. 2001). Consequently, those soil physical, chemical, and biological processes that are affected by soil moisture content can be influenced by the atmospheric $[CO_2]$. For example, Adamsen et al. (2005) observed that inorganic N levels were lower under FACE-grown wheat at various times during the growing season, and that the effect was more pronounced under water-limited conditions. For another example, soil respiration, a microbiological process, has generally been stimulated in plots subjected to $e[CO_2]$ (e.g., Kimball et al. 2002). Greater root respiration and greater microbial respiration from digestion of more root exudates likely play a large role in the stimulation of soil respiration, but at the same time, higher soil water content may also enable increased soil microbial activity, particularly when the water supply is limited and the lower ET rate of $e[CO_2]$-grown plants enables longer activity before the soil moisture supply is exhausted.

17.5 Plant Water Use Efficiency

One of the more universal effects of $e[CO_2]$ on plants is an increase in water use efficiency (WUE = total biomass/total ET; Lawlor and Mitchell 1991; Kimball et al. 2002). As presented previously in the Case Studies chapters in this volume, $e[CO_2]$ has often increased plant biomass production, which is an increase in the numerator in WUE. Also, $e[CO_2]$ has often reduced evapotranspiration (Fig. 17.2), the denominator in WUE. For example, the biomass of sorghum increased about 3 % when supplied with ample water (Fig. 17.3a), while ET was reduced about 13 % (Fig. 17. 2), which represents an increase in WUE of about $[(1 +0.03)/(1 -0.13) -1] \times 100 = 18$ % increase in WUE. Under water stress, the biomass of sorghum increased about 16 % due to growth in $e[CO_2]$ (Ottman et al. 2001) with little change in ET. Thus a 16 % increase in WUE was solely a result of the biomass increase. Thus, for both ample and limited water supply levels, WUE increased due to $e[CO_2]$ but for different reasons.

17.6 Plant Water Relations

The reductions in stomatal conductance and ET with increases in soil water content have important impacts on plant water relations, particularly for water-limited conditions, as reviewed by Wullschleger et al. (2002). Besides (1) an increase in water use efficiency as discussed above, they list impacts (2) on fine root proliferation and whole plant water uptake, (3) on plant water potential, and (4) on solute accumulations and osmotic adjustments. As discussed by Wullschleger et al., reviewed by Kimball et al. (2002), and can be deduced from the Case Studies in this volume, growth at $e[CO_2]$ has generally stimulated root growth, often even more than shoot growth.

However, changes in plant water potential, which is an important measure of the internal water status of plants, have been difficult to detect. As discussed by Kimball et al. (2002), it varies hour by hour through the course of a day; and it decreases (becomes more negative) day by day following a rain or irrigation event that wets the soil. Also, small subtle differences, i.e., that are difficult to detect, can cumulatively affect the way plants grow, which feeds back on their subsequent water potential. For example, suppose one plot of young plants is maintained in a well watered condition, while another is allowed to dry and become stressed for water. Then the dry plot is irrigated and allowed to dry again. During the first cycle, substantial differences in water potential are likely to develop, whereas during the second cycle, the differences will be much smaller – because the second plot now has smaller plants which use water at a slower rate. Therefore, it is difficult to determine meaningful quantitative average changes in plant water potential due to treatments such as $e[CO_2]$.

Nevertheless, in their review, Wullschleger et al. (2002) tabulated the effects of growth in $e[CO_2]$ on water potential from 32 previous studies, mostly from chamber-based experiments. Of these, 19 had a positive response to $e[CO_2]$, ten had no response, and three had a negative response.

FACE experiments have similarly revealed a range of water potential responses, but with more positive than non- or negative responses. For example, cotton, a crop with a relatively large biomass response to $e[CO_2]$ (Fig.17.3a) and little ET response (Fig.17.2), Bhattacharya et al. (1994) found no consistent effect on plant water potential with growth in FACE except near the end of the season. In contrast, grape, a woody C_3 plant like cotton, exhibited a slight improvement in water potential when well watered (Raschi et al. 1996). Wall et al (1994) report that for wheat, which has a somewhat smaller biomass response and larger ET response than cotton, leaves exposed to FACE had slightly, but statistically significant, less negative water potentials compared to $c[CO_2]$-grown plants during most of the daylight period of each day. Sorghum, a C_4 plant with little growth response but a large ET reduction due to growth in elevated CO_2 under ample water and nutrients, had a higher (less negative) average water potential (Wall et al. 2001). Under water stress, water potential improved even more due to the $e[CO_2]$. Thus in these latter FACE experiments, except for the case in which there was no ET reduction, growth under $e[CO_2]$ led to an improvement in plant water potential.

Decreases in ET of $e[CO_2]$-grown plants can lead to some drought avoidance, as discussed in previous sections. However $e[CO_2]$ might also improve plant water relations by increases in drought tolerance, via increases in solutes (i.e., osmotica). Although this hypothesis seems plausible, in their review of the topic, Wullschleger et al. (2002) concluded that osmotic adjustment of leaves and roots is minimal and probably an indirect, secondary response reflecting imbalances between carbohydrate sources and sinks. In contrast, Wall et al. (2001) in their FACE-sorghum experiment believe that increases in specific leaf weight, i.e., greater carbohydrate supply, enhanced the drought tolerance of leaf tissue, which enabled greater growth and yield due to FACE under water-limited conditions.

17.7 Conclusions

Because $e[CO_2]$ causes partial stomatal closure, transpiration from plant leaves is reduced, which has many ramifications for plant water relations, which can be summarized as follows:
- The reduction in leaf transpiration reduces evaporative cooling with a consequential rise in canopy temperatures. Increases of 0.3–1.7 °C at CO_2 con-

centrations of 550 ppm (200 ppm above current concentrations) have been observed, depending on species and conditions. Such canopy temperature changes are likely to cause shifts in the optimum geographic climate areas for growth of crops and other species.

- The reduction in transpiration per unit of leaf area with $e[CO_2]$ generally leads to a reduction in ET per unit of land area. However, the magnitude of such water conservation at $e[CO_2]$ varies with the degree of stimulation of plant growth and the degree of partial stomatal closure. Observed reductions in ET have ranged from near zero for cotton, a woody C_3 species with large growth stimulation, to about 16% for sorghum, a C_4 grass with little growth stimulation. In the absence of global warming, such water conservation will reduce the water requirements of irrigated regions, and with global warming, it will help to keep the requirements from rising as much as the warming alone would cause.
- The reductions in ET with $e[CO_2]$ will also lead to increases in soil moisture content, with consequent effects on numerous soil physical, chemical, and biological processes that are influenced by soil moisture content, such as leaching, mineralization, and soil respiration.
- The reductions in ET and consequent increases in soil moisture can lead to improvements in plant water relations, such as higher plant water potentials. Water conservation with growth in $e[CO_2]$ can enable plants to maintain growth longer into drought cycles.

References

Adamsen FJ, Wechsung G, Wechsung F, Wall GW, Kimball BA, Pinter PJ Jr, LaMorte RL, Garcia RL, Hunsaker DJ, Leavit SW (2005) Temporal changes in soil and biomass nitrogen for irrigated wheat grown under free-air carbon dioxide enrichment (FACE). Agron J 97:160–168

Bhattacharya NC, Radin JW, Kimball BA, Mauney JR, Hendrey GR, Nagy J, Lewin KF, Ponce DC (1994) Leaf water relations of cotton in a free-air CO_2-enriched environment. Agric For Meteorol 70:171–182

Conley MM, Kimball BA, Brooks TJ, Pinter PJJr, Hunsaker DJ, Wall GW, Adam NR, LaMorte RL, Matthias AD, Thompson TL, Leavitt SW, Ottman MJ, Cousins AB, Triggs JM (2001) CO_2 enrichment increases water-use efficiency in sorghum. New Phytol 151:407–412

Ellsworth DS (1999) CO_2 enrichment in a maturing pine forest: are CO_2 exchange and water status in the canopy affected? Plant Cell Environ 22:461–472

Grant RF, Kimball BA, Brooks TJ, Wall GW, Pinter PJJr, Hunsaker DJ, Adamsen FJ, LaMorte RL, Leavitt SW, Thompson TL, Matthias AD (2001) Interactions among CO_2, N, and climate on energy exchange of wheat: model theory and testing with a free air CO_2 enrichment (FACE) experiment. Agron J 93:638–649

Hunsaker DJ, Hendrey GR, Kimball BA, Lewin KF, Mauney JR, Nagy J (1994) Cotton evapotranspiration under field conditions with CO_2 enrichment and variable soil moisture regimes. Agric For Meteorol 70:247–258

Hunsaker DJ, Kimball BA, Pinter PJ Jr, LaMorte RL, Wall GW (1996) Carbon dioxide enrichment and irrigation effects on wheat evapotranspiration and water use efficiency. Trans ASAE 39:1345–1355

Hunsaker DJ, Kimball BA, Pinter PJ Jr, Wall GW, LaMorte RL, Adamsen FJ, Leavitt SW, Thompson TL, Matthias AD, Brooks TJ (2000) CO_2 enrichment and soil nitrogen effects on wheat evapotranspiration and water use efficiency. Agric For Meteorol 104:85–105

IPCC (2001) Climate change 2001: the scientific basis. Contribution from working group I to the third assessment report, inter-governmental panel for climate change (IPCC). Cambridge University Press, Cambridge

Jarvis PG, McNaughton KG (1986) Stomatal control of transpiration: scaling up from leaf to region. Adv Ecol Res 15:1–49

Kammann C, Grunhage L, Gruters U, Janze S, Jager H-J (2005) Response of aboveground grassland biomass and soil moisture to moderate long-term CO_2 enrichment. Basic Appl Ecol (in press)

Kimball BA (2006) Global changes and water resources. In: Lascano RJ, Sojka RE (eds) Irrigation of agricultural crops monograph. American Society of Agronomy, Madison, Wis. (in press)

Kimball BA, LaMorte RL, Pinter PJ Jr, Wall GW, Hunsaker DJ, Adamsen FJ, Leavitt SW, Thompson TL, Matthias AD, Brooks TJ (1999) Free-air CO_2 enrichment and soil nitrogen effects on energy balance and evapotranspiration of wheat. Water Resour Res 35:1179–1190

Kimball BA, Kobayashi K, Bindi M (2002) Responses of agricultural crops to free-air CO_2 enrichment. Adv Agron 77:293–368

Lawlor DW, Mitchell RAC (1991) The effects of increasing CO_2 on crop photosynthesis and productivity: a review of field studies. Plant Cell Environ 14:807–818

Magliulo V, Bindi M, Rana G (2003) Water use of irrigated potato (*Solanum tuberosum* L.) grown under free air carbon dioxide enrichment in central Italy. Agric Ecosyst Environ 97:65–80

Morgan PB, Bollero GA, Nelson RL, Dohleman FG, Long SP (2005) Smaller than predicted increase in aboveground net primary production and yield of field-grown soybean under fully open-air [CO_2] elevation. Global Change Biol 11:1–10

Ottman MJ, Kimball BA, Pinter PJ Jr, Wall GW, Vanderlip RL, Leavitt SW, LaMorte RL, Matthias AD, Brooks TJ (2001) Elevated CO_2 effects on sorghum growth and yield at high and low soil water content. New Phytol 150:261–273

Raschi A, Bindi M, Longobucco A, Miglietta F, Moriondo M (1996) Water relations of *Vitis vinifera* L. plants growing under elevated atmospheric CO_2 concentrations in a FACE setup. In: European Society for Agronomy (ed) Proceedings congress of the European society for agronomy, 7–11 July 1996. European Society for Agronomy, Wageningen, pp 54–55

Samarakoon AB, Gifford RM (1995) Soil water content under plants at high CO_2 concentration and interactions with direct CO_2 effects: a species comparison. J Biogeogr 22:193–202

Sellers PJ, Bounoua L, Collatz GJ, Randall DA, Dazlitch DA, Los SO, Berry JA, Fung I, Tucker CJ, Field CB, Jensen GG (1996) Comparison of radiative and physiological effects of doubled atmospheric CO_2 on climate. Science 271:1402–1406

Tommasi PD, Magliulo V, Dell'Aquila R, Miglietta F, Zaldei A, Gaylor G (2002) Water consumption of a CO_2 enriched poplar stand. Atti del Convegno CNR-ISAFOM, Ercolano

Triggs JM, Kimball BA, Pinter PJ Jr, Wall GW, Conley MM, Brooks TJ, LaMorte RL, Adam NR, Ottman MJ, Matthias AD, Leavitt SW, Cerveny RS (2004) Free-air carbon dioxide enrichment (FACE) effects on energy balance and evapotranspiration of sorghum. Agric For Meteorol 124:63–79

Wall GW, Kimball BA, Hunsaker DJ, Garcia RL, Pinter PJ Jr, Idso SB, LaMorte RL (1994) Diurnal trends in total water potential of leaves of spring wheat grown in a free-air CO_2-enriched (FACE) atmosphere and under variable soil moisture regimes. In: US Water Conservation Laboratory (ed) Annual research report. USDA-ARS, Phoenix, Ariz., pp 73–76

Wall GW, Brooks TJ, Adam NR, Cousins AB, Kimball BA, Pinter PJ Jr, PJ, LaMorte RL, Triggs J, Ottman MJ, Leavitt SW, Matthias AD, Williams DG, Webber AN (2001) Elevated atmospheric CO_2 improved sorghum plant water status by ameliorating the adverse effects of drought. New Phytol 152:231–248

Walter IA, Allen RG, Elliott R, Jensen M E, Itenfisu D, Mecham B, Howell TA, Snyder R, Brown P, Echings S, Spofford T, Hattendorf M, Cuenca RH, Wright JL, Martin D (2000) ASCE's standardized reference evapotranspiration equation. In: Evans RG, Benham BL, Trooien TP (eds) National irrigation symposium, proceedings of the fourth decennial symposium. American Society of Agricultural Engineers, St. Joseph, Mich., pp 209–214

Wullschleger SD, Norby RJ (2001) Sap velocity and canopy transpiration in a sweetgum stand exposed to free-air CO_2 enrichment (FACE). New Phytol 150:489–498

Wullschleger SD, Tschaplinski TJ, Norby RJ (2002) Plant water relations at elevated CO_2 – implications for water-limited environments. Plant Cell Environ 25:319–331

Yoshimoto M, Oue H, Kobayashi K (2005) Responses of energy balance, evapotranspiration, and water use efficiency of canopies to free-air CO_2 enrichment. Agric For Meteorol 133:226–246

18 Biological Nitrogen Fixation: A Key Process for the Response of Grassland Ecosystems to Elevated Atmospheric [CO$_2$]

Ueli A. Hartwig and Michael J. Sadowsky

18.1 Introduction

In addition to carbon (C) and sulphur (S), nitrogen (N) needs to be chemically reduced to enter the biosphere. In the case of N, dintrogen gas (N$_2$) is the most abundant and most chemically stable form. Since the cycling of carbon and nitrogen between their respective abiotic forms and those in the biosphere are to some extent coupled – thus each potentially limiting the flow of the other – nitrogen fixation may be considered as a „sister-process" to photosynthesis. In the long term, a persistent increase in the introduction of C into the biosphere under elevated atmospheric CO$_2$ concentration (e[CO$_2$]) can only occur if either biological (symbiotic) N$_2$ fixation increases, or N cycling changes in a way that a greater proportion of N is turned-over, assuming the ecosystem N use-efficiency does not change much (see Chapter 21).

Increasing attention has been paid to the interaction of CO$_2$ and N availability in terrestrial ecosystems. The e[CO$_2$] is likely to affect C cycling by stimulating photosynthesis and primary productivity in terrestrial ecosystems (see Chapter 14). However, primary productivity in an ecosystem may be limited by other environmental factors, such as water supply, temperature, or the availability of mineral nutrients. The availability of N is one of the key factors that limit plant growth and crop yield. If a greater CO$_2$ availability results in increased plant growth, then e[CO$_2$] ultimately leads to an increased plant demand for N (for a review, see Hartwig 1998). Thus, the extent of the response of plants to e[CO$_2$] may be limited by mineral N availability. Biological, symbiotic N$_2$ fixation is considered to be the main process whereby N is introduced into most terrestrial ecosystems. Since it is assumed that the sequestration of C and N into an ecosystem occurs simultaneously (Granhall 1981; Gifford 1992; Hartwig et al. 1996), introducing greater amounts of C into

Ecological Studies, Vol. 187
J. Nösberger, S.P. Long, R.J. Norby, M. Stitt,
G.R. Hendrey, H. Blum (Eds.)
Managed Ecosystems and CO$_2$
Case Studies, Processes, and Perspectives
© Springer-Verlag Berlin Heidelberg 2006

an ecosystem would ultimately cause a greater demand for N, thereby challenging the rate at which biological N_2 fixation can satisfy the N demand of a
respective ecosystem.

18.2 Elevated Atmospheric [CO_2] Appears Not to Affect the Activity of Symbiotic N_2 Fixation

Numerous studies examining a wide range of N_2-fixing legumes and woody
plant species have shown that there is an increase in total N_2 fixation per plant
under e[CO_2] (for a review, see Hartwig 1998). Given that the growth rate of
the plant increases under e[CO_2], an increase in the rate of N_2 fixation is not
surprising. Likewise, one would expect that N assimilation from mineral
sources increases by the same magnitude. Indeed, data from controlled environment experiments reported by Zanetti et al. (1998) and Almeida et al.
(2000) clearly show that the relative contribution of symbiotic N versus mineral N does not change as the growth rate of white clover increases under
e[CO_2].

It has to be emphasised, however, that these experiments were conducted
under conditions of a continuous supply of mineral N that was not affected by
atmospheric [CO_2]; thus in all these cases, the soil interface, providing a mineral N variable that is very difficult to predict, was lacking (see Chapter 23).
This finding was also confirmed by studies using *Alnus* (Vogel et al. 1997),
Acacia (Schortemeyer et al. 1999) and young *Robinia pseudoacacia* (Feng et
al. 2004). In the long run, nodule development and N_2 fixation showed parallel changes; and specific N_2 fixation was unaffected. However, specific N_2 fixation does appear to increase temporarily in some experiments, especially if
atmospheric [CO_2] suddenly increases while the growth of the nodules cannot keep up with the increased demand for symbiotically fixed N (see Hartwig
1998).

18.3 The Initial Response of Symbiotic N_2 Fixation to Elevated Atmospheric [CO_2] Under Field Conditions is Different From That Under Continuous Nutrient Supply

Under temperate climatic conditions and on fertile soils, such as those found
at the Swiss FACE experiment, the response of symbiotic N_2 fixation to e[CO_2]
was surprisingly different to measured responses seen in the laboratory. Initially, the percentage of fixed N derived from symbiosis in white clover
increased strikingly under both high- and low-N fertilisation conditions, both
in monoculture and in mixtures with perennial ryegrass (*Lolium perenne* L.).

Several studies confirmed that the additional fixed N came from symbiotic N_2 fixation and none from the soil or fertiliser N (Zanetti et al. 1996; Zanetti and Hartwig 1997; Zanetti et al. 1997). This result was later confirmed with alfalfa (*Medicago sativa*) grown under the same conditions (Lüscher et al. 2000) and with white clover (*Trifolium repens*) grown in fertile soil in boxes placed under e[CO_2] (Soussana and Hartwig 1996; Hartwig et al. 2002a). At the Swiss FACE site, the increase in the percentage of N from symbiosis was especially large in the first few years of the Swiss FACE experiment (Richter 2003; Richter et al., unpublished data).

In addition, N nutrition in ryegrass was shown to restrict growth (Zanetti et al. 1997) and thus in many experiments the competitive ability of the legumes in mixed communities increased under e[CO_2] (Newton et al. 1994; Hebeisen et al. 1997; Lüscher et al. 1998; Warwick et al. 1998). Together with the increased clover proportion in mixed swards, symbiotic N_2 fixation increased overall by more than 50 % on a land area basis. This matched the increase in C-fixation through photosynthesis (Ainsworth et al. 2003; Rogers et al. 1998; Isopp et al. 2000), apparently balancing the C:N ratio of the ecosystem (Hartwig et al. 1996, 2000; Soussana and Hartwig 1996).

In addition to changes in the functional response of symbiotic N_2 fixation to e[CO_2], micro-organisms also apparently respond to e[CO_2] in an indirect manner (see Chapter 23). This is presumably due to increased root growth and changes in rhizodeposition rates. For example, Schortemeyer et al. (1996) reported that e[CO_2] resulted in a 2-fold increase in populations of *Rhizobium leguminosarum* bv. *trifolii*, the symbiont of clover, specifically in the rhizosphere of white clover (*Trifolium repens* L.). Similarly, Montealegre et al. (2002) reported that e[CO_2] resulted in changes in microbial community composition in rhizosphere and bulk soil from white clover, and that there was an 85 % increase in total rhizosphere bacteria and a 170 % increase in respiring rhizosphere bacteria when assessed on a land area basis. Likewise, arbuscular mycorrhiza population changed with time, in a way to adapt to the new plant and soil situation as induced by e[CO_2] (Gamper et al. 2004, 2005; Hartwig et al. 2002b). Taken together, these studies indicate that the rhizosphere is a specific and selective habitat for microbial populations, and that bulk soil cannot be overlooked in our assessment of the impacts of e[CO_2] on long-term soil C and N pools.

One possible reason for the lack of correlation between symbiotic N_2 fixation or legume growth and e[CO_2] may be due to the limitation of another essential nutrient. Phosphorous (P) availability, a limiting factor for ecosystem productivity in large areas of the world, has long been known to inhibit both competitive ability and symbiotic N_2 fixation of leguminous plants (see Cadisch et al. 1993). Low P has previously been suggested to restrict legume response to e[CO_2] (Hartwig et al. 1996), and has been impressively shown by Niklaus et al. (1998a), Stöcklin and Körner (1999) and Almeida et al. (1999, 2000). Another issue may be competition for light: legumes have been shown to respond less to e[CO_2] than non-leguminous dicots if they experience shading by the latter as a result of relatively infrequent cutting (Teyssonneyre et al. 2002).

Under e[CO_2], symbiotic N_2 fixation presumably increases more than expected as a result of increased growth of leguminous plants. However, this is apparently true only if there are other competing N sinks that increase under e[CO_2] (see Chapter 21). If such additional or increasing N sinks are missing in the ecosystem, or if additional N sinks become progressively saturated by N, this over-proportional stimulating effect of e[CO_2] on symbiotic N_2 fixation is expected to disappear. This may explain results obtained in an alpine pasture experiment where the percentage of N derived from symbiotic N_2 fixation did not increase under e[CO_2] (Arnone 1999). Since alpine ecosystems are highly N-rich, but relatively unproductive (Arnone 1997; Jacot et al. 2000a, b), e[CO_2] may not have greatly affected the availability of mineral N; thus symbiotic N_2 fixation was not challenged.

18.4 What Are the Possible Reasons For the Differential Responses of Symbiotic N_2 Fixation to Elevated Atmospheric [CO_2] in Laboratory and Field Experiments?

To a large extent, the percentage of N from symbioses clearly depends on the availability of soil N (for a review, see Hartwig 1998; Fig. 18.1). Thus, any processes in the ecosystem which lead to a reduction in the availability of mineral N in the soil – at least in relation to plant growth – are ultimately expected to lead to an increased percentage of N derived from symbioses (see above). Therefore it is assumed that, under field conditions and under e[CO_2], the availability of soil N is affected indirectly (Hartwig et al. 1996). This was proposed by Diaz et al. (1993) and is referred to as a nutrient feedback mechanism. With respect to N availability in the soil, an apparent lack of soil mineral N was indicated by several parameters during the first 3 years of the Swiss FACE experiment. The N nutrition of plants did not increase to the same extent as expected due to the higher plant growth rate under e[CO_2]. There was a strong increase in the apparent root:shoot ratio of perennial ryegrass (Jongen et al. 1995; Hebeisen et al. 1997; Hartwig et al. 2000, 2002a), a strong reduction in the N concentration of above-ground tissue of perennial ryegrass (Zanetti et al. 1997) and a lower index of N nutrition in perennial ryegrass (Zanetti et al. 1997), but no change in the concentration of mineral N, or in the rate of N mineralization in the soil under e[CO_2] (Gloser et al. 2000; Richter et al. 2003; Sowerby et al. 2000). Similar results were obtained by Soussana et al. (1996) and Casella et al. (1996). Another significant factor may be denitrification; it was indeed shown that e[CO_2] may lead to increased rates of denitrification (Baggs et al. 2003a, b; Ineson et al. 1998). However, this issue is

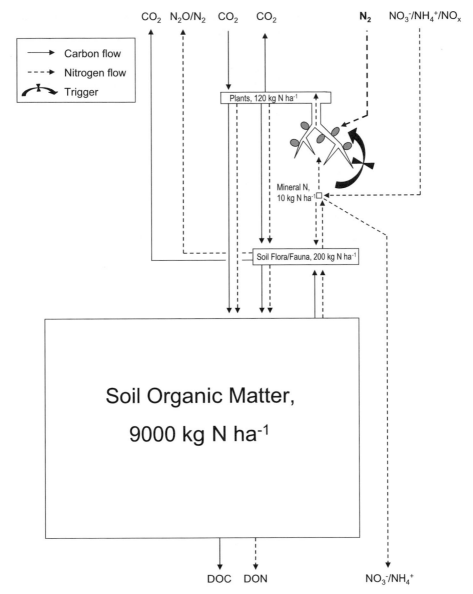

Fig. 18.1 Scheme of N pools and C/N movements in terrestrial ecosystems. The numbers reflect orders of magnitude changes that may be valid for a highly productive grassland ecosystem, such as those in temperate climatic conditions. *DOC* Dissolved organic carbon, *DON* dissolved organic nitrogen

still uncertain; and a recent study showed that e[CO_2] may have little effect on nitrifying and denitrifying enzyme activity in various European grassland soils (Barnard et al. 2004).

These data suggest that, during the initial phase of the Swiss FACE experiment, processes and N pools in the soil were affected by the increased C introduction into the ecosystem (see Chapter 21) in a manner such that mineral N was apparently insufficient for optimal plant growth.

18.5 The Time Component, While Often Suggested, Is Now Evident in the 10-Year Swiss FACE Experiment

Very few experiments examining the influence of e[CO_2] on ecosystem structure and function are carried out for long time periods. Data from the Swiss FACE experiment indicate a progressive saturation of the initial, apparently increasing, N sinks under e[CO_2]. That is, the e[CO_2]-induced increase in the percentage of N assimilated from symbiotic N_2 fixation (%N_{sym}) in white clover appears to decrease with time, but notably only under a high-N fertiliser input regime (Richter 2003; Richter et al., unpublished data; see Chapter 21). This is also consistent with the progressively increasing CO_2 response of *Lolium perenne* after several years in the high fertiliser treatment of the same experiment (Daepp et al. 2000, 2001; Schneider et al. 2004). In contrast, during the entire 10 years under low-N fertiliser input, the responses on N_2 fixation and plant growth remained more or less stable (Daepp et al. 2001; Richter 2003; Richter et al., unpublished data). These data indicate that, in a very productive grassland ecosystem, a continuous N supply by symbiotic N_2 fixation may result in a new steady-state of the N economy under e[CO_2], especially on fertile soil under favourable climatic conditions and with a very high rate of N fertilisation. This change in steady state, however, can take decades under less productive conditions. Given the slow increase in the global atmospheric [CO_2], N cycling in terrestrial ecosystems may have been able to adapt continuously to the gradually increasing atmospheric [CO_2].

These observations at the plant functional level are confirmed by measurements of selected soil processes. Schneider et al. (2004) showed that N supply from mineralization was increased towards the late stage of the Swiss FACE experiment, thereby also promoting growth of non-leguminous plants under e[CO_2]. The apparent significance of symbiotic N_2 fixation, thereby, is supported by the finding that the retention of new N in the soil after several years was increased under e[CO_2] in *Trifolium repens* swards (Hartwig et al. 2002a; Van Groeningen et al. 2003).

The same sequence of events may be shown in the composition of populations of soil micro-organisms involved in N_2 fixation. For example, Montealegre et al. (2000) reported that, in the early periods of the Swiss FACE experiment, e[CO_2] altered the population structure and competitive ability of *Rhizobium leguminosarum* bv. *trifolii* isolates relative to those rhizobia recovered from plants grown under ambient conditions. However, by year 10 of the FACE experiment, the population structure of rhizobia under e[CO_2] was similar to that under control conditions (Stoeber et al., unpublished data). This suggested that the CO_2-induced change to population structure was transient in nature and was most likely influenced by both the N status of the ecosystem as well as the type and concentration of root exudates released by clover plants (see Chapter 23).

18.6 The Significance of Symbiotic N_2 Fixation Under Elevated Atmospheric [CO_2] in Terrestrial Ecosystems: An Attempt to Reach a General Conclusion

A key factor regulating symbiotic N_2 fixation is the availability of mineral N and the legume's N sink (Fig. 18.1). If the production of mineral N is inhibited, or the N is immobilized, then denitrification or leaching is intensified, less mineral N is available for plant growth and, thus, the percentage of N from symbiosis increases under e[CO_2] (Hartwig 1998). This is in fact a very sensitive bioassay for mineral N in the soil. One difficulty in predicting N cycling in terrestrial ecosystems is related to the reduced consumption of water by plants under e[CO_2] (Jackson et al. 1994; Casella et al. 1996; Niklaus et al. 1998b). Under relatively dry conditions, N mineralization may be constrained while, under moderately humid conditions, N mineralization may run at a higher rate and, under rather humid conditions, denitrification may be stimulated, as reflected by the seasonal variability reported for nitrate reductase activity (Deiglmayr et al. 2004).

Another factor is H_2 evolution, a process that is ultimately linked to biological N_2 fixation in systems lacking the Hup+ phenotype (usually the case in temperate symbiotic systems). Since several soil bacteria can use H_2 as a substrate, increased symbiotic N_2 fixation may also lead to a stimulation in the activity of selective microbial populations and, thus, to changes in N and C cycling (Dong and Layzell 2002).

Based on the above, we can predict that, if e[CO_2] leads to an increased N sink anywhere in a ecosystem (see Chapter 21), then symbiotic N_2 fixation will increase. As a result, symbiotic N_2 fixation, as the "sister process" of photosynthesis, will maintain a balanced C:N ratio in the ecosystem. This has also been suggested from a modelling study (Thornley and Cannell 2000). In a long-term experiment like in the 10-year Swiss FACE experiment, on a highly productive grassland ecosystem, an initial striking increase in symbiotic N_2 fixation and a change in the community composition of appropriate symbiotic bacteria may be expected. With progressing experimental duration, however, a readjustment of symbiotic N_2 fixation and an apparent reconstitution of the composition of the appropriate symbiotic bacteria may occur. All these events are associated with an apparent adjustment of N cycling in the system. Finally, the main CO_2-increased N sink appears to be plants themselves (Richter et al. 2003), while the soil processes appear to readjust with time. However, depending on their nature (e.g. magnitude of net primary production, sequestration of organic matter, denitrification, net productivity, mineral nutrition, etc.), such an observed e[CO_2]-induced increase and subsequent decrease in the apparent relative N limitation associated with changes in N_2 fixation (and possibly in the composition of the involved microbial populations) may occur either faster or slower in other ecosystems (Thornley and Cannell 1997).

18.7 Conclusion

Data from our 10 years of studies done at the Swiss FACE site, along with numerous process studies done by ourselves and by others, clarifies the role that biological (symbiotic) N_2 fixation plays in a CO_2-rich world.

- Under e[CO_2], symbiotic N_2 fixation increases as a result of increased plant growth (N demand) and not due to direct CO_2 stimulation leading to greater photosynthate availability.
- Under fertile soil conditions, e[CO_2] apparently caused changes in soil processes and nitrogen status; and, as a result, increases in total symbiotic N_2 fixation (N sink-driven) and changes in the population structure of N_2-fixing soil micro-organisms were detected.
- Under fertile soil conditions and ample water availability, e[CO_2] apparently caused a pronounced N limitation in the initial periods of CO_2 enhancement. However, within a few years, a new N balance was apparently reached (progressive N saturation); but only under conditions of high-N input and not under a low-N input. This was accompanied by a readjustment of symbiotic N_2 fixation capacity in legumes and by further shifts in the population of N_2-fixing soil micro-organisms to a structure seen previous to CO_2 enrichment.
- The integrated nature and interdependence of photosynthesis and biological (symbiotic) N_2 fixation was confirmed from the Swiss FACE experiment.

Acknowledgements.   Our studies were supported by the Swiss National Energy Fund, ETH Zurich, Swiss Department of Science and Education (COST), Federal Office of Agriculture and Federal Office of Energy.

References

Ainsworth EA, Rogers A, Blum H, Nösberger J, Long SP (2003) Variation in acclimation of photosynthesis in *Trifolium repens* after eight years of exposure to free air CO_2 enrichment (FACE). J Exp Bot 54:2769–2774

Almeida JFP, Lüscher A, Frehner M, Oberson A, Nösberger J (1999) Partitioning of P and the activity of root acid phosphatase in white clover (*Trifolium repens* L) are modified by increased atmospheric CO_2 and P fertilisation. Plant Soil 210:159–166

Almeida JPF, Hartwig UA, Frehner M, Nösberger J, Lüscher A (2000) Evidence that P deficiency induces N feedback regulation of symbiotic N_2 fixation in white clover (*Trifolium repens* L). J Exp Bot 51:1289–1297

Arnone JA III (1997) Indices of plant N availability in an alpine grassland under elevated CO_2. Plant Soil 190:61–66

Arnone JA III (1999) Symbiotic N_2 fixation in a high Alpine grassland: effects of four growing seasons of elevated CO_2. Funct Ecol 13:383–387

Baggs EM, Richter M, Cadisch G, Hartwig UA (2003a) Denitrification in grass swards is increased under elevated atmospheric CO_2. Soil Biol Biochem 35:729–732

Baggs EM, Richter M, Hartwig UA, Cadisch G (2003b) Nitrous oxide emissions from grass swards during eight years of elevated atmospheric pCO_2 (Swiss FACE). Global Change Biol 9:1214–1222

Barnard R, Barthes L, Le Roux X, Harmens H, Raschi A, Soussana J-F, Winkler B, Leadley PW (2004) Atmospheric CO_2 elevation has little effect on nitrifying and denitrifying activity in four European grasslands. Global Change Biol 10:488–497

Cadisch G, Sylvester-Bradley R, Boller BC, Nösberger J (1993) Effects of phosphorus and potassium on N_2 fixation (^{15}N-dilution) on field-grown *Centrosema acutifolium* and *C. macrocarpum*. Field Crop Res 31:329–340

Casella E, Soussana J-F, Loiseau P (1996) Long-term effects of CO_2 enrichment and temperature increase on temperate grass sward. Plant Soil 182:83–99

Daepp M, Suter D, Almeida JPF, Isopp H, Hartwig UA, Frehner M, Blum H, Nösberger J, Lüscher A (2000) Yield response of *Lolium perenne* swards to free air CO_2 enrichment increased over six years in a high-N input system. Global Change Biol 6:805–816

Daepp M, Nösberger J, Lüscher A (2001) Nitrogen fertilisation and development stage affect the response of yield, biomass partitioning and morphology of *Lolium perenne* L swards to elevated pCO_2. New Phytol 150:347–358

Deiglmayr K, Philippot L, Hartwig UA, Kandeler E (2004) Structure and activity of the nitrate reducing community in the rhizosphere of *Lolium perenne* and *Trifolium repens* under long-term elevated atmospheric pCO_2. FEMS Microbiol Ecol 49:445–454

Diaz S, Grime JP, Harris J, McPherson E (1993) Evidence of a feedback mechanism limiting plant response to elevated carbon dioxide. Nature 364:616–617

Dong Z, Layzell DB (2002) H_2 oxidation, O_2 uptake and CO_2 fixation in hydrogen treated soils. Plant Soil 229:1–12

Feng Z, Cyckmans J, Flessa H (2004) Effects of elevated carbon dioxide concentration on growth and N_2 fixation of young *Rhobinia pseudoacacia*. Tree Physiol 24:323–330

Gamper H, Peter M, Jansa J, Lüscher A, Hartwig UA, Leuchtmann A (2004) Arbuscular micorrhizal fungi benefit from seven years of free air CO_2 enrichment in well-fertilized grass and legume monocultures. Global Change Biol 10:189–199

Gamper H, Hartwig UA, Leuchtmann A (2005) Mycorrhizas improve nitrogen nutrition of *Trifolium repens* after 8 yr of selection under elevated atmospheric CO_2 partial pressure. New Phytol (in press)

Gifford RM (1992) Interaction of carbon dioxide with growth-limiting environmental factors in vegetation productivity, implications for the global carbon cycle. In: Desjardins RL, Gifford RM, Nilson T, and Greenwood EAN (eds) Advances in bioclimatology, vol 1. Springer, Berlin Heidelberg New York, pp 24–58

Gloser V, Jezikova M, Lüscher A, Frehner M, Blum H, Nösberger J, Hartwig UA (2000) Soil mineral nitrogen availability was unaffected by elevated atmospheric pCO_2 in a four years old field experiment (Swiss FACE). Plant Soil 227:291–299

Granhall U (1981) Biological nitrogen fixation in relation to environmental factors and functioning of natural ecosystems. In: Clark FE, Rosswall T (eds) Terrestrial nitrogen cycles, processes, ecosystem strategies and management impacts, vol 33. SCOPE Ecological Bulletins, Stockholm, pp 131–144

Hartwig UA (1998) The regulation of symbiotic N_2 fixation: a conceptual model of N feedback from the ecosystem to the gene expression level. Perspect Plant Ecol Evol Syst 1:92–120

Hartwig UA, Zanetti S, Hebeisen T, Lüscher A, Frehner M, Fischer B, Van Kessel C, Hendrey GR, Blum H, Nösberger J (1996) Symbiotic nitrogen fixation: one key to understand the response of temperate grassland ecosystems to elevated CO_2? In: Körner C, Bazazz F (eds) Carbon dioxide, populations, and communities. Academic Press, San Diego, pp 253–264

Hartwig UA, Lüscher A, Daepp M, Blum H, Soussana J-F, Nösberger J (2000) Due to symbiotic N_2 fixation, five years of elevated atmospheric pCO_2 had no effect on litter N concentration in a fertile grassland ecosystem. Plant Soil 224:43–50

Hartwig UA, Lüscher A, Nösberger J, Kessel C van (2002a) Nitrogen-15 budget in model ecosystems of white clover and perennial ryegrass exposed for four years at elevated atmospheric pCO_2. Global Change Biol 8:194–202

Hartwig UA, Wittmann P, Braun R, Hartwig-Räz B, Jansa J, Mozafar A, Lüscher A, Leuchtmann A, Frossard E, Nösberger J (2002b) Arbuscular mycorrhiza infection enhances the growth response of *Lolium perenne* to elevated atmospheric pCO_2. J Exp Bot 53:1207–1213

Hebeisen T, Lüscher A, Zanetti S, Fischer BU, Hartwig UA, Frehner M, Hendrey GR, Blum H, Nösberger J (1997) Growth response of *Trifolium repens* L. and *Lolium perenne* L. as monocultures and bi-species mixture to free air CO_2 enrichment and management. Global Change Biol 3:149–160

Ineson P, Coward PA, Hartwig UA (1998) Soil gas fluxes of N_2O, CH_4 and CO_2 beneath *Lolium perenne* under elevated CO_2: the Swiss free air carbon dioxide enrichment experiment. Plant Soil 198:89–95

Isopp H, Frehner M, Long SP, Nösberger J (2000) Sucrose-phosphate synthase responds differently to source–sink relations and to photosynthetic rates: *Lolium perenne* L growth at elevated pCO_2 in the field. Plant Cell Environ 23:597–607

Jackson RB, Sala OE, Field CB, Mooney HA (1994) CO_2 alters water use, carbon gain, and yield for the dominant species in a natural grassland. Oecologia 98:257–262

Jacot KA, Lüscher A, Nösberger J, Hartwig UA (2000a) Symbiotic N_2 fixation of various legume species along an altitudinal gradient in the Swiss Alps. Soil Biol Biochem 32:1043–1052

Jacot KA, Lüscher A, Nösberger J, Hartwig UA (2000b) The relative contribution of symbiotic N_2 fixation and other nitrogen sources to grassland ecosystems along an altitudinal gradient in the Alps. Plant Soil 225:201–211

Jongen MB, Jones MB, Hebeisen T, Blum H, Hendrey GR (1995) The effects of elevated CO_2 concentrations on the root growth of *Lolium perenne* and *Trifolium repens* grown in a FACE system. Global Change Biol 1:361–372

Lüscher A, Hendrey GR, Nösberger J (1998) Long term responsiveness to free air CO_2 enrichment of functional types, species and genotypes of perennial grassland. Oecologia 113:37–45

Lüscher A, Hartwig UA, Suter D, Nösberger J (2000) Direct evidence that in fertile grassland symbiotic N_2 fixation is an important trait for a strong response of plants to elevated atmospheric CO_2. Global Change Biol 6:655–662

Montealegre CM, Kessel C van, Blumenthal JM, Hur H-G, Hartwig UA, Sadowsky MJ (2000) Elevated atmospheric CO_2 alters microbial population structure in a pasture ecosystem. Global Change Biol 6:475–482

Montealegre CM, Kessel C van, Russelle MP, Sadowsky MJ (2002) Changes in microbial activity and composition in a pasture ecosystem exposed to elevated atmospheric carbon dioxide. Plant Soil 243:197–207

Newton PCD, Clark H, Bell CC, Glasgow EM, Campbell BD (1994) Effects of elevated CO_2 and simulated seasonal-changes in temperature on the species composition and growth-rate in pasture turves. Ann Bot 73:53–59

Niklaus PA, Leadley PW, Stöcklin J, Körner C (1998a) Nutrient relations in calcareous grassland under elevated CO_2. Oecologia 116:67–75

Niklaus PA, Spinnler D, Körner C (1998b) Soil moisture dynamics of calcareous grassland under elevated CO_2. Oecologia 111:201–208

Richter M (2003) Influence of elevated atmospheric CO_2 concentration on symbiotic N_2 fixation and availability of nitrogen in grassland ecosystems. PhD thesis 15185, ETH, Zurich, 125 pp

Richter M, Hartwig UA, Frossard E, Cadisch G (2003) Gross fluxes of nitrogen in grassland soil exposed to elevated pCO_2 for seven years. Soil Biol Biochem 35:1325–1335

Rogers A, Fischer BU, Bryant J, Frehner M, Blum H, Raines CA, Long SP (1998) Acclimation of photosynthesis to elevated CO_2 under low-nitrogen nutrition is affected by the capacity for assimilate utilization perennial ryegrass under free-air CO_2 enrichment. Plant Physiol 118:683–689

Schneider MK, Lüscher A, Richter M, Aeschlimann U, Hartwig UA, Blum H, Frossard E, Nösberger J (2004) Ten years of free-air CO_2 enrichment altered the mobilisation of N from soil in *Lolium perenne* L swards. Global Change Biol 10:1377–1388

Schortemeyer M, Hartwig UA, Hendrey GR, Sadowsky M (1996) Microbial community changes in the rhizospheres of white clover and perennial ryegrass exposed to free-air carbon dioxide enrichment (FACE). Soil Biol Biochem 28:1717–1724

Schortemeyer M, Atkin OK, McFarlane N, Evans JR (1999) The impact of elevated atmospheric CO_2 and nitrate supply on growth, biomass allocation, nitrogen partitioning and N_2 fixation of *Acacia melanoxylon*. Aust J Plant Physiol 26:737–747

Soussana JF, Hartwig UA (1996) The effects of elevated CO_2 on symbiotic N_2 fixation: a link between the carbon and nitrogen cycles in grassland ecosystems. Plant Soil 187:321–332

Soussana JF, Casella E, Loiseau P (1996) Long-term effects of CO_2 enrichment and temperature increase on a temperate grass sward II. Plant nitrogen budgets and root fraction. Plant Soil 182:101–114

Sowerby A, Blum H, Gray TRG, Ball AS (2000) The decomposition of *Lolium perenne* in soils exposed to elevated CO_2: comparisons of mass loss of litter with soil respiration and soil microbial biomass. Soil Biol Biochem 32:1359–1366

Stöcklin J, Körner C (1999) Interactive effects of elevated CO_2, P availability and legume presence on calcareous grassland: results of a glasshouse experiment. Funct Ecol 13:200–209

Teyssonneyre F, Picon-Cochard C, Falcimagne R, Soussana J-F (2002) Effects of elevated CO_2 and cutting frequency on plant community structure in a temperate grassland. Global Change Biol 8:1034–1046

Thornley JHM, Cannell MGR (1997) Temperate grassland response to climate change: an analysis using the Hurley pasture model. Ann Bot 80:205–221

Thornley JHM, Cannell MGR (2000) Dynamics of mineral N availability in grassland ecosystems under increased [CO_2]: hypothesis evaluated using the Hurley pasture model. Plant Soil 224:153–170

Van Groeningen KJ, Six J, Harris D, Blum H, Kessel C van (2003) Soil C-13-N-15 dynamics in a N_2-fixing clover system under long exposure to elevated atmospheric CO_2. Global Change Biol 9:1751–1762

Vogel CS, Curtis PS, Thomas RB (1997) Growth and nitrogen accretion of dinitrogen-fixing *Alnus glutinosa* (L) Gaertn. under elevated carbon dioxide. Plant Ecol 130:63–70

Warwick KR, Taylor G, Blum H (1998) Biomass and compositional changes occur in chalk grassland turves exposed to elevated CO_2 for two seasons in FACE. Global Change Biol 4:375–385

Zanetti S, Hartwig UA (1997) Symbiotic N_2 fixation increases under elevated atmospheric pCO_2 in the field. Acta Oecol 18:285–290

Zanetti S, Hartwig U, Lüscher A, Lüscher A, Hebeisen T, Frehner M, Fischer BU, Hendrey GR, Blum H, Nösberger J (1996) Stimulation of symbiotic N_2 fixation in *Trifolium repens* (L) under elevated atmospheric pCO_2 in a grassland ecosystem. Plant Physiol 112:575–583

Zanetti S, Hartwig UA, Kessel C van, Lüscher A, Hebeisen T, Frehner M, Fischer BU, Hendrey GR, Blum H, Nösberger J (1997) Does nitrogen nutrition restrict the CO_2 response of fertile grassland lacking legumes? Oecologia 112:17–25

Zanetti S, Hartwig UA, Nösberger J (1998) Elevated atmospheric CO_2 does not affect per se the preference for symbiotic nitrogen as opposed to mineral nitrogen of *Trifolium repens* L. Plant Cell Environ 21:623–630

19 Effects of Elevated [CO$_2$] and N Fertilization on Interspecific Interactions in Temperate Grassland Model Ecosystems

A. Lüscher and U. Aeschlimann

19.1 Introduction

Grasslands cover 24 % of the terrestrial surface (Sims and Risser 2000) and play an important role in the global carbon cycle. The response of these ecosystems to e[CO$_2$] is therefore of major ecological importance. In humid, temperate regions, agricultural grasslands are often sown and are typically dominated by perennial ryegrass (*Lolium perenne*) and white clover (*Trifolium repens*).

In monocultures, the responses of *L. perenne* to long-term e[CO$_2$] and N fertilization are well documented (Schneider et al. 2004). Elevated [CO$_2$] increases the C:N ratio in the ecosystem, which increases the demand for N in order to maintain greater plant growth. In non-P-limited grasslands containing legumes, the increased N demand at e[CO$_2$] may be met by symbiotic N$_2$ fixation (Zanetti et al. 1996). Biomass production and the species-specific characteristics of resource acquisition were affected by changes in [CO$_2$] and N supply. In *L. perenne*, e[CO$_2$] led to lower canopy height and an increased root biomass (Daepp, Nösberger, & Lüscher 2001), whereas in *T. repens* e[CO$_2$] stimulated stolon branching and reduced the specific leaf area (Ryle and Powell 1992). The response of biomass production to e[CO$_2$] differed between legume and grass species (Lüscher, Hendrey, & Nösberger 1998) and symbiotic N$_2$ fixation was a crucial plant character for this (Lüscher et al. 2000). The altered availability of growth resources is also very likely to influence the benefits and costs of interactions between plant species grown in mixtures. Compared to monocultures, much less is known about the effects of changes in [CO$_2$] and the availability of N when species are grown in mixtures. Mixed swards are inherently more complex than monocultures and the outcome of changes in resource availability is unpredictable. Even without experimental modification of resource availability, clover proportions in mixed swards fluc-

Ecological Studies, Vol. 187
J. Nösberger, S.P. Long, R.J. Norby, M. Stitt,
G.R. Hendrey, H. Blum (Eds.)
Managed Ecosystems and CO$_2$
Case Studies, Processes, and Perspectives
© Springer-Verlag Berlin Heidelberg 2006

tuate widely over time (Kessler and Nösberger 1994; Lüscher at al. 2005). In addition, clover growth and distribution are typically patchy and fluctuate irregularly (Cain et al. 1995). Since growth in a mixture allows interactions between plant species, the results obtained are of high ecological relevance. Depending on the degree of niche separation, the interactions between the species in a mixture can lead either to interspecific competition or to beneficial effects of resource complementarity and facilitation. The benefits or costs of interspecific interactions can be represented by the relative yield, which is defined as the ratio of the yield of a species in a mixture to the yield of the same species in monoculture.

A key element of species dynamics in a grass–legume system is a nitrogen-based trade-off between grass and clover (Schwinning and Parsons 1996). Because clover has the ability to fix atmospheric N_2, it has a competitive advantage when the availability of N in the soil is low. Through the release of N from the clover to the soil, this additional fixed N may become available to the grass, which is the stronger competitor at high levels of soil N. A dynamic equilibrium between competition and facilitation is established between the two species. The changes in the competitive ability of legumes and grasses due to variation in $[CO_2]$ and available N are likely to influence the plant species composition of a sward.

Since *L. perenne* and *T. repens* differ strongly in their resource acquisition characteristics, we expected a higher yield in the mixture than calculated from the monocultures, due to resource complementarity, i.e. a relative yield total >1. This beneficial effect is likely to be affected by $[CO_2]$ and the availability of N. Based on the results obtained from monocultures, we hypothesized that in a mixture with *L. perenne*, *T. repens* will benefit more from $e[CO_2]$ and a limited availability of mineral N.

The Swiss FACE experiment offered a good opportunity to study the long-term effects of elevated atmospheric $[CO_2]$ and varying N application on the interspecific interactions in a managed grassland system maintained as a bi-species mixture of a grass (*L. perenne*) and a legume (*T. repens*). The acting, underlying mechanisms are more readily detected in this simplified bi-species system than in a multi-species community.

19.2 Materials and Methods

19.2.1 Experimental Site

Data on species composition of swards and of interactions between species are presented from the Swiss FACE experiment. The Swiss FACE is located at Eschikon (8°41' E, 47°27' N) near Zurich, at an altitude of 550 m above sea level. The long-term effects of atmospheric [CO$_2$] and N fertilization on fertile grassland ecosystems (pure swards and bi-species mixtures) were examined in the field. The experiment was arranged in three blocks, each consisting of two circular areas (18 m diameter), one CO$_2$-enriched (600 ppm [CO$_2$]) and the other an ambient control (360 ppm [CO$_2$]). CO$_2$ fumigation began in May 1993 and lasted each year for the whole growing season (March to November) during daylight hours. A more detailed description of the FACE setup is given in Section 8.2.

19.2.2 Experimental Treatments

L. perenne cv. Bastion and *T. repens* cv. Milkanova were sown in 5.3 m^2 plots in 1992. The swards were maintained either as monocultures or as mixtures of the two species. The sowing rates for *T. repens* and *L. perenne*, respectively, were 0.8 g m^{-2} and 3.2 g m^{-2} in the monocultures and were 0.4 g m^{-2} and 1.6 g m^{-2} in the mixtures (replacement design). There are different experimental designs for examining interactions between plant species. Simple designs, such as the additive or the replacement design, have limitations (Connolly 1997; Snaydon and Satorre 1989) and new designs have been proposed (Connolly and Wayne 2005; Ramseier et al. 2005). However, certain restrictions for replacement designs are not relevant for the data presented here. The reasons for this are:
1. There was not one final harvest, as the experiment was carried out over ten years, with five harvests each year. Therefore, changes in the yield of each species over time can be addressed.
2. The relative proportion of the species within the mixed swards varied greatly over time (e.g. between 10 % and 85 % of *T. repens* within one year; see Lüscher et al. 2005). Consequently, the initial density of these permanent species with their high potential for vegetative multiplication and/or self-thinning was not decisive for the outcome of the mixture composition.
3. All the swards, with varying N and [CO$_2$] treatments, started with the same proportion of each species. Thus, the differences between treatments discussed here are not due to differences in initial species density or proportion.

Since 1993, the swards were cut five times each year at a height of 5 cm and the harvested material removed. All plots were fertilized with 5.5 g P m^{-2} and 24.1 g K m^{-2} each year, to ensure that these nutrients were non-limiting for plant growth (Daepp et al. 2000). The P content of the leaves of *T. repens* proved that P availability was non-limiting for plant growth (Lüscher et al. 2004). To examine the effects of N availability on the response of the ecosystem to e[CO_2], two levels of N fertilization (14 g N m^{-2} year^{-1} and 56 g N m^{-2} year^{-1}) were applied. The N fertilizer was applied as liquid NH_4NO_3 at the beginning of each regrowth period. The amount applied was divided between the five successive regrowth periods in the proportion 30, 20, 20, 15 and 15 %, with these percentages corresponding to the expected yields at the end of each regrowth period. Each treatment (combinations of sward type × [CO_2] × N fertilization) was repeated twice within each block.

19.2.3 Data Collection and Statistical Analysis

For biomass determination at harvest, the cut plant material was oven-dried at 65 °C for 48 h prior to weighing. A subsample was ground into powder and analyzed for total N content. Relative yield (RY) was calculated for each growing season as the annual crop yield of a species in mixture divided by the annual crop yield of the species in monoculture. Relative yield total (RYT) of the mixture was obtained by adding the RY of both species. Relative N yield (RNY) and relative N yield total (RNYT) were calculated analogously based on the harvested N yields (crop yield multiplied by the %N content of the biomass). The statistical analyses were carried out using the GLM procedure (SAS ver. 8.02; SAS Institute 1999). The model was a split-plot with [CO_2] as the main plot factor.

19.3 Results

19.3.1 Proportion of *T. repens* in Mixture

Elevated [CO_2] increased the proportion of *T. repens* in the mixed sward by more than 30 % at both levels of N supply (Table 19.1; CO_2 $P<0.05$). Compared to low N supply, high N supply decreased the proportion of *T. repens* in the mixture by more than 40 % (Table 19.1; N $P<0.01$), irrespective of [CO_2] (N × CO_2 n.s.).

19.3.2 Biomass and Nitrogen Yield

The sward type had a significant effect on the biomass and N yield derived from *T. repens*, being lower in the mixture than in monoculture (Table 19.1). In contrast, the crop yield of *L. perenne* did not significantly differ between the mixture and monoculture (Table 19.1; sward × species $P<0.0001$; Table 19.2).

Table 19.1 Annual mean values of proportion of clover in mixtures, crop yield, nitrogen yield, relative yield and relative nitrogen yield

| | 14 g N m^{-2} year^{-1} | | 56 g N m^{-2} year^{-1} | | |
	360 ppm CO$_2$	600 ppm CO$_2$	360 ppm CO$_2$	600 ppm CO$_2$	Standard error
Proportion of *T. repens* (%)					
Mixtures	29	39	15	22	3
Crop yield (g m^{-2} year^{-1})					
L. perenne monoculture	717	776	1241	1462	19
T. repens monoculture	830	989	830	990	19
L. perenne mixture	770	709	1142	1199	19
T. repens mixture	309	515	182	295	19
Total mixture	1079	1224	1324	1494	
Nitrogen yield (g m^{-2} year^{-1})					
L. perenne monoculture	14.5	12.1	38.4	38.1	0.6
T. repens monoculture	34.1	38.1	34.8	38.0	0.6
L. perenne mixture	17.9	17.4	34.6	33.7	0.6
T. repens mixture	12.5	19.0	7.7	11.4	0.6
Total mixture	30.4	36.4	42.3	45.1	–
Relative crop yield					
L. perenne	1.06	1.09	0.92	0.79	0.03
T. repens	0.37	0.51	0.24	0.30	0.03
Total	1.43	1.60	1.16	1.09	
Relative nitrogen yield					
L. perenne	1.23	1.43	0.91	0.85	0.03
T. repens	0.37	0.49	0.24	0.29	0.03
Total	1.60	1.92	1.15	1.14	

Table 19.2 Analysis of variance for the species' annual biomass and nitrogen yield

| | | Crop yield | | Nitrogen yield | |
Source	DF	MS	Probability	MS	Probability
Mainplot					
CO$_2$	1	314350	0.003	67	n.s.
Error A	4	7693	–	18	–
Subplot					
Nitrogen	1	1114835	0.0001	1885	0.0001
Sward	1	2498779	0.0001	3290	0.0001
Species	1	3546986	0.0001	47	0.08
CO$_2$ × nitrogen	1	13578	n.s.	2	n.s.
CO$_2$ × sward	1	8398	n.s.	7	n.s.
CO$_2$ × species	1	49293	n.s.	172	0.001
Nitrogen × sward	1	258955	0.0004	337	0.0001
Nitrogen × species	1	2195802	0.0001	3374	0.0001
Sward × species	1	1646798	0.0001	3375	0.0001
Error B	81	19115	–	15	–

Overall, crop yield in both species was increased at e[CO_2], when compared to c[CO_2] (Table 19.1; CO_2 $P<0.003$; CO_2 × species n.s.; Table 19.2). This effect was not significantly different between levels of N supply and sward type (N × CO_2 n.s., sward × CO_2 n.s.; Table 19.2). The effect of e[CO_2] on N yield was positive in *T. repens* but negative in *L. perenne* (Table 19.1; CO_2 × species $P<0.001$; Table 19.2).

At high N supply as compared to low N, the biomass and N yield of *L. perenne* was increased in both monoculture and the mixture (Table 19.1). The high N supply reduced the biomass and N yield of *T. repens* in the mixture, whereas in the monoculture there was no significant effect of N supply on biomass and N yield (Table 19.1; N × species $P<0.0001$, N × sward $P<0.0001$; Table 19.2).

19.3.3 Relative Yield of Biomass and Nitrogen

In both species, the RY of biomass was not affected by [CO_2] (Table 19.3). e[CO_2] increased the RYT at low N supply, but reduced it at high N supply (Table 19.1; CO_2 × N $P<0.01$; Table 19.3). The high N supply reduced the RY and RYT in both species, when compared to low N (Table 19.3; N $P<0.0001$).

At e[CO_2], there was an increase in RNY of *T. repens* at both N levels, and of *L. perenne* at low N supply. There was a decrease in RNY of *L. perenne* at high N supply (Table 19.1; CO_2 $P<0.003$, CO_2 × N $P<0.003$; Table 19.3). This resulted in an increase in RNYT at e[CO_2] only at the low N supply (Table 19.1; CO_2

Table 19.3 Analysis of variance for relative yield and relative yield total of biomass and nitrogen

Source	DF	Relative crop yield MS	Proba-bility	Relative nitro-gen yield MS	Proba-bility	DF	Relative crop yield total MS	Proba-bility	Relative nitro-gen yield total MS	Proba-bility
Mainplot
CO_2	1	0.008	n.s.	0.076	0.003	1	0.014	0.05	0.137	0.003
Error A	4	0.001	–	0.002	–	4	0.002	–	0.003	–
Subplot
Nitrogen	1	0.463	0.0001	1.107	0.0001	1	0.836	0.0001	1.990	0.0001
Species	1	4.498	0.0001	7.043	0.0001	0	–	–	–	–
CO_2 × nitrogen	1	0.005	n.s.	0.239	0.003	1	0.070	0.01	0.137	0.001
CO_2 × species	1	0.040	n.s.	0.077	0.08	0	–	–	–	–
Nitrogen × species	1	0.063	n.s.	0.001	n.s.	0	–	–	–	–
Error B	37	0.023	–	0.024	–	16	0.008	–	0.009	–

$P<0.003$, CO$_2$ × N $P<0.001$; Table 19.3). Compared to low N, the high N supply reduced the RNY of both species and RNYT (Table 19.1; N $P<0.0001$, N × species n.s.; Table 19.3).

19.4 Discussion

19.4.1 Interspecific Differences in the Response to e[CO$_2$] Were Augmented in the Mixed Community When Compared to the Pure Sward

The interspecific differences in the response to e[CO$_2$] between *L. perenne* and *T. repens* observed in monocultures (Table 19.1; Hebeisen et al. 1997) lead to the hypothesis that the interactions among these species in a mixed plant community will be altered when the atmospheric [CO$_2$] increases. In this study, we found that interspecific differences in the yield response of *T. repens* and *L. perenne* to e[CO$_2$] were augmented in bi-species mixtures when compared to the pure swards (Table 19.1), thereby indicating changes in the species' interactions. The yield response of *L. perenne* to e[CO$_2$] in the mixture was smaller (−2 %), when compared to the pure sward (+13 %). With *T. repens*, the response was clearly stronger in the bi-species mixture (+65 %) than in the pure sward (+19 %) (Table 19.1; Hebeisen et al. 1997). Similarly, the yield response of 14 different genotypes of *T. repens* grown in established *L. perenne* swards to e[CO$_2$] reached +99 % when averaged over the three years of the experiment (Lüscher et al. 1998). Consequently, the increase in total yield of the mixture at e[CO$_2$] (13 % on average over the ten years) was mainly due to a strong increase in the yield of *T. repens* (Table 19.1).

The interspecific differences in the response to e[CO$_2$], particularly in the mixture, resulted in a consistent and significant increase in the proportion of *T. repens* in the bi-species mixture at both levels of N supply (Table 19.1). Similarly, in more complex mixtures containing other grass, legume and non-legume dicot species, the proportion of legumes was significantly higher at e[CO$_2$] (Lüscher et al. 1996). This effect was also observed in a diverse permanent plant community at the New Zealand FACE experiment (Edwards et al. 2001; Ross et al. 2004) and seems to be a general phenomenon (Campbell et al. 2000; Lüscher et al. 2005). The level of N fertilization affected the proportion of *T. repens* in the mixture to an even greater extent (proportion of 34 % at low N and 19 % at high N; Table 19.1).Varying the levels of N fertilization may therefore offer the possibility to mitigate any changes in the legume proportion within a mixture, brought about by increased atmospheric [CO$_2$]. These results demonstrate that e[CO$_2$] not only does affect the harvested yield, but also the interspecific interactions and the botanical composition of mixed plant communities. This is important, since changes in species composition may affect ecosystem functioning (e.g. nutrient cycling) and forage quality (Lüscher et al. 2005).

19.4.2 Competitive Ability Depended Strongly on the Species, the N and [CO$_2$] Treatments

Clear differences between the RYs of *L. perenne* and *T. repens* show that the two species were not equally competitive for resources. The RY and RNY of *T. repens* were significantly smaller than 0.5 in three of the four treatments (Table 19.1), but were between 0.79 and 1.43 for *L. perenne*. This provides evidence that white clover suffered adversely from competition with the grass. The grass appeared to compete more successfully for resources and even to profit from synergistic effects (RY >1.0).

In the low N treatment, the low availability of mineral N was the main factor in limiting the growth of *L. perenne* in the pure sward. This is evident from the doubling of the yield and the leaf area index (LAI) of *L. perenne* at high N, when compared to low N fertilization (Table 19.1; Daepp et al. 2001). Daepp et al. (2001) found that in the Swiss FACE system, even the application of 56 g N m^{-2} year^{-1} did not result in unlimited growth of *L. perenne*. At low N supply, the high RY and RNY values for *L. perenne* in the mixture are suggested to be due to its very dense rooting system, which is efficient at exploring the soil and extracting nutrients. The root mass of *L. perenne* was found to be seven times greater than that of *T. repens* (Jongen et al. 1995; Hebeisen et al. 1997). As a result, the amount of mineral N derived from the soil and the fertilizer taken up by *L. perenne* in the mixed sward was similar to that in the monoculture (Zanetti et al. 1997). This was despite the competition from *T. repens* for mineral N and the sowing rate of *L. perenne* in the mixture being half that of the monoculture. Due to the very efficient root system of *L. perenne*, *T. repens* had to meet its N demand primarily (up to 90 %) through symbiotic N$_2$ fixation (Zanetti et al. 1997). In addition, the apparent transfer of symbiotically fixed N from *T. repens* to *L. perenne* contributed up to 40 % of the total N assimilated by *L. perenne* grown in the low N mixture (Soussana and Hartwig 1996; Zanetti et al. 1997). These processes help to explain the clear synergistic effects and the resulting RY >1.0 for *L. perenne* at the low N supply.

Due to its stoloniferous habit and limited petiole length, *T. repens* is ineffective in placing its leaves in the upper levels of mixed canopies to compete successfully for light (Winkler and Nösberger 1985; Schwank et al. 1986). This is most probably the main reason for the reduced competitive ability of *T. repens* in the bi-species mixture under high N fertilization. It is notable, that the level of N fertilization had no effect on the *T. repens* yield when grown in pure swards. Thus, all the effects of N fertilization on *T. repens* observed in the mixtures are indirect effects caused through the doubling of yield and leaf area index and, thus, increased shading by the competitor *L. perenne*. *T. repens* responds to shading by elongating its petioles, thus placing the young leaves in the upper layers of the canopy where there is sufficient radiation for growth (Faurie et al. 1996). Single leaves of *T. repens* can reach a height above 40 cm in the sward (Woledge et al. 1992) but, to do so, *T. repens* must invest a large proportion of assimilates into the growth of the petioles. The proportion of petioles in the harvested yield of *T. repens* was 10 % at a low LAI and increased to 50 % at a high LAI (Blum, unpublished data). Consequently, the relative growth rate (RGR; Hunt and Parson 1994) of *T. repens* was low due to a low leaf area ratio (LAR; Soussana et al. 1995). This is a disadvantage when competing with the companion grass. High LAR is important for a high RGR of plant species from nutrient-rich sites (Lambers and Poorter 1992).

Shading not only reduces the RGR of *T. repens* during the period of shading, but may also have 'long-term' consequences for *T. repens* by reducing stolon branching (Lötscher and Nösberger 1996). As well as reducing radiation, shading in plant communities changes the spectral composition of the light (Sattin et al. 1994). Low red/far-red ratios resulted in reduced branching of *T. repens* stolons (Robin et al. 1994; Lötscher and Nösberger 1997). Elevated [CO₂] increased the competitive ability of *T. repens*. This may be due to reduced shading of *T. repens* by *L. perenne*, because the height of *L. perenne* plants was reduced under e[CO₂] when compared to c[CO₂] (Clark et al. 1995; Daepp et al. 2001); and a higher proportion of radiation was transmitted to the lower layers of the canopy, probably due to the more erect growth habit of *L. perenne* (Suter et al. 2001).

19.4.3 Resource Complementarity Strongly Depended on the N and [CO₂] Treatments

The results show that, for the low N treatment, an extremely high resource complementarity for the two species occurred. This is evident from the exceptionally high values for RYT (between 1.4 and 1.6) and the RNYT (between 1.6 and 1.9) of the mixtures. This was due to the facts that: (i) *L. perenne* was competing very effectively for mineral N in the soil, (ii) a high proportion of the N yield of *T. repens* was derived from symbiotic N_2 fixation (up to 90 %; Zanetti et al. 1997) and (iii) a significant proportion of apparently transferred symbiotic N was found in the harvested N yield of *L. perenne* (up to 40 %; Zanetti et al. 1997). Thus, it is not surprising that the total N yield harvested from the mixture plots over the ten years of the experiment (33 g N m^{-2} year^{-1}, Table 19.1) significantly exceeded the amount of N that was applied as fertilizer (14 g N m^{-2} year^{-1}) or the amount of N harvested from the low-N grass monocultures (13 g N m^{-2} year^{-1}).

Under low N fertilization, e[CO₂] further increased the resource complementarity of the two species, but did not increase competition. This is evident from the increased RY of *T. repens* and the non-reduction of the RY of *L. perenne*. As a result, the RYT of the mixture increased. The same is true for the relative N yields: RNYT increased from 1.6 at $c[CO_2]$ to 1.9 at $e[CO_2]$. The reason for the increased resource complementarity at $e[CO_2]$ when compared to current, is a further increase in the N limitation of the system under $e[CO_2]$. This is evident from: (i) the lower N yield of the *L. perenne* monoculture under elevated than under $c[CO_2]$ (Table 19.1) and (ii) the reduction of the N nutrition index (Lemaire et al. 1989; Soussana et al. 1996) by 36% under $e[CO_2]$ as measured by Zanetti et al. (1997). This reduction occurred even after the critical N concentration for $e[CO_2]$ conditions was corrected for (Soussana et al. 1996). Therefore, this decline cannot be explained by the lower N requirement under $e[CO_2]$, but solely by an increased N limitation of growth.

Under low N fertilization, the increased RYT and RNYT at $e[CO_2]$ were mainly due to a $[CO_2]$-induced increase of the N input through symbiotic N_2 fixation. When legumes were present in the FACE system, the increased uptake of C, as a result of the $[CO_2]$-induced increase in the rate of photosynthesis (Rogers et al. 1998; Ainsworth et al. 2003) was buffered by an increase in symbiotic N_2 fixation (proportion of N derived from symbiosis) in *T. repens* plants (Zanetti et al. 1996; Soussana and Hartwig 1996) and by a greater proportional yield of *T. repens* in the plant community (Table 19.1; Hebeisen et al. 1997). Symbiotic N_2 fixation apparently plays a key role in maintaining the C:N balance in these fertile grassland ecosystems under $e[CO_2]$.

In the high N fertilization treatment, the RYT was close to 1.0, indicating a greatly reduced resource complementarity and a fully competitive system. At high N, the N yield derived from mineral N uptake (from fertilizer and soil) increased in the *L. perenne* monoculture by a factor of 2.9, compared to the low N treatment (Table 19.1). However, in the mixture, the yield of N as a result of symbiotic fixation and apparent N transfer decreased strongly, due mainly to: (i) the smaller proportion of *T. repens* in the mixture and (ii) the smaller proportion of N derived from symbiotic fixation in *T. repens* (Zanetti et al. 1997). The effect of $e[CO_2]$ on RNYT did not indicate any increase in resource complementarity at high N. This suggests that N availability does not strongly limit growth under $e[CO_2]$ in the high N treatment, and is in accord with the increased N availability observed in the pure swards of *L. perenne* after the first three years of the experiment (Schneider et al. 2004).

The extreme resource complementarity (RNYT of up to 1.9) at low N and the complete loss of resource complementarity at high N demonstrate that mineral N availability is the most important limiting factor for determining plant growth and interaction in the low N fertilization treatments of the Swiss FACE experiment, where other nutrients (e.g. P, Section 19.2.2; Lüscher et al. 2004) were applied at rates appropriate for high-yielding grasslands. Such high resource complementarity as reported here is enabled by: (i) functional type mixtures of grass and legume, (ii) N being the limiting factor and (iii) the legume being strongly dependent on N uptake through symbiotic N_2 fixation. Thus, the degree of synergistic effects in a plant community seems not to be a question of biodiversity per se, but is dependent on environmental conditions and appears crucially to rely on the presence of legumes for a high resource complementarity (Hooper and Dukes 2004). If a growth factor, other than N, was limiting, the niche separation between *L. perenne* and *T. repens* would have been much smaller, as the two species explore very similar sources for the other growth factors. This is evident from RYT being close to 1.0 in the high N treatment.

19.5 Conclusions

Elevated [CO₂] and N fertilization influenced markedly yield, species proportion and interspecific interactions in temperate grassland. These changes may significantly affect amount and quality of forage and ecosystem functioning.

- Interspecific differences in the response to e[CO₂] were stronger in the mixed sward (–2 % for *L. perenne* and +65 % for *T. repens*) than in the pure swards (+13 % for *L. perenne* and +19 % for *T. repens*), demonstrating that e[CO₂] does affect not only the yield, but also the interspecific interactions and the species composition of mixed plant communities. Thus, studying the ecosystem response to e[CO₂] needs experiments with mixed plant communities.
- RY and RNY <0.5 for *T. repens* provides evidence that *T. repens* was adversely affected from competition with the grass, while the grass competed more successfully for resources (RY >0.5) or even clearly gained from synergistic effects (RY >1.0).
- The extreme resource complementarity (RNYT of up to 1.9) at low N and the loss of resource complementarity (RYT and RNYT close to 1.0) at high N demonstrate that mineral N availability was the most important limiting factor for plant growth and interspecific interactions in the low N treatment of the Swiss FACE experiment. Thus, this FACE experiment with grasses and legumes provides a good tool to study effects of e[CO₂] on the N cycle of grassland ecosystems under strongly limiting and non-limiting N availability.

Acknowledgements.   Our studies were supported by the Swiss National Energy Fund, ETH Zurich, Swiss Department of Science and Education (COST), Federal Office of Agriculture and Federal Office of Energy.

References

Ainsworth EA, Davey PA, Hymus GJ, Osborne CP, Rogers A, Blum H, Nösberger J, Long SP (2003) Is stimulation of leaf photosynthesis by elevated carbon dioxide concentration maintained in the long term? A test with *Lolium perenne* grown for 10 years at two nitrogen fertilization levels under free air CO_2 enrichment (FACE). Plant Cell Environ 26:705–714

Cain ML, Pacala SW, Silander JA Jr, Fortin MJ (1995) Neighbourhood models of clonal growth in the white clover *Trifolium repens*. Am Nat 145:888–917

Campbell BD, Stafford Smith DM, Ash AJ, Fuhrer J, Gifford RM, Hiernaux P, Howden SM, Jones MB, Ludwig JA, Manderscheid R, Morgan JA, Newton PCD, Nösberger J, Owensby CE, Soussana JF, Tuba Z, ZuoZhong C (2000) A synthesis of recent global change research on pasture and rangeland production: Reduced uncertainties and their management implications. Agric Ecosyst Environ 82:39–55

Clark H, Newton PCD, Bell CC, Glasgow EM (1995) The influence of elevated CO_2 and simulated seasonal-changes in temperature on tissue turnover in pasture turves dominated by perennial ryegrass (*Lolium perenne*) and white clover (*Trifolium repens*). J Appl Ecol 32:128–136

Connolly J (1997) Substitutive experiments and the evidence for competitive hierarchies in plant communities. Oikos 80:179–182

Connolly J, Wayne P (2005) Assessing determinants of community biomass composition in two-species plant competition studies. Oecologia 142:450–457

Daepp M, Suter D, Lüscher A, Almeida JPF, Isopp H, Hartwig UA, Blum H, Nösberger J (2000) Yield response of *Lolium perenne* swards to free air CO_2 enrichment increased over six years in a high-N-input system. Global Change Biol 6:805–816

Daepp M, Nösberger J, Lüscher A (2001) Nitrogen fertilization and developmental stage alter the response of *Lolium perenne* to elevated CO_2. New Phytol 150:347–358

Edwards GR, Clark H, Newton PCD (2001) The effects of elevated CO_2 on seed production and seedling recruitment in a sheep-grazed pasture. Oecologia 127:383–394

Faurie O, Soussana JF, Sinoquet H (1996) Radiation interception, partitioning and use in grass-clover mixtures. Ann Bot 77:35–45

Hebeisen T, Lüscher A, Zanetti S, Fischer BU, Hartwig UA, Frehner M, Hendrey GR, Blum H, Nösberger J (1997) Growth response of *Trifolium repens* L. and *Lolium perenne* L. as monocultures and bi-species mixture to free air CO_2 enrichment and management. Global Change Biol 3:149–160

Hooper DU, Dukes JS (2004) Overyielding among plant functional groups in a long-term experiment. Ecol Lett 7:95–105

Hunt R, Parson IT (1994) A computer program for deriving growth-functions in plant analysis. J Appl Ecol 11:297–307

Jongen M, Jones MB, Hebeisen T, Blum H, Hendrey G (1995) The effects of elevated CO_2 concentrations on the root-growth of *Lolium perenne* and *Trifolium repens* grown in a face system. Global Change Biol 1:361–371

Kessler W, Nösberger J (1994) Factors limiting white clover growth in grass/clover systems. Proc Gen Meet Eur Grassl Fed 15:525–538

Lambers H, Poorter H (1992) Inherent variation in growth-rate between higher plants – a search for physiological causes and ecological consequences. Adv Ecol Res 23:187–261

Lemaire G, Gastal F, Salette J (1989) Analysis of the effect of N nutrition on dry matter yield and optimum N content. Proc Int Grassl Congr 16:179–180

Lötscher M, Nösberger J (1996) Influence of position and number of nodal roots on outgrowth of axillary buds and development of branches in *Trifolium repens* L. Ann Bot 78:459–465

Lötscher M, Nösberger J (1997) Branch and root formation in *Trifolium repens* is influenced by the light environment of unfolded leaves. Oecologia 111:499–504

Lüscher A, Hebeisen T, Zanetti S, Hartwig UA, Blum H, Hendrey GR, Nösberger J (1996) Differences between legumes and nonlegumes of permanent grassland in their responses to free-air carbon dioxide enrichment: its effect on competition in a multispecies mixture. In: Körner C, Bazzaz F (eds) Carbon dioxide, populations, and communities. Academic Press, San Diego, pp 287–300

Lüscher A, Hendrey GR, Nösberger J (1998) Long-term responsiveness to free air CO$_2$ enrichment of functional types, species and genotypes of plants from fertile permanent grassland. Oecologia 113:37–45

Lüscher A, Hartwig UA, Suter D, Nösberger J (2000) Direct evidence that symbiotic N$_2$ fixation in fertile grassland is an important trait for a strong response of plants to elevated atmospheric CO$_2$. Global Change Biol 6:655–662

Lüscher A, Daepp M, Blum H, Hartwig UA, Nösberger J (2004) Fertile temperate grassland under elevated atmospheric CO$_2$ – role of feed-back mechanisms and availability of growth resources. Eur J Agron 21:379–398

Lüscher A, Fuhrer J, Newton PCD (2005) Global atmospheric change and its effect on managed grassland systems. In: McGilloway DA (ed) Grassland: a global resource. Wageningen Academic, Wageningen, pp 251–264

Ramseier D, Connolly J, Bazzaz FA (2005) Carbon dioxide regime, species identity and influence of species initial abundance as determinants of change in stand biomass composition in five-species communities: an investigation using a simplex design and RGRD analysis. J Ecol 93:502–511

Robin C, Hay MJM, Newton PCD, Greer DH (1994) Effect of light quality (red far-red ratio) at the apical bud of the main stolon on morphogenesis of *Trifolium repens* L. Ann Bot 74:119–123

Rogers A, Fischer BU, Bryant J, Frehner M, Blum H, Raines CA, Long SP (1998) Acclimation of photosynthesis to elevated CO$_2$ under low-nitrogen nutrition is affected by the capacity for assimilate utilization. Perennial ryegrass under free-air CO$_2$ enrichment. Plant Physiol 118:683–689

Ross DJ, Newton PCD, Tate KR (2004) Elevated [CO$_2$] effects on herbage production and soil carbon and nitrogen pools and mineralization in a species-rich, grazed pasture on a seasonally dry sand. Plant Soil 260:183–196

Ryle GJA, Powell CE (1992) The influence of elevated pCO$_2$ and temperature on biomass production of continuously defoliated white clover. Plant Cell Environ 15:593–599

SAS Institute (1999) The SAS system for Windows ver 8.02. SAS Institute, Cary, N.C.

Sattin M, Zuin MC, Sartorato I (1994) Light quality beneath field-grown maize, soybean and wheat canopies – red–far red variations. Physiol Plant 91:322–328

Schneider MK, Lüscher A, Richter M, Aeschlimann U, Hartwig UA, Blum H, Frossard E, Nösberger J (2004) Ten years of free-air CO$_2$ enrichment altered the mobilization of N from soil in *Lolium perenne* L. swards. Global Change Biol 10:1377–1388

Schwank O, Blum H, Nösberger J (1986) The influence of irradiance distribution on the growth of white clover (*Trifolium repens* L.) in differently managed canopies of permanent grassland. Ann Bot 57:273–281

Schwinning S, Parsons AJ (1996) Analysis of the coexistence mechanisms for grasses and legumes in grazing systems. J Ecol 84:799–813

Sims PL, Risser PG (2000) Grasslands. In: Barbour MG, Billings WG (eds) North American terrestrial vegetation, Cambridge University Press, New York, pp 323–356

Snaydon RW, Satorre EH (1989) Bivariate diagrams for plant competition data: modifications and interpretation. J Appl Ecol 26:1043–1057

Soussana JF, Hartwig UA (1996) The effects of elevated CO_2 on symbiotic N_2 fixation: a link between the carbon and nitrogen cycles in grassland ecosystems. Plant Soil 187:321–332

Soussana JF, Vertès F, Arregui MC (1995) The regulation of clover shoot growing points density and morphology during short-term clover decline in mixed swards. Eur J Agron 4:205–215

Soussana JF, Casella E, Loiseau P (1996) Long-term effects of CO_2 enrichment and temperature increase on a temperate grass sward. II. Plant nitrogen budgets and root fraction. Plant Soil 182:101–114

Suter D, Nösberger J, Lüscher A (2001) Response of perennial ryegrass to free-air CO_2 enrichment (FACE) is related to the dynamics of sward structure during regrowth. Crop Sci 41:810–817

Winkler L, Nösberger J (1985) Einfluss der Schnitthäufigkeit und N-Düngung auf die Bestandesstruktur und die vertikale Verteilung von Weissklee (*Trifolium repens* L.) in einer Dauerwiese. J Agron Crop Sci 155:43–50

Woledge J, Davidson K, Dennis WD (1992) Growth and photosynthesis of tall and short cultivars of white clover with tall and short grasses. Grass Forage Sci 47:230–238

Zanetti S, Hartwig UA, Lüscher A, Hebeisen T, Frehner M, Fischer BU, Hendrey GR, Blum H, Nösberger J (1996) Stimulation of symbiotic N_2 fixation in *Trifolium repens* L. under elevated atmospheric pCO_2 in grassland ecosystems. Plant Physiol 112:575–583

Zanetti S, Hartwig UA, Van Kessel C, Lüscher A, Hebeisen T, Frehner M, Fischer BU, Hendrey GR, Blum H, Nösberger J (1997) Does nitrogen nutrition restrict the CO_2 response of fertile grassland lacking legumes? Oecologia 112:17–25

20 The Potential of Genomics and Genetics to Understand Plant Response to Elevated Atmospheric [CO₂]

G. Taylor, P.J. Tricker, L.E. Graham, M.J. Tallis, A.M. Rae,
H. Trewin, and N.R. Street

20.1 Introduction

20.1.1 What We Know and What We Need to Know

There is now a pressing need to understand more about long-term adaptation and genetic changes in future CO_2 concentrations, particularly for adaptive traits that are relevant to plant productivity and ecological characteristics that determine survival, fitness, yield and interaction with pests and pathogens (Ward and Kelly 2004). We wish to identify the genes that determine ecological success in future CO_2 environments (Feder and Mitchell-Olds 2003) and plant yields in crop systems (Martin 1989) – a subject defined as *ecological and environmental genomics*. We have an unprecedented opportunity to utilise new genomic and genetic techniques to address these questions in relation to elevated CO_2 using current FACE facilities. We have already quantified changes in the major characteristics determining function, productivity, yield and fitness that are sensitive to elevated CO_2 (Ainsworth and Long 2005), including increased photosynthesis, Rubisco acclimation, decreased plant water use and altered plant canopy architecture and leaf quality, as reported in this volume. The advantages of FACE experiments are clear – they provide realistic environmental conditions to study both short- and long-term responses of a wide range of crop, managed and natural ecosystems. The potential of new technologies is also clear. We are now able to consider the expression of many thousands of genes simultaneously, using microarrays, a technology originally developed for human disease screening (Schena et al. 1995) that has become routine in laboratory studies, but with very few field experiments on plants reported. Plant microarrays were first developed for

Ecological Studies, Vol. 187
J. Nösberger, S.P. Long, R.J. Norby, M. Stitt,
G.R. Hendrey, H. Blum (Eds.)
Managed Ecosystems and CO₂
Case Studies, Processes, and Perspectives
© Springer-Verlag Berlin Heidelberg 2006

the model *Arabidopsis* (Schaffer et al. 2000) and this initially restricted their use, but arrays are now available for a wide range of crop plants, including maize (Fernandes et al. 2002), rice (Wasaki et al. 2003) and soybean (Vodkin et al. 2004), and for a tree, poplar (Andersson et al. 2004). Microarrays allow us to consider *gene expression* – previously only possible with mRNA blots and other forms of difficult differential expression.

Complementary to genomic approaches are those of high-throughput protein identification – proteomics and metabolic profiling (metabolomics). These technologies may be considered together in an approach that has been defined as *systems biology* (Ideker et al. 2005), or *integrative biology*, the guiding principal of which is that all constituents of a cell should be studied at once, in order to obtain a meaningful understanding of networks and controls that are operating at any given time or in response to altered conditions (Blanchard 2004). However, in its widest sense, the science of ecological and environmental genomics may also be used to consider other technologies that allow us to elucidate aspects of the genome responsible for adaptive traits (Cronk 2005). This includes the use of natural genetic variation to identify quantitative trait loci (QTL) explaining a wide range of developmental and adaptive changes that occur in response to an altered environment, as summarised in Table 20.1.

20.1.2 Can an Integrative (Systems) Biology Approach be Useful?

Complex biological systems should be considered as units where billions of molecules interact together to transform energy into life. Such complexity cannot easily be broken down in a reductionist approach, by studying only a single part of the system – one gene and its regulation, for example – but must be considered together and understood by integrating information from gene, protein and metabolite, as illustrated in Fig. 20.1.

Using an integrative biology approach, the FACE experiment may be seen as central to the production of large quantitative datasets at gene, protein and metabolite level (Fig. 20.1, step 1). Linking these to the development of computational models and analysis approaches (step 2), the output (step 3) is then used for hypothesis-driven experiments in controlled environments (step 4), to undertake short-term and rapid experimentation (step 5). Using experimental data (step 6), the hypothesis can then be refined with further second-stage experimentation using FACE facilities (step 7). The ability to manage this large and complex amount of information is at the forefront of current developments in bioinformatics and this is what currently limits our understanding (Thimm et al. 2004). In the remainder of this brief review, we document the major high-throughput technologies available to future FACE scientists, reviewing on-going activity in FACE experimentation and providing a glimpse of future experiments that may be possible in FACE facilities.

Table 20.1 Genomic, genetic and other high-throughput technologies available for, but as yet largely unexploited, in FACE experiments

Technology	Description	Use in FACE experiments?
Genomics 1 Microarrays – global transcript profiling	Glass-based cDNA spotted arrays or oligonucleotide-synthesised arrays, with several thousand probes representing a significant portion of the genes of an organism. Limited in the past by sequence information from species of interest. Cross-species hybridisations may be possible.	Limited use. First data available from soyFACE and POPFACE for cDNA microarrays with several thousand ESTs showing only small numbers of genes appeared sensitive to elevated CO_2, with variability and reproducibility an issue (Miyazaki et al. 2004; Taylor et al. 2005)
2 Natural genetic variation and QTL discovery	The identification of areas of the genome responsible for complex traits (those generally determined by several rather than single genes). Utilises a molecular genetic map saturated with markers, e.g. SSRs and a segregating population (Rae et al. 2005).	No reported use in FACE experiments, although plans currently underway. Open-top chamber study of *Populus* (Ferris et al. 2002; Rae et al. 2005).
3 Association genetics	The utilisation of genetic variation (such as SNPs) in a natural population to find associations with phenotypic variation.	No reported use. Difficulty of placing natural populations into a FACE facility. Possible for *Arabidopsis* ecotypes.
Proteomics	Protein profiling. The use of 2-D gels, ICAT and AQUA for protein. Identification using mass spectrometry and isotopic signatures.	No reported use, but POPFACE samples under analysis. Controlled-environment study identified 13 proteins sensitive to elevated CO_2 (Bae and Sicher 2004).
Metabolomics	Gas chromatography mass spectrometry-based metabolic profiling for the identification of metabolites in tandem with PCA and hierarchical clustering techniques can reveal informative biochemical phenotypes (Fiehn et al. 2000).	No reported studies in FACE.

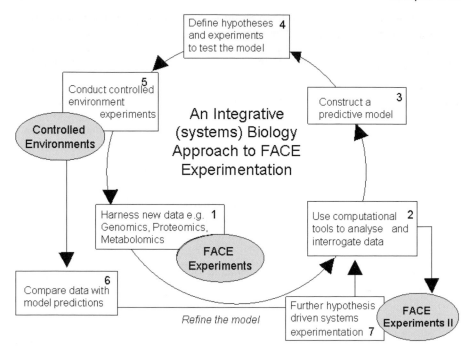

Fig. 20.1 An integrated (systems) biology approach to future experimentation in FACE facilities utilising the high through-put technologies of genomics, proteomics and metabolomics

20.2 Genomics in Field-Grown Plants

20.2.1 Transcript Profiling

The sequencing of plant genomes and the development of genomic techniques has increased our fundamental understanding of plant growth and function. To date, genome studies have been used to discover and classify gene functions with an increasing number of plant *environmental genomic* studies completed, including plant response to stresses such as ozone (Matsuyama et al. 2002).

The introduction of the microarray has allowed realistic *global transcript profiling* to be undertaken in many replicate biological samples. Microarrays allow the parallel screening of gene expression for (potentially) all genes from an organism either at a particular time or in response to a treatment. Microarrays allow speculative investigation of gene expression in the absence of hypothesis-testing (Schrader et al. 2004). This technology is now firmly established as a valuable research tool and its use is becoming routine

(Alba et al. 2004). There are a number of variations in the production of microarrays, primarily related to how the array is spotted or synthesised, as reviewed by Deyholos and Galbraith (2001), but here we focus on the use of cDNA glass-based microarrays, since currently these offer much potential to FACE scientists, being relatively cheap to produce and use, with more and more species of interest available for spotting. The production of cDNA microarrays involves spotting probes onto a solid support. Physically, glass slides are favoured above material filters because they are solid, transparent and have low fluorescence, allowing the target direct access to the probe, and are easy to visualise. The major limitation to field biologists has been the availability of sequence information for the organisms of interest. Ideal probes are fragments of cDNA that have been sequence-validated, annotated and are unique, which show minimal cross-hybridisation to related sequences and collectively represent a comprehensive portion of the expressed genome. Typically gene expression is profiled by the competitive hybridisation of cDNA from two targets, each labelled with a different fluorescent dye (Fig. 20.2). A ratio of expression for a target at a particular probe is the desired result; and therefore the nature of the probe and the amount of the target are key considerations. In order to obtain reliable results, targets must hybridise to probes with a high degree of specificity and sensitivity. In practice, researchers are limited in their choice of probes by the EST libraries available, as the sequencing, resequencing for validation and production of cDNA clones is expensive and time-consuming.

This genome-wide approach has advantages. By assuming that genes with related functions may be co-regulated, candidate genes likely to be involved in the same processes are discovered (Eisen et al. 1998) and classes of functional genes may be identified (Golub et al. 1999). For example, cDNA microarray analysis of poplar leaves during senescence revealed that overall gene expression was up-regulated, demonstrating that senescence involves active processes rather than a down-regulation of earlier activity and that specific groups of functional genes associated with senescence may be identified (Bhalerao et al. 2003; Andersson et al. 2004).

Bioinformatics – "conceptualizing biology in terms of macromolecules and then applying 'informatics' techniques to understand and organize the information associated with these molecules, on a large scale" (Luscombe et al. 2001) – uses information from large microarray datasets to cluster gene expression profiles into broad patterns of biological behaviour and to filter specific genes of interest (Wu 2001), as depicted in Fig. 20.2.

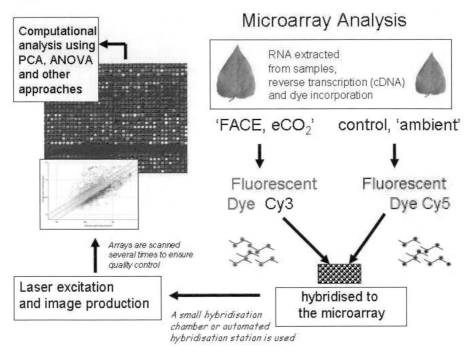

Fig. 20.2 Performing a microarray hybridisation requires the extraction of high quality RNA from field material that is reverse transcribed to form cDNA. Fluorescent dyes, contrasting between treatments ('targets') are incorporated and hybridised to the 'probes' on the array, often a glass-based cDNA array, although other arrays with shorter-sequence probes exist. Following hybridisation in controlled conditions, laser excitation is used to quantify the relative incorporation of the dye with each spot, giving a measure of relative RNA expression in the two samples from *FACE* and *ambient* CO_2. Large datasets are produced that require proprietary software or specialist programming. Parametric statistics, if used, must incorporate routines to avoid type 1 and type 2 statistical errors

20.2.2 Use of Expression Arrays in FACE Experiments

Few attempts have been made to use microarrays in current FACE facilities. In the POPFACE experiment, transcript profiling was completed in years 2, 3, 5 and 6 of exposure to elevated CO_2 (Taylor et al. 2005) and this approach revealed several novel results (summarised in Fig. 20.3). A consistent result was the finding in that the effect of elevated CO_2 on relative transcript expression depended strongly on stage of leaf development. For young leaves, differential expression suggested that genes were up-regulated in elevated compared to expression in current CO_2, whilst the opposite was true for older leaves, as illustrated in Fig. 20.3. A second finding was that rather few transcripts appeared sensitive to elevated CO_2 (a few dozen), but this may be a

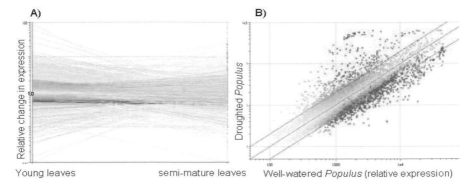

Fig. 20.3 Transcript profiling in *Populus* trees grown in either current (350 ppm) or elevated (550 ppm) CO_2 for 6 years (**A**), or trees exposed to drought (**B**). A *Young leaves* depicts the changes in gene expression for elevated compared to current CO_2, where each *line* is a single spot (transcript) on the array, with *red* for differential up-regulated in elevated CO_2 (>1.0), *green* for down-regulation in elevated CO_2 (<1.0) and *yellow* for no change (relative expression ~1.0). The pattern of response to elevated CO_2 differed depending on leaf age. **B** Each spot on the array is represented by a spot on the figure, with colour notations as in (**A**). A one-to-one and two-fold change in gene expression is depicted by the *parallel lines*. Those genes up-regulated in drought are shown in *red*

reflection of environmental variation. Transcripts of note included that for the small sub-unit of Rubisco, which was most likely to be up- and down-expressed in elevated relative to current CO_2 in young and semi-mature leaves respectively, a result confirmed by quantitative RT-PCR (Taylor et al. 2005). Most studies of Rubisco expression using Northern blot analysis have revealed a down-regulation of this transcript in elevated CO_2, although Moore et al. (1998) also showed that up-regulation was possible, particularly in young leaves. The transcript profiling from POPFACE also revealed significant changes in the expression of xyloglucan endotransglycosylase/hydrolase (XTH) transcripts for a cell wall-loosening enzyme, known to show increased activity in elevated CO_2 (Ranasinghe and Taylor 1996; Ferris et al. 2001), an effect associated with increased leaf cell expansion and leaf area (Taylor et al. 2003). Interestingly, at the ASPENFACE site, transcripts for this enzyme were also up-regulated in elevated CO_2, following a transcript-profiling study using Nylon membrane arrays (Gupta et al. 2005a, b). It would appear that this is an important growth mechanism in elevated CO_2, but perhaps that was predictable before these hybridisations were performed. The real value of transcript profiling is in revealing previously unsuspected genes such as the calcium-dependent protein kinase that appeared to be up-regulated in elevated compared to CO_2; and the RAS-related GTP-binding protein was consistently down-regulated in elevated compared to current CO_2 for semi-mature leaves (Taylor et al. 2005). At the soyFACE site, a preliminary transcript-profiling experiment using *Arabidopsis* was also reported by Miyazaki et al. (2004) and revealed that, for *Arabidopsis* at least, larger differences in gene expression were apparent between controlled environment versus field than between current versus elevated CO_2. Taken together, these three preliminary studies suggest that field-grown plant material is subjected to large environmental variation on both a seasonal and daily basis and it is possible that this, combined with small changes in the expression of many genes rather than large changes in the expression of few genes, means that microarrays are difficult systems to work with in the field.

20.2.3 QTL Discovery for Responsive Traits

Because patterns of plant development, growth and productivity are controlled by many rather than single genes, they can be resolved at the genomic level using population and quantitative genetics, through the identification of QTL. Elucidation of QTL is not only useful for analysing important agronomic traits in crops, but also for understanding fundamental aspects of genetic control in plants, particularly model species, grown under differing conditions. It can provide evidence that a plant characteristic of interest has a genetic component and is a good start-point for future studies on individual genes and genomic regions, or in focusing on the inheritance and evolution of specific traits of interest. To our knowledge, few studies of QTL identification in elevated CO_2 have been published (Ferris et al. 2002). This is surprising, because the approach has yielded valuable insight into plant response to a range of environmental changes, prompting gene-cloning strategies, and identified *Arabidopsis* as a valuable model to understand the ecological significance of genetic variation (Alonso-Blanco et al. 1998).

There is considerable evidence that past changes in atmospheric elevated CO_2 have acted as a selection pressure, leading to altered plant development and adaptation. For example, stomatal numbers have declined since preindustrial and across geological time-scales (Hetherington and Woodward 2003) – an effect attributed to rising atmospheric CO_2. These adaptive changes are likely to have an effect on plant competitive ability and fitness. In order to use the power of quantitative genetics to unravel genetic response to global change, we need first to develop a well characterised segregating population, such as *Arabidopsis* (Lister and Dean 1993) and also *Populus* (Bradshaw et al. 1994), which must be genotyped; and, coupled to this, the whole population (in replicate) is then subjected to the conditions of interest. However, there are some technical difficulties, since the numbers involved in any mapping study tend to be large (Lister and Dean 1993).

By exposing a mapping population to elevated CO_2, we have revealed these responses as well as detected the underlying QTL determining growth and development traits. Elevated [CO_2] resulted in the production of larger trees,

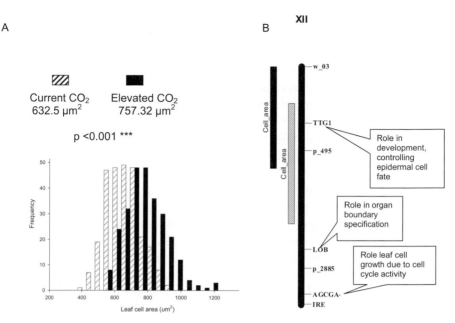

Fig. 20.4 The use of quantitative genetics to elucidate the genomic regions sensitive to elevated CO_2, as yet a technology not utilised in FACE facilities. **A** Here, an open-top chamber experiment on *Populus* reveals that the F_2 progeny of an interspecific cross (*P. trichocarpa* × *P. deltoides*) had significantly larger leaf epidermal cells in elevated CO_2. This information was used to determine several QTL (areas of the genome) responsible for this trait. **B** On linkage group XII of *Populus*, a QTL for cell area was found in both current and elevated CO_2 treatments and at least one candidate gene, taken from the physical sequence of *Populus* (TTG1 – known to determine cell patterning and fate), is found to co-locate to this region. This would be a good candidate gene for further study. The physical sequence of *Populus* may be interrogated at http://genome.jgi-psf.org/Poptr1/Poptr1.home.html

which is in keeping with other literature reporting increased above- and below-ground plant growth and biomass accumulation (Norby et al. 1999). Figure. 20.4 shows how leaf epidermal cell expansion was stimulated by elevated CO_2, identifies a QTL for this trait (several others were present) and co-locates a putative candidate gene that may be contributing to this response. Similar responses to elevated CO_2 were found when the RILs of *Arabidopsis* were exposed to this treatment in four separate experiments over the course of 3 years (Rae et al. 2005).

20.2.4 Association Genetics

QTL discovery in segregating pedigrees relies on linkage between a QTL and a marker generated by hybridisation between the parents of a segregating pedigree. This means that recombination, and therefore mapping resolution, is limited by the number of generations involved to produce the segregating pedigree. In general, these studies have positioned the identified QTL within chromosomal regions spanning 10–40 cM (Kearsey and Farquhar 1998). Although this level of precision may be sufficient for some applications of marker-assisted selection, to more accurately identify candidate genes for the trait of interest, the theory of association mapping has been put forward.

Association genetics utilises linkage disequilibrium (LD) to identify QTL in natural populations. LD is the non-random association between alleles, usually at linked loci (Weir 1990). This occurs when alleles at different loci occur together more often than expected; LD is a statistical measure that quantifies the non-independence of genotypes at several loci.

Interest in the study of LD has increased dramatically in recent years due to genomic technologies enabling rapid identification of haplotypes at many genetic loci, either by DNA sequencing or by high-throughput single nucleotide polymorphisms (SNP) analysis. Two main approaches can be taken in association mapping. One is the whole genome scan. In the presence of significant LD, of the order of tens of kilobases or more, it can be possible to identify genetic regions that are associated with a particular trait of interest by a high-density genome scan of individuals from an existing population. Alternatively, if LD declines rapidly in the population, a candidate gene approach may be taken where variation in genes, that are presumed to be of importance, are analysed and associated with the variation in the trait, i.e. genes are identified that may be responsible for the trait of interest by screening a limited number of candidate genes. Individual SNP haplotypes within a candidate gene are systematically tested for association with the phenotype of interest. In some cases, it should be possible to identify a polymorphism within a gene that is responsible for the difference in alternative phenotypes (Palaisa et al. 2004), sometimes termed quantitative trait nucleotides (QTN; Rafalski and Morgante 2004).

LD is expected to vary greatly in different populations because of the randomness of history, but the average rate of decay of LD (i.e. the genetic or physical distance over which LD can be measured) depends on the demographic history of the population. In particular, the extent of selfing versus outcrossing in plant populations can have a strong effect. It has been shown that LD is extensive in the mainly inbreeding species A. thaliana, but that it is far from genome-wide (Nordborg 2000). Association-mapping studies have been successful in human studies (e.g. Cardon et al. 1994; Fullerton et al. 2002) and livestock (Kirkpatrick and Jarne 2000), but its use is just beginning in plants (Gupta et al. 2005a, b).

Association studies can be used in a similar manner, in FACE experiments, as linkage analysis in segregating pedigrees, but the process differs in that it examines a set of presumably unrelated genotypes containing more allelic diversity and recombination events than in a typical controlled cross, therefore reducing the time and cost of producing the pedigree and increasing the reso-lution and applicability. The predicted effect of the QTL should not be specific to a single family or mapping pedigree. The major limitation of LD mapping is that it provides little insight into the mechanistic basis of the LD detected (e.g. LD may not be due to linkage, but population subdivision and admixture), so that genomic localisation and cloning of genes based on LD may not always be successful. Therefore joint linkage and LD-mapping strategies have been devised (Wu et al. 2002). At present, there are no reported association studies in FACE experiments, but this should be a useful tool for future work.

20.3 Proteomics and Metabolomics in Field-Grown Plants

The term *genomics* is used increasingly to encompass a range of high-throughput investigative methods that examine all components of a biologi-cal state at various scales, principally transcriptomics, proteomics, and metabolomics. The term *functional genomics* describes the use of genomics to ascribe functional roles to genes and their products. Proteomics is an ever-developing technology now widely used in a range of scientific disciplines in order to analyse the proteome (the *prote*in component of the gen*ome*) with the potential to elucidate responses to biotic and abiotic stresses, including elevated CO_2. Such results complement transcriptome data and further aid the understanding of the functional identity of all plant genes existing in the genome. Microarrays are confounded by the fact that they only provide infor-mation on changes occurring at the level of transcription and do not take into account post- translational modifications that may occur before the protein is fully functional. Using proteomics in conjunction with information gained from the transcriptome further aids the understanding of biological processes and mechanisms (Tao and Aebersold 2003). In order to identify and quantify proteins correctly and accurately, appropriate methods are required that are both sensitive and powerful. There are several different approaches that have been developed for quantitative protein profiling. The two-dimen-sional gel electrophoresis technique (hereafter referred to as 2-DE) was pri-marily the most commonly used proteomic procedure. However, it is a limited technology and, like most proteomic technologies, does not identify post-translational modifications that may be of regulatory significance. This is a major limitation. The 2-DE has been used in long-term exposure of *Arabidop-sis* to elevated carbon dioxide levels. In this experiment, 13 proteins were found to differ significantly in response to elevated CO_2 (Bae and Sicher 2004). Six of the proteins were identified by mass spectrometry (MS) and were involved in plant development or stress and photosynthesis (Bae and Sicher 2004). However, the 2-DE approach to proteomics was much criticised, since the technique requires a large amount of technical expertise to produce adequate gels (Quadroni and James 1999) and is also extremely time-consuming. Furthermore, in the experiments conducted by Bae and Sicher (2004), it was difficult to identify hydrophobic proteins or those that were hard to solubilise. It is also apparent that this technique is not appropriate for detecting proteins present in low abundance. In addition to the gel-based 2-DE, isotopically labelling peptides may also be used to identify proteins. One such technique, isotope-coded affinity tags (ICAT) can be used to determine the relative amounts of proteins from cells subjected to two different conditions. The stable isotope tags used in the technique are chemically identical and simply differ in mass. The proteins from one treatment are labelled with 'heavy' tags, whilst those from treatment two are labelled with 'light' tags. The two samples are combined and, after protein digestion, they are run through a column where the tagged peptides are captured. After fractionation, tandem MS is used to identify the labelled peptides. The ICAT approach has been used for quantification and has produced some good results (Ranish et al. 2003). There have however been problems with the reagents used in the ICAT method, which are currently being further developed. Recently, a similar approach was developed, termed iTRAQ (Applied Biosystems; www.appliedbiosystems. com). Similarly to ICAT, it is a non-gel-based approach and, following protein digestion, the peptides are tagged and analysed by tandem MS. There are a number of other non-gel-based approaches, including combined fractional diagonal chromatography (COFRADIC), which was developed in Professor Joël Vandekerckhove's laboratory (University of Ghent). The technique involves separating peptides based upon chromatographic differences and using MS for identification. Appropriate analytical methods are essential for unravelling the complex results produced from proteomic experiments. There are a number of databases constructed to aid the interpretation of the results from MS data, including MASCOT (Perkins et al. 1999), PepFrag (Qin et al. 1997) and MS-Tag (Clauser et al. 1999). Furthermore, a centralized proteomics database was recently created, with the aim of sharing experimental data with other researchers (http://bioinformatics.icmb.utexas.edu/OPD). Similar schemes had already been set up for transcript analysis of microarrays, such as the Stanford Microarray Database. Such apparatus will further encourage the use of proteomics for future research purposes.

Another technique involving isotopic labelling is known as absolute quantification of proteins (AQUA; http://www.proteome.soton.ac.uk/aqua.htm) and measures absolute protein expression (in terms of number of molecules per cell). It is a highly sensitive method which involves using stable isotopically labelled peptides (e.g. ^{13}C) as a reference. Tandem MS is again used to quantify the protein of interest by comparison with the level of the corresponding reference. Applications of this technique however are not well documented.

Proteomic approaches have been used in *Arabidopsis* and alfalfa (*Medicago sativa*) to study cell wall proteins (Chivasa et al. 2002; Watson et al. 2004).

Whilst many named cell wall proteins with known biological function have been identified using such techniques, including expansins, glucanases and peroxidases, a number of previously unnamed proteins have also been identified (Chivasa et al. 2002), thus illustrating the potential for protein discovery. Although not yet utilised extensively for protein profiling in FACE experiments, there is huge potential for this application which may be realised in the next few years. Since it has already been shown that 2-DE can be successfully used to detect in proteins in response to ozone in rice seedlings (*Oryza sativa*; Agrawal et al. 2002) and elevated CO_2 in *Arabidopsis* (Bae and Sicher 2004), it seems likely that future potential of protein profiling is large.

Metabolomics is becoming an important component of the holistic era of genomic investigations, although it has currently received less attention than genomic approaches. In order to understand and model complex plant responses to environment, it is essential to quantify and understand changes in the metabolome, alongside those of genome and proteome. The plant metabolome consists of the low molecular weight molecules present within cells and the study of these metabolites is termed *metabolomics* (Fiehn et al. 2000). Both primary metabolites (e.g. amino acids, fatty acids, carbohydrates) and secondary metabolites (e.g. flavonoids, terpenoids) are present within plant cells and represent end-products of gene expression. Their study alongside transcriptomics and proteomics is therefore critical to attaining an integrated biology understanding of adaptation and development. Many metabolites have critical functions in resistance and stress responses of plants and, commercially, are important as they are constituents of the taste, smell and colour of edible crops and flowers (Bino et al. 2004). The composition of cellular metabolites defines the biochemical state of a cell and the metabolome is tightly linked to the biological functioning of cells. This close relationship between the metabolome and biological function makes metabolomics an essential approach for understanding biochemical adaptation to an altered environment. In the same manner as transcriptomics and proteomics, metabolomic analysis allows patterns of co-expression to be identified and these can indicate regulons that fall under the control of single genes.

Metabolomics can be considered the biological interpretation of data derived from chemical data through the application of complex mathematical techniques. Metabolic profiling requires the identification of low molecular weight compounds, which can be achieved through gas chromatography (GC), high-performance liquid chromatography (HPLC), and nuclear magnetic resonance (NMR). Compounds are chromatographically separated and compared to a well defined sets of calibration standards. Metabolomics aims to identify metabolites present in crude extracts, using NMR, MS (including quadrupole time-of-flight MS; QTOF) and Fourier-transformed infrared spectroscopy (FT-IR). Metabolites are then identified by comparison of spectral data to that available in databases (Wagner et al. 2003) such as NIST (www.nist.gov), Wiley (www.wileyregistry.com) Sigma–Aldrich (http://www.

sigmaaldrich.com/Area_of_Interest/Equipment_Supplies__Books/Key_Reso
urces/Spectral_Viewer.html) and other research databases, such as the Max
Planck Institute of Molecular Plant Physiology (www.mpimp-golm.mpg.
de/mms-library/index-e.html). At present, these databases contain only a
fraction of the total metabolomic pool, representing less than 30% of identified spectral peaks (Bino et al. 2004). It is estimated that a typical *Arabidopsis*
leaf will contain in the order of 5000 primary and secondary metabolites, with
only 10% of these currently having functional annotation data available
(Wagner et al. 2003). These 5000 represent only a fraction of the 200 000
metabolites thought to exist within the plant kingdom (Pichersky and Gang
2000). In order to attain information about the entire metabolome, combinations of the above technologies are required (Hall et al. 2002). The combination of technologies used represents a balance between speed, accuracy and
resolution (Sumner et al. 2003).

20.4 The Importance of Experimental Design and Sampling Strategy in FACE Facilities

A crucial component of genomics is that of experimental design, data handling and data analysis. For such extensive and complicated datasets as those
produced by genomics, the complexity, learning curve and essential role of
data analysis cannot be understated. For FACE experiments, an added consideration is that of environmental heterogeneity, which may be considerable
across both small and large temporal and spatial scales. We already know that
gene expression can be affected by time of day (Michael and McClung 2003),
temperature (Seki et al. 2002) and light environment (Bertrand et al. 2005)
and that all these act to confound treatment effects unless a relatively strin-
gent sampling strategy is applied, for example by 'time of day' or 'season'
(including daylength). It also critical to determine which type of parameters
or 'traits' are likely to be responsive to CO_2 and tractable using these technologies. Most FACE experiments have some differences between rings within
each treatment –'block effects' – and correct statistical treatment of these data
is essential. Biological replicates (individual plants) may be pooled in some
approaches to remove some of this variation, although this should always be
checked against individual replicates. Analysis is the visualisation and identification of responsive components, followed by the functional annotation of
genes, proteins and metabolites and their placement within biochemical
pathways. These last two steps are essential to the formation of a biological
understanding of the functional contribution of metabolites. Another aspect
to analysis involves the handling of quantitative data. Statistical analysis of
the data is required to determine the probability that gene expression, for
example, is significantly affected by the condition of interest or between the

tissues, organs, developmental stages, or genotypes under consideration. Many issues arising from the handling of large-scale datasets have been resolved or are currently being researched in relation to transcriptomics data analysis. Due to the large number of profiles that can be obtained from high-throughput 'omics', analytical methods to reduce the dimensionality of the data are often employed. The most common of these are principal components analysis (PCA), hierarchical cluster analysis (HCA) and K-means clustering. Sumner et al. (2003) offers a brief overview of these. As previously mentioned, examination of the functional classification of genes and their metabolites is of critical importance in order to inform a biological understanding of metabolome responses and their role in plant–environment interactions. To this end, many resources have been made available recently that enable visual interpretation of data from a range of genomics data. One such resource is a software package called MapMan (Thimm et al. 2004; http://gabi.rzpd.de/projects/MapMan/). Figure 20.5 shows an example image from the elevated CO_2 dataset that is freely available from the MapMan website. It gives an overview of cellular response mechanisms. Each block within the categories represents a gene and the expression level of that gene is colour-coded (blue for down-regulation in response to elevated CO_2 and red for up-regula-

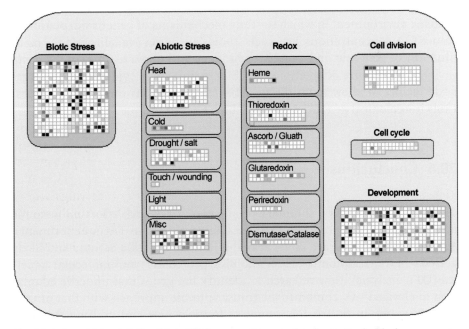

Fig. 20.5 Categories involved in cellular response mechanisms. Each *block* represents genes and the expression of the gene in response to elevated CO_2 is colour-coded, where *blue* represents down-regulation and *red* up-regulation. Source: http://gabi.rzpd.de/projects/MapMan/data.shtml

tion). Other such tools include Aracyc (http://www.arabidopsis.org/), Kyoto encyclopedia of genes and genomes (KEGG; http://www.genome.jp/kegg/pathway.html) and Gene ontology (GO; www.geneontology.org).

As metabolomics represents an end-result of gene expression and proteome expression, its functioning and response are highly coupled to the biological state of the organism. Resolution at the metabolome level may prove to be higher than that of the proteome or the transcriptome. As identification and annotation of previously uncharacterised metabolites and their inclusion in public databases increases, metabolomics will play a central role in contributing to an integrated biology level understanding of plant–environment interactions.

20.5 The Future

The potential of genomics has yet to be fully realised in FACE experiments. There is no doubt that an integrative (systems) Biology approach will be pertinent over the coming years. Crop production will be altered in future climates and there is a need to link Biology to molecular crop breeding and improvement programmes, with a strong emphasis on field as opposed to controlled environment studies. In addition, FACE experiments provide a realistic environment in which to study mechanisms of genetic adaptation to future CO_2 concentrations. Few such studies are as yet available and the combination of genomic and genetic techniques will allow us to gain powerful insights into future adaptations. FACE experiments provide the ideal large-scale facility to allow multi-disciplinary teams to focus on important questions in ecological genomics in future high CO_2 conditions.

20.6 Conclusions

This chapter presents a summary of all current research effort on genomic and other 'omic' approaches that are being utilized in FACE experiments. Surprisingly few data are as yet available, but we describe activity and likely activity in transcriptomics, metabolomics, proteomics and molecular genetics (QTL analysis). New research to identify the genes that underlie adaptation to elevated CO_2 combines the transcriptomic approach with that of QTL discovery and, in future, this promises to yield new and exciting data that can only be collected at field-scale, since large replicated populations are necessary.

- New technologies and resources in molecular genetics and genomics have to date been largely unused in large-scale ecosystem manipulative experiments such as FACE. Few transcriptome, proteome and metabolome studies have been undertaken. Similarly, the use of natural genetic variation to isolate QTL and genomic regions linked to adaptation to elevated CO_2 have been little considered, despite their potential.

- Model species, including poplar, arabidopsis, maize, rice and soybean, can currently be used to elucidate genomic response to elevated CO_2 because they have wide-ranging resources, including physical DNA sequence, microarrays, protein databases and molecular genetic maps. In the future, genomic resources will be extended to a wide range of ecologically relevant species. This promises to provide exciting new insights into long-term ecosystem responses to elevated CO_2 and ozone.
- The first transcriptome studies have revealed a small set of genes that may be sensitive to long-term exposure to elevated CO_2, including genes involved in the control of growth of the plant cell wall and cell size and shape. Microarrays, however, present inherent difficulties for field-grown material which may be highly variable and where experimental design and sampling are critical in obtaining high quality data.
- QTL for adaptive responses to elevated CO_2 have been identified in poplar and arabidopsis; and these provide the first clues to genetic adaptation to elevated CO_2. The genes underlying these QTL should be determined as a matter of priority.
- A systems biology approach to genomics should enable full integration of the environmental factors conferring a given phenotype, overcoming limitation centred on gene expression studies alone and providing novel insight into growth, development and adaptation in elevated CO_2.

References

Agrawal GK, Rakwal R, Yonekura M, Kubo A, Saji H (2002) Proteome analysis of differentially displayed proteins as a tool for investigating ozone stress in rice (*Oryza sativa L.*) seedlings. Proteomics 2:947–959

Ainsworth EA, Long SP (2005) What have we learned from 15 years of free-air CO_2 enrichment (FACE)? A meta-analytic review of the responses of photosynthesis, canopy properties and plant production to rising CO_2. New Phytol 165:351–371

Alba R, Fei ZJ, Payton P, Liu Y, Moore SL, Debbie P, Cohn J, D'Ascenzo M, Gordon JS, Rose JKC, Martin G, Tanksley SD, Bouzayen M, Jahn MM, Giovannoni J (2004) ESTs, cDNA microarrays, and gene expression profiling: tools for dissecting plant physiology and development. Plant J 39: 697–714

Alonso-Blanco C, El-Assal SE, Coupland G, Koornneef M (1998) Analysis of natural allelic variation at flowering time loci in the landsberg erecta and Cape Verde Islands ecotypes of *Arabidopsis thaliana*. Genetics 149:749–764

Andersson A, Keskitalo J, Sjodin A, Bhalerao R, Sterky F, Wissel K, Tandre K, Aspeborg H, Moyle R, Ohmiya Y, Bhalerao R, Brunner A, Gustafsson P, Karlsson J, Lundeberg J, Nilsson O, Sandberg G, Strauss S, Sundberg B, Uhlen M, Jansson S, Nilsson (2004) A transcriptional timetable of autumn senescence. Genome Biol 5:R24

Bae H, Sicher R (2004) Changes in soluble protein expression and leaf metabolite levels in *Arabidopsis thaliana* grown in elevated carbon dioxide. Field Crop Res 90:61–73

Bertrand C, Benhamed M, Li YF, Ayadi M, Lemonnier G, Renou JP, Delarue M, Zhou DX (2005) *Arabidopsis* HAF2 gene encoding TATA-binding protein (TBP)-associated factor TAF1, is required to integrate light signals to regulate gene expression and growth. J Biol Chem 280:1465–1473

Bhalerao R, Keskitalo J, Sterky F, Erlandsson R, Bjorkbacka H, Jonsson Birve S, Karlsson J, Gardestrom P, Gustafsson P, Lundeberg J, Jansson S (2003) Gene expression in autumn leaves. Plant Physiol 131:1–13

Bino RJ, Hall RD, Fiehn O, Kopka J, Saito K, Draper J, Nikolau BJ, Mendes P, Roessner-Tunali U, Beale MH, Trethewey RN, Lange BM, Wurtele ES, Sumner LW (2004) Potential of metabolomics as a functional genomics tool. Trends Plant Sci 9:418–425

Blanchard JL (2004) Bioinformatics and systems biology, rapidly evolving tools for interpreting plant response to global change. Field Crops Res 90:117–131

Bradshaw HD, Villar M, Watson BD, Otto KG, Stewart S, Stettler RF (1994) Molecular genetics of growth and development in *Populus*. III. A genetic linkage map of a hybrid poplar composed of RFLP, STS, and RAPD markers. Theor Appl Genet 89:167–178

Cardon LR, Fulker DW (1994) The power of interval mapping of quantitative trait loci using selected sib pairs. Am J Hum Genet 55:825–833

Chivasa S, Ndimba BK, Simon WJ, Robertson D, Yu XL, Knox JP, Bolwell P, Slabas AR (2002) Proteomic analysis of the *Arabidopsis thaliana* cell wall. Electrophoresis 23:1754–1765

Clauser, K.R., Baker, P., and Burlingame, A.L. (1999) Role of accurate mass measurement (±10 ppm) in protein identification strategies employing MS or MS/MS and database searching. Anal Chem 71:2871–2882

Cronk QCB (2005) Plant eco-devo: the potential of poplar as a model organism. New Phytol 166:39–48

Deyholos MK, Galbraith DW (2001) High-density microarrays for gene expression analysis. Cytometry 43:229–238

Eisen MB, Spellman PT, Brown PO, Botstein D (1998) Cluster analysis and display of genome-wide expression patterns. Proc Natl Acad Sci USA 95:14863–14868

Feder ME, Mitchell-Olds T (2003) Evolutionary and ecological functional genomics. Nature 4:649–655

Fernandes J, Brendel V, Gai XW, Lal S, Chandler VL, Elumalai P, Galbraith DW, Pierson EA, Walbot V (2002) Comparison of RNA expression profiles based on maize expressed sequence tag frequency analysis and micro-array hybridization. Plant Physiol 128:896–910

Ferris R, Sabatti M, Miglietta F, Mills RF, Taylor G (2001) Leaf area is stimulated in *Populus* by free air CO_2 enrichment (POPFACE), through cell expansion and production. Plant Cell Environ 24:305–315

Ferris R, Long L, Bunn SM, Robinson KM, Bradshaw HD, Rae AM, Taylor G (2002) Leaf stomatal and epidermal cell development: identification of putative quantitative trait loci in relation to elevated carbon dioxide concentration in poplar. Tree Physiol 22:633–640

Fiehn O, Kopka J, Dormann P, Altmann T, Trethewey RN, Willmitzer L (2000) Metabolite profiling for plant functional genomics. Nat Biotechnol 18:1157–1161

Fullerton J, Cubin M, Bhomra A, Davidson S, Miller S, Turn M, Dolby C, Mott R, Wang C, Tiwari H, Allison D, Neale M, Fairburn C, Goodwin G, Flint J (2002) Linkage analysis of extremely discordant and concordant sibling pairs identifies QTL that influence variation in a human personality trait. Am J Hum Genet 71[Suppl]:263

Golub TR, Slonim DK, Tamayo P, Huard C, Gaasenbeek M, Mesirov JP, Coller H, Loh ML, Downing JR, Caligiuri MA, Bloomfield CD, Lander ES (1999) Molecular classification of cancer: class discovery and class prediction by gene expression monitoring. Science 286:531–537

Gupta P, Duplessis S, white H, Karnosky DF, Martin F, Podila GK (2005a) Gene expression patterns of trembling aspen trees following long-term exposure to interacting elevated CO_2 and tropospheric O_3. New Phytol (in press)

Gupta PK, Rustgi S, Kulwal PL (2005b) Linkage disequilibrium and association studies in higher plants: present status and future prospects. Plant Mol Biol 57:461–485

Hall R, Beale M, Fiehn O, Hardy N, Sumner L, Bino R (2002) Plant metabolomics: the missing link in functional genomics strategies. Plant Cell 14:1437–1440

Hetherington AM, Woodward FI (2003) The role of stomata in sensing and driving environmental change. Nature 424:901–908

Ideker T, Galitski T, Hood L (2005) A new approach to decoding life: Systems biology. Annu Rev Genomics Hum Genet 2:343–372

Kearsey MJ, Farquhar AGL (1998) QTL analysis in plants; where are we now? Heredity 80:137–142

Kirkpatrick M, Jarne P (2000) The effects of a bottleneck on inbreeding depression and the genetic load. Am Nat 155:154–167

Lister C, Dean C (1993) Recombinant inbred lines for mapping RFLP and phenotypic markers in *Arabidopsis thaliana*. Plant J 4:745–750

Luscombe NM, Greenbaum D, Gerstein M (2001) What is bioinformatics? A proposed definition and overview of the field. Methods Inform Med 40:346–358

Martin GB (1998) Gene discovery for crop improvement. Curr Opin Biotechnol 9:220–226

Matsuyama T, Tamaoki M, Nakajima N, Aono M, Kubo A, Moriya S, Ichihara T, Suzuki O, Saji H (2002) cDNA microarray assessment for ozone-stressed *Arabidopsis thaliana*. Environ Pollut 117:191–194

Michael TP, McClung CR (2003) Enhancer trapping reveals widespread circadian clock transcriptional control in Arabidopsis. Plant J 31:279–292

Miyazaki S, Fredricksen M, Hollis KC, Poroyko V, Shepley D, Galbraith DW, Long SP, Bohnert HJ (2004) Transcript expression profiles of Arabidopsis thaliana grown under controlled conditions and open-air elevated concentrations of CO_2 and O_3. Field Crop Res 90:47–59

Moore BD, Cheng SH, Rice J, Seemann JR (1998) Sucrose cycling, Rubisco expression, and prediction of photosynthetic acclimation to elevated atmospheric CO_2. Plant Cell Environ 21:905–915

Norby RJ, Wullschleger SD, Gunderson CA, Johnson DW, Ceulemans R (1999) Tree responses to rising CO_2 in field experiments: implications for the future forst. Plant, Cell Environ 22:683–714

Nordborg M (2000) Linkage disequilibrium, gene trees and selfing: an ancestral recombination graph with partial self-fertilization. Genetics 154:923–929

Palaisa K, Morgante M, Tingey S, Rafalski A (2004) Long-range patterns of diversity and linkage disequilibrium surrounding the maize Y1 gene are indicative of an asymmetric selective sweep. Proc Natl Acad Sci USA 101:9885–9890

Perkins DN, Pappin DJC, Creasy DM, Cottrell JS (1999) Probability-based protein identification by searching sequence databases using mass spectrometry data. Electrophoresis 20:3551–3567

Pichersky E, Gang DR (2000) Genetics and biochemistry of secondary metabolites in plants: an evolutionary perspective. Trends Plant Sci 5:439–445

Quadroni M, James P (1999) Proteomics and automation. Electrophoresis 20:664–677

Qin J, Fenyo D, Zhao YM, Hall WW, Chao DM, Wilson CJ, Young RA, Chait BT (1997) A strategy for rapid, high confidence protein identification. Anal Chem 69:3995–4001

Rae AM, Graham LE, Street NR, Hughes J, Hanley ME, Tucker J, Taylor G (2005) QTL for growth and development in elevated carbon dioxide in two model plant genera: a novel approach for understanding adaptation to climate change? Global Change Biol (in press)

Rafalski A, Morgante M (2004) Corn and humans: recombination and linkage disequilibrium in two genomes of similar size. Trends Genet 20:103–111

Ranasinghe S, Taylor G (1996) Mechanism for increased leaf growth in elevated CO_2. J Exp Bot 47:349–358

Ranish JA, Yi EC, Leslie DM, Purvine SO, Goodlett DR, Eng J, Aebersold R (2003) The study of macromolecular complexes by quantitative proteomics. Nat Genet 33:349–355

Schaffer R, Landgraf J, Perez-Amador M, Wisman E (2000) Monitoring genome-wide expression in plants. Curr Opin Biotechnol 11:162–167

Schena M, Shalon D, Davis RW, Brown PO (1995) Quantitative monitoring of gene-expression patterns with a complementary-DNA microarray. Science 270:467–470

Schrader J, Nilsson J, Mellerowicz E, Berglund A, Nilsson P, Hertzberg M, Sandberg G (2004) A high-resolution transcript profile across the wood-forming meristem of poplar identifies potential regulators of cambial stem cell identity. Plant Cell 16:2278–2292

Seki M, Narusaka M, Ishida J, Nanjo T, Fujita M, Oono Y, Kamiya A, Nakajima M, Enju A, Sakurai T, Satou M, Akiyama K, Taji T, Yamaguchi-Shinozaki K, Carninci P, Kawai J, Hayashizaki Y, Shinozaki K (2002) Monitoring the expression profiles of 7000 *Arabidopsis* genes under drought, cold and high-salinity stresses using a full-length cDNA microarray. Plant J 31:279–292

Sumner LW, Mendes P, Dixon RA (2003) Plant metabolomics: large-scale phytochemistry in the functional genomics era. Phytochemistry 62:817–836

Tao WA, Aebersold R (2003) Advances in quantitative proteomics via stable isotope tagging and mass spectrometry. Curr Opin Biotechnol 14:110–118

Taylor G, Tricker PJ, Zhang FZ, Alston VJ, Miglietta F, Kuzminsky E (2003) Spatial and temporal effects of free-air CO_2 enrichment (POPFACE) on leaf growth, cell expansion, and cell production in a closed canopy of poplar. Plant Physiol 131:177–185

Taylor G, Street NR, Tricker PJ, Sjödin A, Graham L, Skogström O, Calfapietra C, Scarascia-Mugnozza, Janssen S (2005) The transcriptome of *Populus* in elevated CO_2. New Phytol 167:143–154

Thimm O, Blasing O, Gibon Y, Nagel A, Meyer S, Kruger P, Selbig J, Muller LA, Rhee SY, Stitt M (2004) MAPMAN: a user-driven tool to display genomics data sets onto diagrams of metabolic pathways and other biological processes. Plant J 37:914–939

Vodkin LO, Khanna A, Shealy R, Clough SJ, Gonzalez DO, Philip R, Zabala G, Thibaud-Nissen F, Sidarous M, Stromvik MV, Shoop E, Schmidt C, Retzel E, Erpelding J, Shoemaker RC, Rodriguez-Huete AM, Polacco JC, Coryell V, Keim P, Gong G, Liu L, Pardinas J, Schweitzer P (2004) Microarrays for global expression constructed with a low redundancy set of 27,500 sequenced cDNAs representing an array of developmental stages and physiological conditions of the soybean plant. Genomics 5:73

Wagner C, Sefkow M, Kopka J (2003) Construction and application of a mass spectral and retention time index database generated from plant_GC/EI-TOF-MS metabolite profiles. Phytochemistry 62:887–900

Ward JK, Kelly JK (2004) Scaling up evolutionary responses to elevated CO_2: lessons from *Arabidopsis*. Ecol Lett 7:427–440

Wasaki J, Yonetani R, Shinano T, Kai M, Osaki M (2003) Expression of the OsPI1 gene, cloned from rice roots using cDNA microarray, rapidly responds to phosphorus status. New Phytol 158:239–248

Watson BS, Lei ZT, Dixon RA, Sumner LW (2004) Proteomics of Medicago sativa cell walls. Phytochemistry 65:1709–1720

Weir BS, Basten CJ (1990) Sampling strategies for distances between DNA-sequences. Biometrics 46:551–572

Wu R, Ma CX, Casella G (2002) Joint linakge and linkage disequilibrium mapping of quantitative trait loci in natural populations. Genetics 160:779–792

Wu TD (2001) Analysing gene expression data from DNA microarrays to identify candidate genes. J Pathol 195:53–65

21 The Impact of Elevated Atmospheric [CO$_2$] on Soil C and N Dynamics: A Meta-Analysis

K.-J. van Groenigen, M.-A. de Graaff, J. Six, D. Harris, P. Kuikman, and C. van Kessel

21.1 Introduction

The current rise in atmospheric [CO$_2$], a consequence of human activities such as fossil fuel burning and deforestation, is thought to stimulate plant growth in many ecosystems (Bazzaz and Fajer 1990). Gifford (1994) suggested that the resulting increase in C assimilation by plants and its subsequent sequestration in the soil could counterbalance CO$_2$ emissions. However, higher plant growth rates in a CO$_2$-rich world can only be sustained if the soil supplies plants with additional nutrients (Zak et al. 2000; Luo et al. 2004). Therefore, the effect of elevated (e)[CO$_2$] on soil N availability is of key importance when predicting the potential for C storage in terrestrial ecosystems.

In short-term experiments, soil N availability can decrease (Diaz et al. 1993) or increase (Zak et al. 1993) under e[CO$_2$], depending on the response of the soil microbial community. Moreover, plants under e[CO$_2$] can increase N uptake at the expense of microbial N consumption (Hu et al. 2001). Clearly, the impact of higher [CO$_2$] levels on C and N dynamics in terrestrial ecosystems depends on a set of complex interactions between soil and plants. Also, the establishment of equilibrium between soil organic matter (SOM) input and decomposition can take up to decades or longer. Therefore, we need long-term experiments under realistic field situations to predict changes in ecosystems under future [CO$_2$].

The use of open-top chambers (OTC) and free-air carbon dioxide enrichment (FACE) techniques allowed for CO$_2$ fumigation studies under far more realistic conditions than before (Rogers et al. 1983; Hendrey 1993; Chapter 2). Over the past two decades, many OTC and FACE experiments have been conducted, covering a wide range of terrestrial ecosystems. Soil characteristics

Ecological Studies, Vol. 187
J. Nösberger, S.P. Long, R.J. Norby, M. Stitt, G.R. Hendrey, H. Blum (Eds.)
Managed Ecosystems and CO$_2$
Case Studies, Processes, and Perspectives
© Springer-Verlag Berlin Heidelberg 2006

related to C and N cycling have been studied in most of these experiments, but no clear pattern has emerged that allows us to generalize about CO_2 enrichment effects on SOM dynamics (Zak et al. 2000).

By affecting soil C and N dynamics, ecosystem management practices could influence an ecosystem's response to $e[CO_2]$ (see Chapter 1). For example, the addition of fertilizer N might affect both the input and loss of soil C by affecting the $[CO_2]$ response of plant growth (Oren et al. 2001) and decomposition rates (Rice et al. 1994; Niklaus and Körner 1996). Intensive soil disturbances can also affect soil C and N dynamics, as they disrupt aggregates containing physically protected SOM (Six et al. 2002). Thus, we should take into account differences in ecosystem management when we compare results from CO_2 enrichment studies.

The sensitivity of individual experiments to detect changes in soil C is low because of high spatial variability and the large size of the soil C pool compared to the input of C (Hungate et al. 1996). The statistical power to identify changes in SOM pools across individual experiments might be increased by a quantitative integration of research results. Meta-analytic methods enable placing confidence limits around effect sizes; therefore they provide a robust statistical test for overall $[CO_2]$ effects across multiple studies (Curtis and Wang 1998). Moreover, they enable to test whether there are significant differences in the mean $[CO_2]$ response between categories of studies (Hedges and Olkin 1985).

Meta-analyses have recently been used to summarize the effect of $e[CO_2]$ on plant physiology, litter quality and decomposition rates (Curtis and Wang 1998; Norby et al. 2001; Ainsworth et al. 2002). In comparison, few researchers have used meta-analysis to summarize $[CO_2]$ effects on SOM dynamics (Jastrow et al. 2005). For this review we compiled the available data from FACE and OTC experiments on a number of soil characteristics related to soil C and N cycling. Using meta-analytic techniques, we compared the effect of CO_2 enrichment on these characteristics between several levels of ecosystem management.

21.2 Materials and Methods

21.2.1 Database Compilation

Data were extracted from 65 published studies on SOM dynamics in FACE and OTC experiments (see end of chapter for list of database references). The response variables included in the meta-analysis are listed in Table 21.1. Whenever values were reported in tables, they were taken directly from the publication. Results presented in graphs were digitized and measured to esti-

Table 21.1 List of response variables included in the meta-analysis; and their abbreviations, as used in figures

Parameter abbreviation	Definition
C	Soil C content
N	Soil N content
C:N	Soil C to N ratio
MicC	Microbial C content
MicN	Microbial N content
rCO$_2$	Microbial respiration, measured in short-term (<15 days) incubations
MinN	N mineralization rates, measured in short-term (<30 days) incubations
GNI	Gross N immobilization, measured by ^{15}N pool dilution methods
GNM	Gross N mineralization, measured by ^{15}N pool dilution methods
Cmax	Potential mineralizable C, measured in long-term (>55 days) incubations
Csol	Soluble C

mate values for the particular pool or flux. Data reported on an area base were converted to a weight base, using soil density data whenever available. In all other cases, equal bulk soil density in current (c)[CO$_2$] and e[CO$_2$] treatments was assumed.

To make meaningful comparisons between experiments, a number of restrictions were applied to the data. Data were included for soil layers ranging in depth from 0–5 cm to 0–40 cm. When data were reported for multiple depths, we included results that best represented the 0–10 cm soil layer. As our review focuses on mineral soils, measurements on forest litter layers, marsh and rice paddies were excluded from the database. The e[CO$_2$] levels of the experiments included in the data base ranged from 450 ppm to 750 ppm. Data were not corrected for the degree of CO$_2$ enrichment. When more than one e[CO$_2$] level was included in the experiment, only the results at the level that is approximately twice c[CO$_2$] were included. The duration of the [CO$_2$] treatment had to be at least 100 days (the approximate length of a growing season in the temperate zone). Results from different N treatments, plant species and communities, soils and irrigation treatments within the same experiment were considered independent measurements. These studies were included separately in the database. Reich et al. (2001) reported the effect of e[CO$_2$] on N mineralization under 16 grassland species, which were grown both in monocultures and in mixtures. If results would be included for all separate species, their experiment would dominate the MinN data set. Therefore, only the results for the 16 species mixture plots were

Table 21.2 Categorical variables used to summarize experimental conditions; and the values they could assume in the analysis of between-group heterogeneity

	Level 1	Level 2	Level 3
Method	OTC	FACE	
N fertilization (kg ha^{-1} year^{-1})	<30	30–150	>150
Vegetation (a)	Planted	Natural	
Vegetation (b)	Herbaceous	Woody	
Vegetation (c)	Crops	Non-crops	

used. For OTC experiments, data from the control chambers rather than the non-chamber control plots were included as the results at $c[CO_2]$. In case these were available, data for blower controls in FACE experiments were included as the results at $c[CO_2]$.

Results on C and N fluxes were all based on incubation data (laboratory and in situ). Data for microbial biomass were based on the fumigation– extraction method (Vance et al. 1987) and the substrate induced respiration technique (Anderson and Domsch 1978). Data for soluble C were based on extractions using either cold water, or 0.5 M solutions of KCl or K_2SO_4. For total soil C and N contents, only the most recent data for each study were incorporated. For data on microbial biomass and activities, time series from the most recent year of measurement were included whenever available. In these cases, the average values at $c[CO_2]$ and $e[CO_2]$ were calculated over time. Experimental conditions were summarized by a number of categorical variables: type of exposure facility, N addition and vegetation type (Table 21.2). Vegetation was characterized as either herbaceous or woody, crop or non-crop and planted or natural. Vegetation was considered planted if it was sown or placed into the soil less than 10 years before the start of the CO_2 fumigation. Thus, all experiments on planted vegetation were conducted on physically disturbed soil. The duration of each experiment (i.e. years of CO_2 fumigation) was also included in the database. The number of observations on microbial responses to $e[CO_2]$ was relatively low compared to observations on soil C and N contents. To ensure that each N fertilization class was well represented, we decided to pool the two highest classes for these response variables.

21.2.2 Statistical Analyses

The data set was analyzed with meta-analytic techniques described by Curtis and Wang (1998) and Ainsworth et al. (2002), using the statistical software MetaWin ver. 2.1 (Rosenberg et al. 2000). The natural log of the response ratio (r = response at $e[CO_2]$/response at $c[CO_2]$) was used as a metric for all variables and is reported as the percent change at e[CO$_2$] ([r–1]×100). A mixed model was used for our analysis, based on the assumption that a random variation in [CO$_2$] responses occurred between studies.

The effect of e[CO$_2$] on total soil C content, soil N content and soil C:N ratios was analyzed using a weighted parametric analysis. In this analysis, each individual observation was weighted by the reciprocal of the mixed-model variance, which was the sum of the variance of the natural log of the response ratio and the pooled within-class variance (Curtis and Wang 1998). Results from studies that did not report standard deviation were included conservatively by assigning them the minimum weight calculated from other studies in the data set, according to Norby et al. (2001).

To test whether differences in [CO$_2$] responses could be explained by experimental conditions, results were compared between categories of studies. In the weighted analysis, the total heterogeneity for a group of comparisons (Q_t) was partitioned into within-class heterogeneity (Q_w) and between class heterogeneity (Q_b), according to Curtis and Wang (1998). The impact of experiment duration was tested as a continuous variable. For this analysis, Q_t was partitioned in heterogeneity that was explained by the regression model (Q_m) and the amount of residual error heterogeneity.

For all other response variables, the standard deviations were often not available. Because standard deviations are required for a weighted parametric analysis, an unweighted analysis using resampling techniques was conducted on these variables instead. In the unweighted analysis, bootstrapping techniques were used to calculate confidence intervals on mean effect size estimates for the whole data set and for categories of studies (Adams et al. 1997). To assess the effect of experiment duration on variability in microbial [CO$_2$] responses, we compared short-term (1–2 growing seasons) and long-term (>2 growing seasons) experiments.

In both the unweighted and weighted analyses, the [CO$_2$] effect on a response variable was considered significant if the 95% confidence interval did not overlap 0, and marginally significant if the 90% confidence interval did not overlap 0. Means of categories were considered significantly different if their 95% confidence intervals did not overlap.

21.3 Results

21.3.1 Soil C and N Contents

Total soil C increased significantly by 4.1% at e[CO$_2$] (Fig. 21.1), but the [CO$_2$] response depended on fertilizer N additions (Table 21.3). In experiments that received less than 30 kg N ha^{-1} year^{-1}, soil C contents were unaffected. How-

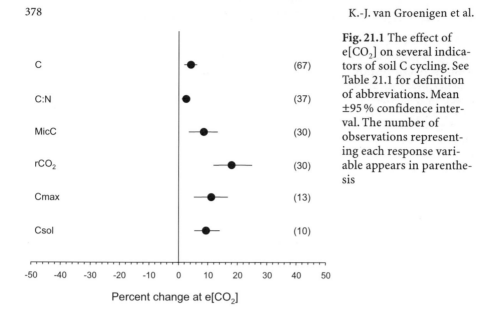

Fig. 21.1 The effect of e[CO_2] on several indicators of soil C cycling. See Table 21.1 for definition of abbreviations. Mean ±95 % confidence interval. The number of observations representing each response variable appears in parenthesis

Table 21.3 Between group heterogeneity (Q_b) for the [CO_2] response of total soil C and N contents and C:N ratio. For the continuous variable time, the heterogeneity explained by the regression model (Q_m) is reported. Response variables are represented by k observations. * $P<0.05$, ** $P<0.01$

Variable	k	Method	N	Vegetation (planted/ natural)	Vegetation (herbaceous/ woody)	Vegetation (crops/ non-crops)	Time
C	67	0.03	7.56*	0.47	2.19	1.99	2.24
N	48	0.00	1.37	0.62	1.03	0.16	0.68
C/N	37	0.40	5.78	11.50**	0.89	5.34*	0.39

ever, soil C accumulation became apparent with increasing inputs of N. Experiments receiving between 30 kg and 150 kg N ha^{-1} year^{-1} showed a significant increase in soil C at e[CO_2] (+4.3 %), whereas experiments receiving >150 kg N ha^{-1} year^{-1} showed a significant [CO_2] response of +8.1 %. Within the N fertilization classes, none of the variables affected the [CO_2] response for soil C.

As the data set was divided, not every categorical variable was represented in each sub-group. The data set of experiments receiving >150 kg N ha^{-1} year^{-1} contained no experiments on natural vegetation. Moreover, this subgroup was heavily biased towards herbaceous plants; only one of the studies was carried out on woody plants. However, across the whole soil C data set, the [CO_2] response did not differ between herbaceous and woody species or between

Fig. 21.2 The effect of e[CO$_2$] on several indicators of soil N cycling. See Table 21.1 for definition of abbreviations. Mean ±95 % confidence interval. The number of observations representing each response variable appears in parenthesis

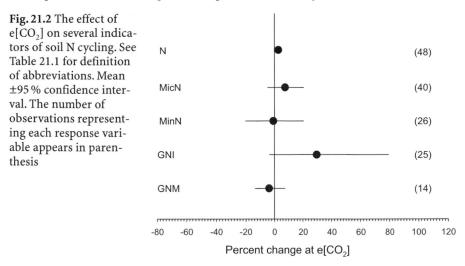

Percent change at e[CO$_2$]

planted and natural vegetation (Table 21.3). None of the other categorical variables, or the continuous variable time affected the [CO$_2$] response for soil C content either.

Soluble C increased by 9.4 % at e[CO$_2$]. Soil N concentrations significantly increased by 2.8 % at e[CO$_2$], with none of the categorical variables affecting the [CO$_2$] response (Fig. 21.2). Averaged over all experiments, soil C:N increased by 2.4 % at e[CO$_2$] (Fig. 21.1). The [CO$_2$] response with respect to C:N differed between experiments on planted and natural vegetation (Table 21.3); only under planted vegetation did soil C:N increase (+4.0 %) at e[CO$_2$]. A significant Q_b was found for the distinction between crops and non-crops. Nevertheless, the [CO$_2$] response for soil C:N did not differ between these categories (i.e. their 95 % confidence intervals overlapped). Experiments on crops were represented by only two studies with highly different [CO$_2$] responses. Thus, of all categorical variables, the distinction between planted and natural vegetation best explained the variation in [CO$_2$] effects on soil C:N.

21.3.2 Microbial Biomass and Activity

Microbial respiration and potential mineralizable C increased by 18.0 % and 11.1 % at e[CO$_2$], respectively. Microbial C also increased, but showed a smaller response (+8.5 %). The [CO$_2$] response for microbial respiration and microbial C tended to be stronger for high N than for low N treatments (Fig. 21.3), but the differences between fertilization classes were not significant. The effect of e[CO$_2$] on microbial N pools and fluxes was characterized by large confidence intervals, indicating large differences in [CO$_2$] responses

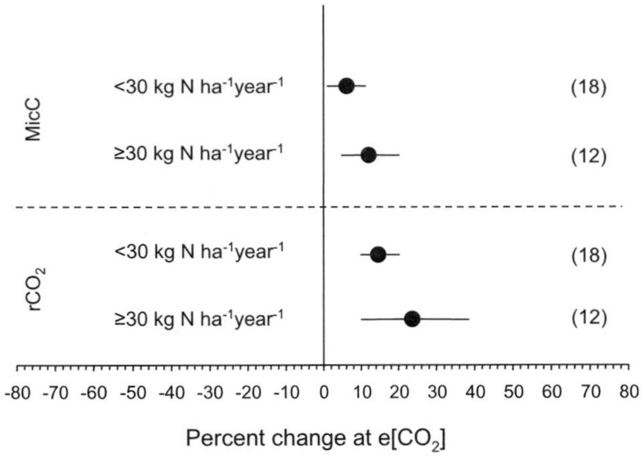

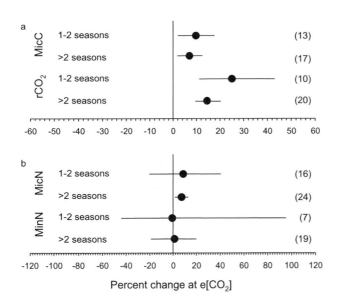

Fig. 21.3 The effect of e[CO$_2$] on microbial C and microbial respiration, as affected by N addition. See Table 21.1 for definition of abbreviations. Mean ±95 % confidence interval. The number of studies representing each response variable appears in parenthesis

Fig. 21.4 The effect of e[CO$_2$] on microbial C and microbial respiration (**a**) and microbial N and N mineralization (**b**) in short- and long-term studies. See Table 21.1 for definition of abbreviations. Mean ±95 % confidence interval. The number of observations representing each response variable appears in parenthesis

between studies (Fig. 21.2). Of all variables relating to soil N dynamics, only gross N immobilization was marginally significantly affected by e[CO$_2$] (+29.8 %). Although CO$_2$ enrichment had no overall effect on microbial N contents, it significantly increased microbial N in long-term studies (Fig. 21.4b). As with microbial C, most of the short-term studies on microbial N involved planted vegetation. The [CO$_2$] response for microbial N, N mineralization, and to a lesser extent, microbial C and microbial respiration showed relatively large confidence intervals for short-term experiments (Fig. 21.4a, b).

21.4 Discussion

21.4.1 Soil C Contents

The overall increase in soil C pools at e[CO$_2$] suggests a potential for soil C sequestration. However, the average 4.1 % increase in total soil C is small, taking into consideration the spatial variability in individual field experiments. Not surprisingly, only a small number of experiments reported significant increases in soil C at e[CO$_2$] (Rice et al. 1994; Wood et al. 1994; Prior et al. 1997, 2004, 2005; Williams et al. 2000; Hagedorn et al. 2001).

Our estimate of the effect of CO$_2$ enrichment on total soil C has several constraints. First, initial soil C contents in c[CO$_2$] and e[CO$_2$] plots might differ, thereby affecting the measured [CO$_2$] effect on soil C contents (Schlesinger and Lichter 2001). In individual experiments, however, the chance of larger initial C contents in c[CO$_2$] plots is equivalent to the chance of larger initial C contents in e[CO$_2$] plots. Pre-existing differences are therefore expected to cancel each other out in large data sets such as the one used in the current meta-analysis.

Secondly, our estimate solely concerns soil C in the top soil layer. As most of the new C enters the soil in the top layer, this is where the [CO$_2$] effect on soil C is expected to be strongest. Also, treatment effects on soil C stocks might differ between soil depths. For instance, Mack et al. (2004) found that N fertilization increased SOM in the top layer of an arctic tundra soil, whereas it caused SOM at lower depths to decline. Thus, great caution is required when extrapolating the [CO$_2$] effect on soil C in the top layer to lower depths.

The increase in soluble C at e[CO$_2$] suggests that leaching of C might increase. Leached C either accumulates at lower depths or ends up in the groundwater, thereby contributing to C sequestration. However, a total C analysis of the top 75 cm of a grassland soil after 10 years of e[CO$_2$] showed no indication of increased precipitation of dissolved organic C (DOC) at lower depths (Van Kessel et al. 2006). In a model forest under e[CO$_2$], losses of C through DOC leaching were small compared to soil C inputs (Hagedorn et al. 2002). These results suggest that losses of DOC do not form a main part of the 'missing carbon sink'.

Soil C contents only increased at e[CO$_2$] in experiments that received 30 kg N ha^{-1} year^{-1} or more. These results corroborate models predicting that additional ecosystem C storage under future [CO$_2$] will be limited by N availability (Hungate et al. 2004a). Moreover, they suggest that previous estimates of soil C sequestration at e[CO$_2$] might be overly optimistic (Jastrow et al. 2005). It should be noted that the initial soil N availability of individual experiments has not been taken into account in our meta-analysis. Differences in initial soil N availability can affect the impact of e[CO$_2$] on plant growth, and thus soil C input. Unfortunately, there is no indicator on the nutritional status of the soil that was available for all experiments.

As plant growth under $e[CO_2]$ is often limited by nutrient availability (Curtis and Wang 1998; Oren et al. 2001), we argue that the dependence of soil C storage on N fertilization is mainly caused by its effect on soil C input. Nitrogen additions can increase soil C stabilization (Neff et al. 2002), thereby exacerbating increases in soil C following a rise in soil C inputs. However, to our knowledge, not one study has found a positive interaction between elevated CO_2 and N fertilization for stabilization of new C. Thus, we found no evidence that suggests the positive effect of N fertilization on soil C sequestration under elevated CO_2 is related to its effect on C stabilization per se.

The positive effect of $e[CO_2]$ on potential mineralizable C is relatively strong compared to its effect on soil C contents, suggesting that the rise in soil C is mainly due to expanding labile C pools. As these pools are typically small and have high turnover rates, their contribution to soil C sequestration is limited. A substantial increase in soil C requires that additional C entering the soil is stabilized in long-lived pools. Thus, the protective capacity of soils largely determines the potential to sequester C under $e[CO_2]$ in the long term (Six et al. 2002). Ongoing soil disturbance decreases the physical protection of new SOM, thereby reducing the stabilization of C (Paustian et al. 2000). Yet, crops and non-crops did not differ in the $[CO_2]$ response for soil C (Table 21.3), suggesting that ongoing physical soil disturbance did not interact with the effect of CO_2 enrichment. One explanation for the lack of such an interaction is that CO_2 fumigation studies are too short for differences in the physical stabilization of C to affect total C contents. Also, most experiments on crops are heavily fertilized. The positive effect of N fertilization on soil C input could potentially compensate for the negative effect of soil disturbance on soil C accumulation.

As soil C input generally increases at $e[CO_2]$, its effect on soil C contents is expected to grow over time in most individual experiments. Therefore, we hypothesized that the effect of $e[CO_2]$ on soil C was stronger in long-term studies. In fact, all studies that found a significant increase in soil C at $e[CO_2]$ were at least 2 years old. Yet, experiment duration did not affect the $[CO_2]$ response for soil C in the total data set, nor in any of the N fertilization classes (Table 21.3). Apparently, differences between experiments are too large to detect a time $\times$ $[CO_2]$ interaction.

Several meta-analyses suggest that the effect of $e[CO_2]$ on plant growth is stronger for trees than for herbaceous plants (Ainsworth et al. 2005; De Graaff et al. 2006). Our data show that this difference does not translate into a stronger $[CO_2]$ response for soil C under woody plants. A possible explanation for the lack of difference between plant life forms is that the residence time of assimilated C is relatively short for herbaceous plants (Schlesinger 1997). Thus in this category an increase in plant growth could rapidly result in increases in soil C input and thereby promote soil C sequestration. Given that

CO$_2$ fumigation studies typically last only a few years, a relative large increase in soil C inputs for herbaceous plants might compensate for a relative small increase in plant growth.

21.4.2 Microbial Biomass and Activity

In our meta-analysis, CO$_2$ enrichment significantly increased soil microbial C contents and microbial respiration as measured by incubation. An increase in [CO$_2$] also stimulated microbial respiration in an intact plant–soil system (Hungate et al. 1997a) and enzymatic activities under natural grassland (Ebersberger et al. 2003), all pointing to enhanced microbial activity. Numerous studies found increases in soil C input at e[CO$_2$] (Hungate et al. 1997a; Hoosbeek et al. 2004; Pendall et al. 2004). As soil microorganisms are generally C-limited (Anderson and Domsch 1978), an increase in C availability most likely contributed to the rise in microbial activity. In water-limited ecosystems, an increase in [CO$_2$] can also enhance microbial activity through soil moisture feedbacks (Ebersberger et al. 2003; Pendall et al. 2003), due to an increase in water use efficiency of plants.

In nutrient-poor grasslands and prairies, the effect of e[CO$_2$] on microbial respiration is generally small, but strongly increases when fertilizer N is added (Rice et al. 1994; Niklaus and Körner 1996; Hungate et al. 1997a). Nonetheless, our meta-analysis did not reveal significant differences in the [CO$_2$] response of microbial respiration between N fertilizer classes. We argue that the lack of significance is due to differences in initial soil N availability between experiments, which are obscuring the effect of N additions. However, the significant increase in soil C contents in fertilized ecosystems under e[CO$_2$] suggests that a possible positive effect of N additions on the [CO$_2$] response for microbial respiration does not outweigh the increase in soil C input.

The confidence intervals associated with the [CO$_2$] responses for microbial biomass and activity are relatively large in short-term studies (Fig. 21.4a, b). In these studies, soil C input largely consists of rhizodeposition. Whereas rhizodeposition generally increases under e[CO$_2$] (see Chapter 22), the magnitude of the [CO$_2$] response differs strongly between species (e.g. Paterson et al. 1996), possibly contributing to the variety of microbial responses. Also, short-term studies mostly involved planted vegetation and were thus conducted on physically disturbed soil. Soil disturbance stimulates the release of physically protected native SOM, a process that might affect microbial responses to e[CO$_2$]. Thus, we do not know whether it is the short duration of experiments per se or soil disturbance that broadens the array of microbial responses. Nonetheless, our results suggest that one should be cautious when using results from short-term experiments on planted vegetation for long-term predictions on the ecosystem level.

21.4.3 Soil N Dynamics

The significant increase in soil C:N ratios at e[CO_2] suggests a potential decrease in soil N availability. However, the large variability in [CO_2] responses of microbial N contents and N transformation rates makes it hard to predict how future increases in [CO_2] will affect soil N cycling. On the one hand, the marginally significant increase in gross N immobilization at e[CO_2] suggests a rise in microbial N demand. On the other hand, Mikan et al. (2000) found that gross N immobilization and plant N uptake by *Populus tremuloides* increased under e[CO_2], without affecting microbial N contents. Their results suggest that the effect of increased gross N immobilization on soil N availability can be compensated by enhanced turnover of microbial N. However, in our meta-analysis, microbial N contents increased under e[CO_2] in long-term experiments. This implies that, over time, a rise in [CO_2] will increase the amount of N immobilized in the microbial biomass.

These findings are in line with Luo et al. (2004), who suggested that, when e[CO_2] stimulates biomass production in unfertilized ecosystems, the resulting increase in soil C inputs will gradually reduce N availability. This process is called progressive nitrogen limitation (PNL). The negative effect of PNL might be temporarily alleviated by increased efficiency of plant N uptake, due to increased fine root production (Mikan et al. 2000) or increased mycorrhizal colonization of roots (Rillig et al. 2000). However, the rise in plant growth and soil C input resulting from these adaptations increases the soil C:N ratio, thereby further enhancing microbial competition for N. Therefore, [CO_2]-induced mechanisms that increase plant N uptake without a net ecosystem gain of N are self-limiting (Hungate et al. 2004a).

In theory, PNL can be postponed or alleviated in ecosystems where e[CO_2] increases biological N_2 fixation and/or increases N retention (Luo et al. 2004; Chapter 18). The significant increase in soil N contents at e[CO_2] implies that one or both of these processes is indeed stimulated. To our knowledge, a sustained increase in symbiotic N_2 fixation under e[CO_2] has only been measured in fertilized systems (Zanetti et al. 1996; Ross et al. 2004). In an unfertilized scrub oak community, the initial doubling of N_2 fixation was followed by a decrease in following years, which was attributed to Mo limitation (Hungate et al. 2004b). In unfertilized natural grasslands, N_2 fixation did not increase at e[CO_2] due to P limitations (Niklaus et al. 1998). These results suggest that increases in symbiotic N_2 fixation under e[CO_2] might have a limited effect on preventing PNL.

Soil N leaching in a number of forest ecosystems decreased at e[CO_2] (Hungate et al. 1999; Hagedorn et al. 2000; Johnson et al. 2004), suggesting a positive effect on N retention. The effect of e[CO_2] on trace N gas losses is less clear. The emission of N_2O in a natural desert ecosystem was not affected by e[CO_2] (Billings et al. 2002), but NH_3 volatilization sporadically increased. Hungate et al. (1997b) found that NO emissions from natural grassland decreased at e[CO$_2$]. Similarly, Mosier et al. (2003) found that NO emissions from a fertilized shortgrass steppe were lower in plots that had previously been subjected to e[CO$_2$]. In both cases, the reduction in gaseous N losses was explained by an increase in microbial immobilization of N.

Williams et al. (2001) found an increased ^{15}N retention in SOM under a tallgrass prairie at e[CO$_2$], which was explained by enhanced microbial N demand. In the same experiment, soil C contents increased significantly under e[CO$_2$] (Williams et al. 2000). Conversely, Niklaus and Körner (2004) found that ^{15}N retention in a native grassland was not affected by e[CO$_2$]. In the same experiment, a relative small stimulation of soil C input and limited stabilization of new C prevented increases in soil C contents at e[CO$_2$]. From these results, we conclude that microbial N immobilization is a main mechanism involved in N retention in nutrient poor soils under e[CO$_2$]. However, an increase in N retention at e[CO$_2$] requires an increase in soil C. Effectively, the additional C sequestered at e[CO$_2$] is used to retain more N, so that the N cycle tracks the C cycle (Thornley and Cannell 2000).

It is important to note that all experimental studies finding PNL applied a step-increase in atmospheric [CO$_2$]. In the real world, a gradual increase in atmospheric [CO$_2$] would allow more time to adjust the N status of the soil, possibly decreasing the severity of PNL (Luo et al. 2004). In unfertilized systems, the gain of N through increased retention under e[CO$_2$] is limited by the amount of atmospheric N deposition. In these systems, the net gain of N will thus be small (Thornley and Cannel 2000). It may take decades to centuries for ecosystems to reach a new equilibrium, where gain of N through retention has compensated for the [CO$_2$] induced decrease in soil N availability and factors other than N availability are limiting the ecosystem's response to e[CO$_2$].

21.5 Future Research Needs

Recently, several papers discussed future research needs regarding SOM dynamics under e[CO$_2$] (Luo et al. 2004; Pendall et al. 2004; Jones and Donnelly 2004). One of the general conclusions was that the extrapolation of results from field experiments to long-term predictions for actual ecosystems continues to be a main challenge. Longer durations of experiments and combined experimental and modeling studies will contribute to more accurate predictions on the fate of SOM in a CO$_2$-rich world.

More experiments on natural ecosystems are needed, as they are currently underrepresented. This is especially the case for forest ecosystems: to date, no data are available on the effect of e[CO$_2$] on SOM dynamics in mature, natural forests. The recently developed WebFACE technique (Pepin and Körner 2002) enables to study the effect of e[CO$_2$] on tall canopy trees. This technique is currently used to study the effect of e[CO_2] on plant physiology (Körner et al. 2005), but may enable studying SOM dynamics in the future.

Our meta-analysis strongly suggests that additional N is needed to sequester C under e[CO_2]. Thus, in unfertilized ecosystems, a mechanistic understanding of N supply processes is essential to predict the effect of e[CO_2] on SOM dynamics. Whereas available data suggest that the effect of e[CO_2] on N_2 fixation will be limited in natural ecosystems (Hungate et al. 2004b; Niklaus et al. 1998), more experiments on a wider number of ecosystems are needed before definitive conclusions can be drawn. The effect of e[CO_2] on N retention has been studied by following the fate of isotopically labeled inorganic N in grassland and prairie soils (Williams et al. 2001; Niklaus and Körner 2004; Van Kessel et al. 2006). The same technique might be used to study N retention in other ecosystems.

We also need more insight into the effect of e[CO_2] on soil C stabilization mechanisms. Soil aggregation has been found to increase under e[CO_2] (Rillig et al. 1999; Six et al. 2001; Prior et al. 2004), thereby potentially increasing physical SOM stabilization. An increase in [CO_2] might also affect chemical SOM stabilization by its effect on litter quality (e.g. Parsons et al. 2004). The long-term impact of these feedback mechanisms on C sequestration is unclear and requires more experimental data for verification. Finally, our meta-analysis suggests that physical soil disturbances can influence the effect of e[CO_2] on SOM dynamics. To our knowledge, the combined effect of physical soil disturbance and increased atmospheric [CO_2] has only been studied in one experiment (Prior et al. 2005). Such experiments might enable us to reveal the relative importance of atmospheric [CO_2] and ecosystem management practices for soil C sequestration.

21.6 Conclusions

Field experiments are a valuable tool for predicting the effect of future [CO_2] on soil organic matter (SOM) dynamics. Using meta-analytic techniques, we reviewed the effect of CO_2 enrichment on SOM dynamics under field conditions. Our analysis summarized the effects of 65 studies and covered results from both OTC and FACE experiments.

- Averaged over all studies, soil C contents increased by 4.1%. A relatively strong increase in potential mineralizable C (+11.1%) and soluble C (+9.4%) suggests that the rise in soil C is largely due to expanding labile C pools.
- The effect of CO_2 enrichment on soil C contents depended on N availability; soil C contents only increased under e[CO_2] in experiments that received 30 kg N ha^{-1} year^{-1} or more.

- Short-term experiments on planted vegetation showed a larger range of soil microbial responses to $e[CO_2]$ than long-term experiments on natural vegetation.
- Microbial C contents and microbial respiration increased by 8.5 % and 18.0 % at $e[CO_2]$, respectively. The effect of CO_2 enrichment on microbial activity tended to be higher for fertilized studies. However, since soil C storage at $e[CO_2]$ depends on N additions, we conclude that at high N fertilization rates the $[CO_2]$ response for soil C input outweighs that of microbial respiration.
- Stimulation of gross N immobilization at $e[CO_2]$ was 29.8 %, whereas gross N mineralization rates remained unaffected. These results, together with a 7.2 % increase in microbial N contents at $e[CO_2]$ in long-term studies, suggest that higher $[CO_2]$ levels enhanced microbial N demand.
- An increase in microbial N demand at $e[CO_2]$, together with the dependency on N additions for soil C storage at $e[CO_2]$ are in line with the progressive nitrogen limitation theory.

Database References.   Allen et al. (2000) Ecological Applications 10:437–448. Ambus and Robertson (1999) Plant and Soil 209:1–8. Andrews et al. (2000) Soil Biology and Biochemistry 32:699–706. Barnard et al. (2004) Global Change Biology 10:488–497. Billings et al. (2002) Soil Biology and Biochemistry 34:1777–1784. Billings et al. (2004) Global Biochemical Cycles 18:GB1011. Calabritto et al. (2002) Journal of Mediterranean Ecology 3:23–27. Cardon et al. (2001) Soil Biology and Biochemistry 33:365–373. Carnol et al. (2002) Global Change Biology 8:590–598. De Graaff et al. (2004) Global Change Biology 10:1922–1935. Dijkstra et al. (2005) Plant Soil 272:41–52. Ebersberger et al. (2003) Soil Biology and Biochemistry 35:965–972. Finzi and Schlesinger (2003) Ecosystems 6:444–456. Hagedorn et al. (2001) European Journal of Soil Science 52:619–628. Hagedorn et al. (2002) Soil Biology & Biochemistry 34:355–366. Hagedorn et al. (2003) Global Change Biology 9:862–872. Holmes et al. (2003) Global Change Biology 9:1743–1750. Hoosbeek et al. (2004) Global Biogeochemical Cycles 18:GB1040. Hu et al. (2001) Nature 409:188–191. Hungate et al. (1996) Ecology 77:2505–2515. Hungate et al. (1997a) Nature 388:576–579. Hungate et al. (1997b) Biogeochemistry 37:89–109. Hungate et al. (1997 c) Oecologia 109:149–153. Hungate et al. (1999) Global Change Biology 5:781–789. Ineson et al. (1996) Plant and Soil 187:345–350. Islam et al. (2000) Global Change Biology 6:255–265. Jäger et al. (2003) Journal of Applied Botany 77:117–127. Jastrow et al. (2005) Global Change Biology 11:2057–2064. Johnson et al. (1994) Plant and Soil 165: 129–138. Johnson et al. (2000) Plant and Soil 224:99–113. Johnson et al. (2002) Nature 416:82–83. Johnson et al. (2003) Ecological Applications 13: 1388–1399. Johnson et al. (2004) Biogeochemistry 69:379–403. Leavitt et al. (2001) New Phytologist 150:305–314. Lichter et al. (2005) Ecology 86:1835–1847. Moscatelli et al. (2001) Soil Use and Management 17:195–202. Niklaus and Körner (1996) Plant and Soil 184:219–229. Niklaus (1998) Global Change Biology 4:451–458. Niklaus et al. (1998) Oecologia 116:67–75. Niklaus et al. (2001) Oecologia 127:540–548. Niklaus et al. (2003) Global Change Biology 9:585–600. Pendall et al. (2004) New Phytologist 162:447–458. Prior et al. (1997) Journal of Environmental Quality 26:1161–1166. Prior et al. (2004) Soil Science 169:434–439. Prior et al. (2005) Global Change Biology 11:657–665. Reich et al. (2001) Nature 410:809–812. Rice et al. (1994) Plant and Soil 165:67–74. Richter et al. (2003) Soil Biology and Biochemistry

35:1325–1335. Ross et al. (2004) Plant and Soil 260:183–196. Schortemeyer et al. (1996) Soil Biology and Biochemistry 28:1717–1724. Schortemeyer et al. (2000) Global Change Biology 6:383–391. Sinsabaugh et al. (2003) Applied Soil Ecology 24:263–271. Søe et al. (2004) Plant and Soil 262:85–94. Sonnemann and Wolters (2005) Global Change Biology 11:1148–1155. Torbert et al. (2004) Environmental Management 33:132–138. Van Kessel et al. (2000) Global Change Biology 6:435–444. Van Kessel et al. (2006) Global Change Biol (in press). Wang et al. (1998) Acta Botanica Sinica 40: 1169–1172. Williams et al. (2003) Global Biochemical Cycles 17:GB 1041. Williams et al. (2000) Plant and Soil 227:127–137. Williams et al. (2001) SSSAJ 65:340–346. Williams et al. (2004) SSSAJ 68:148–153. Wood et al. (1994) Agricultural and Forest Meteorology 70:103–116. Zak et al. (1993) Plant and Soil 151:105–117. Zak et al. (2000b) Ecological Applications 10:47–59. Zak et al. (2003) Ecological Applications 13:1508–1514

Acknowledgements. We would like to thank Mark Williams, Donald R. Zak, Sharon Billings, Adrien Finzi, William Holmes, Des Ross, Diana Ebersberger, Frank Hagedorn, Marcel Hoosbeek, Steve Prior, Shuijin Hu, Michael Richter, Cristina Moscatelli, Paolo de Angelis, Pascal A. Niklaus, John Lichter, Peter Reich, Feike Dijkstra, Dale Johnson and Zoe Cardon for sharing their raw data with us. Thanks to Lisa Ainsworth for her advice on the meta-analytic procedure. Finally, we would like to thank Bruce Hungate for his useful comments on an earlier version of the manuscript. This study was funded by the National Science Foundation, Division of Environmental Biology (grant number DEB-0120169), and the Dutch DWK Research Programme on Nutrient Management from the Ministry of Agriculture, Nature Management and Fisheries.

References

Adams DC, Gurevitch J, Rosenberg MS (1997) Resampling tests for meta-analysis of ecological data. Ecology 78:1277–1283

Ainsworth EA, Davey PA, Bernacchi CJ, Dermody OC, Heaton EA, Moore DJ, Morgan PB, Naidu SL, Yoo RA HS, Zhu XG, Curtis PS, Long SP (2002) A meta-analysis of elevated CO_2 effects on soybean (*Glycine max*) physiology, growth and yield. Global Change Biol 8:695–709

Ainsworth EA, Long SP (2005) What have we learned from 15 years of free-air CO_2 enrichment (FACE)? A meta-analytic review of the responses of photosynthesis, canopy properties and plant production to rising CO_2. New Phytol 165:351–372

Anderson JPE, Domsch KH (1978) A physiological method for the quantitative measurement of microbial biomass in soil. Soil Biol Biochem 10:215–221

Bazzaz FA, Fajer ED (1990) Plant life in a CO_2-rich world. Sci Am 266:68–74

Billings SA, Schaeffer SM, Evans RD (2002) Trace N gas losses and N mineralization in Mojave desert soils exposed to elevated CO_2. Soil Biol Biochem 34:1777–1784

Curtis PS, Wang XZ (1998) A meta-analysis of elevated CO_2 effects on woody plant mass, form and physiology. Oecologia 113:299–313

De Graaff MA, Van Groenigen KJ, Six J, Hungate BA, Van Kessel C (2006) Interactions between plant growth and soil nutrient cycling under elevated CO_2: a meta-analysis. Global Change Biol (in press)

Diaz S, Grime JP, Harris J, McPherson E (1993) Evidence of a feedback mechanism limiting plant response to elevated carbon dioxide. Nature 363:616–617

Ebersberger D, Niklaus PA, Kandeler E (2003) Long term CO$_2$ enrichment stimulates N-mineralisation and enzyme activities in calcareous grassland. Soil Biol Biochem 35:965–972

Gifford RM (1994) The global carbon cycle: a viewpoint on the missing sink. Aust J Plant Physiol 21:1–15

Hagedorn F, Bucher JB, Tarjan D, Rusert P, Bucher-Waillin I (2000) Responses of N fluxes and pools to elevated atmospheric CO$_2$ in model forest ecosystems with acidic and calcareous soils. Plant Soil 224:273–286

Hagedorn F, Maurer S, Egli P, Blaser P, Bucher JB, Siegwolf R (2001) Carbon sequestration in forest soils: effect of soil type, atmospheric CO$_2$ enrichment, and N deposition. Eur J Soil Sci 52:619–628

Hagedorn F, Blaser P, Siegwolf F (2002) Elevated atmospheric CO$_2$ and increased N deposition effects on dissolved organic carbon – clues from δ^{13}C signature. Soil Biol Biochem 34:355–366

Hedges LV, Olkin I (1985) Statistical methods for meta-analysis. Academic, New York

Hendrey GR (1993) Free-air carbon dioxide enrichment for plant research in the field. Smoley, Boca Raton, Fla.

Hoosbeek MR, Lukac M, Van Dam D, Godbold DL, Velthorst EJ, Biondi FA, Peressotti A, Cotrufo MF, Angelis P de, Scarasscia-Mugnozza G (2004) More new carbon in the mineral soil of a poplar plantation under free air carbon enrichment (PopFACE): cause of increased priming effect? Global Biogeochem Cycles 18:GB1040

Hu S, Chapin FS, Firestone MK, Field CB, Chiariello NR (2001) Nitrogen limitation of microbial decomposition in a grassland under elevated CO$_2$. Nature 409:188–191

Hungate BA, Jackson, RB, Field CB, Chapin FS III (1996) Detecting changes in soil carbon in CO$_2$ enrichment experiments. Plant Soil 187:15–145

Hungate BA, Holland EA, Jackson RB, Chapin FS III, Mooney HA, Field CB (1997a) The fate of carbon in grasslands under carbon dioxide enrichment. Nature 388:576–579

Hungate BA, Lund CP, Pearson HL, Chapin FS III (1997b) elevated CO$_2$ and nutrient addition alter soil N cycling and N trace gas fluxes with early weason wet-up in a California annual grassland. Biogeochemistry 37:89–109

Hungate BA, Dijkstra P, Johnson DW, Hinkle CR, Drake BG (1999) Elevated CO$_2$ increase N$_2$ fixation and decreases soil nitrogen mineralization in Florida scrub oak. Global Change Biol 5:781–789

Hungate BA, Dukes JS, Shaw R, Luo Y, Field CB (2004a) Nitrogen and climate change. Science 302:1512–1513

Hungate BA, Stiling PD, Dijkstra P, Johnson DW, Ketterer ME, Hymus GJ, Hinkle CR, Drake BG (2004b) CO$_2$ elicits long-term decline in nitrogen fixation. Science 304:1291–1291

Jastrow JD, Miller RM, Matamala R, Norby RJ, Boutton TW, Rice CW, Owensby CE (2005) Elevated atmospheric carbon dioxide increases soil carbon. Global Change Biol 11:2057–2064

Johnson DW, Cheng W, Joslin JD, Norby RJ, Edwards NT, Todd DE (2004) Effects of elevated CO$_2$ on nutrient cycling in a sweetgum plantation. Biogeochemistry 69:379–403

Jones MB, Donnelly A (2004) Carbon sequestration in temperate grassland ecosystems and the influence of management, climate and elevated CO$_2$. New Phytol 164:423–439

Körner C, Asshoff R, Bignucolo O, Hattenschwiler S, Keel SG, Pelaez-Riedl S, Pepin S, Siegwolf RTW, Zotz G (2005) Carbon flux and growth in mature deciduous forest trees exposed to elevated CO$_2$. Science 309:1360–1362

Luo Y, Su B, Currie WS, Dukes JS, Finzi A, Hartwig U, Hungate B, McMurtrie RE, Oren R, Parton WJ, Pataki DE, Shaw MR, Zak DR, Field CB (2004) Progressive nitrogen limitation of ecosystem responses to rising atmospheric carbon dioxide. BioScience 54:731–739

Mack MC, Schuur EAG, Bret-Harte MS, Shaver GR, Chapin III FS (2004) Ecosystem carbon storage in arctic tundra reduced by long-term nutrient fertilization. Nature 431:440–443

Mikan CJ, Zak DR, Kubiske ME, Pregitzer KS (2000) Combined effects of atmospheric CO_2 and N availability on the belowground carbon and nitrogen dynamics of aspen mesocosms. Oecologia 124:432–445

Mosier AR, Pendall E, Morgan JA (2003) Effect of water addition and nitrogen fertilization on the fluxes of CH_4, CO_2, NO_x, and N_2O following five years of elevated CO_2 in the Colorado shortgrass steppe. Atmos Chem Phys 3:1703–1708

Neff JC, Townsend AR, Gleixner G, Lehman SJ, Turnbull J, Bowman WD (2002) Variable effects of nitrogen additions on the stability and turnover of soil carbon. Nature 419:915–917

Niklaus PA, Körner C (1996) Responses of soil microbiota of a late successional alpine grassland to long term CO_2 enrichment. Plant Soil 184:219–229

Niklaus PA, Körner C (2004) Synthesis of a six-year study of calcareous grassland responses to in situ CO_2 enrichment. Ecol Monogr 74:491–511

Niklaus PA, Leadly PW, Stocklin J, Körner C (1998) Nutrient relations in calcareous grassland under elevated CO_2. Oecologia 116:67–75

Norby RJ, Cotrufo MF, Ineson P, O'Neill EG, Canadell JG (2001) Elevated CO_2, litter chemistry, and decomposition: a synthesis. Oecologia 127:153–165

Oren R, Ellisworth DS, Johnson KH, Phillips N, Ewers BE, Maier C, Schafer KVR, McCarthy H, Hendrey G, McNulty SG, Katul GG (2001) Soil fertility limits carbon sequestration by forest ecosystems in a CO_2 enriched atmosphere. Nature 411:466–469

Parsons WFJ, Lindroth RL, Bockheim JG (2004) Decomposition of *Betula papyrifera* leaf litter under the independent and interactive effects of elevated CO_2 and O_3. Global Change Biol 10:1666–1677

Paterson E, Rattray EAS, Killham K (1996) Effect of elevated atmospheric CO_2 concentration on C-partitioning and rhizosphere C-flow for three plant species. Soil Biol Biochem 28:195–201

Paustian K, Six J, Elliott ET, Hunt HW (2000) Management options for reducing CO_2 emissions from agricultural soils. Biogeochemistry 48:147–163

Pendall E, Del Grosso S, King JY, LeCain DR, Milchunas DG, Morgan JA, Mosier AR, Ojima DS, Parton WA, Tans PP, White JWC (2003) Elevated atmospheric CO_2 effects and soil water feedbacks on soil respiration components in a Colorado grassland. Global Biogeochem Cycles 17:GB1046

Pendall E, Mosier AR, Morgan JA (2004) Rhizodeposition stimulated by elevated CO_2 in a semiarid grassland. New Phytol 162:447–458

Pepin S, Körner C (2002) Web-FACE: a new canopy free-air CO_2 enrichment system for tall trees in mature forests. Oecologia 133:1–9

Prior SA, Torbert HA, Runion GB, Rogers HH, Wood CW, Kimball BA, LaMorte RL, Pinter PJ, Wall GW (1997) Free-air carbon dioxide enrichment of wheat: soil carbon and nitrogen dynamics. J Environ Qual 26:1161–1166

Prior SA, Runion GB, Torbert HA, Rogers HH (2004) Elevated atmospheric CO_2 in agroecosystems: soil physical properties. Soil Sci 169:434–439

Prior SA, Runion GB, Rogers HH, Torbert HA, Reeves DW (2005) Elevated atmospheric CO_2 effects on biomass production and soil carbon in conventional and conservation cropping systems. Global Change Biol 11:657–665

Reich PB, Knops J, Tillman D, Craine J, Ellsworth D, Tjoelker M, Lee T, Wedink D, Naeem S, Bahauddin D, Hendrey G, Jose S, Wrage K, Goth J, Bengston W (2001) Plant diversity enhances ecosystem responses to elevated CO_2 and nitrogen deposition. Nature 410:809–812

Rice CW, Garcia FO, Hampton CO, Owensby CE (1994) Soil microbial response in tall-grass prairie to elevated CO$_2$. Plant Soil 165:67–74

Rillig MC, Wright SF, Allen MF, Field CB (1999) Rise in carbon dioxide changes soil structure. Nature 400:628–628

Rillig MC, Hernandez GY, Newton PCD (2000) Arbuscular mycorrhizae respond to elevated atmospheric CO$_2$ after long-term exposure: evidence from a CO$_2$ spring in New Zealand supports the resource balance model. Ecol Lett 3:475–478

Rogers HH, Heck WW, Heagle AS (1983) A field technique for the study of plant-responses to elevated carbon-dioxide concentrations. J Air Pollut Control Assoc 33:42–44

Rosenberg MS, Adams DC, Gurevitch J (2000) MetaWin, statistical software for meta-analysis, ver 2. Sinauer Associates, Sunderland, Mass.

Ross DJ, Newton PCD, Tate KR (2004) Elevated [CO$_2$] effects on herbage production and soil carbon and nitrogen pools and mineralization in a species-rich, grazed pasture on a seasonally dry sand. Plant Soil 260:183–196

Schlesinger WH (1997) Biogeochemistry: an analysis of global change, 2nd edn. Academic, San Diego, Calif.

Schlesinger WH, Lichter J (2001) Limited carbon storage in soil and litter of experimental forest plots under increased atmospheric CO$_2$. Nature 411:466–469

Six J, Carpentier A, Kessel C van, Merckx R, Harris D, Horwath WR, Luscher A (2001) Impact of elevated CO$_2$ on soil organic matter dynamics as related to changes in aggregate turnover and residue quality. Plant Soil 234:27–36

Six J, Conant RT, Paul EA, Paustian K (2002) Stabilization mechanisms of soil organic matter: implications for C-saturation of soils. Plant Soil 241:155–176

Thornley JHM, Cannell MGR (2000) Dynamics of mineral N availability in grassland ecosystems under increased [CO$_2$]: hypotheses evaluated using the Hurley pasture model. Plant Soil 224:153–170

Vance ED, Brookes PC, Jenkinson DS (1987) An extraction method for measuring soil microbial biomass-C. Soil Biol Biochem 19:703–707

Van Kessel C, Boots B, de Graaff MA Harris D, Blum H, Six J (2006) Soil C and N sequestration in a grassland following 10 years of free air CO$_2$ enrichment. Global Change Biol (in press)

Williams MA, Rice CW, Owensby CE (2000) Carbon dynamics and microbial activity in tallgrass prairie exposed to elevated CO$_2$ for 8 years. Plant Soil 227:127–137

Williams MA, Rice CW, Owensby CE (2001) Nitrogen competition in a tallgrass prairie ecosystem exposed to elevated carbon dioxide. Soil Sci Soc Am J 65:340–346

Wood CW, Torbert HA, Rogers HH, Runion GB, Prior SA (1994) Free-air CO$_2$ enrichment effects on soil carbon and nitrogen. Agric For Meteorol 70:103–116

Zak DR, Pregitzer KS, Curtis PS, Teeri JA, Forgel R, Randlett DL (1993) Elevated atmospheric CO$_2$ and feedback between carbon and nitrogen cycles. Plant Soil 151:105–117

Zak DR, Pregitzer KS, King JS, Holmes WE (2000) Elevated atmospheric CO$_2$, fine roots, and the response of soil micro organisms: a review and hypothesis. New Phytol 147:201–222

Zanetti S, Hartwig UA, Luscher A, Hebeisen T, Frehner M, Fisher BU, Hendrey GR, Blum H, Nosberger J (1996) Stimulation of symbiotic N$_2$ fixation in Trifolium repens L under elevated atmospheric pCO$_2$ in a grassland ecosystem. Plant Physiol 112:575–583

22 The Influence of Elevated [CO$_2$] on Diversity, Activity and Biogeochemical Functions of Rhizosphere and Soil Bacterial Communities

S. Tarnawski and M. Aragno

22.1 Introduction

Does the elevation of atmospheric [CO$_2$] directly influence soil microbiota? If the present rate of atmospheric [CO$_2$] increase is maintained, it may be predicted that a doubling in concentration will occur after about one century. [CO$_2$] values around 600 ppm could be reached during the second half of the twenty-first century. Due mainly to the respiratory activities of inhabiting micro-organisms, soils atmosphere contains between 2000 ppm and 38 000 ppm CO$_2$ (Gobat et al. 2004). Therefore, soil acts as a source for atmospheric CO$_2$; and a doubling of this latter's concentration would not significantly affect its content in soil.

Extremely high [CO$_2$] may directly affect the soil microflora. In soils exposed to CO$_2$ concentrations of 14.6–65.2 % as a consequence of seismic activity in the Mammoth Mountain (California), Salmassi et al. (2003) observed a decrease in overall bacterial diversity, a strong decrease in the abundance of *Proteobacteria* and an increase in the proportion of *Acidobacteria*, compared to similar soils not subjected to e[CO$_2$]. However, such concentrations are far higher (1–2 orders of magnitude) than CO$_2$ concentrations expected in normal soils.

An indirect effect of e[CO$_2$] may be hypothesized through the intermediary of plants and roots (Hu et al. 1999). Indeed, CO$_2$ fixation by C$_3$ plants, with an apparent k_m value around 450 ppm, should strongly respond to a [CO$_2$] increase between 300 ppm and 600 ppm. This contrasts with C$_4$ plants: with their higher affinity (apparent k_m around 70 ppm), CO$_2$ fixation is almost saturated by the present atmospheric concentration and should respond only slightly to such an increase. Qualitative, as well as quantitative effects of e[CO$_2$] on C$_3$ plants should manifest altogether on litter formation, root pro-

Ecological Studies, Vol. 187
J. Nösberger, S.P. Long, R.J. Norby, M. Stitt,
G.R. Hendrey, H. Blum (Eds.)
Managed Ecosystems and CO$_2$
Case Studies, Processes, and Perspectives
© Springer-Verlag Berlin Heidelberg 2006

duction and plant activities (e.g. rhizodeposition, water and ions uptake, respiration). Soil microbiota active in litter degradation and those living under the influence of the root (that is, in the rhizosphere) are then suitable to respond indirectly to such an increase. In the case of meadow plants, up to 30–50 % of the total assimilated carbon during photosynthesis can be translocated to the soil (Kuzyakov 2001; Gobat et al. 2004). The rate of net photosynthesis and carbon allocation to the roots are enhanced under e[CO_2]. A significant part of this carbon is released by plants roots (rhizodeposition) in the rhizosphere. The C flow may be increased and its composition altered under e[CO_2]. Consequently, plants modify their surrounding soil and play a major role in the structuration of rhizosphere microflora (Paterson 2003).

Therefore, the present paper mainly focuses on the rhizosphere microbiota.

22.2 Interactions Between Soil Microbiota and Rhizosphere Conditions

The rhizosphere is under the reciprocal influence of the root, the soil and the microflora. The root activity comprises the production of organic compounds, as well as water and ion uptake and respiratory oxygen consumption. The microflora is characterized by its numerous activities, including respiration, dissolution and the uptake of minerals and organic materials, the degradation of plant mucilages and the production of secretions, as well as a number of plant growth-promoting and plant-protecting properties. Some bacteria and fungi are also root parasites or deleterious.

Microbiota will respond in different ways to the high nutrient flow in the rhizosphere. One has first to keep in mind that an increase in bacterial biomass will result in an increase in bacterial grazing, mainly by protozoans, nematodes and rotifers. This establishes a nutrient chain which limits the bacterial biomass, allowing part of the soil inorganic nutrients to be kept for plant nutrition. This generates a rapid turnover of the bacteriomass and therefore favours fast-growing r-strategists, in comparison to the slow-growing K-strategists which dominate in the bulk soil. Indeed, rDNA clones isolated from rhizosphere fractions showed a higher proportion of sequences identified to cultivated species than clones isolated from bulk soil (Aragno 2005). In the latter, most clones were unidentifiable with described species, or identified only with "environmental" sequences, some of which probably related to "viable but not culturable" organisms, that is, extreme oligotrophs.

Contrary to what was often supposed, there is as a rule a decreasing diversity from bulk soil to root bacterial communities. This probably results from the above mentioned "election" of a lower number of fast-growing populations (r-strategists), from the possible selection operated by the bacteria-grazing microfauna, and from possible selective effects operated by signals from the root. This appeared clearly in the rhizospheres of white clover and ryegrass (Fig. 22.1; Marilley et al. 1998; Marilley and Aragno 1999). The "rhizosphere effect" on bacterial communities is therefore altogether a decrease in diversity and a shift toward "root competent" populations (e.g. *Pseudomonas*) responding positively to the presence of the root.

Rhizosphere microbiota affect the soil oxygen, carbon and nitrogen cycles. Combined with root respiration, oxygen consumption by rhizodeposition-utilizing aerobes and by chemoautotrophs oxidizing mineralization products (e.g. nitrifiers) creates a negative gradient of oxygen concentration towards the root. In turn, micro-oxic conditions favour processes requiring low-oxygen partial pressures, like N$_2$ fixation, denitrification or hydrogen cyanide production. Root respiration and heterotrophic microbial activities also result in a strong CO$_2$ production. Many rhizosphere bacteria synthesize highly stable exopolymers, particularly exopolysaccharides, which accumulate in the soil (bacterial humin) where together they represent a structural agent (agglomeration of soil particles, mucigel formation around the root in replacement of labile, plant-derived polysaccharides) and a carbon sink. Bacteria and fungi also produce phenol-oxidases (peroxidases, polyphenol-oxidases) which activate plant-derived phenolic compounds in prelude to their spontaneous polymerization into humic acids and insolubilization humin, which belong to the stablest forms of organic carbon in soils, with life-times which may be expressed in centuries or millenia (Gobat et al. 2004). This humin then associates with clays to form ion-absorbing complexes.

Depending on the N-content of rhizodeposition, its mineralization in the rhizosphere may result in opposite effects on soil N (Fig. 22.2). If the rhizode-

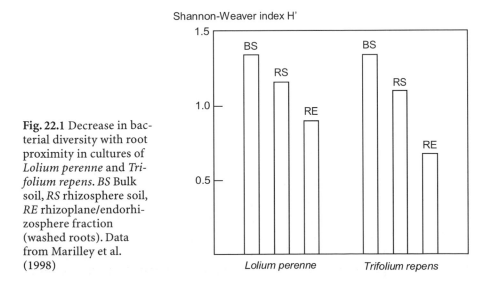

Fig. 22.1 Decrease in bacterial diversity with root proximity in cultures of *Lolium perenne* and *Trifolium repens*. *BS* Bulk soil, *RS* rhizosphere soil, *RE* rhizoplane/endorhizosphere fraction (washed roots). Data from Marilley et al. (1998)

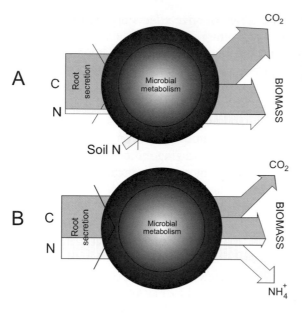

Fig. 22.2 Opposite effects on soil nitrogen of the mineralization of nitrogen-poor (**A**) and nitrogen-rich (**B**) rhizodeposition

position C/N ratio is lower than the N balance needed for bacterial biomass formation, there is a net release of NH_4^+ through ammonification. In contrast, with a low N-content of root production, there is an uptake of soil N by the bacterial biomass. So, changes in the N-content of rhizodeposition may affect soil N, particularly in conditions where N is a limiting factor in soil.

Under limiting N concentrations, N_2 fixation should be favoured in the rhizosphere, due to simultaneous high energy input (through rhizodeposition), low oxygen and low nitrogen. Hamelin et al. (2002) have shown the presence of an abundant and diverse N_2-fixing microflora in the rhizosphere of a perennial, oligonitrophilic grass growing in N-poor soils. Under high N, denitrification should be favoured in the rhizosphere under low $[O_2]$.

Rhizosphere bacterial communities do not result simply from the casual encounter of soil bacteria with the root. A number of root-competent bacterial populations in the rhizosphere live in mutualistic interactions with the plant. They benefit from rhizodeposition-derived nutrients and in some cases from other root-derived factors, like micro-oxic conditions (e.g. for N_2 fixation), growth factors, attracters or even inducers of enzyme activities. In return, these populations may exhibit properties favouring plant growth and productivity: they are then termed "plant growth-promoting rhizobacteria" (PGPR). Mutualism is a highly sophisticated process implying co-adaptation of all partners; and this is only conceivable through a long co-evolution. Among the PGP properties identified, let us mention: solubilization of minerals through acidolysis or chelation, secretion of mineralizing enzymes, N_2 fixation, synthesis of exopolymers and production of phytohormones. Another type of positive interaction between the rhizosphere bacteria and plants is the antagonistic effect of bacteria against plant parasites.

Other mutualistic interactions include a third partner, the mycorrhizal fungus. There are mycorrhiza-helper bacteria (Garbaye 1994). They may interact with the mycorrhiza itself (in the mycorrhizosphere), e.g. by favouring colonization by the fungus, or with the extraradical mycelium (in the hyphosphere), at a distance from the rhizosphere. In this case, they may participate in the solubilization or liberation of ions (e.g. phosphate, iron) in prelude to their translocation by the fungus to the plant, or facilitate the germination of spores (in the case of arbuscular mycorrhizal fungi).

22.3 Effect of e[CO$_2$] on Rhizodeposition

It is generally accepted that an increase in [CO$_2$] alters root secretions quantitatively and qualitatively, particularly in conditions of N- or P-limitation. An increase in C-flux and a higher C/N ratio of rhizodeposition was often noticed, even when no increase in plant biomass was observed (Paterson et al. 1997; Hu et al. 1999). In other words, plants under e[CO$_2$] decrease the allocation of N-rich metabolites and increase the allocation of C-rich metabolites to root secretions.

Studies on rhizodeposition in the rhizosphere, particularly on its main part, the secretion of low-molecular-weight soluble organic compounds, presents a major difficulty: it is performed either in hydroponic solutions, which differ completely from the rhizosphere conditions, or in artificial soils, which do not constitute a reservoir of rhizosphere-adapted micro-organisms. Analysis of root secretions in natural soils is hampered by the fact that rhizosphere micro-organisms continuously take up easily metabolizable compounds, so that their actual concentration results from the flux between secretion and uptake, which itself depends on the biomass, affinity and consumption rate of consumers (Hu et al. 1999). Therefore, the instantaneous concentration and composition of root secretion does not rely on the actual flows of secreted compounds. The quantitative and qualitative characteristics of root secretion may also depend on the external concentrations of these compounds, which themselves depend on microbial uptake rate. For example, the re-absorption of exudates by the root, often observed in hydroponic conditions, could be insignificant in a real rhizosphere. Caution should therefore be given to the interpretation of measurements on the response of root secretion to a [CO$_2$] increase.

22.4 Responses of Microbial Biomass, Cell Number and Activity

A direct estimation of bacterial biomass in soil is impossible. Only indirect measurements, such as the analysis of a cell component (ATP, phospholipid fatty acids – PLFA), the decomposition of killed biomass (chloroform fumigation) or total cell count (often difficult to apply), allow a relative measurement of the "amount of bacteria". But comparisons are possible, provided the same method is applied. The number of cultivable bacteria has quite a different significance, linked to growth strategy. Indeed, a low proportion of the living bacterial cells in a soil are actually cultivable on conventional (and rich) laboratory media. These are adapted to high nutrient flows and may be considered as r-strategists. The "non-cultivable" bacteria appear rather as K-strategists adapted to extremely low nutrient levels.

There is no general tendency towards an increase or not in rhizosphere microbial biomass under $e[CO_2]$ (Hu et al. 1999; Zak et al. 2000b). After 3 months of fumigation with 700 ppm $^{14}CO_2$, a 42 % increase in microbial biomass was measured in Lolium perenne soil (van Ginkel and Gorissen 1998; van Ginkel et al. 2000). An increase in bacterial biomass was also noticed in fertile pastures after five years under $e[CO_2]$ (Hu et al. 2001). In contrast, in soils under L. perenne and Trifolium repens, Richter et al. (2003) did not detect any change in bacterial biomass after 7 years fumigation with 600 ppm CO_2 compared to current conditions. Montealegre et al. (2002) did not detect any effect of $[CO_2]$ increase on total bacterial populations in rhizosphere soil of the same grass. No changes in microbial biomass were detected in the rhizosphere and soils of poplars (Lussenhop et al. 1998; Kelly et al. 2003).

In a tallgrass prairie system exposed to $e[CO_2]$ for 8 years, Williams et al. (2000) showed a slight increase (not significant) in microbial C and N biomass, whereas there was a significant increase in overall microbial activity, indicating a probable increase in the specific activity of microbial biomass. Similarly, an increase in microbial activity, together with unchanged microbial biomass, was noticed in sandstone grasslands (Hungate et al. 2000). Under L. perenne, the amount of photosynthetic carbon allocated to bacterial biomass was not influenced by $[CO_2]$, whereas the amount of non-microbial carbon in the rhizosphere increased by a factor of 2.6 in 28-day cultures (Griffiths et al. 1998). This may result from an increase in the turnover of C and microbial biomass, that is, an increase in biomass specific activity. Indeed, actual bacterial numbers and biomass result from (i) bacterial growth and (ii) grazing by the microfauna (mainly protozoans, nematodes and rotifers). Therefore, an increase in bacterial growth may be followed by an increase in grazing, resulting in a higher turnover without biomass and/or cell number increase. Grazing then results in the recycling of nutrients from the microbial biomass in favour of the plant. So, through a combination of the highly efficient solubilization, mineralization and ion concentration capabilities of many microbial populations, such a system would increase the flux of nutriments from directly unavailable pools to the plant. This stresses the importance of the higher trophic levels in the functioning of the rhizosphere, as was demonstrated by Clarholm (1985, 1989). Her experiments on wheat cultures in unfertilized sterile soil inoculated with soil bacteria with or without the addition of amoebae, showed an increase in plant N biomass of 3–4 times in the presence of these latter. In this sense, several authors, e.g. Hungate et al. (2000), Williams et al. (2000) and Yeates et al. (1997), noticed an overall increase in the activity of bacteria and fungi and their grazer protozoans and microarthropods in a rhizosphere under e[CO$_2$].

Due to increased rhizodeposition, there may also be shifts among bacterial populations linked to growth strategies. An increase in cultivable bacteria and a simultaneous decrease in total bacterial cells (with a dominance of oligotrophic, "non-cultivable" K-strategists) under e[CO$_2$] were observed by Hodge et al. (1998) in the rhizosphere of *L. perenne* and by Insam et al. (1999) in artificial tropical ecosystems. The number of cultivable bacteria was significantly higher in *L. perenne* rhizosphere after 2 years under 600 ppm CO$_2$ as compared to current conditions, whereas no change was observed in the distant soil (Marilley et al. 1999). This was confirmed 7 years later on the same plots by Fromin et al. (2005). Except for a short period in spring, no such change occurred under *T. repens*. This may be related to the fact that the C/N ratio of roots did not change with e[CO$_2$] in clover, whereas it increased significantly in ryegrass (Jongen et al. 1995).

As a whole, it appears that, under N-limitation, there is a higher carbon allocation to rhizodepositon, which in turn favours cultivable, fast-growing r-strategists adapted to feed on easily metabolizable substrates. This results in a decreased soil N availability (Hartwig et al 1996), due to higher sequestration by the plant and, depending on the conditions, a decreased bacterial biomass (Hu et al. 2001). Then, the slow-growing K-strategists, more adapted to degrade less labile substrates, are hampered and these substrates accumulate, increasing the soil organic matter content. Another hypothetical mechanism may be suggested for the shift from K- to r-strategists: increased growth of these latter is followed by an increase in grazer populations, which however would not feed specifically on the fast-growers and then would decrease the relative numbers of slow-growing K-strategists. Anyhow, the consequence of increased soil organic matter is a sink in atmospheric CO$_2$ in response to a [CO$_2$] increase! How such a tendency would maintain in the long term and function as an effective negative feed-back regulation of atmospheric [CO$_2$] needs yet to be evaluated.

22.5 Effects on Soil Structure and Enzyme Activities

There is a lack of knowledge regarding the response to $e[CO_2]$ of rhizosphere and soil extracellular enzyme activities, particularly those related to ligninolysis and humification (e.g. phenol-oxidases). Most authors considered the accumulation of soil organic matter only globally, without relating it to the specific humification processes. Fog (1988) has shown that, in the presence of high concentrations of available N, the degradation of labile compounds (e.g. cellulose) is stimulated, whereas ligninolytic enzymes are inhibited. Therefore, N sequestration under $e[CO_2]$ results in increased ligninolytic activities, leading to increased polycondensation humification. Hu et al. (1999) noticed a difference in the accumulation of phenolic compounds, which was higher in forest soils than under herbaceous plants under $e[CO_2]$. In artificial tropical ecosystems, a higher degree of humification, as revealed by extractable optical density, was observed under 610 ppm CO_2 (Insam et al. 1999).

In the presence of arbuscular mycorrhizal fungi (AMF), Rillig et al. (1999) noticed a modification of soil aggregate structure under $e[CO_2]$. There was an increase in the number of small aggregates (0.25–1.0 mm), whereas the number of larger ones (1.0–2.0 mm) was not modified. This was related to the increased production of glomalin, a protein secreted by the AMF.

22.6 Responses of Bacterial Community Structure to $e[CO_2]$

The evaluation by different authors of the effects of an increase in atmospheric $[CO_2]$ on the structure of soil and rhizosphere microbial communities led to contrasting conclusions. This may result from the different systems studied, but possibly also from the different methods applied to assess community structure and diversity.

Hence, Griffiths et al. (1998) did not observe significant differences in the structure of microbial communities associated with *L. perenne* under $e[CO_2]$, as revealed by a global characterization of soil DNA (thermal denaturation and G+C content).Using PLFA profiles, Kandeler et al. (1998) and Zak et al. (1996) did not observe consistent effects of $e[CO_2]$. However, by the same approach, Montealegre et al. (2002) detected changes in white clover communities in response to CO_2 enrichment. T-RFLP profiles of bacterial communities from 16S rDNA were significantly changed in response to $e[CO_2]$ in soils under *Populus tremelloides* (Kelly et al. 2003). Marilley et al. (1998), using a cloning–restriction procedure from PCR-amplified 16S ribosomal genes showed a considerable increase in the ratio of the *Rhizobium leguminosarum* population in the *T. repens* rhizoplane/endorhizosphere fraction under $e[CO_2]$. This corresponded to a decrease in the *Pseudomonas* spp. guild in both rhizosphere soil and washed root fractions. In contrast, under *L. perenne*, the same authors showed a remarkable increase in the *Pseudomonas* spp. guild in the rhizosphere soil under e[CO$_2$]. The relative increase was less marked in the rhizoplane/endorhizosphere fraction, which was already considerably enriched in *Pseudomonas* under c[CO$_2$] compared to the rhizosphere soil. Under both clover and ryegrass, the effects of e[CO$_2$] on bulk soil were insignificant.

Jossi et al. 2006 studied the response of root and soil-inhabiting bacterial communities to e[CO$_2$] by molecular fingerprinting of total (revealed by DNA-based approach) and active (revealed by RNA-based approach) bacterial communities associated with the two perennial grasses: *L. perenne* and *Molinia coerulea*. She observed an influence of e[CO$_2$] on global community structure, especially in the root fraction. e[CO$_2$] (600 ppm) had a greater influence on active bacterial communities than on total present communities. It also affected specific bacterial populations. For instance, *Actinobacteria* populations (known to be soil engineers) were especially active in distant soil, but were little affected by e[CO$_2$]. In contrast, δ-*Proteobacteria* (mostly polysaccharide-hydrolysing *Myxobacteria*) were stimulated by e[CO$_2$] in the vicinity of the root. These results confirm the indirect effect of e[CO$_2$] on soil microbiota via the plant. The stimulation of polysaccharide utilisers could be related to an increase in root mucilage secretion.

Regarding "functional" or "metabolic" communities, Insam et al. (1999) did not observe changes in an artificial tropical ecosystem, using both community level physiological profiles with Biolog plates and PLFA profiles. However, Hodge et al. (1998) and Hamelin (2004) noticed a faster utilization of Biolog C-sources in rhizosphere soil of perennial grasses under e[CO$_2$]. They linked this observation to the higher concentration of cultivable bacteria. Elhottova et al. (1997) showed changes in the composition of bacterial nutritional groups under e[CO$_2$]. Under white clover, Montealegre et al. (2000) showed a shift in the genetic composition of *R. leguminosarum* bv. *trifolii* populations (symbiotic N$_2$ fixators) under e[CO$_2$]. When *Rhizobium* strains isolated from e[CO$_2$] were put in competition with strains isolated from c[CO$_2$] for clover nodulation, there was a 17 % increase in nodule occupancy by the former in cultures under e[CO$_2$].

22.7 Elevated [CO$_2$] and Nitrogen Cycle in Soil and Rhizosphere

22.7.1 N-pools, Uptake and Mineralization

Nitrogen is a major bioelement, typical of most of the active biomolecules. It is also a major limiting factor for plant growth, along with phosphorus and iron. So, great attention has been given to the consequences of e[CO$_2$] on N pools and N cycle-related activities. Indeed, due to the generally observed increase in C-rhizodeposition as a consequence of e[CO$_2$], several responses may be hypothesized which imply N-related pools and activities:

1. Increased C/N of the rhizodeposition may imply increased N immobilization in bacterial and plant biomass (Fig. 22.2) and therefore a decrease in soil available N, leading to decreased microbially mediated organic C degradation and consequently to higher accumulation of soil organic matter (SOM).
2. With low available soil N and under lower [O$_2$] resulting from increased microbial respiration due to increased root C secretion (also an energy source), an increased associative N$_2$ fixation could occur in the rhizosphere.
3. With high available soil N concentrations and unhampered nitrification, the lower [O$_2$] and higher organic C in the rhizosphere could induce more denitrification, resulting in a net loss of N in the ecosystem.

A significant, although moderate increase in total soil N was observed in an unfertilised tallgrass prairie exposed to e[CO$_2$] for 8 years (Williams et al. 2000). In the same experiment, the concentration of soil inorganic N was significantly higher with e[CO$_2$] during the growing season (May–June), indicating a probable greater N-mineralization linked to root secretions.

Neither soil organic N mineralization nor NH$_4^+$ consumption responded to a previous exposure to 600 ppm CO$_2$ for 7 years in an artificial grassland cultivated with either *L. perenne* or *T. repens* (Richter et al. 2003). Contrary to the above-mentioned experiments, however, this grassland was fertilised with 14 g m^{-2} year^{-1} N/NH$_4$NO$_3$.

Studying experimental cultures of poplars in reconstituted soils for 2.5 growing seasons, Zak et al. (2000a) did not observe significant changes in soil N-cycling under 357 ppm and 707 ppm CO$_2$. The authors put forward an interesting hypothesis to explain these different responses to e[CO$_2$]: in young, developing open ecosystems, where the roots have not yet fully colonized the soil, microbial metabolism would be mainly influenced by the relatively large pool of soil organic matter, as compared to the input from rhizodeposition. In contrast, significant effects would be expected in conditions of full soil occupation by the roots, which would then alter microbial activities through their

productions and activities. We could also speak of a higher degree of "rhizos-
phericity" of the soil in the latter situation than in the former.

22.7.2 N$_2$ Fixation

Except in the case of symbiotic nodulation (see Chapter 18), there are very few
reports on the effect of e[CO$_2$] on N$_2$ fixation and N$_2$-fixing bacteria, particularly in the rhizosphere. This results probably from the fact that most studies
dealt with fertilised and artificially planted systems, where associative N$_2$ fixation would not play a significant role. However, a significant N$_2$-fixing activity was measured in the roots of plants which may show a luxuriant growth in
soils with extremely low inorganic N concentrations, such as soils subjected to
nitrification/denitrification cycles in zones of water table fluctuation. This
was the case with *Phragmites australis*, the common reed (M. Arreguit, personal communication) and with *M. coerulea* (Hamelin et al. 2002).

Marilley et al. (1999) showed a strong increase in the ratio of clones related
to *R. leguminosarum* in the rhizoplane/endorhizosphere fraction of white
clover roots under e[CO$_2$], which could result from an increased production
of specific attracters (e.g. flavonoids) in the root secretion.

22.7.3 Nitrification

Nitrification is a strictly aerobic process performed by two guilds of bacteria
that act in synergy: the ammonia-to-nitrite oxidizers and the nitrite-to-
nitrate oxidizers. They are neutrophilic to slightly alkalophilic, so nitrification
would be hampered in acidic soils. They are also strict chemolitho-auto-
trophs, with ammonia and nitrite, respectively, serving as electron donors for
lithotrophic respiration. Hence, nitrification does not depend on the availability of organic compounds, but only on the presence of ammonia, oxygen and
carbon dioxide. Therefore, we do not agree with the assumption that "a microbial community dominated by bacteria favours nitrification" (Hu et al. 1999, p.
435). Nitrification is per se an acidifying process (oxidation of ammonia to
nitric acid). However, nitrate uptake by plants increases the pH. In the absence
of plants, such acidification, combined with the high solubility of nitrate salts,
facilitates the leaching of minerals. So, direct effects of increased rhizodeposition on nitrification should not be expected. However, indirect effects, such as
a decrease in ammonium and/or oxygen concentrations could hamper it,
whereas it would be favoured through increased N mineralization (ammonification).

In most systems studied, nitrification activity was decreased under e[CO$_2$].
Hence, a 41 % decrease in gross nitrification rate was measured in ryegrass
swards under 600 ppm [CO$_2$] in comparison with current conditions (Baggs et

al. 2003). N_2O emissions from nitrification were seven times higher under current conditions than under $e[CO_2]$, whereas N_2O from denitrification did not change significantly. In monospecific grassland mesocosms (Barnard et al. 2004), nitrification activity was strongly decreased under $e[CO_2]$ under *Holcus lanatus* plants, but was not affected under *Festuca rubra* plants. Nitrate production (i.e. N mineralization + nitrification) in soils from *Pinus sylvestris* cultures under $e[CO_2]$ was doubled when incubated in the laboratory under $e[CO_2]$, whereas no changes were measured between high and current $[CO_2]$ cultures if incubated under $c[CO_2]$. $[CO_2]$ during incubation did not alter nitrate production in soils from cultures under $e[CO_2]$. This could be explained by an adaptation of the composition of denitrifying guild to $e[CO_2]$. In soils under poplars, after 2.5 growth seasons, Zak et al. (2000a) did not detect significant changes in gross and net nitrification rates due to $e[CO_2]$.

22.7.4 Denitrification

Denitrification sensu stricto is an anaerobic respiratory process involving the reduction of nitrate to gaseous N, mainly N_2 and N_2O. The N_2O/N_2 ratio may vary according to environmental conditions, such as redox potential, so N_2O production alone should not be taken as a measure of denitrifying activity, unless the measurement is performed in the presence of an inhibitor of N_2O-reductase, such as acetylene. Moreover, N_2O is also formed through the nitrification process (Baggs et al. 2003). However, the use of ^{15}N isotopes allows the distinction to be made between N_2O produced by nitrification and that by denitrification (Baggs and Blum 2004). Other processes of dissimilatory nitrate reduction include nitrate-to-nitrite reduction, followed by nitrite accumulation in pure cultures, and nitro-ammonification, that is reduction to ammonia, which is more likely to occur in permanently anoxic environments. Most denitrifying bacteria are otherwise aerobes (e.g. *Pseudomonas* spp., *Paracoccus* spp., *Ralstonia eutropha*) which have the ability to shift from aerobic to nitrate respiration when oxygen becomes limiting. Oxygen behaves generally as a repressor of denitrification, although there are exceptions (e.g. *Thiosphaera pantotropha*). So, the presence of denitrifyers in a given environment may be related to completely other reasons than the occurrence of oxygen limitation, although frequent and long-lasting oxygen depletion periods may result in an enrichment of denitrifying populations.

The conditions prevailing in the rhizosphere, particularly the low oxygen concentration and the high flow of easily metabolizable carbon sources might favour denitrification and denitrifying bacteria. As these parameters would be still more marked under $e[CO_2]$, this could result in increased denitrification. Indeed, Baggs et al. (2003) observed that total denitrification (measured as $N_2 + N_2O$ production) increased more than three-fold in a *L. perenne* sward under 600 ppm CO$_2$, as compared with current concentration. This increase resulted from an increase in N$_2$ production, whereas N$_2$O production through denitrification did not change significantly. Possibly, a decrease in O$_2$ concentration under e[CO$_2$] could explain this increase in the N$_2$/N$_2$O ratio, as well as the simultaneous decrease in nitrification. However, in longer-term experiments, Baggs and Blum (2004) showed a clear increase in N$_2$O production through denitrification, confirming the results obtained earlier by Ineson et al. (1998) on the same swards after 2 years of fumigation.

The response of bacterial denitrifying populations to the rhizosphere conditions gave contradictory results. For example, Clays-Josserand et al. (1995, 1999) and Delorme et al. (2003) showed a higher proportion of denitrifying *Pseudomonas* in the rhizosphere than in the distant soil. However, Roussel-Delif et al. (2005) showed a decrease in the frequency of nitrate-to-nitrite and denitrifying *Pseudomonas* with root proximity in swards of *L. perenne* with a low N fertilisation (14 g N m^{-2} year^{-1}). In *M. coerulea* on its native, low-N soil, the proportion of nitrate-to-nitrite reducers was lower in the root fraction, whereas the proportion of strict denitrifyers increased. It must be stressed, however, that the former authors studied flax and tomato, two annual plants cultivated in artificial soils, whereas the latter worked on dense swards of perennial grasses, that is, dense and mature versus open and young ecosystems. Moreover, in the latter case, N was a limiting factor of plant growth. It was assumed that the decrease in the ratio of denitrifying/total pseudomonads with root proximity was due to competition between plants and nitrate-dissimilating bacteria for the scarcely available nitrate (Fromin et al. 2005).

The ratio of denitrifying/total *Pseudomonas* clearly increased in the root fraction of *L. perenne* cultivated with low N fertilisation (14 g N m^{-2} year^{-1}) under e[CO$_2$], as compared to current concentration (Roussel-Delif et al. 2005). A slight increase was observed in the rhizosphere soil, whereas no change occurred in the bulk soil. This verifies the initial postulate that the main effect of e[CO$_2$] occurs through the intermediation of the root. However, no clear effect was detected in *Molinia*.

With high N fertilisation (56 g N m^{-2} year^{-1}), the above-postulated competition for N between denitrification and plants is not likely to occur. Indeed, the N requirement of *L. perenne* was estimated at 29 g N m^{-2} year^{-1} (Richter 2003). In these conditions, the frequency of denitrifying peudomonads was similar in all fractions and did not respond to e[CO$_2$] (Fromin et al. 2005).

The significance of nitrate-to-nitrite reducers in the rhizosphere, which greatly dominate the true denitrifyers, is not clear. Indeed, nitrite does not accumulate. It might be further reduced by true denitrifyers, or used as an electron sink by fermentative bacteria, improving their energy yield. But another process, which was not studied so far in the rhizosphere, utilizes nitrite: the anammox reaction (Mulder et al. 1995), that is, anaerobic ammonium oxidation. Bacteria belonging to a well defined phylogenetic group, the order *Planctomycetales*, are able to perform the following reaction:

$$NH_4^+ + NO_2^- \rightarrow N_2 + 2H_2O$$

This thermodynamically favourable reaction of "nitrifying denitrification" provides energy for an anaerobic autotrophic metabolism, so anammox bacteria do not rely on organic carbon sources. Anammox bacteria would benefit from the low oxygen concentration in the rhizosphere environnement, stressed under $e[CO_2]$. This would increase N losses in the ecosystem.

22.8 Plant-Growth Promoting Rhizobacteria

Although rhizosphere bacteria, and particularly pseudomonads, are likely to interfere with plants, often in a favourable manner (PGPR; Lugtenberg and Dekkers 1999), very little is known about the response of these organisms and their plant-related traits to $e[CO_2]$. Tarnawski (2004) studied a collection of 1228 *Pseudomonas* strains isolated from distant and rhizosphere soil and from root fractions of two perennial grassland systems (*L. perenne* and *M. coerulea*) under both current (360 ppm) and elevated (600 ppm) $[CO_2]$. The response of siderophore producers to $e[CO_2]$, was positive in *Lolium* and negative in *Molinia*. In both plants, $e[CO_2]$ induced a negative response in the proportion of HCN-producing strains, considered as potential inhibitors of root-parasitic fungi. However, both populations were in a higher proportion in the root fraction relative to bulk soil. No effect of root proximity or $e[CO_2]$ was observed on the proportion of auxin producers. Siderophore producers are likely to interfere in several ways with plants: they present a competitive advantage in rhizosphere competence and root colonization (Lugtenberg and Dekkers 1999), whereas siderophore production may favour plant nutrition and altogether compete with root parasitic fungi (Elad and Baker 1985; Sharma and Johri 2003).

22.9 Discussion and Perspectives

A critical aspect in $e[CO_2]$ studies was put forward by Hu et al. (1999) in their excellent review: the need for long-term experiments. Even after 10 years fumigation, it may be difficult to extrapolate the results to foresee the evolution over the next century. Moreover, many experiments were performed on a much shorter term, which could better reflect an instantaneous situation than a long-term evolution. The same authors stressed that many of the experiments performed consisted of studying young, open ecosystems in conditions where the plants were not subjected to competition with others and the microbiota responded more to soil resources than to the supplement of nutrients brought by the increased rhizodeposition. There is a huge difference between a tomato plant in a rhizobox and a multi-species, unfertilised prairie! In this latter case, the extreme complexity, competitivity and stability of such an ecosystem would result in a very slow evolution, again requiring long-term experimentations. In most CO$_2$ fumigation experiments, the studied ecosystems are submitted from one day to the next to e[CO$_2$] which is supposed to occur progressively over the following century. This might induce, over the short term, unrealistic stresses in the response of the plants and of their ecosystems (Klironomos et al. 2005). The resilience times related to such stresses are not known, so it is impossible to affirm whether an observed response to e[CO$_2$] is just a reaction to a stress or will persist forever.

However, long-term effects of e[CO$_2$] in dense, perennial plant communities could lead to the accumulation in the bulk soil of factors (e.g. SOM) or microbial populations, primarily synthesized or stimulated in the root or rhizosphere fractions. Indeed, in such soils, the root system is renewed year after year, dead roots being subjected to humification whereas new roots are growing in the nearby bulk soil. However, the activity of certain microbial populations may require direct contact with the root webs (endophytic or rhizoplane habitat). In this case, although present, the populations in the rhizosphere or bulk soils would function mainly as reservoirs for the re-inoculation of new roots. So, the presence of a given population does not indicate per se that it actually fulfils the function which one would expect from its properties. In this case, a RT-PCR procedure, allowing detection of the transcription of related genes, should be applied.

Future studies on the effects of e[CO$_2$] on soil and rhizosphere microbiota should give more attention to:

- the use of modern, molecular tools for the description of microbial communities, at the functional as well as phylogenetical levels, as well as the use of statistical/numerical tools to analyse and synthesize these results (Fromin et al. 2002);
- the response of root-competent populations, particularly PGPR;
- the nature of the carbon compounds accumulated in response to e[CO$_2$], the biological and enzymatic processes involved in their formation and degradation, as well as their turnover rates and long-term evolution;
- the effect on secondary food-chains starting from rhizodeposition, including bacterial and fungal grazers;
- the importance of associative N$_2$ fixation, particularly in perennial, unfertilised natural ecosystems.

22.10 Conclusions

During the past 12 years, a huge number of results have been obtained by researchers from many different disciplines. However, trying to put all this information together results in a "patchwork", making an integrated view difficult. It is somewhat like a jigsaw puzzle in which most of the pieces are missing and the overall picture does not yet appear. Indeed, although outstanding research was performed with the best adapted modern methodology, most groups worked intra-, rather than interdisciplinarily. We are therefore far from able to draw general conclusions about the full effect of e[CO_2] on the Earth's ecosystem. Will it, according to the Gaia hypothesis, respond by negative feed-back regulations in a homeostatic manner? Will it lead, through positive feed-backs, to a global catastrophe? Besides the highly disputed climatic changes through the increase in greenhouse effect, the more discrete, but perhaps equally important effect on photosynthesis could modify in a considerable manner the relationships between plants in their ecosystems. Many of these relationships are under the control of the hidden face of the plant ecosystem, that is, the soil microbiota. In the present state of our knowledge, some tendencies appear regarding soil microbiotic responses to e[CO_2], among which:

- The main effects of atmospheric e[CO_2] on soil microbiota occur via plant metabolism (particularly in C3 plants) and root secretions. These effects increase with root proximity and are maximal in the root (rhizoplane and endorhizosphere) fraction, whereas there is little or no effect in the bulk soil.
- e[CO_2] in perennial plant communities may lead to accumulation in the bulk soil of soil organic matter, due to a decreased N-availability and a stimulation of the utilization of low-molecular-weight organic compounds by rhizosphere microbiota, relative to more recalcitrant compounds.
- There is no clear effect of e[CO_2] on the bacterial biomass and/or cell numbers in the rhizosphere. However, the biomass specific activity and turnover are in general increased, as well as grazing by protozoa. In such conditions, r-strategists are favoured.
- e[CO_2] increases, at least over the short and medium term, the ratio of root-competent populations (e.g. pseudomonads) in the rhizosphere.
- N cycle-related microbial populations, particularly denitrifyers, respond mainly to e[CO_2] under low nitrogen input, i.e. under N-limitation conditions.

We feel however that much has still to be investigated to understand the fate of macrobiota in their interaction with microbiota, in response to an evolution in part provoked, but not controlled, by mankind.

Acknowledgments. The research performed in the authors' laboratory was supported by grant 3100-068208.02 from the Swiss National Science Foundation. We are thankful to the "FACE" group of this laboratory, which was co-ordinated by Dr. Laurent Marilley and Dr. Nathalie Fromin, with the enthusiastic participation of Drs. Sylvie Teyssier-Cuvelle, Jérôme Hamelin, David Roesti and Pierre Rossi, and that of Maïté Arreguit, Maryline Jossi and Ludovic Roussel-Delif. Prof. Josef Nösberger and the "FACE" group in the Institute of Plant Science at ETH-Z, Eschikon, introduced us to the problem of e[CO₂], invited us to collaborate with the Swiss FACE experiment and gave their assistance for maintaining our *Molinia* cultures under e[CO₂]. We acknowledge in particular the friendly help of Ueli Hartwig and Herbert Blum.

References

Aragno M (2005) The rhizosphere: a hot spot of bacterial diversity. In: Satyanarayana T, Johri BN (eds) Microbial diversity: current perspectives and potential applications. I.K. International Publishing House, New Delhi

Baggs EM, Blum H (2004) CH_4 oxidation and emissions of CH_4 and N_2O from *Lolium perenne* swards under elevated atmospheric CO_2. Soil Biol Biochem 36:713–723

Baggs EM, Richter M, Cadish G, Hartwig UA (2003) Denitrification in grass swards is increased under elevated atmospheric CO_2. Soil Biol Biochem 35:729–732

Barnard R, Barthes L, Le Roux X, Leadley PW (2004) Dynamics of nitrifying activities, denitrifying activities and nitrogen in grassland mesocosms as altered by elevated CO_2. New Phytol 162:365–376

Clarholm M (1985) Interactions of bacteria, protozoa and plants leading to mineralization of soil nitrogen. Soil Biol Biochem 17:181–187

Clarholm M (1989) Effects of plant–bacterial–amoebal interactions on plant uptake of nitrogen under field conditions. Biol Fertil Soils 8:373–378

Clays-Josserand A, Lemanceau P, Philippot L, Lensi R (1995) Influence of two plant species (flax and tomato) on nitrogen dissimilative abilities within fluorescent *Pseudomonas* spp. Appl Environ Microbiol 61:1745–1749

Clays-Josserand A, Ghiglione JF, Philippot L, Lemanceau P, Lensi R (1999) Effect of soil type and plant species on the fluorescent peudomonads nitrate dissimilating community. Plant Soil 209:275–282

Delorme S, Philippot L, Edel-Hermann V, Deulvot C, Mougel C, Lemanceau P (2003) Comparative genetic diversity of the *narG*, *nosZ* and 16S rRNA genes in fluorescent pseudomonads. Appl Environ Microbiol 69:1004–1012

Elad Y, Baker R (1985) The role of competition for iron and carbon in suppression of chlamydospore germination of *Fusarium oxysporum* by *Pseudomonas* spp. Phytopathology 75:1053–1059

Elhottova D, Tríska J, Santruckova H, Kveton J, Santrucek J, Simkova M (1997) Rhizosphere microflora of winter wheat plants cultivated under elevated CO_2. Plant Soil 197:251–259

Fog K (1988) The effect of added nitrogen on the rate of decomposition of organic matter. Biol Rev 63:433–462

Fromin N, Hamelin J, Tarnawski S, Roesti D, Jourdain-Miserez K, Forestier N, Teyssier-Cuvelle S, Gillet F, Aragno M, Rossi P (2002) Statistical analyses of denaturing gel electrophoresis (DGE) fingerprinting patterns. Environ Microbiol 4:634–643

Fromin N, Tarnawski S, Roussel-Delif L, Hamelin J, Baggs EM, Aragno M (2005) Nitrogen fertilisation rate alters the frequency of nitrate-dissimilating *Pseudomonas* spp. in the rhizosphere of *Lolium perenne* grown under elevated pCO_2 (Swiss FACE). Soil Biol Biochem 37:1962–1965

Garbaye J (1994) Helper bacteria – a new dimension to the mycorrhizal symbiosis. New Phytol 128:197–210

Ginkel JH van, Gorissen A (1998) In situ decomposition of grass roots as affected by elevated atmospheric carbon dioxide. Soil Sci Soc Am J 62:951–958

Ginkel JH van, Gorissen A, Polci D (2000) Elevated atmospheric carbon dioxide concentration: effects of increased carbon input in a *Lolium perenne* soil on microorganisms and decomposition. Soil Biol Biochem 32:449–456

Gobat JM, Aragno M, Matthey W (2004) The living soil: fundamentals of soil science and soil biology. Science Publishers, Enfield

Griffiths BS, Ritz K, Ebblewhite N, Paterson E, Killham K (1998) Ryegrass rhizosphere microbial community structure under elevated carbon dioxide concentrations, with observations on wheat rhizosphere. Soil Biol Biochem 30:315–321

Hamelin J (2004) PhD thesis, University of Neuchâtel, Neuchâtel

Hamelin J, Fromin N, Tarnawski S, Teyssier-Cuvelle S, Aragno M (2002) NifH gene diversity in the bacterial community associated with the rhizosphere of *Molinia coerulea*, an oligonitrophilic perennial grass. Environ Microbiol 4:477–482

Hartwig UA, Zanetti S, Hebeisen T, Lüscher A, Frehner M, Fischer B, Kessel C van, Hendrey GR, Blum H, Nösberger J (1996) Symbiotic nitrogen fixation: one key to understanding the response of temperate grassland ecosystem to elevated CO_2? In: Körner C, Bazzaz FA (eds) Community, population and evolutionary responses to elevated CO_2 concentration. Academic Press, New York, pp 253–264

Hodge A, Paterson E, Grayston SJ, Campbell CD, Ord BG, Killham K (1998) Characterisation and microbial utilisation of exudate material from the rhizosphere of *Lolium perenne* grown under CO_2 enrichment. Soil Biol Biochem 30:1033–1043

Hu S, Firestone M, Chapin FS (1999) Soil microbial feedbacks to atmospheric CO_2 enrichment. Tree 14:433–437

Hu S, Chapin, FS, Firestone MK, Field CB, Chiariello NR (2001) Nitrogen limitation of microbial decomposition in a grassland under elevated CO_2. Nature 409:188–191

Hungate BA, Jaeger CH, Gamara G, Chapin FS, Field CB (2000) Soil microbiota in two annual grasslands: responses to elevated atmospheric CO_2. Oecologia 124:589–598

Ineson P, Coward PA, Hartwig UA (1998) Soil gas fluxes of N_2O, CH_4 and CO_2 beneath *Lolium perenne* under elevated CO_2: the Swiss free air carbon dioxide enrichment experiment. Plant Soil 198:89–95

Insam H, Baath E, Berreck M, Frostegard A, Gerzabek MH, Kraft A, Schinner F, Schweiger P, Tschuggnall G (1999) Responses of the soil microbiota to elevated CO_2 in an artificial tropical ecosystem. J Microbiol Methods 36:45–54

Jongen M, Jones MB, Hebeisen T, Blum H, Hendrey GR (1995) The effects of elevated CO_2 concentrations on the root growth of *Lolium perenne* and *Trifolium repens* grown in a FACE system. Global Change Biol 1:361–371

Jossi M, Hamelin J, Tarnawski S, Gillet F, Kohler F, Aragno M, Fromin N (2006) How does elevated pCO_2 modify total and metabolicaly active bacterial communivites in the rhizosphere of two perrennial grasses grown in field conditions? FEMS Microbiol Ecol 55:339–350

Kandeler E, Tscherko D, Hobbs PJ, Kampichler V, Jones TH (1998) The response of soil microorganisms and roots to elevated CO_2 and temperature in a terrestrial model ecosystem. Plant Soil 202:251–262

Kelly JJ, McCormack J, Janus LR, Angeloni N, Rier ST, Tuchman NC (2003) Elevated atmospheric CO_2 alters belowground micobial communities associated with quaking aspen roots. Abstr Gen Meet Am Soc Microbiol 103:N-017

Klironomos JN, Allen MF, Rillig MC, Piotrowski J, Makvandi-Nejad S, Wolfe BE, Powell JR (2005) Abrupt rise in atmospheric CO$_2$ overestimates community response in a model plant–soil system. Nature 433: 621–624

Kuzyakov Y (2001) Tracer studies of carbon translocation by plants from the atmosphere into the soil (a review). Eurasian Soil Sci 34, 28–42

Lugtenberg BJ, Dekkers LC (1999) What makes Pseudomonas bacteria rhizosphere competent? Environ Microbiol 1:9–13

Lussenhop J, Treonis A, Curtis PS, Teeri JA, Vogel CS (1998) Response of soil biota to elevated atmospheric CO$_2$ in poplar model systems. Oecologia 113:247–251

Marilley L, Aragno M (1999) Phylogenetic diversity of bacterial communities difffering in degree of proximity of Lolium perenne and Trifolium repens. Appl Soil Ecol 13:127–136

Marilley L, Vogt G, Blanc M, Aragno M (1998) Bacterial diversity in the bulk soil and rhizosphere fractions of Lolium perenne and Trifolium repens as revealed by PCR restriction analysis of 16s rDNA. Plant Soil 198:219–224

Marilley L, Hartwig UA, Aragno M (1999) Influence of an elevated atmospheric CO$_2$ content on soil and rhizosphere bacterial communities beneath Lolium perenne and Trifolium repens under field conditions. Microb Ecol 38:39–49

Montealegre CM, Kessel C van, Blumenthal JM, Hur HG, Hartwig UA, Sadowsky MJ (2000) Elevated atmospheric CO$_2$ alters microbial population structure in pasture ecosystem. Global Change Biol 6:475–482

Montealegre CM, Kessel C van, Blumenthal JM, Hur HG, Hartwig UA, Sadowski MJ (2002) Changes in microbial activity and composition in a pasture ecosystem exposed to elevated atmospheric carbon dioxide. Plant Soil 243:187–207

Mulder A, Vandegraaf AA, Robertson LA, Kuenen JG (1995) Anaerobic ammonium oxidation discovered in a denitrifying fluidized-bed reactor. FEMS Microbiol Ecol 16, 177–183

Paterson E (2003) Importance of rhizodeposition in the coupling of plant and microbial productivity. Eur J Soil Sci 2003:54

Paterson E, Hall JM, Rattray EAS, Griffiths BS, Ritz K, Killham K (1997) Effect of elevated CO$_2$ on rhizosphere carbon flow and soil microbial processes. Global Change Biol 3:363–377

Richter M (2003) PhD thesis, ETH-Z, Zürich

Richter M, Hartwig UA, Frossard E, Noesberger J, Cadisch G (2003) Gross fluxes of nitrogen in grassland soils exposed to elevated atmospheric pCO$_2$ for seven years. Soil Biol Biochem 22:1325–1335

Rillig MC, Wright SF, Allen MF, Field CB (1999) Rise in carbon dioxide changes soil structure. Nature 400:628

Roussel-Delif L, Tarnawski S, Hamelin J, Philippot L, Aragno M, Fromin N (2005) Frequency and diversity of nitrate reductase genes among nitrate-dissimilating Pseudomonas in the rhizosphere of perennial grasses grown in field conditions. Microb Ecol 49:63–72

Salmassi TM, Ayala EA, Barco RA, Becerra C, Chakhalyan MM, Connor KM, Cuevas G, Dolmajian HK, Hirun MS, Lam RG, Martinez AE, Martinez JG, Mendosa RY, Moshkani S, Nguyen JQ, Ovasapyan K, Shin CS, Sok D, West KL, Wong NJ, Wu RC, Monterosa A, Khachikian CS (2003) Effects of elevated CO$_2$ on soil bacterial diversity: a comparative study at Mammoth Mountain, Ca. Abstr Gen Meet Am Soc Microbiol 103:N-116

Sharma A, Johri BN (2003) Growth promoting influence of siderophore-producing Pseudomonas strains GRP3A and PRS$_9$ in maize (Zea mays L.) under iron limiting conditions. Microbiol Res 158:243–248

Tarnawski S (2004) PhD thesis, University of Neuchâtel, Neuchâtel

Williams MA, Rice CW, Owensby CE (2000) Carbon dynamics and microbial activity in tallgrass prairie exposed to elevated CO_2 for 8 years. Plant Soil 227:127–137

Yeates GW, Tate KR, Newton PCD (1997) Response of the fauna of a grassland soil to doubling of atmospheric carbon dioxide concentration. Biol Fertil Soils 25:307–315

Zak DR, Ringelberg DB, Pregitzer KS, Randlett DL, White DC, Curtis PS (1996) Soil microbial communities beneath *Populus grandidentata* Michx. grown under elevated atmospheric CO_2. Ecol Appl 6:257–262

Zak DR, Pregitzer KS, Curtis PS, Holmes WE (2000a) Atmospheric CO_2 and the composition and function of soil microbial communities. Ecol Appl 10:47–59

Zak DR, Pregitzer KS, King JS, Holmes WE (2000b) Elevated atmospheric CO_2, fine roots and the response of soil microroganisms: a review and hypothesis. New Phytol 147:201–222

23 Increases in Atmospheric [CO$_2$] and the Soil Food Web

D. A. PHILLIPS, T. C. FOX, H. FERRIS, and J. C. MOORE

23.1 Introduction

Organic matter deposited in soil by plants is the energy source for a complex web of functionally and nutritionally interconnected species. Bacteria and fungi, the initial consumers of soil organic matter, are themselves substrates for a multitude of tiny predators and grazers, including protozoa, nematodes, and arthropods, which comprise the soil food web (Brussaard et al. 1997). Dead plant tissue (i.e. litter), from both aboveground and belowground sources, is the dominant pathway by which plant carbon (C) moves to soil (Schlesinger and Lichter 2001), but living roots also transfer C to soil through turnover of fine roots (Jackson et al. 1997; Matamala et al. 2003) or living cells (Hawes 1990) and as soluble exudates (Rovira 1991). Direct herbivory of roots by certain nematodes (Ferris 1982) and other parasites constitutes another channel for C movement to soil. Plant C transferred to mycorrhizal fungi can be viewed as exudation because these organisms are separated from plant cells by membranes. Exuded compounds released from living roots may be more important than previously recognized because they are dynamically linked to plant growth (Farrar et al. 2003) and can be influenced both passively (Owen and Jones 2001; Jones et al. 2005) and actively (Phillips et al. 2004) by soil microorganisms. Thus, if either plant production of exudates or microbial pilfering of these compounds increases, one can foresee major effects on soil food web organisms and associated C storage. The importance of understanding how plant growth and microbial productivity are linked is widely recognized (Paterson 2003), but the specific mechanisms that control those connections are poorly understood.

Ecological Studies, Vol. 187
J. Nösberger, S.P. Long, R.J. Norby, M. Stitt,
G.R. Hendrey, H. Blum (Eds.)
Managed Ecosystems and CO$_2$
Case Studies, Processes, and Perspectives
© Springer-Verlag Berlin Heidelberg 2006

23.1.1 Soil Food Webs: The Concept

Soil food webs are assemblages of diverse, interdependent species. Although clear trophic levels often exist in aboveground ecosystems, the interdependence of microorganisms (bacteria and fungi), microfauna (protozoa and nematodes), mesofauna (mites, collembolans, and enchytraeids), and macrofauna (earthworms, ants, termites, and herbivorous insects) makes it difficult to distinguish trophic levels in belowground systems (Brussaard et al. 1997). As a result, soil organisms frequently are assigned to "functional" groups that share particular ecosystem roles, such as root colonization or predation (Fig. 23.1). Studies in which individual species have been removed show that not all species in a functional group are required for the basic operation of an ecosystem (Laakso and Setälä 1999). The exact niche of each species can be unclear, but temporal and spatial differences in species activity are often revealed by detailed observation (Gunapala et al. 1998; Ferris and Matute 2003).

One fundamental characteristic of soil food webs is that they are primarily heterotrophic assemblages which depend ultimately on autotrophic plants for a continuing supply of C resources. This fact suggests selection pressures may have favored survival of mutualistic interactions that stimulate plant growth while promoting an immediate or ultimate transfer of plant C to the soil organisms (Wall and Moore 1999). Existing ecological data support this idea.

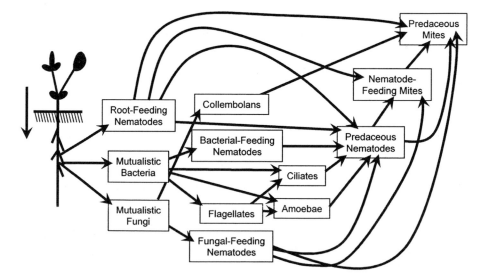

Fig. 23.1 Carbon flow in a soil food web associated with plant roots. Autotrophic plants supply C to root-colonizing organisms, which transfer C to various interconnected grazers and predators. (Moore and de Ruiter 1991)

For example, plant growth is stimulated when predators, such as nematodes and protozoa, consume soil bacteria and fungi (Clarholm 1985; Ingham et al. 1985; Wardle 1999; Wardle et al. 2004). Defining the myriad molecular mechanisms by which such interactions regulate plant growth, however, remains an important task (Phillips et al. 2003; Bonkowski 2004).

23.2 Effects of Elevated [CO$_2$] on Soil Organic Matter and the Food Web

The fate of newly derived soil organic matter interests scientists concerned with problems caused by increasing levels of atmospheric [CO$_2$]. Greater soil C storage, for example, may offer a short-term solution until longer-term methods and policies are developed for reducing CO$_2$ emissions or increasing CO$_2$ sequestration (Batjes 1998; Lal 2003). While elevated [CO$_2$] can promote plant productivity, including root growth, when sufficient mineral nutrients are available (Williams et al. 1981; Curtis et al. 1990; Owensby et al. 1993; Jongen et al. 1995), little of the additional C remains sequestered in soil for any length of time (Van Kessel et al. 2000; Schlesinger and Lichter 2001), possibly because soil microorganisms respire the new C rapidly. Such results emphasize the importance of understanding how soil microorganisms interact with increasing levels of [CO$_2$], but reviews of studies examining that topic show varied effects (Sadowsky and Schortemeyer 1997; Kampichler et al. 1998; Zak et al. 2000).

One meta-analysis of 47 studies evaluating responses of soil microorganisms to elevated [CO$_2$] found that soil respiration (root + microbial respiration) under plants exposed to higher [CO$_2$] generally, but not always, increased (Zak et al. 2000). The extent of changes in soil respiration varied from a 10 % decline to a 162 % increase, but mean increases across all reports were 51 % (±52 % standard deviation, SD) for grasses, 49 % (±24 % SD) for herbaceous dicots, and 42 % (±24 % SD) for woody plants (Zak et al. 2000). Calculated values for the microbial respiration component of soil respiration in the same report found mean increases from elevated [CO$_2$] of 34 % (±35 % SD) for grasses, 34 % (±18 % SD) for herbaceous dicots, and 20 % (±23 % SD) for woody plants. Estimates of microbial biomass showed increases with elevated [CO$_2$] of 17 % (±86 % SD) for grasses, 29 % (±29 % SD) for herbaceous dicots, and 19 % (±46 % SD) for woody plants. Other studies of these and related parameters, which were not included in that meta-analysis, show similarly divergent results (Zak et al. 1996; Cheng and Johnson 1998; Inubushi et al. 1999; Ball et al. 2000; Bruce et al. 2000; Montealegre et al. 2000; Sowerby et al. 2000; King et al. 2001; Wiemken et al. 2001; Phillips et al. 2002; Ronn et al. 2002, 2003). No attempt will be made here to rationalize these differences, but common sources of variation, including biological species present and mineral nutrient status, undoubtedly influenced the results. One overall conclusion may be that elevated [CO_2] increases soil respiration, microbial respiration, and microbial biomass, if other factors are not limiting, but increases in any particular case will be difficult to predict and probably are a function of environmental conditions, the community structure, and metabolic rates of individual organisms.

Studies of elevated [CO_2] effects on soil food web structure show other incompletely explained changes. Data indicate that increased [CO_2] frequently stimulates mycorrhizal fungi (Langley et al. 2003; Rillig and Field 2003; Treseder et al. 2003; Olsrud et al. 2004); and such effects could alter soil food webs. In one study where fungal grazers were examined, the number of collembolans, which prefer non-mycorrhizal fungi, increased with additional N and elevated [CO_2] as their favored fungal food sources proliferated more than the mycorrhizal fungi (Klironomos et al. 1997). In other experiments, elevated [CO_2] altered the mixture of species present in assemblages of nematodes (Hoeksema et al. 2000; Yeates et al. 2003), protozoans (Treonis and Lussenhop 1997; Yeates et al. 2003), and collembolans (Jones et al. 1998). In most cases, the changes in species present occurred without altering the total abundance of organisms. Such changes may not always reflect varying availability of C resources because other parameters, such as soil moisture, could also have changed.

Soil food webs frequently are dominated by either bacteria or fungi, which are viewed as separate channels for energy flows (Moore and Hunt 1988; De Ruiter et al. 1993). Measurements at the ETH FACE site showed clear effects of elevated [CO_2] and N fertilization on these two energy channels (J.C. Moore and H. Ferris, unpublished data). Saprobic and mycorrhizal fungi declined with increased N, while protozoa feeding in the bacterial channel increased with N additions and decreased under elevated [CO_2]. Responses of bacterial-feeding, fungal-feeding, and plant-feeding nematodes were not definitive, but omnivorous nematodes, which are supported by both energy channels, declined with N fertilization, which is consistent with their sensitivity to mineral fertilizers (Tenuta and Ferris 2004). These changes are consistent with shifts toward the bacterial energy channel, which occur with increasing amounts of N derived from mineralization (Moore et al. 2003). They also suggest shifts toward the fungal energy channel under elevated [CO_2], in which case the potential for C storage and N immobilization may increase.

Given uncertainty over changes in total soil C under elevated [CO_2], attention has focused on how altered [CO_2] affect particular C inputs from plants. In forest ecosystems, elevated [CO_2] increased plant litter (Schlesinger and Lichter 2001) and living fine roots (Matamala and Schlesinger 2000). At the same time, however, the accumulation of both litter and dead fine roots was restricted by their turnover rates (Matamala and Schlesinger 2000; Schlesinger and Lichter 2001; Matamala et al. 2003). Such results suggest that the degradation capacity of soil food webs exceeded any incremental C inputs produced by elevated [CO$_2$]. A complete understanding of the forces operating in such experiments, however, requires an analysis of soluble root exudates.

Separating soluble root exudates from degradation of organic compounds in complex ecosystems is difficult; and for this reason data that assess changes in dissolved organic carbon under elevated [CO$_2$] (Jones et al. 1998; Uselman et al. 2000) cannot be interpreted as measures of root exudation. The possibility that root exudation rises with elevated [CO$_2$] is supported by an increase in oxalate outside the root under higher [CO$_2$] (Delucia et al. 1997), but data describing the effects of elevated [CO$_2$] on other key root exudates, such as amino acids and sugars (Fan et al. 2001), have not been reported. Because bacteria and fungi often stimulate root exudation (Meharg and Killham 1995), predators and grazers on these microorganisms could influence exudation and their interactions may affect soil C storage. For these reasons, a direct examination of how elevated [CO$_2$], microorganisms and the soil food web affect root exudation is justified.

23.3 Root Exudation and the Effects of Elevated [CO$_2$]

Root exudation of soluble compounds, such as amino acids, is a multi-faceted process involving both efflux and influx components (Fig. 23.2; Jones and Darrah 1994). Thus amino acid "exudation" is more properly viewed as a *net* efflux. Simple sugars move in and out of plant roots by mechanisms similar to those controlling amino acid fluxes (Jones and Darrah 1996), but dicarboxylic acids move primarily out of the root, and little influx has been detected at ecologically relevant concentrations (Jones and Darrah 1995). In biochemical terms, the passive efflux of amino acids is driven primarily by large differences in concentration between the inside (e.g. 10 mM) and the outside (e.g. 0.1–10 μM) of root cells, while influx involves proton-pumping ATPases that maintain an electrochemical potential difference across the plasma membrane to support uptake into the plant by proton-coupled amino acid transporters (Farrar et al. 2003). Other materials released from roots, such as proteins, complex carbohydrates, and insoluble cellular debris, are often mediated by physical processes, including herbivory by nematodes, enchytraeids, and insects, which are not addressed here.

Early work established that the presence of microorganisms around roots could increase photosynthate released by the plant into the soil (Meharg and Killham 1995). For amino acids, that increase occurs because amino acids present in soil at low concentrations, such as those coming from root exudation, are taken up more effectively by microorganisms than by roots (Jones et al. 2005). Obviously any amino acids used by microorganisms would not be available for reabsoprtion by the plant (Owen and Jones 2001). Though com-

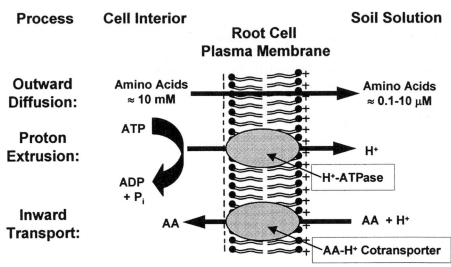

Fig. 23.2 Plant processes involved in amino acid (*AA*) exudation from roots. Amino acid exudation represents the net efflux resulting from separate efflux and influx processes (Jones and Darrah 1994). Exudation of most simple sugars also follows the principles shown here (Jones and Darrah 1996), but the dicarboxylic acid citrate is lost through efflux without being reabsorbed (Jones and Darrah 1995)

pletely correct, this traditional view of microorganisms as passive receptacles for amino acids lost from roots lacks the complexity suggested by data indicating that compounds produced by bacteria and fungi can actively increase the net efflux of amino acids (Phillips et al. 2004). That work, which quantified both efflux and influx components for 16 amino acids, showed that treating roots of four species of axenic plants with 100 μM 2,4-diacetylphloroglucinol (DAPG), increased net efflux 3- to 20-fold. DAPG is a common product of *Pseudomonas* bacteria (Cook et al. 1995), and 100 μM can be an ecologically relevant concentration (Bonsall et al. 1997). Other microbial compounds, including phenazine and phenazine-1-carboxylate from *Pseudomonas* (Taraz et al. 1990) and zearalenone from *Fusarium* fungi (Jimenez et al. 1996), also enhanced amino acid exudation in the absence of microorganisms (Phillips et al. 2004).

Additional experiments have now quantified influx and efflux of amino acids in maize, annual ryegrass, and medic seedlings treated with 425 ppm or 850 ppm [CO_2] (Phillips et al. 2006). Those results show that elevated [CO_2] probably can enhance rhizodeposition by two mechanisms. First, in C-3 wheat and medic, higher [CO_2] promoted root growth without altering amino acid efflux rate (nmol g^{-1} root fresh weight), and thus a larger root surface area would allow more exudation. Second, in C-4 maize elevated [CO_2] did not stimulate root or shoot growth, but there was a 44 % increase in the total efflux rate of 16 amino acids, which was associated with a significant ($P=0.05$) increase in efflux rates of six individual amino acids. These studies used axenic seedlings to examine the innate efflux and influx capacities of plants growing in the absence of culturable microorganisms. Roots of the three plant species studied, under both ambient and elevated [CO_2], took up the 16 amino acids at rates 94–374 % higher than they were effluxed, but in soil, adsorption of amino acids to soil particles before they were recovered through influx to the root should increase rhizodeposition under elevated [CO_2]. These striking results emphasizes the important role of microorganisms as both passive (Owen and Jones 2001; Jones et al. 2005) and active (Phillips et al. 2004) promoters of root exudation.

23.4 Linking Plants to Soil Food Webs under Changing [CO_2]

Soil food webs clearly can stimulate plant growth (Bonkowski 2004); and increased exudation under elevated [CO_2] has the potential to promote this effect by nurturing food web activities. Understanding these interactions and how they relate to progressive N limitations (Luo et al. 2004) are requirements for predicting the effects of climate change on soil food web functions. Predation of microorganisms by nematodes, protozoans, and/or arthropods contributes significantly to plant growth. In quantitative terms, predation can increase growth of a perennial grass by 145 % when nematodes feed on bacteria (Ingham et al. 1985). Mineralization of limiting nutrients (e.g. N) is one component of the growth stimulation (Ferris et al. 1997; Laakso and Setälä 1999; Wardle 1999; Wardle et al. 2004) but not a total explanation. For example, bacterial-feeding protozoa promoted biomass accumulation in woody tree seedlings by 55 % even when a complete, N-containing nutrient solution was supplied every 2 h (Jentschke et al. 1995). Other analyses suggest that additional benefits of predation, including the release of particular organic products (Phillips et al. 1999) and the promotion of root colonization by beneficial bacteria (Bonkowski and Brandt 2002), also contribute to plant growth. Thus two key processes underlying the promotion of plant growth by soil food webs must be explained: (1) root colonization by microorganisms and (2) predation of bacteria and fungi by nematodes, protozoa, and arthropods.

Root colonization reflects direct activities of microorganisms, as well as the indirect effects of their predators. Direct microbial interactions with roots through adhesion (Matthysse and McMahan 2001), biofilm formation (O'Toole et al. 2000), responses to their own quorum-sensing compounds (von Bodman et al. 2003) or plant-derived quorum-sensing mimics (Teplitski et al. 2000), and competitive exclusion of competing fungi (Cook et al. 1995) are topics of active investigation, but detailed connections to the soil food web are poorly understood. Indirect effects include the physical transport of bacteria to new resources by nematodes (Ingham et al. 1985; Brown et al. 2004). Some bacteria are transferred on the outer surface of nematodes, but many survive passage through the nematode intestine. Amoebae predators, in contrast, reduce the number of bacteria on roots while simultaneously increasing the proportion of auxin-producing bacteria (Bonkowski and Brandt 2002). One result of this population shift is an increase in lateral root length. Such plant-growth-promoting bacteria occur commonly on roots and may enhance growth through multiple mechanisms (Ryu et al. 2003). Any mechanism that promotes growth of the root or root-colonizing microorganisms obviously has the potential to benefit the larger soil food web.

Predation, the other key process involved with the soil food web promotion of plant growth, is well characterized at the organismic level, but molecular mechanisms are poorly understood (Phillips et al. 2003). Nematodes show preferences for certain bacterial species (Moens et al. 1999), which may be based on attraction, repellence or a combination of both, and the potential benefit of such preferences is evident in the fact that nematode growth rates differ as much as 12-fold when they are supplied with various bacterial species as food sources (Venette and Ferris 1998). Little is known about how nematodes locate bacteria, insect larvae, or other nematodes, but such complex behaviors clearly involve receptors and neurotransmissions (Chao et al. 2004). One can reasonably hypothesize that nematode selection of bacteria or detection of the root involves responses to particular compounds. Examining this hypothesis requires careful tests that measure two forms of nematode movement, kinesis and taxis (Young et al. 1998; Rodger et al. 2003). Chemotactic responses of nematodes are generally assumed to result from differences in signal perceived at the amphid neurons on either side of the head, although laser ablation of chemosensory neurons in one amphid does not prevent chemotaxis in *C. elegans* (Bargmann and Mori 1997). Studies of chemotactic behavior are often performed on agar surfaces on which many nematodes are oriented with their lateral surfaces at right angles to the agar and the signal source. In that case, the dorso-ventral movements of the nematode body probably do not expose the amphids to differences in signal strength. A three-dimensional matrix, although less tractable observationally, is probably a more realistic environment for studies of chemotactic behavior of soil nematodes (Perry and Aumann 1998; Lee 2002).

Nematodes move toward increasing $[CO_2]$ (Klingler 1965; Robinson 1995; Lee 2002), but because they distinguish between CO_2-producing roots and insect larvae, they must sense additional factors (Ruhm et al. 2003). While they are attracted to plant roots (Prot 1977) and root exudates (Viglierchio 1961; Riddle and Bird 1985), no plant-specific compounds that specifically produce the response have been identified. We hypothesize that bacterial-feeding nematodes commonly found near plant roots respond positively to plant signature compounds and both positively and negatively to bacterial compounds, which they use to select particular bacterial species. Current experiments are testing these concepts.

Several regulatory molecules involved in communication between plants and soil food web organisms have been found in recent studies. For example, plant roots release chemical factors, which regulate bacterial quorum-sensing genes normally responding to N-acyl homoserine lactone (AHL) signals from other bacteria (Teplitski et al. 2000). These genes control key processes involved in root colonization, including motility, biofilm formation, and antibiotic production (von Bodman et al. 2003). Other examples include lumichrome, a riboflavin breakdown product, which increases plant growth at low concentrations (5–50 nM; Phillips et al. 1999), and homoserine lactone, an AHL degradation product that can increase stomatal opening and transpiration when supplied to roots at 10 nM (Joseph and Phillips 2003). Also, the bacterial product DAPG not only enhances amino acid exudation from roots (Phillips et al. 2004), but, in certain cases, it promotes plant growth (De Leij et al. 2002). Current experiments are detecting DAPG effects on gene transcription related to growth (T.C. Fox and D.A. Phillips, unpublished data). Other studies have already shown that treating roots with 10 nM AHLs alters the accumulation of over 150 plant proteins inside the treated region (Mathesius et al. 2003). Thus, there is mounting evidence that low external concentrations of key compounds produced by food web organisms have major effects on the functioning of both plants and the associated soil organisms. This evidence emphasizes that soil food webs have evolved to function in an environment with myriad active chemical factors. The processes of root colonization and predation undoubtedly reflect the effects of many such compounds, but whether elevated [CO$_2$] generally affects the production of secondary metabolites in plants that might reach soil food web organisms is unclear. For example, several phenolic compounds in roots and shoots of *Plantago maritima* increased under higher [CO$_2$] (Davey et al. 2004), but no changes in total phenolics were detected in several tree species exposed to elevated [CO$_2$] (Hamilton et al. 2004).

We doubt that a doubling of atmospheric [CO$_2$] will disrupt interactions between plants and the soil food web. Increases in plant litter and root exudation with rising [CO$_2$] may elevate soil food web activity until the autotrophic plant community restricts exudation for some reason, such as a progressive limitation of available mineral N (Luo et al. 2004). Atmospheric [CO$_2$] levels are currently approaching 400 ppm. When estimates of atmospheric [CO$_2$] levels based on stomatal abundance in fossils (Retallack 2001) are related to probable evolutionary interactions between terrestrial plants and soil organisms, it is evident that roots and soil food webs have co-evolved through multiple periods when atmospheric [CO$_2$] exceeded 2000 ppm (Fig. 23.3). It seems logical, therefore, that a natural balance or buffering of biochemical and physiological processes will help plants and their associated soil food webs survive [CO$_2$] much higher than current conditions.

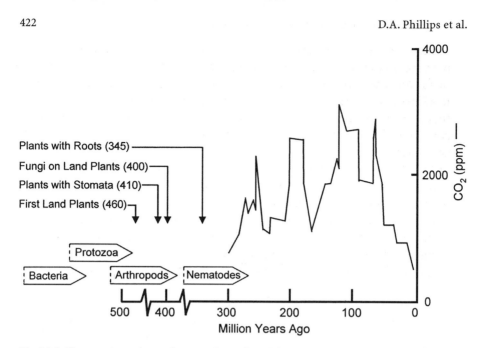

Fig. 23.3 Changes in estimated atmospheric [CO_2], based on stomatal abundance on leaf fossils (Retallack 2001) and their relationship to possible milestones in the evolution of soil food webs. Fossil evidence for many soft-bodied organisms present in soil food webs is limited (Poinar 1983), but numerous highly evolved individuals, morphologically similar to modern arthropods, are present in fossils from 400 million years ago

23.5 Conclusions

Organic inputs to soil are comprised largely of plant debris and root exudation, which is responsible for rhizodeposition. Increases in organic matter inputs from plants growing under elevated [CO_2] affect soil microorganisms and a limited set of conclusions can be drawn.

- Bacterial and fungal communities in soil ecosystems use such plant materials as resources to support multiple levels of tiny grazers and predators, which comprise soil food webs.
- Ten years of elevated [CO_2] at the ETH FACE site produced data on soil protozoa and nematodes that are consistent with adjustments predicted for availability of soil bacteria and fungi.
- Disparate changes in soil microorganisms and complex adjustments in food web structure reported under higher [CO_2] in a multitude of other experiments suggest that a better understanding of C resource availability is needed.
- Increases in living root mass under elevated [CO_2] could affect soil food webs through additional exudation, but limited information is available on

changes in root exudation under such conditions. We summarize here a new, more complex, understanding of root exudation that includes mechanisms by which microorganisms, and possibly their predators within the food web, can actively enhance root exudation. Initial experiments indicate that higher $[CO_2]$ can increase root exudation of amino acids under axenic conditions by two separate mechanisms and these could result in more rhizodeposition.

- Little is known about how elevated $[CO_2]$ levels alter predation, another key connection between the soil food web and the plant, but reductionist studies are beginning to support the concept that specific molecules affect predation and influence many organismic interactions in the root zone.
- Because the fossil record suggests soil food webs were exposed to widely varied levels of $[CO_2]$ for long periods, a certain stability of these interactions should be expected as global atmospheric $[CO_2]$ increases.

Acknowledgements. This work was supported by NSF grant DEB-0120169 and by award US-3353-02 from BARD, the US–Israel Binational Agricultural Research and Development Fund.

References

Ball AS, Milne E, Drake BG (2000) Elevated atmospheric-carbon dioxide concentration increases soil respiration in a mid-successional lowland forest. Soil Biol Biochem 32:721–723

Bargmann CI, Mori I (1997) Chemotaxis and thermotaxis. In: Riddle DL, Blumenthal T, Meyer BJ, Preiss JR (eds) *C. elegans* II. Cold Spring Harbor Laboratory Press, Cold Spring Harbor, pp 717–737

Batjes NH (1998) Mitigation of atmospheric CO_2 concentrations by increased carbon sequestration in the soil. Biol Fertil Soils 27:230–235

Bodman S von, Bauer W, Coplin D (2003) Quorum sensing in plant-pathogenic bacteria. Ann Rev Phytopath 41:455-482

Bonkowski M (2004) Protozoa and plant growth: the microbial loop in soil revisited. New Phytol 162:617–631

Bonkowski M, Brandt F (2002) Do soil protozoa enhance plant growth by hormonal effects? Soil Biol Biochem 34:1709–1715

Bonsall RF, Weller DM, Thomashow LS (1997) Quantification of 2,4-diacetylphloroglucinol produced by fluorescent *Pseudomonas* spp in vitro and in the rhizosphere of wheat. Appl Environ Microbiol 63:951–955

Brown DH, Ferris H, Fu S, Plant R (2004) Positive feedback in a model food web. Theor Popul Biol 65:143–152

Bruce KD, Jones TH, Bezemer TM, Thompson LJ, Ritchie DA (2000) The effect of elevated atmospheric carbon dioxide levels on soil bacterial communities. Global Change Biol 6:427–434

Brussaard L, Behan-Pelletier VM, Bignell DE, Brown VK, Didden W, Folgarait P, Fragoso C, Freckman DW, Gupta V, Hattori T, Hawksworth DL, Klopatek C, Lavelle P, Malloch

DW, Rusek J, Soderstrom B, Tiedje JM, Virginia RA (1997) Biodiversity and ecosystem functioning in soil. Ambio 26:563–570

Chao MY, Komatsu H, Fukuto HS, Dionne HM, Hart AC (2004) Feeding status and serotonin rapidly and reversibly modulate a *Caenorhabditis elegans* chemosensory circuit. Proc Natl Acad Sci USA 101:15512–15517

Cheng WX, Johnson DW (1998) Elevated CO_2, rhizosphere processes, and soil organic matter decomposition. Plant Soil 202:167–174

Clarholm M (1985) Interactions of bacteria, protozoa, and plants leading to mineralization of soil nitrogen. Soil Biol Biochem 17:181–187

Cook RJ, Thomashow LS, Weller DM, Fujimoto D, Mazzola M, Bangera G, Kim D (1995) Molecular mechanisms of defense by rhizobacteria against root disease. Proc Natl Acad Sci USA 92:4197–4201

Curtis PS, Balduman LM, Drake BG, Whigham DF (1990) Elevated atmospheric CO_2 effects on belowground processes in C_3 and C_4 estuarine marsh communities. Ecology 71:2001–2006

Davey MP, Bryant DN, Cummins I, Ashenden TW, Gates P, Baxter R, Edwards R (2004) Effects of elevated CO_2 on the vasculature and phenolic secondary metabolism of *Plantago maritima*. Phytochemistry 65:2197–2204

De Leij F, Dixon-Hardy JE, Lynch JM (2002) Effect of 2,4-diacetylphloroglucinol-producing and non-producing strains of *Pseudomonas fluorescens* on root development of pea seedlings in three different soil types and its effect on nodulation by *Rhizobium*. Biol Fertil Soils 35:114–121

De Ruiter P, Moore J, Zwart K, Bouwman L, Hassink J, Bloem J, De Vos J, Marinissen J, Didden W, Lebbink G, Brussaard L (1993) Simulation of nitrogen mineralization in the belowground food webs of two winter-wheat fields. J Appl Ecol 30:95–106

Delucia EH, Callaway RM, Thomas EM, Schlesinger WH (1997) Mechanisms of phosphorus acquisition for ponderosa pine seedlings under high CO_2 and temperature. Ann Bot 79:111–120

Fan TW-M, Lane AN, Shenker M, Bartley JP, Crowley D, Higashi RM (2001) Comprehensive chemical profiling of gramineous plant root exudates using high-resolution NMR and MS. Phytochemistry 57:209–221

Farrar J, Hawes M, Jones D, Lindow S (2003) How roots control the flux of carbon to the rhizosphere. Ecology 84:827–837

Ferris H (1982) The role of nematodes as primary consumers. In: Freckman DW (ed) Nematodes in soil ecosystems. University of Texas, Austin, pp 3–13

Ferris H, Matute M (2003) Structural and functional succession in the nematode fauna of a soil food web. Appl Soil Ecol 23:93–110

Ferris H, Venette RC, Lau SS (1997) Population energetics of bacterial-feeding nematodes: carbon and nitrogen budgets. Soil Biol Biochem 29:1183–1194

Gunapala N, Venette RC, Ferris H, Scow KM (1998) Effects of soil management history on the rate of organic matter decomposition. Soil Biol Biochem 30:1917–1927

Hamilton JG, Zangerl AR, Berenbaum MR, Pippen J, Aldea M, DeLucia EH (2004) Insect herbivory in an intact forest understory under experimental CO_2 enrichment. Oecologia 138:566–573

Hawes MC (1990) Living plant cells released from the root cap: A regulator of microbial populations in the rhizosphere? Plant Soil 129:19–27

Hoeksema JD, Lussenhop J, Teeri JA (2000) Soil nematodes indicate food web responses to elevated atmospheric CO_2. Pedobiologia 44:725–735

Ingham RE, Trofymow JA, Ingham ER, Coleman DC (1985) Interactions of bacteria, fungi and their nematode grazers: effects on nutrient cycling and plant growth. Ecol Monogr 55:119–140

Inubushi K, Cheng WG, Chander K (1999) Carbon dynamics in submerged soil microcosms as influenced by elevated CO$_2$ and temperature. Soil Sci Plant Nutr 45:863–872

Jackson RB, Mooney HA, Schulze ED (1997) A global budget for fine root biomass, surface area, and nutrient contents. Proc Natl Acad Sci USA 94:7362–7366

Jentschke G, Bonkowski M, Godbold DL, Scheu S (1995) Soil protozoa and forest tree growth – non-nutritional effects and interaction with mycorrhizae. Biol Fertil Soils 20:263–269

Jimenez M, Manez M, Hernandez E (1996) Influence of water activity and temperature on the production of zearalenone in corn by three *Fusarium* species. Int J Food Microbiol 29:417–421

Jones DL, Darrah PR (1994) Amino-acid influx at the soil-root interface of *Zea Mays* L. and its implications in the rhizosphere. Plant Soil 163:1–12

Jones DL, Darrah PR (1995) Influx and efflux of organic acids across the soil–root interface of *Zea mays* L. And its implications in rhizosphere C flow. Plant Soil 173:103–109

Jones DL, Darrah PR (1996) Re-sorption of organic compounds by roots of *Zea mays* L. and its consequences in the rhizosphere III. Characteristics of sugar influx and efflux. Plant Soil 178:153–160

Jones DL, Shannon D, Junvee-Fortune T, Farrar JF (2005) Plant capture of free amino acids is maximized under high soil amino acid concentrations. Soil Biol Biochem 37:179–181

Jones TH, Thompson LJ, Lawton JH, Bezemer TM, Bardgett RD, Blackburn TM, Bruce KD, Cannon PF, Hall GS, Hartley SE, Howson G, Jones CG, Kampichler C, Kandeler E, Ritchie DA (1998) Impacts of rising atmospheric carbon dioxide on model terrestrial ecosystems. Science 280:441–443

Jongen M, Jones MB, Hebeisen T, Blum H, Hendrey G (1995) The effects of elevated CO$_2$ concentrations on the root growth of *Lolium perenne* and *Trifolium repens* grown in a face system. Global Change Biol 1:361–371

Joseph CM, Phillips DA (2003) Metabolites from soil bacteria affect plant water relations. Plant Physiol Biochem 41:189–192

Kampichler C, Kandeler E, Bardgett RD, Jones TH, Thompson LJ (1998) Impact of elevated atmospheric CO$_2$ concentration on soil microbial biomass and activity in a complex, weedy field model ecosystem. Global Change Biol 4:335–346

King JS, Pregitzer KS, Zak DR, Kubiske ME, Holmes WE (2001) Correlation of foliage and litter chemistry of sugar maple, *Acer Saccharum*, as affected by elevated CO$_2$ and varying N availability, and effects on decomposition. Oikos 94:403–416

Klingler J (1965) On the orientation of plant nematodes and of some other soil animals. Nematologica 11:14–18

Klironomos JN, Rillig MC, Allen MF, Zak DR, Kubiske M, Pregitzer KS (1997) Soil fungal-arthropod responses to *Populus tremuloides* grown under enriched atmospheric CO$_2$ under field conditions. Global Change Biol 3:473–478

Laakso J, Setälä H (1999) Sensitivity of primary production to changes in the architecture of belowground food webs. Oikos 87:57–64

Lal R (2003) Global potential of soil carbon sequestration to mitigate the greenhouse effect. Crit Rev Plant Sci 22:151–184

Langley JA, Dijkstra P, Drake BG, Hungate BA (2003) Ectomycorrhizal colonization, biomass, and production in a regenerating scrub oak forest in response to elevated CO$_2$. Ecosystems 6:424–430

Lee DL (2002) Behaviour. In: Lee DL (ed) The biology of nematodes. Taylor and Francis, London, pp 369–387

Luo Y, Su B, Currie W, Dukes J, Finzi A, Hartwig U, Hungate B, McMurtrie R, Oren R, Parton W, Pataki D, Shaw M, Zak D, Field C (2004) Progressive nitrogen limitation of ecosystem responses to rising atmospheric carbon dioxide. Bioscience 54:731–739

Matamala R, Schlesinger WH (2000) Effects of elevated atmospheric CO_2 on fine root production and activity in an intact temperate forest ecosystem. Global Change Biol 6:967–979

Matamala R, Gonzalez-Meler MA, Jastrow JD, Norby RJ, Schlesinger WH (2003) Impacts of fine root turnover on forest NPP and soil C sequestration potential. Science 302:1385–1387

Mathesius U, Mulders S, Gao MS, Teplitski M, Caetano-Anolles G, Rolfe BG, Bauer WD (2003) Extensive and specific responses of a eukaryote to bacterial quorum-sensing signals. Proc Natl Acad Sci USA 100:1444–1449

Matthysse A, McMahan S (2001) The effect of the *Agrobacterium tumefaciens attR* mutation on attachment and root colonization differs between legumes and other dicots. Appl Environ Microbiol 67:1070–1075

Meharg AA, Killham K (1995) Loss of exudates from the roots of perennial ryegrass inoculated with a range of micro-organisms. Plant Soil 170:345–349

Moens T, Verbeeck L, de Maeyer A, Swings J, Vincx M (1999) Selective attraction of marine bacterivorous nematodes to their bacterial food. Mar Ecol Progr Ser 176:165–178

Montealegre CM, Van Kessel C, Blumenthal JM, Hur H-G, Hartwig U, Sadowsky MJ (2000) Elevated atmospheric CO_2 alters microbial population structure in a pasture ecosystem. Global Change Biol 6:475–482

Moore JC, Hunt HW (1988) Resource compartmentation and the stability of real ecosystems. Nature 333:261–263

Moore JC, Ruiter PC de (1991) Temporal and spatial heterogeneity of trophic interactions within below-ground food webs. Agric Ecosyst Environ 34:371–397

Moore JC, McCann K, Setälä H, Ruiter PC de (2003) Top-down is bottom-up: Does predation in the rhizosphere regulate aboveground dynamics? Ecology 84:846–857

O'Toole G, Kaplan HB, Kolter R (2000) Biofilm formation as microbial development. Annu Rev Microbiol 54:49–79

Olsrud M, Melillo JM, Christensen TR, Michelsen A, Wallander H, Olsson PA (2004) Response of ericoid mycorrhizal colonization and functioning to global change factors. New Phytol 162:459–469

Owen AG, Jones DL (2001) Competition for amino acids between wheat roots and rhizosphere microorganisms and the role of amino acids in plant N acquisition. Soil Biol Biochem 33:651–657

Owensby CE, Coyne PI, Ham JM, Auen LM, Knapp AK (1993) Biomass production in a tallgrass prairie ecosystem exposed to ambient and elevated CO_2. Ecol Appl 3:644–653

Paterson E (2003) Importance of rhizodeposition in the coupling of plant and microbial productivity. Eur J Soil Sci 54:741–750

Perry RN, Aumann J (1998) Behaviour and sensory responses. In: Perry RN, Wright DJ (eds) The physiology and biochemistry of free-living and plant-parasitic nematodes. CAB International, Wallingford, pp 75–102

Phillips DA, Joseph CM, Yang GP, Martínez-Romero E, Sanborn JR, Volpin H (1999) Identification of lumichrome as a *Sinorhizobium* enhancer of alfalfa root respiration and shoot growth. Proc Natl Acad Sci USA 96:12275–12280

Phillips DA, Ferris H, Cook DR, Strong DR (2003) Molecular control points in rhizosphere food webs. Ecology 84:816–826

Phillips DA, Fox TC, King MD, Bhuvaneswari TV, Teuber LR (2004) Microbial products trigger amino acid exudation from plant roots. Plant Physiol 136:2887–2894

Phillips DA, Fox TC, Six J (2006) Root exudation (net efflux of amino acids) may increase rhizodeposition under elevated carbon dioxide. Global Change Biol 12:561–567

Phillips RL, Zak DR, Holmes WE, White DC (2002) Microbial community composition and function beneath temperate trees exposed to elevated atmospheric carbon dioxide and ozone. Oecologia 131:236–244

Poinar GO (1983) The natural history of nematodes. Prentice–Hall, Englewood Cliffs, N.J.

Prot JC (1977) Amplitude et cinétique des migrations du nématode *Meloidogyne javanica* sous l'influence d'un plant de tomate. Cah ORSTOM Ser Biol 11:157–166

Retallack GJ (2001) A 300-million-year record of atmospheric carbon dioxide from fossil plant cuticles. Nature 411:287–290

Riddle DL, Bird AF (1985) Responses of the plant parasitic nematodes *Rotylenchus reniformis*, *Anguina argostis* and *Meloidogyne javanica* to chemical attractants. Parasitology 91:185–195

Rillig MC, Field CB (2003) Arbuscular mycorrhizae respond to plants exposed to elevated atmospheric CO$_2$ as a function of soil depth. Plant Soil 254:383–391

Robinson AF (1995) Optimal release rates for attracting *Meloidogyne incognita*, *Rotylenchulus reniformis*, and other nematodes to carbon dioxide in sand. J Nematol 27:42–50

Rodger S, Bengough AG, Griffiths BS, Stubbs V, Young IM (2003) Does the presence of detached root border cells of *Zea mays* alter the activity of the pathogenic nematode *Meloidogyne incognita*? Phytopathology 93:1111–1114

Ronn R, Gavito M, Larsen J, Jakobsen I, Frederiksen H, Christensen S (2002) Response of free-living soil protozoa and microorganisms to elevated atmospheric CO$_2$ and presence of mycorrhiza. Soil Biol Biochem 34:923–932

Ronn R, Ekelund F, Christensen S (2003) Effects of elevated atmospheric CO$_2$ on protozoan abundance in soil planted with wheat and on decomposition of wheat roots. Plant Soil 251:13–21

Rovira AD (1991) Rhizosphere research – 85 years of progress and frustration. In: Keister DL, Cregan PB (eds) The rhizosphere and plant growth. Kluwer Academic, Dordrecht, pp 3–13

Ruhm R, Dietsche E, Harloff HJ, Lieb M, Franke S, Aumann J (2003) Characterisation and partial purification of a white mustard kairomone that attracts the beet cyst nematode, *Heterodera schachtii*. Nematology 5:17–22

Ryu CM, Farag MA, Hu CH, Reddy MS, Wei HX, Pare PW, Kloepper JW (2003) Bacterial volatiles promote growth in *Arabidopsis*. Proc Natl Acad Sci USA 100:4927–4932

Sadowsky MJ, Schortemeyer M (1997) Soil microbial responses to increased concentrations of atmospheric CO$_2$. Global Change Biol 3:217–224

Schlesinger WH, Lichter J (2001) Limited carbon storage in soil and litter of experimental forest plots under increased atmospheric CO$_2$. Nature 411:466–469

Sowerby A, Blum H, Gray TRG, Ball AS (2000) The decomposition of *Lolium perenne* in soils exposed to elevated CO$_2$: Comparisons of mass loss of litter with soil respiration and soil microbial biomass. Soil Biol Biochem 32:1359–1366

Taraz K, Schaffner EM, Budzikiewicz H, Korth H, Pulverer G (1990) 2,3,9-Trihydoxyphenazin-1-carbonsäure – ein unter Berylliumeinwirkung gebildetes neues Phenazinderivat aus *Pseudomonas fluorescens*. Z Naturforsch 45b:552–556

Tenuta M, Ferris H (2004) Relationship between nematode life-history classification and sensitivity to stressors: ionic and osmotic effects of nitrogenous solutions. J Nematol 36:85–94

Teplitski M, Robinson JB, Bauer WD (2000) Plants secrete substances that mimic bacterial *N*-acyl homoserine lactone signal activities and affect population density-dependent behaviors in associated bacteria. Mol Plant Microbe Interact 13:637–648

Treonis AM, Lussenhop JF (1997) Rapid response of soil protozoa to elevated CO$_2$. Biol Fertil Soils 25:60–62

Treseder KK, Egerton-Warburton LM, Allen MF, Cheng YF, Oechel WC (2003) Alteration of soil carbon pools and communities of mycorrhizal fungi in chaparral exposed to elevated carbon dioxide. Ecosystems 6:786–796

Uselman SM, Qualls RG, Thomas RB (2000) Effects of increased atmospheric CO_2, temperature, and soil N availability on root exudation of dissolved organic carbon by a N-fixing tree (*Robinia pseudoacacia* L.). Plant Soil 222:191–202

Van Kessel C, Horwath WR, Hartwig U, Harris D, Lüscher A (2000) Net soil carbon input under ambient and elevated CO_2 concentrations: isotopic evidence after 4 years. Global Change Biol 6:435–444

Venette RC, Ferris H (1998) Influence of bacterial type and density on population growth of bacterial-feeding nematodes. Soil Biol Biochem 30:949–960

Viglierchio DR (1961) Attraction of parasitic nematodes by plant root emanations. Phytopathol 51:136–142

Wall DW, Moore JC (1999) Interactions underground: soil biodiversity, mutualism and ecosystem processes. BioScience 49:109–117

Wardle DA (1999) How soil food webs make plants grow. Trends Ecol Evol 14:418–420

Wardle DA, Bardgett RD, Klironomos JN, Setälä H, Putten WH van der, Wall DH (2004) Ecological linkages between aboveground and belowground biota. Science 304:1629–1633

Wiemken V, Ineichen K, Boller T (2001) Development of ectomycorrhizas in model beech-spruce ecosystems on siliceous and calcareous soil: A 4-year experiment with atmospheric CO_2 enrichment and nitrogen fertilization. Plant Soil 234:99–108

Williams LE, DeJong TM, Phillips DA (1981) Carbon and nitrogen limitations on soybean seedling development. Plant Physiol 68:1206–1209

Yeates GW, Newton PCD, Ross DJ (2003) Significant changes in soil microfauna in grazed pasture under elevated carbon dioxide. Biol Fertil Soils 38:319–326

Young IM, Griffiths BS, Robertson WM, McNicol JW (1998) Nematode (*Caenorhabditis elegans*) movement in sand as affected by particle size, moisture and the presence of bacteria (*Escherichia coli*). Eur J Soil Sci 49:237–241

Zak DR, Ringelberg DB, Pregitzer KS, Randlett DL, White DC, Curtis PS (1996) Soil microbial communities beneath *Populus grandidentata* crown under elevated atmospheric CO_2. Ecol Appl 6:257–262

Zak DR, Pregitzer KS, King JS, Holmes WE (2000) Elevated atmospheric CO_2, fine roots and the response of soil microorganisms: A review and hypothesis. New Phytolog 147:201–222

Part D   Perspectives

24 FACE Value: Perspectives on the Future of Free-Air CO₂ Enrichment Studies

A. ROGERS, E. A. AINSWORTH, and C. KAMMANN

24.1 The Value of FACE Experiments

Free-air CO_2 enrichment (FACE) studies are the ultimate test bed for hypotheses that seek to explain how plants respond to rising $[CO_2]$; and they provide the most realistic conditions for simulating the impact of future elevated (e)$[CO_2]$ levels (see Chapter 2). FACE studies have many benefits over controlled environment and open-top chamber (OTC) experiments. FACE allows the investigation of an undisturbed ecosystem and does not modify the vegetation's interaction with light, temperature, wind, precipitation, pathogens and insects (Long et al. 2004). This, in combination with the large size of FACE plots, allows the integrated measurement of many plant and ecosystem processes simultaneously in the same plot, avoids many of the problems associated with edge effects prevalent in OTCs (Long et al. 2004), enables significantly more plant material to be harvested without compromising the experiment, and allows plants to be studied throughout their life cycle, including trees that have enough space to develop to canopy closure.

FACE experiments are not without their problems (see Chapter 2). In particular, there has been criticism that the application of a step increase in $[CO_2]$ to ecosystems imposes a strong perturbation (Luo and Reynolds 1999; Newton et al. 2001). Luo and Reynolds (1999) modelled the effects of a step increase in $[CO_2]$ and found that the demand for N caused by the additional carbon influx from the step increase is much greater than that caused by a gradual increase in $[CO_2]$, resulting in an initial overestimation of carbon sequestration in experiments using a step increase in $[CO_2]$. This should be a minor problem in annual agro-ecosystems where plants have spent their whole life-cycle under FACE. However, in natural ecosystems, or perennial agro-ecosystems, a step increase in $[CO_2]$ is more likely to have a significant effect. Continued support is crucial for many existing FACE studies if short-

Ecological Studies, Vol. 187
J. Nösberger, S.P. Long, R.J. Norby, M. Stitt,
G.R. Hendrey, H. Blum (Eds.)
Managed Ecosystems and CO₂
Case Studies, Processes, and Perspectives
© Springer-Verlag Berlin Heidelberg 2006

term disturbance effects are to be distinguished from long-lasting, permanent effects.

Another major problem for FACE experiments is that replication is limited by the costs of operation. FACE experiments typically have only three replicated treatment plots, and therefore, limited statistical sensitivity. As a result, it can be difficult to significantly detect small but physiologically important differences in the response of plants and ecosystems to rising $[CO_2]$. Deployment of FACE experiments is also limited by cost and the difficulty of deploying experiments in areas with poor infrastructure. This is more relevant for unmanaged ecosystems where fully replicated FACE experiments have yet to be deployed in tropical forests, and tropical savannas and grasslands, which together account for over half of the worlds net primary production (Prentice 2001).

24.2 What Have We Learnt From FACE?

Over 15 years of FACE experiments now provide enough data for quantitative integration of results and perhaps the most realistic predictions of how plants will respond to future atmospheric $[CO_2]$. Do the data from FACE experiments support early predictions of plant responses to $e[CO_2]$? This question has been addressed recently by a number of reviews (Ainsworth and Long 2005; Kimball et al. 2002; Long et al. 2004; Nowak et al. 2004). While FACE experiments confirmed many of the results from earlier chamber experiments, there were some surprising differences between fully open-air CO_2 enrichment studies and chamber experiments.

24.2.1 Photosynthesis and Aboveground Productivity

The response of photosynthesis to $e[CO_2]$ has been reported for over 40 species at different FACE experiments (Ainsworth and Long 2005) and is discussed in more detail in Chapter 14. Stomatal conductance decreased 20% under elevated $[CO_2]$, but the ratio of intercellular $[CO_2]$ to external $[CO_2]$ did not change (Ainsworth and Long 2005). Apparent quantum yield of light-limited C_3 photosynthesis increased by 12% under $e[CO_2]$ (Ainsworth and Long, 2005) and C_3 light-saturated photosynthesis and diurnal carbon assimilation increased significantly at $e[CO_2]$, providing the basis for increased aboveground dry matter production (DMP) and crop yield (Fig. 24.1). The increase in DMP and grain yield under $e[CO_2]$ was less than the potential indicated by the increases in carbon uptake (Fig. 24.1). This suggests that there may be bottlenecks downstream of carbon acquisition that are limiting DMP, e.g. the supply of N and other nutrients. Leaf-area index (LAI) increased in trees and

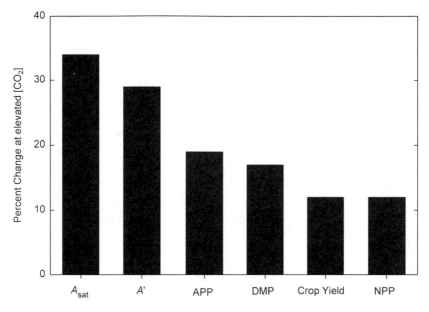

Fig. 24.1 Meta-analysis of FACE effects on light-saturated photosynthesis (A_{sat}), daily integral of carbon uptake (A'), aboveground primary production (APP), dry matter production (DMP), crop yield, and net primary production (NPP). A_{sat}, A', DMP, and crop yield were adapted from Ainsworth and Long (2005) and APP and NPP were adapted from Nowak et al. (2004). In FACE experiments, the response of photosynthesis to e[CO$_2$] is greater than the response of dry matter production, which in turn is greater than the yield response

woody species, but differences between developing and closed canopy forests should be noted (see Chapters 10–13). LAI did not change in herbaceous species, which limited the stimulation of aboveground production and crop yield under FACE. Results from rice and wheat grown under FACE suggest that current productivity models (e.g. Izaurralde et al. 2003) are overestimating future yields and therefore providing overly optimistic world food production projections (Ainsworth and Long 2005).

24.2.2 Photosynthetic Acclimation

Reduced photosynthetic capacity (acclimation) of leaves grown under e[CO$_2$] has been shown to maintain a balance in N and other resources allocated to photosynthetic reactions (Drake et al. 1997; Rogers and Humphries 2000). In the field, foliar carbohydrates accumulated under e[CO$_2$], while N content, Rubisco content, and the maximum rate of carboxylation ($V_{c,max}$) declined (Fig. 24.2). The occurrence of photosynthetic down-regulation was both growth form- and environment-specific (Nowak et al. 2004). In FACE experi-

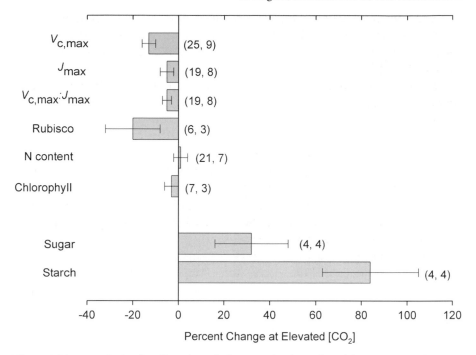

Fig. 24.2 Meta-analysis of acclimation of photosynthesis to e[CO_2] in FACE experiments, adapted from Ainsworth and Long (2005). The percent change in e[CO_2] ±95 % confidence intervals is shown. The numbers of species and FACE sites represented by each mean are shown in parentheses. Rubisco content, N content, chlorophyll content, sugar, and starch were all measured on a leaf area basis

ments, down-regulation of photosynthesis was observed under low-nutrient conditions, in old, but not young needles of evergreens, under cold temperatures late in the growing season, and in wet years, but not dry years (for reviews, see Ainsworth and Long 2005; Nowak et al. 2004). However, in relatively young trees grown under FACE, photosynthetic down-regulation did not occur (Nowak et al. 2004).

24.2.2.1 Response of Different Functional Groups

Analysis of different functional groups has revealed some generalities, but results do not always fit anticipated patterns (Nowak et al. 2004). Generally, C_4 species were less responsive than C_3 species to the increases in [CO_2] provided by FACE experiments (Ainsworth and Long 2005). On average, light-saturated photosynthesis increased marginally (10 %), while diurnal carbon assimilation and dry matter production did not change in C_4 species exposed to e[CO_2] (Ainsworth and Long 2005). However, when C_4 plants were water-stressed, improved water status from reduced stomatal conductance led to a significant stimulation of photosynthesis (Leakey et al. 2004). Clearly, it is the interaction between species and the environment that determines responsiveness to $e[CO_2]$.

To date, trees have been more responsive to $e[CO_2]$ in FACE studies than other functional groups, namely grasses, forbs and legumes (Ainsworth and Long 2005; Nowak et al. 2004). For 12 tree species measured at five different FACE sites, photosynthetic CO_2 uptake increased by 47 %, branch number increased by 25 %, stem diameter was 9 % larger, and LAI increased by 21 % (Ainsworth and Long 2005). Trees also showed a large stimulation in dry matter production and cotton, a woody crop, showed significant and large increases in yield (Ainsworth and Long 2005). This observation is contrary to expectations from earlier studies; however, most of the trees grown under FACE have been young and rapidly growing. Our understanding of the response of mature, closed-canopy forests to $e[CO_2]$ is improving (see Chapter 11), but long-term observations will be required to resolve many of the most basic questions about forested ecosystems. Given the importance of old-growth forests as a carbon pool and potential sink (Carey et al. 2001; Schulze et al. 2000), the value of a future, large, well replicated, long-term FACE experiment in an old-growth forest is clear.

24.2.2.2 Belowground Responses

Effect of C:N on Decomposition. The observation of lower N concentrations in plant tissue grown at $e[CO_2]$ (Cotrufo et al. 1998) led to the hypothesis that the decomposition of plant leaf litter with a higher C:N ratio would be slower at $e[CO_2]$, and possibly limit ecosystem productivity in a CO_2-enriched world (Strain and Bazzaz 1983). Although some experimental evidence has been found for this, the majority of studies failed to support this initial hypothesis (Norby and Cotrufo 1998). Differences in C:N ratios in green tissue were often absent or insignificant when leaves were allowed to become senescent in situ (Norby et al. 2001), but recent studies in the AspenFACE and in the POP/EUROFACE experiments are reporting increased C:N ratios in litter from $e[CO_2]$ plots (see Chapters 10 and 12). However, potential species shifts caused by growth at $e[CO_2]$ may play a greater role in changing the rates of litter decomposition and subsequent nutrient cycling than the direct effects of $e[CO_2]$ on tissue quality (Allard 2004; Kammann et al. 2005).

Primary Productivity. Nowak et al. (2004) recently reviewed belowground primary productivity (BPP) in FACE experiments located on forest, bog, grassland, and desert sites. BPP increased on average by 32 % in $e[CO_2]$ and varied from a 70 % increase in forests to a 7 % decrease in desert systems. However, these differences in ecosystem responses to $e[CO_2]$ were not significant, largely due to the difficulty in measuring BPP accurately (Nowak et al. 2004). New methods for quantifying BPP are urgently needed.

Carbon and Nitrogen Sequestration. In their review, Zak et al. (2000) reported large increases and decreases in soil microbial biomass and microbial C and N in response to growth at e[CO_2]. However, total microbial C and N pools were unaltered and it was changes in microbial community composition and function that were important (Montealegre et al. 2000; Lukac et al. 2003). Groenigen et al. (Chapter 21) reviewed the effects of e[CO_2] on soil C and soil N. Briefly, soil C increased in FACE systems in an N-dependent manner (Fig. 24.3). Under high N fertilization conditions, there was a 9 % increase in soil C, but under medium and low N fertilization, there was no significant change in soil C. Soil N increased slightly under FACE (2.8 %) and the C:N ratio increased in experiments with soil disturbance (Fig. 24.3).

Evidence for N Limitation Feedback on Carbon Acquisition. The above findings suggest that N availability did not decrease, nor did it diminish the response of vegetation to e[CO_2]. Further support of this claim comes from the 10-year Swiss FACE experiment, where photosynthetic stimulation did not change over the course of that experiment, suggesting that carbon acquisition by *Lolium perenne* monocultures did not become N-limited (Ainsworth et al. 2003). In contrast, Oren et al. (2001) attributed the loss of the initial stimulation in the annual carbon increment of *Pinus taeda* grown at e[CO_2] in the Duke Forest prototype experiment to a possible nutrient limitation. Following application of fertilizer, the CO_2-induced biomass carbon increment returned to the levels observed at the start of the experiment, suggesting that N limita-

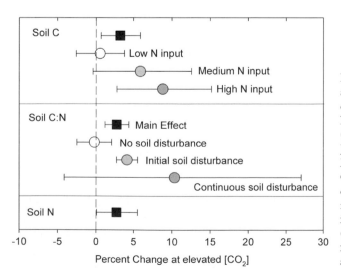

Fig. 24.3 A meta-analysis of FACE effects on soil C, soil N, and the ratio of C:N (data provided by Kees Jan van Groenigen and Marie Anne De Graaff). The percent change at e[CO_2] ±95 % confidence intervals is shown ($n=57$ for soil C, $n=32$ for soil C:N, $n=39$ for soil N)

tions had constrained the response to e[CO$_2$] (Oren et al. 2001). In the adjacent replicated Duke Forest FACE experiment, net primary production (NPP) was correlated with net N mineralization rate but was also related to inter-annual variation in rainfall and degree-days, (see Chapter 11). Whilst an insufficient N supply may limit the response of plants to elevated [CO$_2$], more often than not, in managed ecosystems, N limitation of carbon acquisition has not resulted from a step change in [CO$_2$].

24.3 What Is Missing From Current FACE Research and What Are the Gaps in Understanding?

24.3.1 Additional Treatments

The predicted increase in [CO$_2$] from 540 ppm to 970 ppm by 2100 (Prentice 2001) will not occur in isolation and the importance of studying the effect of rising [CO$_2$] in conjunction with other predicted global change is clear. Of particular importance are the predicted 62 % increase in tropospheric ozone concentration [O$_3$] (Prather and Ehhalt 2001), the predicted increase of between 1.4 °C to 5.8 °C in mean global temperature, and less certain predicted regional variations in precipitation (Cubasch and Meehl 2001). Water supply was added as a split-plot treatment as part of the FACE experiment at the Maricopa Agricultural Research Center, Ariz., and a rainfall displacement study was piloted at the SoyFACE experiment (Kimball et al. 1995; ADB Leakey, personal communication). Two FACE facilities currently include interactions with e[O$_3$] (Karnosky et al. 1999; Morgan et al. 2004); and free-air temperature increase was used to elevate canopy temperature as part of the SwissFACE project (Nijs et al. 1996). Given that the response to the combination of two treatments is not always predictable from the response of individually applied treatments, it is clear that more field studies that include interactions between predicted global change variables are needed. This is particularly important because current process-based terrestrial models that are used to model terrestrial carbon cycling incorporate interactions between changes in [CO$_2$] and climate that have yet to be validated in fully open-air field experiments (Prentice 2001); and of particular importance is the inclusion of physiological acclimation. Hanson et al. (2005) demonstrated the importance of including physiological acclimation in C-cycle models. When included, estimates of annual net ecosystem exchange in an upland-oak forest increased from a 29 % reduction under a multiple global change scenario to a 20 % stimulation. Nesting small drought or temperature treatments within larger CO$_2$ enrichment plots does not provide the best prediction of the response of plants and ecosystems to multiple atmospheric and climatic changes. Larger experiments with fully replicated designs are required.

The value of FACE experiments is their ability to mimic the predicted increase in [CO_2] and [O_3] in as realistic an environment as possible. Care must be taken to ensure that future experiments including interactions with precipitation, and in particular temperature, are designed to reproduce, as closely as possible, the predicted global change scenarios and that they are largely free of artifacts. If this is not done, the value of applying these treatments in the field is significantly reduced.

Simulating global warming in field studies is difficult, and to date, OTCs may offer the best platform for field temperature increases, but as discussed above, OTCs have significant drawbacks. Given what we already know about the response of plants to temperature (Long and Woodward 1988), it is clear that a fully open-air experimental design should include warming of the plant and soil. However, available technologies, namely the use of infrared lamps to heat plant canopies (Nijs et al. 1996) and alternative treatments that warm plant roots (Melillo et al. 2002), have problems. With infrared lamps, the canopy is not warmed uniformly. With low-stature grasses this problem is less serious, but as LAI increases, preferential warming of the upper canopy becomes a problem. Infrared lamps warm and dry the air, increasing the vapor pressure deficit. Global warming is not predicted to be associated with drier air, nor will global temperature increases occur through an exclusive warming of the soil. Even defining the nature of the temperature treatment is challenging. Over the past 50 years, warming on land has been significantly more pronounced at night and mid-to high-northern latitude winters have become warmer, whilst summers have shown little change (Cubasch and Meehl 2001). In addition to diurnal and seasonal patterns, the frequency, timing and duration of temperature extremes will also significantly impact plant responses to global warming (Morison and Lawlor 1999). It is possible that simple ambient +2 °C-type treatments may fail to provide the information we need to predict the response of our ecosystems to future global change scenarios.

24.3.2 Future Challenges

To date, FACE studies have been limited to just a few species, in managed ecosystems, and only one or two varieties of each cultivar. We therefore have little understanding of the variation in germplasm responses to e[CO_2]. Studies have been conducted mostly in temperate ecosystems and these studies have usually only taken place at one location for ecosystems that cover many degrees of latitude. Aboveground research has focused on the response of recently matured leaves at the top of the canopy, but much less is known about developing or senescing foliage. However, on the whole, observation of aboveground plant physiological processes in FACE experiments is well advanced and our future challenge will be to provide mechanistic explanations for the commonly observed physiological responses to growth at e[CO$_2$].

In order to understand the mechanisms that underlie the response of plants and ecosystems to global change, it will be necessary to identify the mechanistic links between levels of biological organization, from changes in gene expression profiles to physiological responses and ultimately to changes in crop yield and the structure and function of ecosystems. This will require large, multidisciplinary, research teams capable of examining responses at macromolecular to ecosystem scales.

Plants and ecosystems sense and interact with rising [CO$_2$] through increased photosynthesis and reduced transpiration (Long et al. 2004); and, given that the [CO$_2$] within the soil is 1–3 orders of magnitude greater than atmospheric [CO$_2$], it is extremely unlikely that soil microorganisms will respond directly to a ca. 200 ppm increase in atmospheric [CO$_2$]. However, changes in nutrient cycling via soil microbial, fungal, and faunal communities have the potential to modulate leaf-level responses to e[CO$_2$], so understanding subsurface processes is an essential part of a full and complete understanding of the response of plants and ecosystems to rising [CO$_2$]. Of particular relevance to managed ecosystems is the need to understand how growth at e[CO$_2$] may impact soil fertility and C sequestration.

Many important questions about how soil processes, particularly C and N cycling, will respond to rising [CO$_2$] remain unanswered and continue to be a challenge. This challenge is in part due to the problem of understanding and quantifying soil processes. To meet this challenge, new approaches are needed to overcome the difficulties associated with the tremendous heterogeneity of soil and the problem of quantifying small and slow acting processes relative to extremely large C and N pool sizes. Several key challenges remain.

24.3.3 What Is the Fate of C Partitioned Belowground?

Closing the carbon budget has been a focus of FACE experiments; and the fate of carbon partitioned belowground has proved a major challenge. A key piece of this puzzle is soil respiration; the magnitude and direction of the response of soil respiration to e[CO$_2$] varies with soil type, climate, seasonality, stand development, and species composition (King et al. 2004; Zak et al. 2000). The contribution of deep roots may be a small, unaccounted-for carbon sink in deeper soils under e[CO$_2$] (Lukac et al. 2003; Marchi et al. 2004); and a proportional increase in deep roots under e[CO$_2$] may have several impacts on soil organic matter (SOM) degradation and trace gas production. Indeed, the question of how much of the C partitioned belowground ultimately remains in the soil as SOM under e[CO$_2$] is an important one for ecologists and policy makers (see Chapter 21)

24.3.3.1 N Cycling

Early predictions that increased N immobilization would limit the response of plants to $e[CO_2]$ have been tempered by whole-ecosystem studies (Norby et al. 2001; Strain and Bazzaz 1983). Symbiotic and free N fixation may increase at $e[CO_2]$ but the impact of a possible P limitation on N fixation is not certain. To date, experiments have been conducted in plants grown in pots in controlled environments but the response of field-grown N-fixers to P limitation at $e[CO_2]$ is unknown (Almeida et al. 1999, 2000; Niklaus and Körner 2004). The response of net nitrification rates to $e[CO_2]$ is unclear, with both increases and decreases reported (Barnard et al. 2004; Ross et al. 2000). Gross N mineralization rates were either unchanged or increases were not significant under FACE (Finzi et al. 2002; Richter et al. 2003). It is unclear if increases in gross N mineralization, combined with rising N-use efficiency (Drake et al. 1997), will meet the rising N demand under $e[CO_2]$ and sustain increases in NPP.

24.3.3.2 Soil Faunal Food Webs and Soil Structure

How will root life span, architecture, and exudation change at $e[CO_2]$ and impact soil food webs? Root colonization by ectomycorrhizal and arbuscular mycorrhizal fungi usually increases under $e[CO_2]$ (Lukac et al. 2003; Rillig et al. 2001). Arbuscular mycorrhizal fungi produce glomalin, which stabilizes soil aggregates (Wright and Anderson 2000). Since glomalin concentration is often increased at $e[CO_2]$ (Rillig et al. 2001) and declines in water repellence have been also been reported (Newton et al. 2004), it is clear that the physical property of the soil could change at $e[CO_2]$. Such changes may impact erosion, fertility, protection of organic matter in aggregates, and hence C sequestration, but little is currently known. Soil faunal community services (e.g. earthworms) play a crucial role in soil fertility and nutrient cycling, but the responses of soil fauna to $e[CO_2]$ are inconsistent (Coûteaux and Bolger 2000). Effects often reflect increases in net BPP, e.g. root-feeding nematode species increased in abundance at $e[CO_2]$ (Yeates et al. 2003). This is particularly important for managed ecosystems, since many of these species are pests/pathogens of economical importance. Also, the effect that $e[CO_2]$-induced changes in soil faunal food webs may have on the performance and subsequent decomposition of herbicides, fungicides, and pesticides is unknown.

24.3.3.3 Trace Gases

The effect of $e[CO_2]$ on fluxes of the greenhouse gases nitrous oxide (N_2O) and methane (CH_4) have rarely been investigated in FACE experiments.

Methane emissions from wetlands are usually stimulated at $e[CO_2]$ (Dacey et al. 1994; Inubushi et al. 2003; Saarnio et al. 2000). On the few occasions they were measured, CH_4 oxidation rates (uptake) were mostly reduced at $e[CO_2]$ (Ineson et al. 1998; Phillips et al. 2001a). Overall, there is evidence for a positive feedback of $e[CO_2]$ on CH_4 fluxes, but this relationship is poorly understood. Nitrous oxide emissions increased at $e[CO_2]$ due to denitrification when nitrate was readily available (Baggs et al. 2003b; Ineson et al. 1998). Nitrous oxide fluxes due to nitrification and denitrification were also dependant on soil moisture content and showed seasonal variation (Phillips et al. 2001b). Clearly, it is important to calculate annual budgets for N_2O emissions to understand whether the overall feedback of $e[CO_2]$ on N_2O emissions will be positive or negative. This is especially important for managed ecosystems where N fertilization and irrigation are part of management practices.

24.4 Technologies for Future FACE Science

24.4.1 The Use of Stable Isotopes

The isotopic signatures of carbon and nitrogen pools provide the means of tracking the transformation of carbon and nitrogen as they flow through plant and soil compartments and the possibility of separately investigating individual soil processes. Two uses of stable isotope technology have emerged, one using small, mostly natural ^{13}C and ^{15}N approaches and the other using high-enrichment, pulse-chase techniques by applying ^{15}N fertilizer, or a [13]CO_2 or [14]CO_2 pulse that is incorporated into the plant–soil system by photosynthesis (Pendall 2002).

The supply of CO_2 used for fumigation in FACE experiments is usually ^{13}C-depleted. This provides the means to distinguish between carbon pools formed prior to CO_2 enrichment and those formed post-fumigation (Hungate et al. 1997). Unfortunately, these measurements are restricted to the elevated CO_2 plots, but root in-growth cores containing soil with an isotopically distinct ^{13}C signature have been used to allow the quantification of C inputs in the control, as well as elevated CO_2 plots (van Kessel et al. 2000). Future FACE experiments on C_3 plants could perhaps be sited on long-term C_4 soils (e.g. *Zea mays* or *Sorghum* fields). Such an approach has been used successfully to quantify carbon inputs due to forest re-growth on an afforested maize field (Del Galdo et al. 2003). However, the advantages afforded by such an approach must be weighed against growing C_3 plants on non-natural or unsuitable soils where the responses may differ from those occurring in their natural environment.

Soil microbial biomass discriminates against [15]N (Robinson 2001) and therefore the ^{15}N abundance in plant material may offer a promising tool to better understand plant–microbial interactions and gross N transformation processes at e[CO_2]. Observed enrichment of ^{15}N in plant material grown at e[CO_2] may indicate a possible increase in microbial activity and an improved root–microbial exploration of old, recalcitrant N pools (Billings et al. 2004). A recent synthesis of $\delta^{15}N$ depletion in foliage grown at e[CO_2], mostly in FACE experiments, concluded that e[CO_2] strongly influenced the N cycle but it was unclear where in the N cycle that the effects of e[CO_2] were manifested (BassiriRad et al. 2003).

In recent years, isotopic pulse labeling of a certain C and N fraction with a high isotopic enrichment of up to 99 % above natural abundance has become a valuable tool in the investigation of C and N cycling in ecosystems. For example, Staddon et al. (2003) used a ^{14}C pulse-labelling approach to determine that the extraradical hyphae of arbuscular mycorrhizal fungi live on average 5–6 days and hence comprise a large, but rapidly turned-over C pool in the soil C cycle. By use of ^{13}C pulse-chase techniques, the short-term gross flow of ^{13}C through various soil C pools, processes and soil organisms might readily be determined in FACE experiments (Ostle et al. 2000; Radajewski et al. 2000). Likewise, labelling with ^{15}N applied as fertilizer to the soil enables N transformation and sequestration processes to be followed in various soil compartments (Baggs et al. 2003a; Richter et al. 2003). The combination of ^{15}N labelling and tracing techniques with new process-based modelling approaches (Müller et al. 2004) may provide new insights into N transformation processes in soils at e[CO_2].

24.4.2 Genomic Technologies and Tools in FACE

Advances in 'omics' technologies have produced an explosion of biological information at the level of transcripts, proteins, protein modifications, and metabolites.

High-throughput technologies allow examination of thousands of genes, proteins and metabolites in parallel (Aharoni and Vorst 2002; Kersten et al. 2002; Kopka et al. 2004; Weckwerth 2003). Systems biology uses understanding of molecular networks and biological modules from functional genomics approaches to comprehend the basis of complex phenotypes (Blanchard 2004; Kitano 2002). Predictive models of biological systems that incorporate biochemical and genetic data are a goal of systems biology and analogous to the approach of global change biology, where advances in technology have spurred realistic experimental platforms for data collection, which provide information for predictive computational models. The challenge is to incorporate large data sets on spatial and temporal scales of genes, metabolites and proteins into models that enable explanation of plant responses to predicted global change events. One step towards this goal will be the development of diagnostic markers for given physiological conditions, e.g. N limitation or oxidative stress. These markers would need to include a temporal dimension to separate short-term and long-term responses. Early steps toward identifying robust markers for C starvation and N assimilation have been made (Foyer et al. 2003; Gibon et al. 2004) and suggest that the identification of diagnostic markers for other physiological processes will also be possible.

The 2004 special issue of *Field Crops Research* (vol 90, issue 1) highlighted the subject area: *Linking functional genomics with physiology for global change research*. Only one case-study from a FACE experiment (*Arabidopsis* grown at SoyFACE) was included in this special issue; and it discussed some of the challenges of applying transcript analysis to field-grown material (Miyazaki et al. 2004). A general conclusion from the study was that results revealed a snapshot of time- and weather-dependent transcripts, superimposed on responses to e[CO$_2$] and e[O$_3$], and repeated profiles of expression would be needed to determine FACE-specific patterns (Miyazaki et al. 2004). This study warned that application of functional genomics to global change science will certainly be challenging, but advances in genomic technologies for other species (including poplar, soybean, maize, and rice) will aid in merging the fields. Integration of molecular data into models should improve our predictions of plant responses to global change.

24.5 A Potential Problem for Long-Running FACE Experiments?

It is clear that, in order to answer many of the questions associated with fundamental belowground processes, operation of FACE experiments must be continued for a significant length of time. This is particularly important for experiments where species composition changes may be slow to appear, where a step increase in [CO$_2$] has been applied to a mature ecosystem, and in experiments on forest systems where it may take several years to reach canopy closure. Currently the longest-running CO$_2$ enrichment experiment is the OTC experiment on *Scirpus olneyi* (Rasse et al. 2005) that began CO$_2$ fumigation in 1987. Due to seasonal variation in meteorological conditions, the effect of e[CO$_2$] over time is often assessed by comparing the ratio of a response at e[CO$_2$] to the response of the c[CO$_2$] control; and this could lead to a misinterpretation of the CO$_2$ response. Consider a hypothetical 25-year FACE experiment running from 1990 to 2015 with a treatment [CO$_2$] set point of 550 ppm. Over the course of the experiment the control [CO$_2$] would have risen from 354 ppm to 392 ppm (Prentice 2001), whilst the treatment [CO$_2$] would have remained unchanged at 550 ppm. If we assume that the relationship between net CO$_2$ assimilation (*A*) and intercellular [CO$_2$] (c_i) remained constant over the course of the experiment and we use the equations of Farquhar et al. (1980) to calculate photosynthetic stimulation based on three categories of

Table 24.1 Reduction in stimulation of A in FACE plots relative to current $[CO_2]$ control plots due to rising $[CO_2]$ over a hypothetical 25-year FACE experiment starting in 1990 when the $[CO_2]$ was 354 ppm. The increase in $[CO_2]$ over the course of this hypothetical experiment is based on the IPCC-predicted increases of 1.5 mmol mol^{-1} year^{-1} (Prentice et al. 2001)

[CO$_2$] (ppm)	Modeled assimilation[a] (µmol m^{-2} s^{-1})		
	C$_3$ crop	Hardwood	Conifer
354	18.03	8.89	4.21
392	19.84	9.83	4.72
550	22.19	13.30	5.43
588	22.58	14.03	5.54

Year stimulation modeled, with [CO$_2$] (ppm)	Percent stimulation in A at elevated [CO$_2$]		
	C$_3$ crop	Hardwood	Conifer
1990	33	50	29
2015 (set point, 550)	21	35	15
2015 (ambient + 196)	23	43	17

[a] A was modeled using the equations of Farquhar et al. (1980). PPI = 1400 µmol m^{-2} s^{-1}, temperature = 25 °C, c_i/c_a = 0.7, RH = 90%. Values for $V_{c,max}$ and J_{max} were taken from Wullschleger (1993). $V_{c,max}$ = 90, 47 and 25 µmol m^{-2} s^{-1} and J_{max} = 171, 104 and 40 µmol m^{-2} s^{-1} for a C$_3$ crop, hardwood, and conifer, respectively

plant: C$_3$ crop, hardwood, and conifer (Wullschleger 1993), the mean relative stimulation in A across these three types of vegetation would be 37% at the beginning of the experiment and only 24% at the end of the experiment (Table 24.1). This problem would not be solved by adopting an ambient +196 ppm fumigation protocol rather than a set point of 550 ppm because the relationship between photosynthesis and $[CO_2]$ is not linear. Due to the shape of the A/c_i plot, long-term FACE experiments over hardwood forests may, to some degree, be able to ameliorate this confounding factor by the use of an ambient +196 ppm fumigation protocol (Table 24.1). Since almost all responses of plants and ecosystems to rising $[CO_2]$ occur downstream of photosynthesis, there is potential to misinterpret the consequences of an apparent reduction in carbon acquisition in long-running experiments.

24.6 Conclusion

Free-air CO$_2$ enrichment studies have been a valuable tool for the investigation of plant and ecosystem responses to rising CO$_2$ levels. The challenges for the next phase of FACE research are clear.

- Multidisciplinary teams of investigators must take advantage of emerging technologies to significantly increase our mechanistic understanding of the responses that FACE experiments have confirmed will take place during the next century.
- If we seek the ability to predict and understand how our managed, and natural, ecosystems will respond to the predicted multiple and concurrent changes in our environment, more interactions with other global change factors must be included in future experiments. To meet these challenges, future FACE experiments will need to be larger to accommodate multiple environmental changes.

Acknowledgements. A.R. was supported by the US Department of Energy Office of Science contract No. DE-AC02-98CH10886 to Brookhaven National Laboratory (BNL). E.A.A. was supported by the Alexander von Humboldt Foundation and the Juelich Research Center, ICG-III.

References

Aharoni A, Vorst O (2002) DNA microarrays for functional genomics. Plant Mol Biol 48: 99–118

Ainsworth EA, Davey PA, Hymus GJ, Osborne CP, Rogers A, Blum H, Nosberger J, Long SP (2003) Is stimulation of leaf photosynthesis by elevated carbon dioxide concentration maintained in the long term? A test with *Lolium perenne* grown for 10 years at two nitrogen fertilization levels under free air CO$_2$ enrichment (FACE). Plant Cell Environ 26: 705–714

Ainsworth EA, Long SP (2005) What have we learned from 15 years of free-air CO$_2$ enrichment (FACE)? A meta-analytic review of responses to rising CO$_2$ in photosynthesis, canopy properties and plant production. New Phytol 165:351–371

Allard V, Newton PCD, Lieffering M, Soussana JF, Grieu P, Matthew C (2004) Elevated CO$_2$ effects on decomposition processes in a grazed grassland. Global Change Biol 10:1553–1564

Almeida JPF, Lüscher A, Frehner M, Oberson A, Nösberger J, (1999) Partitioning of P and the activity of root acid phosphatase in white clover (*Trifolium repens* L.) are modified by increased atmospheric CO$_2$ and P fertilization. Plant Soil 210:159–166

Almeida JPF, Hartwig UA, Frehner M, Nösberger J, Lüscher A, (2000) Evidence that P deficiency induces N feedback regulation of symbiotic N$_2$ fixation in white clover (*Trifolium repens* L.). J Exp Bot 51:1289–1297

Baggs EM, Richter M, Cadisch G, Hartwig UA (2003a) Denitrification in grass swards is increased under elevated atmospheric CO$_2$. Soil Biol Biochem 35:729–732

Baggs EM, Richter M, Hartwig UA, Cadisch G (2003b) Nitrous oxide emissions from grass swards during the eighth year of elevated atmospheric pCO$_2$ (Swiss FACE). Global Change Biol 9:1214–1222

Barnard R, Barthes L, Le Roux X, Leadley PW (2004) Dynamics of nitrifying activities, denitrifying activities and nitrogen in grassland mesocosms as altered by elevated CO$_2$. New Phytol 162:365–376

BassiriRad H, Constable JVH, Lussenhop J, Kimball BA, Norby RJ, Oechel WC, Reich PB, Schlesinger WH, Zitzer S, Sehtiya HL, Silim S (2003) Widespread foliage d^{15}N depletion under elevated CO_2: inferences for the nitrogen cycle. Global Change Biol 9:1582–1590

Billings SA, Schaeffer SM, Evans RD (2004) Soil microbial activity and N availability with elevated CO_2 in Mojave desert soils. Global Biogeochem Cycles 18:1–11

Blanchard JL (2004) Bioinformatics and systems biology, rapidly evolving tools for interpreting plant response to global change. Field Crops Res 90:117–131

Carey EV, Sala A, Keane R, Callaway RM (2001) Are old forests underestimated as global sinks? Global Change Biol 7:339–344

Cotrufo MF, Ineson P, Scott A (1998) Elevated CO_2 reduces the nitrogen concentration of plant tissues. Global Change Biol 4:43–54

Coûteaux M-M, Bolger T (2000) Interactions between atmospheric CO_2 enrichment and soil fauna. Plant Soil 224:123–134

Cubasch U, Meehl GA (2001) Projections of future climate change. In: Houghton JT, Ding Y, Griggs DJ, Noguer M, Linden PJ van der, Dai X, Maskell K, Johnson CA (eds) Climate change 2001: the scientific basis. Cambridge University Press, Cambridge, pp 527–582

Dacey JWH, Drake BG, Klug MJ (1994) Stimulation of methane emissions by carbon dioxide enrichment of marsh vegetation. Nature 370:47–49

Del Galdo I, Six J, Peressotti A, Cotrufo MF (2003) Assessing the impact of land-use change on soil C sequestration in agricultural soils by means of organic matter fractionation and stable C isotopes. Global Change Biol 9:1204–1213

Drake BG, Gonzàlez-Meler MA, Long SP (1997) More efficient plants: a consequence of rising atmospheric CO_2? Annu Rev Plant Physiol Plant Mol Biol 48:609–639

Farquhar GD, von Caemmerer S, Berry JA (1980) A biochemical model of photosynthetic CO_2 assimilation in leaves of C3 species. Planta 149:78–90

Finzi AC, Delucia EH, Hamilton JG, Richter DD, Schlesinger WH (2002) The nitrogen budget of a pine forest under free air CO_2 enrichment. Oecologia 132:567–578

Foyer CH, Parry M, Noctor G (2003) Markers and signals associated with nitrogen assimilation in higher plants. J Exp Bot 54:585–593

Gibon Y, Blaesing OE, Hannemann J, Carillo P, Höhne M, Hendriks JHM, Palcios N, Cross J, Selbig J, Stitt M. (2004) A robot-based platform to measure multiple enzyme activities in *Arabidopsis* using a set of cycling assays: comparison of changes of enzyme activities and transcript levels during diurnal cycles and in prolonged darkness. Plant Cell 16:3304–3325

Hanson PJ, Wullschleger SD, Norby RJ, Tschaplinski TJ, Gunderson CA (2005) Importance of changing CO_2, temperature, precipitation, and ozone on carbon and water cycles of an upland-oak forest: incorporating experimental results into model simulations. Global Change Biol 11:1402–1423

Hungate BA, Holland EA, Jackson RB, Chapin FS, Mooney HA, Field CB (1997) The fate of carbon in grasslands under carbon dioxide enrichment. Nature 388:576–579

Ineson P, Coward PA, Hartwig UA (1998) Soil gas fluxes of N_2O, CH_4 and CO_2 beneath *Lolium perenne* under elevated CO_2: The Swiss free air carbon dioxide enrichment experiment. Plant Soil 198:89–95

Inubushi K, Cheng WG, Aonuma S, Hoque MM, Kobayashi K, Miura S, Kim HY, Okada M (2003) Effects of free-air CO_2 enrichment (FACE) on CH_4 emission from a rice paddy field. Global Change Biol 9:1458–1464

Izaurralde RC, Rosenberg NJ, Brown RA, Thomson AM (2003) Integrated assessment of Hadley Center (HadCM2) climate-change impacts on agricultural productivity and irrigation water supply in the conterminous United States. Agric For Meteorol 117:97–122

Kammann C, Grünhage L, Grüters U, Janze S, Jäger H-J (2005) Response of aboveground grassland biomass and soil moisture to moderate long-term CO_2 enrichment. Basic Appl Ecol 6:351–365

Karnosky DF, Mankovska B, Percy K, Dickson RE, Podila GK, Sober J, Noormets A, Hendrey GR, Coleman MD, Kubiske M, Pregitzer KS, Isebrands JG (1999). Effects of tropospheric O$_3$ on trembling aspen and interaction with CO$_2$: Results from an O$_3$-gradient and a FACE experiment. J Water Air Soil Pollut 116:311–322

Kersten B, Buerkle L, Kuhn EJ, Giavalisco P, Konthur Z, Lueking A, Walter G, Eickhoff H, Schneider U (2002) Large-scale plant proteomics. Plant Mol Biol 48:133–141

Kessel C van, Horwath WR, Hartwig U, Harris D, Lüscher A (2000) Net soil carbon input under ambient and elevated CO$_2$ concentrations: isotopic evidence after 4 years. Global Change Biol 6:435–444

Kimball BA, Kobayashi K, Bindi M (2002) Responses of agricultural crops to free-air CO$_2$ enrichment. Adv Agron 77:293–368

Kimball BA, Pinter PJ, Garcia RL, LaMorte RL, Wall GW, Hunsaker DJ, Wechsung G, Wechsung F, Kartschall T (1995) Productivity and water use of wheat under free-air Co$_2$ enrichment. Global Change Biol 1:429–442

King JS, Hanson PJ, Bernhardt E, DeAngelis P, Norby RJ, Pregitzer KS. 2004. A multi-year synthesis of soil respiration responses to elevated atmospheric CO$_2$ from four forest FACE experiments. Global Change Biol 10:1027–1042

Kitano H (2002) Systems biology: a brief overview. Science 295:1662–1664

Kopka J, Fernie A, Weckwerth W, Gibon Y, Stitt M (2004) Metabolite profiling in plant biology: platforms and destinations. Genome Biol 5:109

Leakey ADB, Bernacchi CJ, Dohleman FG, Ort DR, Long SP (2004) Will photosynthesis of maize in the US Corn Belt increase in future [CO$_2$] rich atmospheres? An analysis of diurnal courses of CO$_2$ uptake under free-air concentration enrichment (FACE). Global Change Biol 10:951–962

Long SP, Ainsworth EA, Rogers A, Ort DR (2004) Rising atmospheric carbon dioxide: plants FACE the future. Annu Rev Plant Biol 55:591–628

Long SP Woodward FI (1988) Plants and temperature. Symposium of the Society for Experimental Biology, Cambridge University Press, Cambridge

Lukac M, Calfapietra C, Godbold DL (2003) Production, turnover and mycorrhizal colonization of root systems of three *Populus* species grown under elevated CO$_2$ (POP-FACE). Global Change Biol 9:838–848

Luo Y, Reynolds JF (1999) Validity of extrapolating field CO$_2$ experiments to predict carbon sequestration in natural ecosystems. Ecology 80:1568–1583

Marchi S, Tognetti R, Vaccari FP, Lanini M, Kaligaric M, Miglietta F, Raschi A (2004) Physiological and morphological responses of grassland species to elevated atmospheric CO$_2$ (concentrations in FACE-systems and natural CO$_2$ springs. Funct Plant Biol 31:181–194

Melillo JM, Steudler PA, Aber JD, Newkirk K, Lux H, Bowles FP, Catricala C, Magill A, Ahrens T, Morrisseau S (2002) Soil warming and carbon-cycle feedbacks to the climate system. Science 298:2173–2176

Miyazaki S, Fredricksen M, Hollis KC, Poroyko V, Shepley D, Galbraith DW, Long SP, Bohnert HJ (2004) Transcript expression profiles of Arabidopsis thaliana grown under controlled conditions and open-air elevated concentrations of CO$_2$ and of O$_3$. Field Crops Res 90:47–59

Montealegre CM, van Kessel C, Blumenthal JM, Hur HG, Hartwig UA, Sadowsky MJ (2000) Elevated atmospheric CO$_2$ alters microbial population structure in a pasture ecosystem. Global Change Biol 6:475–482

Morgan PB, Bernacchi CJ, Ort DR, Long SP (2004) An in vivo analysis of the effect of season-long open-air elevation of ozone to anticipated 2050 levels on photosynthesis in soybean. Plant Physiol 135:2348–2357

Morison JIL, Lawlor DW (1999) Interactions between increasing CO$_2$ concentrations and temperature on plant growth. Plant Cell Environ 22:659–682

Müller C, Stevens RJ, Laughlin RJ (2004) A ^{15}N tracing model to analyse N transformations in old grassland soil. Soil Biol Biochem 36:619–632

Newton PCD, Carran RA, Lawrence EJ (2004) Reduced water repellency of a grassland soil under elevated atmospheric CO_2. Global Change Biol 10:1–4

Newton PCD, Clark H, Edwards GR, Ross DJ (2001) Experimental confirmation of ecosystem model predictions comparing transient and equilibrium plant responses to elevated atmospheric CO_2. Ecol Lett 4:344–347

Niklaus PA, Körner C (2004) Synthesis of a six-year study of calcareous grassland responses to in situ CO_2 enrichment. Ecol Monogr 74:491–511

Nijs I, Teughels H, Blum H, Hendrey G, Impens I (1996) Simulation of climate change with infrared heaters reduces the productivity of Lolium perenne L. in summer. Environ Exp Bot 36:271–280

Norby RJ, Cotrufo MF (1998) Global change – a question of litter quality. Nature 396:17–18

Norby RJ, Cotrufo MF, Ineson P, O'Neill EG, Canadell JG (2001) Elevated CO_2, litter chemistry, and decomposition: a synthesis. Oecologia 127:153–165

Nowak RS, Ellsworth DS, Smith SD (2004) Functional responses of plants to elevated atmospheric CO_2 – do photosynthetic and productivity data from FACE experiments support early predictions? New Phytol 162: 253–280

Oren R, Ellsworth DS, Johnsen KH, Phillips N, Ewers BE, Maier C, Schäfer KVR, McCarthy H, Hendrey G, McNulty SG, Katul GG (2001) Soil fertility limits carbon sequestration by forest ecosystems in a CO_2-enriched atmosphere. Nature 411:469–472

Ostle N, Ineson P, Benham D, Sleep D (2000) Carbon assimilation and turnover in grassland vegetation using an in situ [13]CO_2 pulse labelling system. Rapid Commun Mass Spectrom 14:1345–1350

Pendall E (2002) Where does all the carbon go? The missing sink. New Phytol 153:207–210

Prather M, Ehhalt D (2001) Atmospheric chemistry and green house gases. In: Houghton JT, Ding Y, Griggs DJ, Noguer M, Linden PJ van der, Dai X, Maskell K, Johnson CA (eds) Climate change 2001: the scientific basis. Cambridge University Press, Cambridge, pp 241–287

Prentice IC (2001) The carbon cycle and atmospheric carbon dioxide. In: Houghton JT, Ding Y, Griggs DJ, Noguer M, Linden PJ van der, Dai X, Maskell K, Johnson CA (eds) Climate change 2001: the scientific basis. Cambridge University Press, Cambridge, pp 183–238

Phillips RL, Whalen SC, Schlesinger WH (2001a) Influence of atmospheric CO_2 enrichment on methane consumption in a temperate forest soil. Global Change Biol 7:557–563

Phillips RL, Whalen SC, Schlesinger WH (2001b) Influence of atmospheric CO_2 enrichment on nitrous oxide flux in a temperate forest ecosystem. Global Biogeochem Cycles 15:741–752

Radajewski S, Ineson P, Parekh NR, Murrell JC (2000) Stable-isotope probing as a tool in microbial ecology. Nature 403:646–649

Rasse DP, Peresta G, Drake BG (2005) Seventeen years of elevated CO_2 exposure in a Chesapeake Bay wetland: sustained but contrasting responses of plant growth and CO_2 uptake. Global Change Biol 11:369–377

Richter M, Hartwig UA, Frossard E, Nösberger J, Cadisch G (2003) Gross fluxes of nitrogen in grassland soil exposed to elevated atmospheric pCO_2 for seven years. Soil Biol Biochem 35:1325–1335

Rillig MC, Wright SF, Kimball BA, Pinter PJ, Wall GW, Ottman MJ, Leavitt SW (2001) Elevated carbon dioxide and irrigation effects on water stable aggregates in a Sorghum field: a possible role for arbuscular mycorrhizal fungi. Global Change Biol 7:333–337

Robinson D (2001) δ^{15}N as an integrator of the nitrogen cycle. Trends Ecol Evol 16:153–162

Rogers A, Humphries H (2000) A mechanistic evaluation of photosynthetic acclimation at elevated CO$_2$. Global Change Biol 6:1005–1011

Ross DJ, Tate KR, Newton PCD, Wilde RH, Clark H (2000) Carbon and nitrogen pools and mineralization in a grassland gley soil under elevated carbon dioxide at a natural CO$_2$ spring. Global Change Biol 6:779–790

Saarnio S, Saarinen T, Vasander H, Silvola J (2000) A moderate increase in the annual CH$_4$ efflux by raised CO$_2$ or NH$_4$NO$_3$ supply in a boreal oligotrophic mire. Global Change Biol 6:137–144

Schulze ED, Wirth C, Heimann M (2000) Managing forests after Kyoto. Science 289:2058–2059

Staddon PL, Ramsey CB, Ostle N, Ineson P, Fitter AH (2003) Rapid turnover of hyphae of mycorrhizal fungi determined by AMS microanalysis of ^{14}C. Science 300:1138–1140

Strain BR, Bazzaz FA (1983) Terrestrial plant communities. In: Lemon E (ed) CO$_2$ and plants: the response of plants to rising levels of atmospheric carbon dioxide. AAAS Selected Symposium 84. AAAS, Washington, D.C., pp 177–222

Weckwerth W (2003) Metabolomics in systems biology. Annu Rev Plant Biol 54:669–689

Wright SF, Anderson RL (2000) Aggregate stability and glomalin in alternative crop rotations for the central Great Plains. Biol Fertil Soils 31:249–253

Wullschleger SD (1993) Biochemical limitations to carbon assimilation in C$_3$ plants – a reterospective analaysis of the A/C$_i$ curves from 109 species. J Exp Bot 44:907–920

Yeates GW, Newton PCD, Ross DJ (2003) Significant changes in soil microfauna in grazed pasture under elevated carbon dioxide. Biol Fertil Soils 38:319–326

Zak DR, Pregitzer KS, King JS, Holmes WE (2000) Elevated atmospheric CO$_2$, fine roots and the response of soil microorganisms: a review and hypothesis. New Phytol 147:201–222

Subject Index

Ecological Studies
Volumes published since 2001